Handbook
of
Dimensional
Measurement
Third Edition

Handbook
of
Dimensional
Measurement

Third Edition

Francis T. Farago, Ph.D.
Formerly Senior Research Engineer
General Motors Corporation

Mark A. Curtis, Ed.D.
Professor, Engineering Technology
Ferris State University

Industrial Press Inc.

Library of Congress Cataloging-in-Publication Data

Farago, Francis T.
   Handbook of dimensional measurement / Francis T. Farago, Mark A.
Curtis.—3rd ed.
     p. 608 19.7 x 25.4 cm.
   Includes index.
   ISBN 0-8311-3053-9
   1. Mensuration. 2. Measuring instruments. I. Curtis, Mark A.,
1951–   . II. Title.
T50.F36 1994
681′.2—dc20   94-13703
                                     CIP

Industrial Press Inc.

200 Madison Avenue
New York, NY 10016-4078

Third Edition

First Printing

Copyright © 1994 by Industrial Press Inc., New York, New York. Printed in the United
States of America. All rights reserved. This book, or parts thereof, may not be reproduced,
stored in a retrieval system, or transmitted in any form without the permission of the pub-
lisher.

Composition and manufacturing by The Maple-Vail Book Manufacturing Group

10 9 8 7 6 5 4 3 2

Dedication

The third edition of this book is dedicated to the late **Dr. Francis T. Farago** for providing this foundation, the standard in the field of dimensional measurement.

Contents

List of Tables

Preface to Third Edition

During the twelve years between the publication of the second and third editions of this handbook, many significant developments affecting the field of dimensional measurement have taken place. Increased global competition and ever-increasing demands by consumers for higher-quality products has grown into what is being called a "quality" revolution. This emphasis on quality has naturally focused importance on improving methods of dimensional measurement. Simultaneously, a second revolution involving computer and video technology, as well as digital electronics, was taking place, touching virtually every facet of modern life. Inevitably, these two revolutions met in the manufacturing arena and led to the application of computer, video and electronics technology to the instruments and machines used in dimensional measurement. The result, the latest technology as it applies to metrology, has been carefully integrated into this third edition.

Prior to the computer/electronics revolution of late, the term electronic gage referred to a relatively small portion of the dimensional measurement field. However, today virtually every type of inspection instrument and machine has an electronic or computer version. Therefore, in this edition of the *Handbook of Dimensional Measurement*, the natural evolution of an inspection device is explored within the chapter in which it is first introduced. This organiza-tional concept can be demonstrated using the traditional line graduated vernier caliper introduced in Chapter 2. It became the dial caliper and more recently evolved into the form known as the digital electronic caliper; both are pictured and discussed in Chapter 2. Although the dial caliper has an indicator in addition to line graduations (which could have placed it in Chapter 5 on indicators) and the digital caliper is electronic (which could have placed it in Chapter 7 on electronic gages), the evolutionary form of organization for each major type of device was decided upon.

Over two hundred changes and additions, including seventy-five new figures, have been made in this edition. All of the referenced standards have been updated and much in the way of completely new technology is introduced.

A special thank you must go to the many fine gage manufacturing companies who generously provided information and photographs for inclusion in the handbook. Also, I would like to acknowledge the support of my colleagues at Ferris State University, especially Professor Clare Cook, Department of Electronics, throughout this revision effort. In addition, I very much appreciate the support and patience shown to me by Woodrow Chapman, Editorial Director at Industrial Press, during the completion of this project.

Preface to Second Edition

In the years following the original publication of this book, several objectives have become increasingly important in industrial production and, more specifically, in metalworking. Manufacturing methods introduced recently, and particularly those expected to be applied in the course of this decade, reflect growing emphasis on:

Assured ease of assembly and interchangeability of the majority of product components;

Highest degree of performance reliability;

Compact overall dimensions without affecting extended service life;

Substantial reduction of rejectable parts, even closely approaching "zero-defects"; and

Automation in every area of production, including the inspection of manufactured parts and assembled products.

These trends in the present, and predictably to an even greater extent in the future, will have a significant effect on the choice of the applicable methods of dimensional measurement. While the basic principles and methods discussed in the original chapters that appeared in the first edition are still valid and will continue to be used widely throughout the industry, the effects on dimensional measurements of these new developments must be recognized. Accordingly, four new chapters have been added in this second edition. The general contents of those new chapters, as well as their relationship to the new directions of technological progress, are outlined in the following.

Screw Thread Gaging and Measurement. Since the principles of screw thread measurements are similar to those of general length and profile mea-

surements, that subject was not discussed separately in the first edition. However, the growing emphasis on ease of assembly, prompted also by the expanding application of automated assembly processes, together with the need for improved functional performance of threaded components, often designed in reduced sizes, have widened the requirements for rigorously toleranced product screw threads. Consequently, the application of special screw thread measuring processes has progressed significantly with regard to both the volume of products and the degree of accuracy. Dimensions controlling the functional adequacy of screw threads are being inspected more generally, and very tightly toleranced screw threads are being used in a rapidly growing number of applications.

In view of the expanding role and significant upgrading of screw thread measurements in many areas of industrial production, a thorough discussion of that field of dimensional measurements has been provided in Chapter 16.

Measurement of Gears. The measurement of gears requires particular techniques and instruments, which have been used in machinery production for a long time. More recent advances toward the wide scale use of precision gears, assuring quieter operation, dependable rates of motion transmission, and extended service life, have expanded substantially the application of gear measurements with regard to the number of products inspected, as well as to the range of characteristics considered critical, thereby increasing the demand for specialized instruments. Many advanced types of gear measuring instruments are designed to supply data on various gear parameters in the form of displays, charts or alphanumeric printouts, all facilitating rapid and dependable evaluation.

The continuously growing importance of gear inspection, both functional and analytical, has necessitated the addition of Chapter 17, which provides a systematic description of gear measurements as a distinct branch of modern industrial metrology.

Process Control Gaging. The concept, "The best way to prevent rejects is by avoiding the manufacture of faulty parts," translates into practice as the continuous monitoring of the production process for assuring that each single part produced is in compliance with the pertinent design specifications. The product parameters most subject to variations are the tightly toleranced dimensions, consequently, process monitoring generally requires the continual measurement of critical part dimensions during the actual processing. The commonly used designation of such measurements is *in-process gaging* or, when combined with controlling functions, *process control gaging.*

Again, the general developmental trends of industrial production brought the importance and expanded applications of process control gaging into the foreground, as important contributing means in accomplishing:

Maintenance of rigorous dimensional tolerances in the production process;
Assurance of defect-free products, specifically with respect to controlled geometric dimensions; and
Realizing of significant savings by eliminating the need for subsequent part inspection for the purpose of segregation, rework, and, most importantly, rejection of manufactured parts.

In-process gaging, or even its advanced application as process control gaging, is a dimensional measuring method of not very recent origin, but it definitely belongs to a category of rapid technological advances. Actually, general developments in methods and equipment of modern machinery production, characterized by the increasing use of tightly toleranced parts, highly accelerated production rates of up-to-date machine tools, often designed for unattended operation, combined with the growing emphasis on defect-free products, have resulted in a rapidly expanding use of process control gaging. Consequently, a detailed discussion of the pertinent processes and instruments is given in the new Chapter 18.

Automated Dimensional Measurements. The term automation is used in manufacturing to designate operations of mechanical devices that perform power-actuated functions without requiring human control. In the application of automation for dimensional measurements, usually two major categories of self-controlled actions are performed in a mutually coordinated manner:

Positioning of the sensing member and/or of the specimen; and
Signaling and/or recording of the determined dimensional conditions, commonly in relation to a preset nominal size.

Complementary functions of automated dimensional measuring equipment often include the disposition of the workpiece by its physical transfer, after measurement, to a given receptacle or location in accordance with the determined dimensional conditions. Such dispositional functions may even be the principal purpose of the equipment's operation, other actions having supporting roles only.

While automated processes for dimensional measurements have developed in conjunction with similar trends in the operation of production equipment, the rate of progress in the field of measurement has further been accelerated by the need for more comprehensive inspection, in compliance with endeavors toward assuring upgraded product quality.

Automation is widely applied to modern inspection operations for the purpose of sorting and segregation, by measuring a single or a few significant dimensions. Special equipment for the automatic inspection of workpieces with a considerable number of critical dimensions are also in use in various branches of modern industry, typically, but not exclusively, that of automotive engines produced in large quantities with tightly toleranced components. The evaluation function of the inspection process may also be automated, such as the performance of statistical analyses. The wide availability of multifunction calculators and compact computers has opened new vistas for the expanded and highly efficient applications of coordinate measuring machines; these have become valuable complements of the most advanced numerical controlled (NC) machine tools, besides many other applications for the dependable dimensional measurement of complex components in a wide range of sizes. Finally, various functions of optically aided measurements have also been automated, thereby raising the efficiency of that category of dimensional measurement.

During the past few years significant advances

have taken place in all the above-indicated fields of automated measurements. A comprehensive review of that development in methods and equipment is presented in the new Chapter 19.

The structure and contents of the new chapters were designed to be in harmony with the fifteen chapters of the original edition. A few changes were needed in the original text, to assure compliance with new and revised national standards brought out since the book's original publication date. References are made in the new chapters to the preceding ones, where such ties contributed to combining the original, as well as the new chapters of the book into an organic whole.

Preface to First Edition

Dimensional measurements are a vital link between the designer's intent and the actual product. Thus, design engineering provides detailed specifications in the form of dimensioned and toleranced drawings. Manufacturing engineering converts these into finished products with the aid of a wide variety of dimensional measurements; and quality engineering maintains a continuing control on the product output by still more dimensional measurements carried out with the aid of more or less complex instrumentation.

Phenomenal progress has been achieved during the past decades in the fields of product design, manufacturing and quality control. New and improved materials, better machines and more sophisticated inspection methods and techniques are all contributing to technological changes that are taking place with almost bewildering rapidity. To keep pace with the demands from industry for greater accuracy, repeatability, reliability, ease of use and continuous process control, new and improved dimensional measuring techniques and equipment have had to be developed.

As in most present-day fields of science and technology, however, it is extremely difficult for the practitioner engaged in everyday work to keep abreast of the new methods and equipment in the many ramified areas of dimensional measurement. It may be even more difficult for the beginner, the fledgling engineer or technician, who has not been exposed to the current practices and equipment of a modern manufacturing plant or laboratory, to collect and absorb the huge amount of know-how on which advanced dimensional measurement technology rests. One of the major purposes of this handbook is to provide a readily accessible reference for both the neophyte and the experienced engineer and technician.

To facilitate the acquisition and application of needed information on methods and equipment out of the myriads of facts and data available, ample usage has been made of a highly efficient method of topic discussion: presentation of significant aspects in tabular form. Over eighty tables with diagrammatic illustrations and brief synopses are provided for rapid scanning and review. Thus, important topics of dimensional measuring techniques and equipment can be quickly located and evaluated by means of the comparative analyses arranged for synoptic observation both in words and pictures. More detailed information, together with elaboration of the concise statements of the tables, can then be found in the adjoining text portions of the book.

Emphasis has been laid on the discussion of several important, yet only recently improved systems of measurement which either have not been described at all or have been mentioned only tangentially in other books on dimensional metrology. In addition to such subjects and the current state-of-the-art topics, even a brief discussion of methods and tools which are considered rudimentary, might appear unnecessary. However, aside from the author's attempt toward completeness, the covering of such an extended range of levels is held to be warranted because of the many common fundamentals throughout. Both the basic and the most advanced methods of dimensional measurements rely on similar principles in such aspects as, e.g., part staging datum definition, specimen orientation, referencing, geometric interrelationship, etc. An understanding of the fundamentals of a routine operation obtained by reviewing generally known and practiced methods, can assist both new and experienced engineers in improving their analytical approach toward the more complex measurement problems which, too, must be examined and resolved as geometrical propositions.

Many reputable manufacturers of measuring instruments were helpful in supplying photographic material for illustrating this handbook. The author wishes to express his thanks for their cooperation. Unfortunately, space requirements do place definite limits on the inclusion of illustrations in any handbook. Therefore, while the contributors of the book's illustrations are prominent in the discussed lines of products, neither the selection nor the omission of a particular instrument or manufacturer should be construed as an indication of product evaluation.

In the light of the considerable extent of the subject, the attempt to embrace in a single book the whole field of dimensional measurement would appear a far too ambitious undertaking. Recognizing the physical limitations of a handbook, the emphasis was laid consistently on principles and general characteristics, which are applicable to whole families of processes and instruments. While completeness, even within judiciously set limits can only be approached, it is hoped that this book, by providing access to many principles and practices of modern dimensional measurements, will prove of value to its readers at many different levels of individual experience and responsibility.

Handbook
of
Dimensional
Measurement
Third Edition

I.
Introduction

In common industrial usage, the term *dimensional measurements* designates processes for determining the linear and angular magnitudes of technical parts or of their specific features. These magnitudes are measured and expressed by standard units of length and angle. In an extended usage, dimensional measurements also refer to processes for assessing pertinent geometric characteristics that can affect both the reliability of the basic dimensional measurement and the functional adequacy of the part. Such measurements are concerned with deviations from an ideally regular form or from a surface that is theoretically smooth and has no distinguishable texture.

The basic purpose of dimensional measurements in production is to assure and to verify the agreement of the product with the specifications of the design. Although dimensional measurements in industrial applications may serve other purposes as well, for example, product and production engineering research or development, in the final analysis all these uses are contributory to a common goal: the economical production of parts in compliance with design specifications that are devised for assuring the proper function and service of the manufactured product.

For that reason a handbook on dimensional measurement prepared for engineers and technicians engaged in various areas of industrial production will serve its objectives best when it reflects product-mindedness by considering the measuring processes from the viewpoint of the ultimate purpose and practical applications.

Considering the continued advances in modern industrial production toward greater precision and more reliable function of its products, it is obvious that such pragmatic goals will not detract from the high degree of professional knowledge that dimensional measurements for production purposes often require. The effect of such progressive trends on measuring requirements in industrial production is also reflected by the term *metrology*. This term is also widely used in industry for designating the scientifically based dimensional measurements as distinguished from the empirically developed earlier practices.

The trend toward applying measurement science to industrial operations introduced another specialist into the engineering profession—the engineering metrologist. A few characteristics of that professional will be discussed later.

Various Fields of Engineering Concerned with Dimensional Measurements

In metalworking and in related branches of industry, most of the manufacturing efforts are directed toward producing parts of specific forms and dimensions. Dimensional measurements being the means by which these efforts are guided, coordinated and approved, familiarity with or even intimate knowledge of these means are important elements in the general know-how of engineers engaged in many different sectors of manufacturing industries.

Randomly selected examples, described in general terms and associated with a few important fields of engineering activities in manufacturing, are mentioned to illustrate the point.

Product engineering is guided in the conception of the major or critical part dimensions by stress analyses and/or by empirically determined data that frequently originate from physical testing. In either approach, the general knowledge of dimensional measurement technology is a valuable asset for considering, as the basis of calculations and evaluations, realistic dimensional values whose dependable measurement is technically feasible.

Advances in machine tool building give added substance to the adage: "As long as you can measure it, you can make it." Consequently, knowing what can be measured and to what degree of accuracy can prove beneficial in many areas of creative engineering work.

Designers must be familiar with the capabilities of gaging equipment and with the principles of dimensional measurement in order to apply appropriate dimensioning and tolerancing on engineering drawings, as well as to select practicable datum elements. Only dimensions whose dependable measurement, in locations and relative positions as conceived by the designer, is practicable will actually control the manufacture of products in compliance with the designer's intent.

Manufacturing engineering may be considered the prime user of dimensional measuring equipment and processes. "Make it right the first time" is a concept of particular importance in modern manufacturing, where rework is often uneconomical and the release of defective parts must be scrupulously avoided. In-process gaging, feedback for automatic sizing and tighter tolerances to eliminate the need for selective assembly are three examples of emphasis being placed on the importance of refined measuring methods in the manufacturing process.

Production engineering research, as well as process engineering and facility engineering, are still other areas connected with manufacturing that offer many applications for dimensional measuring processes, frequently even at rather high levels of sophistication. Machine tool acceptance tests with optical tooling and laser beams, process capability studies involving measurements with much finer discrimination than needed in regular production and comparative measurement of the geometric conditions and surface texture of parts machined with alternate methods are three randomly selected examples for illustrating the character of measurements required in these areas.

Tool and gage making, although auxiliaries of the basic manufacturing process, must keep ahead of it in several technological aspects, particularly with regard to dimensional measuring capabilities. The use of comparator type gages is predominant in modern manufacturing, calling for a growing number of master gages with a desirable accuracy level ten times higher than the product tolerance. The need for measurements on the order of the microinch is now common in industrial gage making and control, a trend reflected by the rapidly growing number of industrial metrology laboratories built to provide rigorously controlled environmental conditions.

Quality control is the traditional sector of dimensional measurements that, particularly in the metalworking industry, constitutes a predominant portion of the total inspection work. Quality control departments are generally the final arbiters of product acceptance within the manufacturing plant. Consequently, the methods and equipment used by this sector must be at least equal and preferably superior to those applied in areas whose products are judged by quality control. An intimate knowledge of the principles as well as of the technology and equipment of up-to-date dimensional measurements is also beneficial for the role that quality control engineers may have in designing inspection equipment and in educating inspection personnel.

Product reliability represents a function of relatively recent origin, yet of widely accepted importance in the reliability conscious branches of industry. For defining the measurable parameters of products that are expected to perform in a reliable manner, dimensional conditions have a predominant role in most types of mechanical parts and assemblies. The selected methods of dimensional measurements as well as their actual implementation are important factors of the confidence that can be assigned to correlations between reliable performance and the dimensional characteristics of the product.

Industry's Need for Higher Accuracy in Dimensional Measurements

The progress of industrial measurements toward higher accuracy has been continuous since before the Civil War. The accomplished advances can be considered to represent one order of magnitude, the 1/1000th part and the 1/10,000th part of an inch representing characteristic milestones.

Industry's need for higher accuracy has risen, however, at a much more rapid pace in the decades since World War II. That rise, which has occurred during a period of less than fifty years, progressed from the ten thousandth to the millionth part of an inch, representing an increase in the accuracy level by two orders of magnitude. These values are characteristic of the most sensitive types of measuring instruments in industrial applications and serve to indicate the pace of a progress that deserves thorough consideration when discussing the technology of dimensional measurements.

Global competition has spawned a quality revolution within the manufactured products industry. This fact, when coupled with technical advances in various areas of industrial production, leads to the need for increased accuracy in dimensional measurement. The following list of trends exemplifies this need.

1. Dimensioning of components closer to the theoretical size limits of the required strength, with manufacturing tolerances at a minimum, in order to achieve *material and weight savings* for operational and economic reasons.

2. Reduced envelope dimensions for elements and for complete products, represented by trends known as *compacting* and *miniaturization*, particularly in applications such as computers and other complex electronic devices.

3. *Stress and strain compensating design* for heavily loaded elements whose precisely calculated and built-in form deviations will assure optimum performance under elastic deformations caused by operational loads.

4. *Greater operational precision* and other improvements of the performance of such components as gears, ball and roller bearings, guiding and sealing elements and so forth. To ensure that these objectives are being met, tighter tolerances for size, geometric form and surface texture are being specified.

5. *Automatic assembly,* often at speeds beyond the limits of visual perception, exclude a reliance on fitting, adjustment and segregation. Such processes may require components whose level of accuracy is even higher than that needed for operational purposes.

6. *General interchangeability* for the prompt replacement of component parts or modules, in order to improve the maintainability of assemblies and entire pieces of equipment on a national or even worldwide scope.

7. *Reliable operation* of machinery and other types of equipment is often predicated on parts manufactured and assembled consistently within very close limits of size and form.

8. *Accurate reporting of dimensional information* is being demanded of most industrial suppliers for use in the statistical analysis of their various processing capabilities by original equipment manufacturers (OEMs).

9. *Quality assurance systems,* such as ISO 9000 developed by the International Standards Organization, are being developed and used on a worldwide basis to guarantee that manufactured goods meet certain design, performance and quality specifications.

From the foregoing short list of typical trends it is evident that the continued striving toward more accurate measurements is by far not a mere exercise in expertise. It constitutes a necessary progress in order to satisfy actual needs by providing that degree of dimensional information without which technological advances could not materialize.

That industrywide need for an up-to-date background in measurement science was constantly kept in mind during the development of this book. The various methods and instruments discussed in the text were examined with regard to their applications in industry and from the viewpoint of those primarily responsible for the planning and execution of industrial measurements.

Advances in Dimensional Measuring Instruments

The progress in dimensional measurement technology could not have been accomplished without substantial improvements in the existing types and the introduction of entirely new systems of measuring instruments. A few highlights of such advances will be mentioned for the purpose of visualizing the great potentials for improved measuring techniques that are offered by modern metrological instruments.

Electronic gages were developed parallel with the phenomenal progress in the field of electronics. Measuring instruments operating by electronic amplification, which were originally reserved for laboratories, found many uses in actual manufacturing processes. Electronic gages brought the microinch within the reach of production indirectly, by using setting masters calibrated to millionth of an inch accuracy, and also directly, in such areas as surface-texture, straightness and roundness measurements.

Pneumatic gages grew out of the experimental stage and currently represent the backbone of in-process gaging, where noncontact sensing, small probe dimensions, high amplifications and easily implemented multiple circuitry prove particularly advantageous.

Optical measuring instruments offering means for magnified observation in microscopes and optical projectors, or operating by the amplifying effect of "optical levers," such as in aligning, collimating and angle measuring devices, possess unique measuring potentials that are now being used frequently.

Fringe producing devices operating by means of the interference effects of light beams, by moiré fringes or discrete inductive pulses and by interference fringes derived from laser beams are adaptable for dimensional measurements either by visual fringe count or by means of electronic fringe counting and phase-shift analyzing instruments. Such instrumentation, which translates physical phenomena into industrially applicable tools of extreme precision and stability, opened new vistas for measurements in very fine increments, even over substantial distances, with great speed, and often complemented with digital readouts capable of displaying up to a million, or more, discrete units.

Signal processing, reporting and utilizing devices that operate in conjunction with various types of dimensional measuring instruments are finding diversified applications. Such devices are operated either directly by the variations of current or voltage provided by electrical or electronic instruments, or indirectly, by converting other energies into electrical variables with the aid of pressure or phototransducers, such as in pneumatic or optical gages.

Examples of such devices are: averaging and integrating systems; chart recorders and digital printers; light or sound emitting signals; relays, solenoids, magnetostrictive and other types of actuating devices for machine tool controls, part sorting, segregation and matching. Most of these are essential elements of automatic manufacturing, assembly and inspection equipment.

There should be no doubt regarding the important role that many traditional types of measuring instruments still play in present day industrial production. For that reason no book on dimensional measurement technology would be sufficiently comprehensive without giving due attention to the long since customary means and methods, even though such are discussed in various other books on similar subjects.

Particular emphasis, however, is on measuring technologies and equipment that originate from relatively recent developments, and are not discussed in comparable detail in most other publications on dimensional metrology. In treating such subjects the selection of topics and the extent of the discussion were guided by the relative frequency of applica-

tions and the industrial significance attributed to such newer systems of dimensional measurement.

Dimensional Measurement Concepts

In order to fulfill its purpose, the measurement must be correct to a degree commensurate with the specified tolerances and the required functional service of the product. Two major factors should be considered when evaluating the correctness of dimensional measurements on manufactured parts: (a) the gaging accuracy and (b) the proper location of the measured dimension on the physical part.

The concept of *accuracy* is of common knowledge: it designates the degree of agreement of the measured size with its true magnitude as expressed in standard units of measurement.

The other factor, the physical *location of the dimension being measured* on the part is less commonly realized in its full importance. Actually, improperly located dimensions, not uncommon in industrial practices, can often have a greater effect on the correctness of dimensional measurements than errors due to insufficient indicating accuracy; this factor is an inherent and consequently readily evaluated characteristic of the applied measuring instrument.

The term *precision* is used quite often, and is sometimes incorrectly substituted for the word accuracy. Precision expresses the degree of repeatability of the measuring process. Precision designates how closely identical values are obtained when repeating the same measurement at various intervals, or duplicating them by means of different instruments. It is obvious that the precision of the measurement can be affected by limitations in either of the basic requirements, namely, the accuracy of the instrument and the proper location of the gaging points that determine the dimension being measured on the physical part.

The properly dimensioned engineering drawing must clearly define the location and orientation of the dimension to be measured on the part. However, some degree of discrepancy can hardly be avoided between the designer's intent as expressed on the conveniently developed two-dimensional representation of the three-dimensional part, and the actual transfer of those concepts to the physical surface. To reduce the extent of disagreement between the designer's specifications and the actually measured dimension, the engineering drawing must either define or indicate by distinctly understood inferences the *datum elements* that are positionally related to the specified dimensions and are physically present on the part.

Based on such specifications, the *staging* of the part for measurements can be devised in a manner to assure a good agreement of the dimension being measured with that spelled out on the engineering drawing. The importance of proper staging as a means of accomplishing the correct *location* and *orientation* of the dimensional measurements is repeatedly referred to in the discussions of this book.

In some cases, the staging may also provide the *referencing* of the measurement, by assuring the coincidence of one boundary point of the dimension on the part, with that point, line or plane within the instrument in relation to which the distance is measured, or the size variations are indicated. An example of combined staging and referencing is the diameter measurement of a cylinder placed on the flat stage of a comparator instrument. However, for most technical parts, the referencing points or planes must be mechanically separated from yet conceptually interrelated with the locating surface that is used in the staging phase of the measuring process.

The capability of a measuring instrument to detect and to faithfully indicate even small variations of the measured dimension from a reference value, often called the nominal size, is an important performance characteristic. The measure of that capability is the least size variation that is indicated or displayed by the instrument distinctly and in its true value. For indicating type instruments that operate a moving pointer, the term *sensitivity* is generally used to express this property, whereas for instruments using extended scales observed against index marks and for those with incremental display, for example, digital readout, the terms *resolution* or *discriminating capacity* are more current.

In certain applications, such as in-process gaging or automatic sorting, as well as in moving-stylus type instruments, the time span between input, that is, the sensing of the size variations, and the distinct indication of its value can be an important factor. The respective capacity of the instrument is often designated as its *response speed*.

The indications of the measuring instrument must be consistent and its calibration should be retained over a considerable period of time. The term *stability*, often expressed as the maximum drift over a specific time, is the designation of the instrument's capability for consistent indications. For mechanical gages, and also for more complex instruments whose operation involves several potential sources of error, the term *repeat accuracy* is used, by expressing the numerical equivalent of that condition as the maximum deviation of repeat.

Conditions and Practices Related to Dimensional Measurements

In the planning, implementation and evaluation of dimensional measurements, proper consideration must often be given to conditions that may affect their execution or results. Also deserving attention are practices that were developed for the assurance of better efficiency and higher reliability in the measuring process. Several examples are pointed out and are accompanied by brief descriptions.

The term *material condition* is used in engineering design for relating the specified dimension to the "envelope size" of the part or feature. That imaginary envelope is considered to represent the geometrically perfect form of the examined feature (for example, a cylinder for a round hole) in a size that just touches the most protruding (maximum) or the deepest receding (minimum) elements of the physical surface.

Maximum material condition, as specified and more fully explained in the American National Standards Institute document ANSI Y14.5M, is a designation applied to a given feature (hole, thread, external profile, and so forth) or part, to indicate that its dimension or dimensions are such as to result in the maximum amount of material possible within the allowed dimensional limits, assuming a perfect form. Thus, the maximum material condition for a nominally round hole is when its diameter is at the specified minimum and the hole is a geometrically perfect cylinder. Conversely, the maximum material condition for an external screw thread is when its form is perfect and its diametral dimensions are at their specified maximum.

Maximum material limit is the maximum limit of size of an external dimension or the minimum limit of size of an internal dimension.

The significance of geometric conditions for reliable dimensional measurements. In engineering drawings, the dimensioning of manufactured parts is predicated on geometric forms and surface textures whose degree of regularity is commensurate with the specified dimensional tolerances.

When the required degree of regularity of geometric form and surface is not present, it is to be expected that the reliability of the dimensional measurement will be affected.

Most of such deficiencies that occur may, for the purpose of this review, be assigned to one of the following categories:

a. The geometric forms and interrelations of surfaces defined by such concepts as straightness, flatness, perpendicularity, parallelism, symmetry and so forth;

b. The microgeometric variations of the surface from its conceptually regular form, commonly defined as departures from straightness and circularity of characteristic surface elements; and

c. The microtexture of the surface, assessed as roughness, designating deviations from a theoretically perfect smoothness.

The functional effect of geometric deficiencies is receiving industrywide attention, partly based on empirical findings in service and partly as the result of studies and tests in such areas as lubrication and wear, stress distribution and fatigue and so forth. Higher requirements for consistent regularity also originate from the area of manufacturing in its move toward unconditional interchangeability and automatic assembly. Perhaps less general is the realization of the harmful effects that such deficiencies may have on the accuracy of basic dimensional measurements.

In view of the importance of each of these aspects, namely, the functional adequacy, the manufacturing feasibility and the metrological reliability, particular emphasis is on the measurement of said geometric conditions, considering them as complementary processes of the more familiar methods of linear and angular measurements.

The *reliability of dimensional measurements* may be regarded as a function of the occurring errors and of their assessments. Errors in measurements and of the measuring instruments are, in common metrological practice, assigned to either of two basic categories: random and systematic.

(a) *Random* errors are primarily, yet not exclusively, associated with the human element and are considered as resulting from inaccurate scale reading, improper specimen staging, mistakes in recording, and so forth. Because such errors can neither be predicted nor entirely eliminated, their probable magnitude and relative frequency are evaluated statistically for establishing the probable precision of the measuring process.

The concept of *standard deviation* as a measure of the range of scatter in the measured values from a mean is applied for qualifying the precision of the process or of the instrument. The magnitude of the standard deviation is established mathematically, on the basis of an appropriate number of measurements,

in such a manner that about 68% of all readings should fall within a zone that extends by one standard deviation in both plus and minus directions from a mean value. The Greek letter sigma (σ) is used to designate the standard deviation. In order to assure with a greater probability that the reported measurement values are true, the precision of the process is often appraised by considering an uncertainty zone that extends over $\pm 2\sigma$, comprising theoretically 95.45%, or $\pm 3\sigma$, comprising theoretically 99.73% of all readings.

Assuming that the variation of the individual measurements follows the pattern of a normal distribution curve, the following formula is often used to determine the value of the standard deviation:

$$\sigma = \sqrt{\frac{\Sigma (x_n - \bar{x})^2}{n}}$$

where

$\bar{x}_n =$ the individual readings;
$\bar{x} \; =$ the mean of all readings;
$n \; =$ the total number of (x) readings.

(b) *Systematic* errors are those associated with the measuring instrument's capabilities, environment and other extraneous conditions, which are generally controllable or, at least, measurable.

These two categories, although accepted in general practice, should not be regarded as rigidly delineated. There are error sources, for example, the performance pattern of instruments, that might represent borderline cases and that are appraised by combining fixed magnitudes with the pertinent sigma number as a qualifier.

The capabilities of measuring instruments. When the process requirements of the gaging are adequately assured, it is the capability of the instrument that will primarily determine the reliability of the measurements. This reliability is further influenced by the concepts of *gage repeatability and reproducibility* (gage R&R). Within the context of reliability, the *repeatability* of a gage is measured by having one person measure several different parts at least twice each. These measurements are recorded and the range across the readings for each part is calculated and then used to determine gage repeatability. Therefore, the size of the repeatability error of a gage relates to how close the two separate readings taken with a single gage on the same part by the same person will be. The term *reproducibility* is used to identify the difference in a gage reading (error) when a single part is checked with the same gage by two or more persons. As such, the accuracy of any gage used in a production setting is subject to the errors of repeatability and reproducibility. As a basic rule, the measuring instrument's consistent indicating accuracy should be at least equal to but possibly several times better than the smallest graduation of the readout element, such as the scale or the dial.

Manufacturer's specifications or standards when available, for example, those for dial indicators, define the degree of indicating accuracy expected from the instrument. Nevertheless, the verification of the actual performance accuracy, particularly in the course of usage, is advisable. Such verifications are mandatory for critical measurements and/or when the nominal accuracy level of the instrument is close to the value required from the measuring process.

The verification of the measuring instrument's indicating accuracy, both in its original state and during usage at appropriate intervals, is accomplished by means of properly devised and implemented calibration processes. Various methods of instrument calibration for the purposes of verifying and assuring the reliable accuracy of the dimensional measurements are discussed in several chapters of this book.

The indicating accuracy of a measuring instrument expresses the degree of proportionality with which deviations from an arbitrarily selected reference size are indicated in the scale of the instrument. The absolute size of the reference dimension to which the instrument was set must be established by comparing it to a basic reference standard of length, such as a gage block. The calibration of gage blocks, both in the user's plant, or by outside laboratories, as well as the concept of traceability, designating the controlled relationship of such industrial reference standards with those maintained by the National Bureau of Standards, are also explained.

The effect of environment on dimensional measurements. Absolute stability is not assured even in such seemingly settled substances as rocks; temperature changes and variations in humidity are continuously affecting the geometry of granite surface plates. The effect of temperature variations on the size and on the shape of parts, as well as on the mating conditions of assemblies, are commonly known, although not always properly appraised and considered.

Variations in the environment always affect the results of dimensional measurements. Even for gaging on the production floor, a reasonable degree of care in avoiding extreme variations and, particularly, sudden changes of temperature is indicated. In the measurement of critical dimensions at high resolution, the environmental, particularly temperature, variations can seriously affect the reliability of the obtained results.

For that reason such measurements are carried out in areas where the temperature is maintained at a constant level, usually at 68°F (equivalent to 20°C, the internationally agreed reference temperature for length measurements), with variations on the order of ± 0.5°F, or even less. Such enclosed areas, often designated as metrology laboratory rooms, provide continued exchange of air in a uniformly distributed manner (laminar flow) and at a rate permitting prompt correction of temperature variations that might occur. The introduced air is filtered and dehumidified (relative humidity maintained at a level of 35% to 45%). The air pressure inside the room is kept slightly higher than that of the surrounding areas, from which access is provided through double doors and special pass-through windows. The lighting of these rooms must be uniform, substantially shadowless, of sufficient intensity for delicate work and arranged in a manner to reduce the effects of radiated heat. Although the location of the metrology laboratory is selected so as to reduce the exposure to vibrations, it is often necessary to build a separate foundation for the entire room and/or for the particularly sensitive types of equipment, such as interference microscopes and other categories of optical instruments.

Standardization. Dimensional metrology is a characteristic field where standardization is apt to prove highly beneficial to that wide range of endeavors that must rely on the dependable measurement of geometric conditions. Standardization of dimensional measuring processes and equipment can comprise any or several of the following major aspects:

a. Instrument capacities, envelope and connecting dimensions;

b. Instrument sensitivity, precision, stability and other characteristics related to operation and dependability;

c. Indicating accuracy and methods of verification, known as calibration procedures; and

d. The measuring process, including instrument selection, the applicable techniques, recommended magnifications for accomplishing the required resolution, evaluation of the indications and the assessment of the measured conditions.

The choice of the appropriate process and instrument. For any measuring task and set of conditions there is usually one method or a limited number of methods that can be considered best suited technologically and appropriate economically. For recognizing that particular method and equipment, the review of potential alternatives is an obvious course to follow by the engineering metrologist.

In order to aid in the selection of the optimum method and of the adequate tools for its implementation, synoptic surveys in tabulated form have been prepared for many categories of measuring processes and instruments. No attempts were made to encompass in these tables all possible alternatives. Although providing information pertinent to a meaningful comparison, the data listed were limited to the essentials, in order to avoid a large amount of detail that could defeat the purpose of rapid scanning for comparative evaluation.

Where illustrations are supplied in the tables, they are diagrammatic and are not intended to complement the descriptions. Their purpose is to convey visually the dominant characteristics of the listed types and systems, thereby contributing to the effectiveness of the decision-making mental process.

Engineering Metrology

With the rise of the required accuracy level of dimensional measurements and with the growing importance attributed to the reliability of the reported results, the term metrology has become part of general technical usage. For that reason it seems appropriate to examine the meaning of that term, as well as the functions it requires from its practitioner, the engineering metrologist.

Metrology, in a broader sense, designates the science of all measurements that are made by comparing the dimensionally measurable conditions of solids, or of diverse physical phenomena to generally accepted units of measurement.

In a narrower but widely used interpretation of the term, metrology is a branch of technology concerned with the measurement of geometrically defined dimensions of technical parts. This latter concept, known by the distinctive designation of *dimensional metrology,* or *engineering metrology,* is the subject to be discussed in the following.

Originally, the term metrology referred specifically to the measurement of basic reference standards, such measurements being carried out mainly by the different national metrological laboratories. The combined effect of two concurrent trends: the rapidly increasing industrial need for high precision measurements and the wide availability of measuring instruments with a level of sensitivity and accuracy previously required by only a few laboratories, resulted in the expanded application of metrological techniques, as well as in the more general usage of the term metrology.

Metrology, when designating dimensional measurements in industry, comprises a wide spectrum of measuring processes that are distinguished by more scientific planning, execution and assessment than is applied in common measuring practices. The present-day responsibilities of most engineers in progressive manufacturing industries usually require a far more than superficial understanding of dimensional measurements because the engineers represent an essential link between the dimensioned and toleranced engineering drawing and the actual product. It is this fundamental relationship between the specifications spelled out on the engineering drawing and the actual products produced by production processes that makes engineering metrology an indispensable element in modern manufacturing. Production parts are measured and individual measurements are statistically analyzed to determine the process capability of a particular manufacturing setup. For any given dimension, the process capability is accepted to be plus or minus three standard deviations from the mean, or a total of six standard deviations. The process capability is then compared to the total tolerance allowed on the engineering drawing for the dimension under study (that is, upper dimensional limit minus the lower dimensional limit). By dividing the process capability (that is, six standard deviations) into the total tolerance specified on the engineering drawing, a number called the *capability index* (Cp) is calculated. A Cp of 1.33 is considered to be the minimum acceptable quotient.

Apart from the understanding of dimensional metrology, there is a need for the specialized engineer, who actually develops, directs and evaluates the metrological process. The appropriate designation of such a specialized professional is metrologist or metrology engineer.

As a carryover from earlier practices, the title metrologist seems to have, in general technical usage, a specific connotation, that is, activity associated with the measurement of reference standards.

Although the application of measurement sciences to the development and calibration of basic and reference standards constitutes one of the pillars on which modern technology rests, a high degree of

TABLE 1-1. Comparison of the Responsibilities and Objectives of the Two Branches of Dimensional Metrology

Reference Standard Metrologist	Engineering Metrologist
Concerned with the absolute—the maximum degree of attainable accuracy.	Concerned with a qualified optimum—the consistent accuracy at a degree required for a specific case.
Measures a particular, distinctly specified dimension, regardless of other dimensions of the same body.	Considers the assurance of a functional purpose resulting from the interrelation of several dimensions.
Requires technical parts made expressly with the purpose of embodying a specific dimension—expects excellent workmanship.	Handles parts whose workmanship level is dependent on the functional need and employs a measuring process accommodated to this need.
Measures well-defined dimensions only where no doubt can exist regarding the location on the part of the dimension to be measured.	The dimensions specified are a means to convey the designer's intent regarding the functionally meaningful form and size of the part and may need interpretation for assuring correct measurement.
Requires optimum conditions and the highest level of professional skill.	Concerned with practicality—must adapt himself and the methods employed to conditions and personnel.
The consistent observance of clearly spelled-out procedural instructions is an essential guarantee for repeatability to the last significant digit. Uncompromising adherence to standard methods characterizes the measurement of dimensional reference bodies.	Methods of measurement must often be developed to cope with unusual configurations of the object or locations of the critical dimensions, also considering the functionally meaningful interrelations. Inventiveness and decision-making capacity are important attributes of the engineering metrologist.
Goal: Continuously striving toward a better approach to perfection.	**Goal:** Assuring consistently reliable information at the required level of accuracy, with due regard to the economical aspects.

technical knowledge is also required for adequately transferring the potentials of dependable standards to the manufactured products. That judicious transfer of accuracy potentials calls for a specialist who, distinct from the metrologist in charge of standards, may be called an *engineering metrologist.*

The functions of these two categories of metrologists are not distinctly delineated; actually, they are branches of dimensional metrology, complementary in their objectives, and both are needed to accomplish effectively the dependable dimensional accuracy of manufactured products.

To use the degree of the employed measuring sensitivity as the basis of distinction could lead to erroneous appraisal of the relative values of these two areas of metrological activities. While the reference standard metrology measures in finer units, generally on the order of a microinch, industrial metrology is faced with many other technical problems

whose satisfactory solution depends on the reliability of the measurements that are being carried out in a specific sensitivity range.

As is the case for most classifications related to human activities, the proposed categories are arbitrary and their primary purpose is to differentiate the functions and the objectives of the engineering metrologist from those of the reference standard metrologist. To enable the reader to better visualize the differences in the major responsibilities of these two categories of metrologists, as well as the goals and attitudes that ensue, a tabulated comparison is presented in Table 1-1 above.

Concluding Notes

From the preceding concise survey of the various pertinent aspects of dimensional measurements, it may appear evident that they, on the level to which the designation engineering metrology applies, cover a much wider scope of activities than the proper use of the measuring instruments and the conscientious recording of the obtained indications.

The well planned and executed dimensional measurement can often be considered an *engineered process* based on a thorough understanding of several significant factors, such as the following:

a. The purpose of the measurement, including the desired level of accuracy and degree of confidence.

b. The functional goal that the dimensioning and tolerancing of the engineering drawing of the part is intended to assure.

c. The operational requirements of the measuring process with regard to speed, handling, maintainability, indication display, signaling, recording, data processing and control actuation.

d. The technology of the measuring process, including the correct use of the applied measuring instruments and auxiliary equipment.

e. The capabilities of the process and of the measuring instrument, as well as the verification and assurance of the required performance level.

For purposes of easier information retrieval by the reader, the contents of most individual chapters follow the familiar pattern of classification by instrument categories. However, this concession for the sake of convenient presentation and use must not cause the reader to overlook the priority assigned to determining the appropriate measuring process and defining its requirements. The selection of suitable instruments and accessories, although essential for the proper implementation of the process, constitutes a secondary phase in the engineering of dimensional measurements. Evaluation tables of instrument categories, listing various aspects in which they may offer significant advantages, should prove helpful in comparing alternate types for the selection of that particular system that is best suited to meet specific process requirements.

The authors' intent is to assist in developing an engineering approach in the planning, devising and implementation of dimensional measurements. It is believed that there is a growing need for engineered dimensional measuring practices in order to keep pace with the developments of other industrial and scientific domains in this age of rapid technological advances.

2.
Line Graduated Measuring Instruments

ESSENTIAL PROPERTIES

Line graduated geometric bodies, with graduation spacings representing known distances, are the bases for all direct measurements of specific distances. It follows that instruments having line graduated elements as integral members may be considered the only mechanical means capable of carrying out direct measurements without the assistance of complementary equipment or processes.

The essential property of instruments with line graduated elements is their capability to measure *any* distance within their capacity range. This distinction of line graduated instruments is valid independently of the achievable sensitivity or accuracy of the measurement.

The sensitivity of the measurement will depend primarily on the instrument's basic design. This is most obviously displayed, although not necessarily truthfully represented, by the least distance between the individual graduations of the basic gage element and of the supplementary readout device, when the readout device is part of the instrument.

Factors affecting the measuring accuracy of a line graduated instrument are the original accuracy of the graduation, the level of resolution of the graduation lines and of the readout members, the instrument's design and the general workmanship exercised in its manufacture. The precision of measurements made with line graduated instruments will also be a function of how truly the actual distances to be measured are associated with the corresponding instrument graduations.

Conditions related to instrument accuracy and the potential precision of measurement will be analyzed subsequently for several of the more generally used categories of line graduated measuring instruments.

Categories of Line Graduated Measuring Instruments

Line graduated elements are used in many, widely varying types of measuring instruments. Differences exist in various essential respects, such as intended use, designed sensitivity, accuracy of execution, level of sophistication with regard to readout and many others. The line graduated member can be the instrument itself, without auxiliary devices (for example, the graduated rule). For many important types of measuring instruments, such as vernier calipers and micrometers, the line graduated element is complemented by contact and transfer members, and also is equipped with devices for the proportional evaluation of graduation intervals. Line graduations commonly represent linear distances, but can display angular spacing as well, as discussed in more detail in Chapter 10. Finally, line graduated reference elements are used in many types of optical measuring instruments. In this latter application of line graduated elements, the optical portion and the intended special use of the instruments are considered to be the dominant features; accordingly, detailed discussion of these types of graduated element applications follow in Chapters 8, 9 and 12.

For the convenience of the following discussion, line graduated measuring instruments have been assigned to categories, according to the systems by which the basic principle of line graduated reference elements is put to use for the purpose of length or angle measurement.

TABLE 2-1.   SYSTEMS OF LINE GRADUATED MEASURING INSTRUMENTS – 1

CATEGORY AND SYSTEM OF OPERATION	DIAGRAM OF TYPICAL APPLICATION	DESCRIPTION OF SYSTEM	MEASURING SENSITIVITY IN GENERAL
Line Graduated Rules and Tapes Direct length comparison	Line graduated rule	Steel rules and tapes are available in widely different degrees of accuracy to suit diverse requirements for plain length measuring tools. Graduated rules are made from non-metallic materials too, for subordinate uses. Auxiliary devices extend the field of potential applications.	Commonly fractional, 1/32nd or 1/64th inch; made also with decimal graduations for 1/50th inch smallest division.
Line Graduated Bar Standards Direct length comparison with auxiliary alignment devices	Reference standard bar Graduation lines	The essential characteristic is the sustained accuracy of the graduations, and not the least distance between them. Intervals may be subdivided by other equipment used in conjunction with the reference standard.	High resolution is not the purpose because it is intended as a reference standard with a limited number of distances. For accuracy, see Table 2-2.
Caliper Gages Positive contact transferred to graduated scale	Vernier caliper	Jaws establish the length to be measured by positive contact. One of the jaws is integral with the graduated beam, the other has markings to indicate the corresponding scale position. Vernier improves the measuring sensitivity.	0.001 inch with vernier

TABLE 2-1.   SYSTEMS OF LINE GRADUATED MEASURING INSTRUMENTS — 2

CATEGORY AND SYSTEM OF OPERATION	DIAGRAM OF TYPICAL APPLICATION	DESCRIPTION OF SYSTEM	GENERAL MEASURING SENSITIVITY
Micrometers Direct reading scale complemented with translated motion for subdivision of intermediate distances	Outside micrometer	Rotation of the threaded spindle causes controlled advance of the spindle's contact face relative to an integral reference face. The axial spindle traverse is displayed on a linear scale with intervals equal to the screw pitch, while fractional advance, resulting from partial spindle rotation, is shown on a circumferential scale, frequently read by vernier.	0.001 inch with plain reference mark; 0.0001 inch with vernier graduation
Micrometer Screw Applications Controlled traverse motion	Cross slide with micrometer heads	The micrometer principle is employed in machine tools and instruments to produce controlled traverse motion, either in a single direction or two, mutually perpendicular, directions (cross slide). Frequently micrometer heads act against spring urged slides, the micrometer filling the double function of traversing the member and indicating its displacement.	0.001 inch or 0.0001 inch (with vernier or direct reading with large diameter thimble)
Line Graduated Master Scales Extra fine graduation lines optically magnified for observation and alignment	Projected image of a section of the master scale Main scale Auxiliary scale 88 86 84	The operation of optical instruments for absolute length measurements is based on precise line graduated master scales. Increased resolution of the scale graduation for the purpose of fine incremental subdivision is accomplished through optical magnification of the master scale and the projection of the image on a screen that is equipped with a subdividing device.	0.000025 inch (with the aid of auxiliary scale)

TABLE 2-1.   SYSTEMS OF LINE GRADUATED MEASURING INSTRUMENTS – 3

CATEGORY AND SYSTEM OF OPERATION	DIAGRAM OF TYPICAL APPLICATION	DESCRIPTION OF SYSTEM	GENERAL MEASURING SENSITIVITY
Diffraction Gratings Graduation lines of known spacing but without assigned numerical values.	Basic and superimposed diffraction gratings producing interference fringes	Diffraction gratings are densely spaced parallel lines scribed on plane optical glass surfaces. A superimposed grating with slightly inclined lines will produce interference fringes, with location dependent on the relative position of the gratings. The operation of certain length measuring machines is based on the optical observation of the fringe variations.	0.000250 inch (with photoelectric sensor and electronic counter)
Line Graduated Angle Measuring Instruments Positions of extension arms associated with angular scale graduations.	Universal bevel protractor	Complete circles or segments with scales graduated in degrees, optionally complemented with vernier for fractional values in minutes of arc, are the essential elements of most types of industrial angle measuring instruments.	Direct reading to 1 degree; with vernier, 5 minutes (on stationary instruments also to 1 minute of arc)

By applying these categories, Table 2-1 presents a survey of the more important types of measuring instruments whose operation is based on the use of line graduated reference elements.

LINE GRADUATED BARS, RULES AND TAPES

The measurement of length being a process of comparing the distance to be measured to a known length dimension, the most obvious way for carrying out the measuring operation is by direct comparison.

Straight bodies of suitable length, with graduations in known subdivisions of the basic unit of length measurement, are, in principle, well adapted for the purpose of such comparisons. Starting from a fixed zero point on the scale, which is aligned with one end of the distance to be measured, the graduation line that is in a position corresponding to the other end of the distance indicates the length. Scale

**TABLE 2-2.   ACCURACY REQUIREMENTS FOR DIFFERENT
GRADES OF LINE GRADUATED MEASURING BARS AND RULES**

(Values are indicative only; actual accuracy, particularly for certain grades of reference bars, may substantially differ from those listed.)

MEASURING LENGTH, INCHES	PERMISSIBLE MAX. GRADUATION ERROR (DEVIATION) IN INCHES			
	REFERENCE BARS	INSPECTION BARS	WORKING RULES GRADE I.	WORKING RULES GRADE II.
4	±0.00022	±0.00044	±0.00088	±0.0022
8	±0.00024	±0.00048	±0.00096	±0.0024
20	±0.00030	±0.00060	±0.00120	±0.0030
40	±0.00040	±0.00080	±0.00160	±0.0040
General Formula: L = Total length of graduation	$\pm\left(0.0002 + \dfrac{L}{200,000}\right)$	$\pm\left(0.0004 + \dfrac{L}{100,000}\right)$	$\pm\left(0.0008 + \dfrac{L}{50,000}\right)$	$\pm\left(0.002 + \dfrac{L}{20,000}\right)$

numbers and graduation lines of different height assist in ascribing a numerical value to the corresponding graduation line.

Limitations of Length Measurements by Line Graduated Rules

The obvious simplicity of the line graduated rules measurement process must not conceal the diverse limitations inherent in length measurements by direct comparison. When using line graduated measuring tools for direct measurement of lengths, without the support of auxiliary equipment or instrument members, inaccuracies are to be expected due to several factors:

Instrument limitations

Geometric deficiencies resulting from flatness and parallelism errors, or caused by deflection;

Inaccuracies of scale graduations;

Excessive thickness or poor definition of the graduation lines;

Sensitivity limited by the least increment of the scale graduation.

Observational errors

Alignment deficiencies due to improper coincidence of the distance boundaries with the selected scale graduations;

Parallax errors when observing the scale and the object in a direction not perfectly normal to the surface being measured.

Accuracy of Line Graduated Bars and Rules

Whereas the observational errors can be reduced by means of auxiliary devices that improve the alignment of viewing and by keeping the measuring operation under control, the instrument errors are inherent inaccuracies of the process. The line graduated elements must, therefore, possess an original accuracy that is commensurate with the requirements of the particular measuring process. Although the accuracy requirements of this type of measuring instrument are not yet covered by American standards, the information presented in Table 2-2 (using DIN and other foreign standards) will indicate the instrument accuracies at different levels. These, or similar, accuracy levels may be considered as grades or classes, whose distinct statement on the instrument could be of guidance in evaluating the attainable precision of measurements made with that particular measuring tool.

Line Graduated Reference Bars

Although the concept of line graduated measuring instruments is generally associated with graduated rules, a tool of undisputable usefulness for many shop applications, it is not always realized that there are line graduated measuring bars that represent a very high level of accuracy. As a matter of fact, the basis of all precise length measurements for many decades, the international meter bar, is also essentially a line graduated measuring bar, although without the intermediate graduation.

Metrology laboratories serving scientific and industrial institutions have potential applications for *reference bars*. An important purpose that these reference standards are intended to serve is the calibrations of line graduated inspection tools and of line graduated master scales of measuring instruments. In such calibration processes the direct comparison of lengths by the unaided eye is, of course, not compatible with the required accuracy. The auxiliary devices that utilize the high inherent accuracy of graduated reference bars can be of different designs with a microscope tube as an essential element to provide precisely aligned and sensitive observation of the graduations.

In line graduated measuring bars belonging to a high class of accuracy, deflections can be a source of major errors, unless restrained by proper design. Line graduated reference bars are usually made in modified H, U or X cross-sectional forms, and the graduation lines are applied to that portion of the surface that lies in the neutral plane of the cross-section.

The graduations must not extend to the ends of the reference bar on either side. The surface carrying the graduation should be flat within a few ten thousandths of an inch and the graduation lines must be precisely parallel to each other. Special precautions must be observed in supporting these reference bars to avoid the effect of sag.

Line Graduated Inspection Bars

The primary purpose of line graduated inspection bars is to inspect the accuracy of common line graduated measuring rules, such as those used by toolmakers and mechanics. Inspection bars are usually made in square cross-sections, the side length of the cross-section varying as a function of the total bar length.

The actual length of the bar must extend on both ends beyond the lines of the graduation. The lines are scribed so that they reach the edge of the bar on one of the lengthwise sides. The graduation numbers are erect in that position of the bar where the graduation lines are on the farther edge. This arrangement will facilitate the comparison of the graduations by bringing the edges of the inspection bar and of the rule to be verified to bear closely against each other.

During the measuring process the inspection bar and the rule to be calibrated must rest on a surface plate of inspection grade flatness to avoid major deflections.

Line Graduated Working Rules

These tools serve for references when transferring length dimensions (for example, by means of a divider or a compass), or for direct measurements with the edge of the rule contacting the object to be measured. The cross-sectional form of these rules is rectangular, with a thickness-to-width ratio of about 1:5. According to their primary purpose of use, line graduated working rules are frequently made in two different grades of accuracy, as indicated in Table 2-2.

Note: Working rules comparable to those termed here as Grade II are, by no means, at the bottom of the quality ladder. There are in use many varieties of less accurate graduated rules, with stamped, etched, or printed graduations, made of metals or nonmetallic materials. These commercial types of graduated measuring tools, however, are not intended for precise measurements, and will not be discussed in this book.

The *graduations* are usually applied in a manner that makes the rules best suited for the kind of service they are intended to provide. For the purpose of referencing length dimensions, the graduations of Grade I rules start inside from the end of the rule body. Grade II rules, which are primarily used for direct measurements, have the zero point of the graduation lines coinciding with the end of the rule. This arrangement of the graduations, when applied systematically, also permits the grade to which the accuracy of a particular rule belongs to be recognized at a glance.

The engraving of the graduations on controlled accuracy steel rules is carried out on special graduating machines in order to match, within established tolerances, the master bars of the manufacturer. The master bars should be traceable to the reference stan-

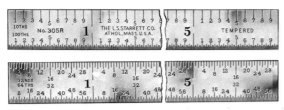

The L. S. Starrett Co.

Fig. 2-1. Steel rule with graduations on four edges—front side and reverse side. Decimal graduations: 10ths and 100ths. Fractional graduations: 32nds and 64ths.

dards of the National Bureau of Standards, in Washington, D.C.[1]

Most models of the industrial type graduated rules have more than a single graduation. Usually, both edges on a side, or on both sides, carry graduations, providing two or four full-length scales (see Fig. 2-1). Frequently used graduations are in 32nds and 64ths of an inch for the fractional scales, and in 10ths and 50ths, exceptionally, in 100ths of an inch for the decimal scales.

Graduated Rule Accessories

When using graduated rules for direct length measurements, which is a common toolmaker and machinist practice, the proper positioning of the rule in relation to the workpiece or the feature to be measured has a great effect on the final reliability of the measurement. There are various accessories available that can be of great help in improving the accuracy of the rule positioning.

Hooks, either of fixed length or with adjustable extension (Fig. 2-2), provide a reversed extension of

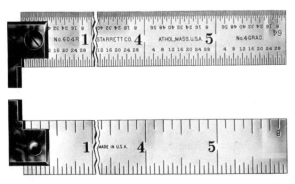

The L. S. Starrett Co.

Fig. 2-2. Reversible and removable hooks mounted on steel rules, for aligning the starting point of the rule graduations with the edge of the object.

---

[1] See Chapter 4 (p. 74), regarding the concept of traceability.

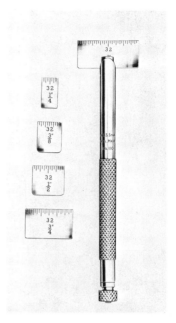

The L. S. Starrett Co.

Fig. 2-3. Clamping shaft for using short rules in confined locations.

the end face of the rule. Hooks permit a good alignment of the zero point on the scale with the borderline of an object surface, and also help to keep the rule in a position substantially normal to the edge of that surface.

Clamping shafts (Fig. 2-3) to hold short rule sections permit direct measurements in cramped areas, where the space confined within the walls of the feature does not provide access for regular size rules.

Parallel clamps (Fig. 2-4) make the rule applicable for use as guides in scribing on cylindrical surfaces lines that are parallel to the object axis and of specific lengths as measured with the aid of the rule graduations.

Foot rests permit graduated steel rules to be held in a substantially vertical position with the starting point of the scale graduations on the support level of the foot block.

The L. S. Starrett Co.

Fig. 2-4. Parallel clamps for aligning steel rules with the axis of cylindrical shafts.

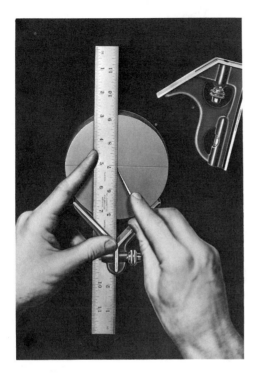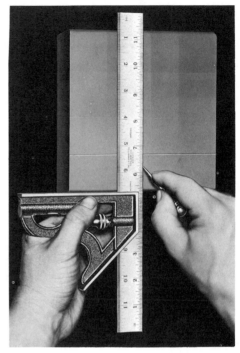

The L. S. Starrett Co.

Fig. 2-5. (Left) Center finders on steel rules permit scribing center lines on the face of round objects. (Right) Square head aligns rule normal to the edge of a straight-sided object.

Square heads and *center finders* (Fig. 2-5) are useful in layout work when the rule is used as a guide for scribing lines, or for measuring distances, square to the workpiece edge or through the center of a circular surface.

Steel Measuring Tapes

The most widely used method of measuring lengths extending over several feet or yards is by means of measuring tapes. The industrial quality steel measuring tapes—as distinguished from the commercial qualities—are made of special tape steel strips, heat treated and tempered, to ensure the consistent flexibility and dimensional stability that is generally required for measuring tools.

Styles, sizes and graduations. For the sake of conveniences of handling, and also for protecting the measuring tape, these instruments are commonly supplied in cases as integral units. The cases have a built-in reel mechanism, and the unwinding of the tape for measurement, as well as its retraction, is carried out by cranking the reel spindle. The small size pocket tapes usually have spring acting retraction mechanisms.

The industrial measuring tapes are currently made in 25, 50, 75 and 100 foot lengths; the pocket tapes are usually 6 and 8 feet long, although other sizes are not uncommon. The graduations are made $\frac{1}{8}$ of an inch on most industrial tapes, and in $\frac{1}{16}$ of an inch on pocket tapes. The inches are numbered consecutively, but start at each foot distance (continuous inch numbering is used for the shorter pocket type tapes). The number of feet preceding the current inch graduations is also clearly marked (Fig. 2-6). The graduations and figures may be applied by different methods, such as printing, embossing or etching, whichever is deemed to be best suited for the intended use of the tape, with particular regard to easy reading and durability.

Accuracy of measuring tapes. The National Bureau of Standards has excellent means and procedures for verifying and certifying master tapes,

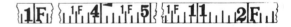

The L. S. Starrett Co.

Fig. 2-6. Sections of a steel tape showing system of graduation.

which then can be used as bench standards by the manufacturers of measuring tapes. Because of the relatively great lengths involved, the accuracy of steel measuring tapes is sensitive (due to expansion or contraction) to the variations of the ambient temperature from the standard 68°F level.

Measuring tapes, when not supported along the used length, will naturally sag, with adverse effect on the correctness of the measured dimension. For precise measurements the tape must be supported over the entire effective length on a substantially flat surface. Furthermore, a moderate force should be applied for stretching the tape. The standard tension for steel tapes up to 100 feet long, positively supported in the horizontal plane, is 10 pounds. Under these conditions steel measuring tapes of controlled quality can be expected to be accurate within 1/10,000th part of their nominal length.

LINE GRADUATED CALIPER GAGES

Essential Elements and Functions

When using line graduated rules to measure length dimensions on physical objects, the major single source of potential error lies in the alignment of the distance boundaries with the corresponding graduations of the rule.

The effect of this potential error is greatly reduced by means of the positive contact members, the jaws, of the caliper gages.

The caliper gage consists essentially of the following elements:

a. The beam, which is actually a line graduated rule of appropriate cross-section;

b. The fixed jaw, or base or cross beam, commonly integral with the graduated beam, and serving as the datum of the measurement; and

c. The sliding jaw, guided along the graduated beam by a well-adjusted slide. This slide has reference marks that are applied in a manner permitting precise alignment with the scale graduations on the beam. The coincidence of the slide marks with specific graduation lines will indicate the sliding jaw's position in relation to the datum.

Other elements of the various types of caliper gages that add to the versatility of their application and raise the accuracy of the measurements, will be discussed later.

The elements of the caliper gages listed before permit substantial improvements in the commonly used measuring techniques over direct measurements with line graduated rules. These improvements stem particularly from the following basic conditions:

1. The datum of the measurement can be made to coincide precisely with one of the boundaries of the distance to be measured;

2. The movable or sliding jaw will achieve positive contact with the object boundary at the opposite end of the distance to be measured (when this is identical with the physical boundary of the object, a condition that is typical for the majority of industrial measurements);

3. The closely observable correspondence of the reference marks on the slide with a particular scale value significantly reduces the extent of readout alignment errors.

Although these improvements of the gaging accuracy substantially extend the field of potential applications for line graduated scales, the essential advantages of reading length dimensions directly over a wide range are retained, with the only limitation arising from the capacity of the instrument.

These advantages make the caliper gages useful instruments of rapid measurements either for the purpose of final length determination, or to achieve a close approximation. In the latter type of applications, the value obtained with the aid of a caliper gage then serves as a guide for setting up more sensitive, but limited range, comparator type instruments for the final measurement at a higher level of accuracy.

Measuring Sensitivity and Accuracy of Vernier Type Caliper Gages

The measuring sensitivity of vernier type caliper gages may be evaluated on the basis of their capacity to associate the gaging position of the jaws with the scale graduation that is in correspondence with the actual size of the dimension to be measured. Slide position reference marks, by being scribed on a beveled surface of the slide window, will end at essentially equal level with the beam graduations, thereby reducing the incidence of parallax errors.

With few exceptions, all caliper gages used in industry have vernier graduations for reference marks on their slides. While the basic graduation of the beam is 1/20th or 1/40th of an inch (0.05 and 0.025 inch, respectively) (Fig. 2-7), the 50 or 25 graduation lines of the vernier, in effect, subdivide the individual beam graduations, resulting in an indirect readout resolution of 1/1000th of an inch (0.001 inch).

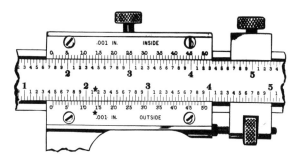

The L. S. Starrett Co.

Fig. 2-7. The slide of a vernier caliper gage for outside and inside measurements. The beam is graduated at 0.050-inch intervals and the vernier plate carries 50 division lines over the space of 49 beam graduations. When in a measuring position the zero mark on the slide is not exactly in line with a division line on the beam; the graduation value just preceding the zero mark must be supplemented with the value of the vernier indications. This latter value is displayed as the number of that single vernier line that exactly coincides with a graduation line on the beam. In the illustration, indicated by stars, that line of the vernier is the fourteenth, resulting in a combined indication value of 1.464 inches.

The actual power of resolution of a vernier caliper also depends on the width and definition of the graduation lines. These are usually heavier on shop grade calipers (0.002 to 0.004 inch wide), and finer on the inspection grades of instruments (0.0015 to 0.0020 inch).

The satin chromium finish of the better grade caliper gages eliminates glare, and stainless steel material reduces corrosion, thereby retaining the original clarity of the graduations.

The capacity of the caliper gages for carrying out accurate measurements is primarily contingent on the accuracy of the scale graduations, although the design of the slide guide ways, the parallelism of the jaw contact faces as well as their perpendicularity to the beam and, of course, the general workmanship of the instrument are also important factors.

The measuring accuracy of the instrument can best be expressed by the maximum indicating error for a specific range of measurements. As a guide, it can be expected that measuring errors due to the instrument, for any length within 4 inches, may be less than 0.001 inch for the inspection grade of vernier caliper gages, but could be three times as much for lower grades.

The Design of Vernier Caliper Gages

Caliper gages are made in different lengths, starting with a useful range of about 5 inches, and are available in lengths of up to 48 inches, or, exceptionally, even longer. The depth of the jaw also var-

ies, although not in a linear relationship with the measuring range.

The form of the jaws may be designed for measuring external surfaces only, or can be provided with features permitting internal measurements as well. Measurements on internal surfaces are made either by inside gaging jaw tips (Fig. 2-8) or directly, by knife edge auxiliary jaws (Fig. 2-9) that can pass each other along a common mating plane. When measuring inside distances with the jaw tips, the combined thickness of the two tips must be added to the indicated dimension, unless the beam has separate graduations for outside and inside measurements (see Fig. 2-7).

Vernier calipers intended for general applications are made with plain sides, usually equipped with *internal locking springs* (Fig. 2-9), whereas for precision measurements the combination type slides are preferred. The latter slides (Fig. 2-7) have a sliding clamp that can be locked in position and serves as the nut for a *fine adjusting screw* acting on the measuring slide.

Some caliper gages have a *depth gage tongue* (Fig. 2-9). This feature has some usefulness for rapid measurements, although the precision of depth gaging by this means is not considered to be equivalent to measurements between the jaws of the same instrument.

The purpose of the *locking screw* on the slide is to maintain a fixed jaw opening for the sequential measurement of parts with the same nominal size, using the caliper as a snap gage. Calipers must not,

The L. S. Starrett Co.

Fig. 2-8. (Left) On external surfaces. (Right) On an internal feature using the offset tip of the jaw.

Etalon/Alina Corp.

Fig. 2-9. Universal vernier caliper gage with knife edges for internal measurements and tongue for depth measurements.

however, be considered equivalent substitutes for snap gages, primarily because the setting accuracy of a vernier caliper is inferior to the gage maker tolerances to which snap gages are made. The locking screw should be on the top side of the slide in order to draw the slide against the lower edge of the beam. The end of the locking screw must not act directly on the beam, but should transmit the locking force through an appropriate insert strip.

Caliper Height Gages

The basic design principles of vernier caliper gages are also applied in caliper height gages. The primary use of caliper height gages is in the field of surface plate work as a layout tool, for marking off vertical distances and for measuring height differences between steps at various levels.

Vernier height gages differ from caliper gages in that they have a single jaw, because the surface plate on which the instrument base rests functions as the reference plane. Vernier gages usually have offset jaws whose contact surfaces can be brought to coincide with the reference plane when the slide position indicates zero height.

The nonintegral reference plane requires great locating stability and internal rigidity from height gages in order to maintain a beam position that is substantially perpendicular to the surface plate. Consequently, with the exception of the less accurate models, vernier height gages are made with wide bases and with bars of strong cross-section; the bars carry the same kind of graduations found on the beams of vernier calipers (Fig. 2-10).

For marking off purposes, scribers can be attached to the contact jaw. These scribers are designed to have the edge substantially at equal level with the contact surface of the jaw, in order to make the height of the scribed line coincident with the dimension indicated by the gage.

Gear Tooth Vernier Calipers

A further application of the vernier caliper principle is found in the gear tooth calipers. These measuring tools are used to check the pitch line thickness of gear teeth by measuring the tooth chord at a specific distance (chordal addendum) from the top of the gear tooth. Although possible errors in the gear blank size, as well as other factors, reduce the ultimate precision of these measurements below the level that could be expected from the 1/1000th-inch sensitivity of the vernier indication, this method of gear inspection still has useful applications in general shop practice.

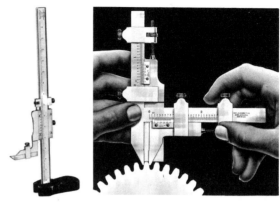

The L. S. Starrett Co.

Fig. 2-10. (Left) Vernier height gage with relief base for referencing measurement from the supporting plane of the base. A scriber is attached to the gage jaw.

Fig. 2-11. (Right) Application of a gear tooth vernier caliper with the gage tongue resting on the top of the gear tooth.

The gage (Fig. 2-11) consists of two independently actuated vernier calipers, each having its own movable slide, but the beams and the stationary jaws are made of a common single piece. One of the slides has the form of a plate, called the tongue of the instrument, which contacts the top of the gear tooth. By moving this slide, the gage can be adjusted to operate at the desired addendum distance. The second slide, integral with the movable jaw, carries out the actual chordal thickness measurement at the pitch line.

Vernier Depth Gages

Operating on the principles of vernier caliper gages, vernier depth gages (Fig. 2-12) differ in the design application of the basic concept, resulting in the reversal of the usual process:

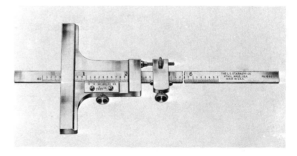

The L. S. Starrett Co.

Fig. 2-12. The vernier depth gage constitutes a modified version of the regular vernier caliper, substituting the face of the beam for the fixed jaw, and employing an element known as the head for the slide.

1. The slide is connected with the cross beam of the instrument, which, by contacting a reference plane on the object surface, establishes the datum of the measurement.

2. The end face of the beam takes over the function of the movable jaw, in contacting the object element whose distance from the datum is to be measured.

The use of the vernier depth gage is not limited to actual depth measurements, although these constitute the major application for the instrument. Distances that are referenced from and are normal to a flat surface of the object, can also be measured conveniently with the aid of vernier depth gages. An example of this latter type of use is measuring the coordinate distances of toolmaker buttons in relation to the edges of a rectangular workpiece.

Caliper Advancements

The dial caliper (Fig. 2-13) resembles and is used in much the same way as the traditional vernier caliper. However, the dial caliper is equipped with a dial indicator in place of a vernier scale. The dial version of the caliper is more easily read than the vernier type. As such, measurement reading errors are minimized and speed of use is increased.

More recently, the caliper has entered the electronic age with the introduction of the digital caliper (see Fig. 2-14). The digital caliper again resembles the traditional universal vernier caliper. However, this newest caliper type has no line graduated measuring scale. The reading is electronically generated and digitally displayed. For further discussion on electronic gages refer to Chapter 7. This digital technology also has been adapted to other common caliper types, such as the large universal caliper and the depth caliper (see Figs. 2-15 and 2-16).

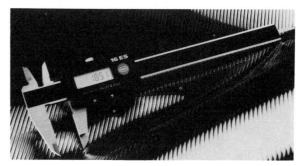

Mahr Gage Co. Inc.

Fig. 2-14. Digital caliper.

The digital caliper, powered by a standard 1.5 volt silver-oxide button battery, offers several advantages over the traditional vernier and newer dial calipers. These advantages include the following: (a) the measurement is read in a single step, (b) a zero

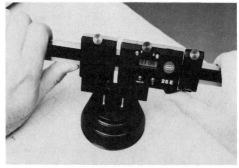

Mahr Gage Co. Inc.

Fig. 2-15. Universal digital caliper.

Mahr Gage Co. Inc.

Fig. 2-16. Digital depth caliper.

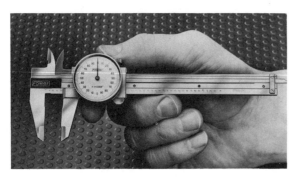

Fred V. Fowler Co. Inc.

Fig. 2-13. Dial caliper.

Mahr Gage Co. Inc.

Fig. 2-17. Digital caliper with minicomputer.

setting from any measuring position is possible, (c) English unit/metric changeover takes place with the touch of a button, and (d) the electronic tying of the digital caliper to a minicomputer or microcomputer for purposes of data collection and statistical evaluation is also possible (see Fig. 2-17).

MICROMETERS

Just a few decades ago micrometers were considered the ultimate in precision for length measurements. Even the measuring machines used at those times in gage laboratories operated on the principles of micrometers.

More sensitive measuring instruments were developed to satisfy the requirements for higher accuracy, and they replaced the use of micrometers in many applications. Most of the higher sensitivity instruments, however, are of the comparator type, which require setting masters for referencing, possess only a reduced measuring range and are, to some extent, generally stationary. These, and many other characteristics, are definite disadvantages in comparison to the portable micrometer, with its capability of measuring actual lengths by direct indication over a substantial measuring range. In addition to these inherent advantages of the micrometer system and the relatively low price of the instruments, many recent improvements in the design and manufacture of micrometer gages contribute to assuring a continuing important role to this category of measuring tools in the field of length measurements.

Operating Principles and Mechanism

The essential element of measuring instruments operating on the micrometer principle is a screw with precisely controlled lead, having a pitch of usually, but not necessarily, 40 threads per inch. The screw of the micrometer is integral with the mea-

suring spindle, whose face establishes the measuring contact with the object. The distance of that contact face from a fixed datum constitutes the measuring length, which is then displayed by the scale graduations of the micrometer.

The micrometer has two basic scales. One is a linear scale to measure directly the axial advance of the spindle in increments large enough to provide distinct reading of the graduation lines. The graduations of this scale are usually identical with the pitch of the micrometer screw. The other is a circumferential scale around a hollow cylinder locked to the micrometer screw to indicate the amount of partial rotation when the last turn of the screw during the gaging process is less than a complete revolution. Micrometers designed for measurement in finer increments also have a vernier scale to permit the evaluation of fractions of circumferential graduations.

Figure 2-18 shows a sectional view of an outside micrometer to illustrate the manner in which these operating principles are actually applied in the design of the instrument. The illustration also shows the nomenclature of the essential elements of the instrument. Correlating the denominations of this particular instrument model with the more general terms used in describing the principles of the mea-

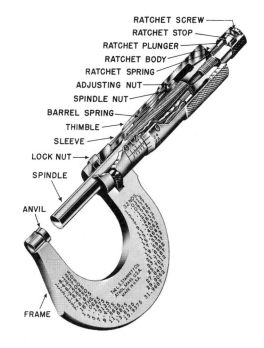

The L. S. Starrett Co.

Fig. 2-18. Sectional view of a micrometer caliper.

suring process we find that: the anvil performs the role of the datum; the sleeve, a fixed member, carries the scale of the axial displacements, as well as the reference marks (single line or vernier) for the circumferential graduation. The reference marks are scribed on the thimble, termed a hollow cylinder in the general description.

Other elements of this type of micrometer instrument, which are shown in the sectional view and have auxiliary functions, are the following:

The *ratchet mechanism* which serves to ensure a uniform measuring force; that would be difficult to maintain just by feel, considering the great mechanical advantage resulting from a fine pitch screw. The ratchet controls the torque that can be transmitted to the threaded spindle, thereby limiting the measuring force to a preset value, for example, two pounds. In some other models of micrometers a friction screw is used instead of the ratchet.

The *adjusting nut*, engaged with the threaded spindle, serves to eliminate play should it develop after prolonged use of the micrometer.

The *sleeve*, maintained in position by a barrel spring, provides means of adjustment should the zero line of the thimble not coincide with the reference mark on the sleeve when the measuring surfaces are in contact with each other.

The *lock nut*, which permits the locking of the spindle at the desired position as determined by the scale reading, and then using the micrometer in a manner comparable to a snap gage.

The Precision of Micrometer Readings

Considering the wide use of screw micrometers, it is desirable to know the degree of confidence that the results of micrometer measurements deserve. The repetitive precision of measurements with a screw micrometer depends on two sets of factors: the *inherent accuracy* of the measuring instrument, and the combined effect of *process errors*.

The *accuracy of the micrometer* will be governed primarily by the following two factors:

The *degree of calibration* of the spindle movement, which will be affected by the lead errors of the screw; the effect is a usually cumulative, and increases with the length of the spindle travel. (*Note*: The aggregate effect of inaccuracies originating from screw lead errors can be reduced by "balanced calibration," that is, by adjusting the thimble to produce error-free reading in the middle of the total, or of the most frequently used section of the spindle traverse.)

The *linearity* of the spindle movement, requiring that any fractional rotation of the screw should result in a proportional advance of the measuring spindle; "drunken" thread, or stick-and-slip condition of the screw in the nut, will have an adverse effect. Deficient linearity will become particularly harmful when superimposed on major calibration errors.

The instrument accuracy is substantially improved by manufacturing the spindles of well stabilized material, precisely grinding the screw thread after hardening, using lapped nuts and applying in general a high degree of workmanship in the manufacturing process of these instruments.

The *calibration process* is a dependable means for assessing the accuracy of the micrometer indications. The process consists of the sequential measurement of gage blocks of known size with the micrometer. The blocks to be measured are selected to represent distances over which the spindle travels for a full or a half turn of the screw. In each step the dimension indicated by the micrometer is recorded on a calibration chart, as shown in Fig. 2-19. The zero line of the chart indicates the nominal size for each reading, and digressions of the actual indications are plotted in this sense and by the scale of the chart ordinate.

The chart recording will usually have a highest crest and a lowest valley, representing the points of the maximum deviation from the nominal level. As a second step, the measurements should then be repeated over a full screw rotation covering a distance that is symmetrically distributed on both sides of the

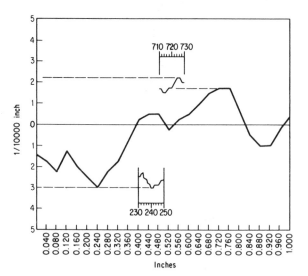

Fig. 2-19. Calibration chart, prepared for a Grade I, outside micrometer with zero- to one-inch measuring range.

originally detected peak deviations in either direction, as shown by the two small inserts in Fig. 2-19. This repeat measurement must be made in increments five to ten times smaller than the steps of the original calibration. Intermediate positions explored by the second measurement could reveal that the actual deviation peak is even greater than the originally charted point.

For the evaluation of the micrometer accuracy, the spread of the deviations in the direction of the ordinate axis may be considered the significant dimension. The errors recorded on the calibration chart will comprise the combined effect of all factors that are related to the measuring accuracy of the micrometer. The more important of these additional factors are as follows:

The flatness and parallelism of the measuring surfaces. The precise method for inspecting this condition is by means of an optical flat (Fig. 2-20). As a general rule, the number of visible interference lines under monochromatic light must not exceed the following values: two fringes for flatness (when checking any one of the measuring surfaces) and six fringes for flatness and parallelism combined (making simultaneous contact with both measuring surfaces and using an optical flat whose faces are plane and parallel).

Deflection of the frame. The applied measuring force will cause a deflection of the frame, resulting in the separation of the measuring surfaces. This effect can be reduced by appropriate frame design and

The Van Keuren Co.

Fig. 2-20. The use of an optical flat for inspecting the flatness and parallelism of the measuring surfaces of a micrometer.

by limiting the applied measuring force, with the aid of ratchet or friction screw, to about two pounds. When kept under proper control, the potential measuring error caused by frame deflection can be kept within about 50 microinches for the one-inch size outside micrometer. The amount of deflection will be larger for micrometers of greater frame size or having wider measuring ranges.

Considering the various factors that affect the measuring accuracy of the frame type outside micrometers, a total spread of potential errors not exceeding 0.000150 inch will be indicative of a good quality of micrometer in the one-inch size range. As a guide, the following formula may be considered for assessing the expected measuring accuracy of a precision grade outside micrometer:

$$\Delta = (150 + 10 \, L) \text{ microinches}$$

where

Δ = aggregate measuring accuracy over the total measuring range of the micrometer;

L = nominal size (maximum measuring length) of the micrometer, in inches.

The *process* errors of micrometer measurement can be caused by heat transfer while holding the instrument, reading errors, inadequate alignment or stability in the mutual positioning of object and measuring tool, wear and many other circumstances.

Various design improvements serve to reduce the incidence of these errors in micrometer measurements. Particular design features directed at increasing the dependability of the measuring process by reducing the effect of some of these potential errors, are discussed below. The design features are listed in groups according to the particular source of potential inaccuracy that these improvements primarily control.

Heat transfer can be reduced by employing plastic insulating grips on the frame. With few exceptions, the micrometer frames are made of steel forgings, a material with practically the same coefficient of thermal expansion as most of the parts to be measured. Because of the differential in the rate of thermal expansion, aluminum frames, although light in weight, are seldom used for micrometers.

Reading errors are substantially reduced by such design features as follows:

a. Satin chrome finish to eliminate glare;

b. Distinct graduation lines applied on a beveled thimble surface to facilitate reading with a minimum of parallax error. Some models of precision micrometers have the graduated surface of the thimble and sleeve mutually flush.

c. A particular type of micrometer has window openings on the thimble, where the hundredths and thousandths values of the measured dimension appear in digits, and only the tenths and the ten-thousandths of an inch values must be determined by reading the graduation lines;

d. Large diameter thimbles for direct reading of the ten-thousandths by graduation lines coinciding with a single reference mark, thus eliminating the vernier as a potential source of reading error.

Alignment and holding stability can be improved for measurements of small parts by using a stand that rests on the bench to clamp the hand micrometer instead of holding it in the free hand. For the repetitive measurement of light parts the use of bench micrometers can provide definitive advantages.

The *applied measuring force* is usually limited by friction screw or ratchet. When precise control of the static measuring force is required, indicator micrometers may be used. These will be discussed further on in the chapter.

Wear will most commonly occur on the measuring faces because of their direct contact with the workpiece. Carbide-faced anvils and spindle tips greatly reduce the wear on these surfaces, thus maintaining over a prolonged period of use the original setting of the micrometer, as well as the parallelism of the contact surfaces. After prolonged use of the micrometer, wear will occur in the threaded members that can affect the original setting and measuring accuracy of the micrometer. *Resetting* the thimble position to the original calibration (Fig. 2-21) and *readjusting* the clearance of the spindle movement by tightening the nut (Fig. 2-22) will usually improve the functioning of the instrument to a level equal or comparable to its original accuracy.

The L. S. Starrett Co.

Fig. 2-21. (Left) Resetting the thimble position of a micrometer to original calibration.

Fig. 2-22. (Right) Readjusting the spindle nut of a micrometer to eliminate harmful play between the members.

Micrometers for Outside Measurements

The most common application for micrometers is the measurement of length dimensions between two parallel end surfaces on the outer side of an object or feature. This is currently known as *external* or *outside measurement.*

The micrometer is a convenient instrument for outside measurements because (1) it supplies promptly the desired size information; (2) it has a relatively wide range of measurement (for example, 10,000 times the least increment of its scale); (3) its use does not require specialized skill; and (4) the measuring tool is easily adaptable to diverse object forms. These and many other advantages account for the wide application of micrometers in production and inspection.

It is a logical consequence of this extended field of potential applications that outside micrometers are required and made in many varieties of sizes, forms and accuracy grades.

The *size* of a micrometer refers to the limits of its measuring range. The most common size is the one-inch size, which permits measurements over a range from zero to one inch in length. The same range of one inch can also be applied to other limit dimensions by mounting the basic screw mechanism on other frame sizes. Accordingly, outside micrometers are also made in sizes of 1 to 2 inches, from 2 to 3 inches, and so on to a 24-inch maximum limit, exceptionally even larger. Because micrometers whose lower range limit is greater than zero cannot have direct contact between spindle face and anvil, setting masters must be used for calibration purposes. These setting gages, also known as standards, must be a size equal to the lower limit of the micrometer's measuring range, and are usually procured together with the micrometer.

The larger size micrometers are available in sets, complete with standards, generally in cases with individual compartments, permitting micrometer measurements over a wide range of lengths, from zero up to the maximum size of the largest instrument in the set.

Although the basic form of the outside micrometer frame is a bow, with the spindle at one end and the anvil at the other end, variations in design are frequent, particularly in the larger sizes. The common micrometers have fixed anvils, limiting the useful range of measurements to the length of the screw travel; however, here are models with adjustable or interchangeable anvils, resulting in a widely extended measuring range (Fig. 2-23).

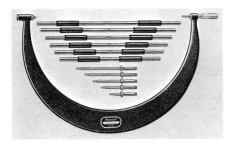

The L. S. Starrett Co.

Fig. 2-23. Outside micrometer for extended measuring range with interchangeable anvils and setting standards.

While on most standard micrometers the *contact surfaces* on the spindle and on the anvil are the flat faces of these basically cylindrical members, micrometers are also made for special measurements where the contact tips have particular forms. Examples are the following:

a. The *disc type micrometer* for thickness measurements on features that provide a narrow clearance only for the penetration of the gage contact elements (Fig. 2-24);

b. The *blade type micrometer* to measure diameters at the bottom of narrow grooves (Fig. 2-25);

c. The *screw thread micrometer* with vee-formed anvil and conical spindle tip, to measure pitch diameters;

d. The *pointed contact micrometer,* with conical tips having a very small flat land on the spindle and on the anvil, to measure inside recesses where regular tips could not penetrate; and

e. The *ball contact micrometer* for measuring the wall thickness of tubes and other cylindrical bodies. Usually the ball anvil is supplied as an attachment for standard outside micrometers. The attachment

The L. S. Starrett Co.

Fig. 2-24. The disc type micrometer is designed for measuring the thickness of closely spaced sections.

The L. S. Starrett Co.

Fig. 2-25. Blade type micrometers can penetrate into the bottom of narrow grooves. The spindle is of nonrotating design.

can easily be snapped on the end of either the anvil or the spindle. The diameter of the ball, commonly 0.200 inch, must be subtracted from the value of the micrometer reading.

It is customary to assign the *grade* or quality level of micrometers on the basis of the least increment of the indicated dimension. Although most micrometers have screws with 40 threads per inch, resulting in a 0.025-inch advance per turn of the screw, the graduations of the thimble can vary, providing different subdivisions of that advance for each complete turn. A plain 25 graduation on the thimble circumference observed by a single line mark on the sleeve will permit the reading of each 0.001-inch spindle advance, this type of micrometer being commonly designated the "thousandths" grade. Substituting a vernier scale for the single reference mark on the sleeve will permit the subdivision of each thimble graduation interval into ten parts, resulting in a "ten-thousandths" micrometer. Similar results can be accomplished by retaining a single line reference mark, but subdividing each of the 25 basic thimble graduations into 10 intermediate increments. This arrangement will allow the direct reading of the ten-thousandths, thereby eliminating the potential errors connected with the reading of the vernier. For practical reasons, however, the direct subdivision of the basic thimble graduations is only feasible for oversize thimbles, with 2-inch or larger diameter.

Although the least graduation is not equivalent to accuracy, it is a common practice with the manufacturers of micrometers to assign instruments with the finer scale increments to a higher grade of overall accuracy, supported by suitable design and appropriate workmanship.

Micrometers for Inside Measurements

The micrometer type instruments have many useful applications for inside measurements as well.

These comprise the measurement of an object feature bounded by walls having parallel elements in the gaging plane, such as the diameter of a cylindrical bore or the width of a parallel-sided groove.

Inside measurements, particularly of bore diameters, are affected by more variables than the measurement of length between outside surfaces. Small bore sizes prevent the introduction of inside micrometers. The depth of the bore, when its diameter must be measured at a greater distance from the open end, can cause difficulties for micrometer applications. Finally, the requirement of measuring a cylindrical feature across its axis and in a plane normal to the axis, calls for conditions that may be accomplished only partially when relying solely on the skill of the operator, this being the case when using a plain inside micrometer.

The particular aspects of inside measurements created the need for a variety of inside measuring instruments operating on the micrometer principle, or used in conjunction with a regular outside micrometer.

The *inside micrometer* consists of a head part comprising the micrometer screw, and of independent measuring rods. The rods are made in different lengths, in steps of one inch, can be assembled with the head part by means of a threaded connection and are accurately positioned on a shoulder. The micrometer screw of the smaller size heads has a $\frac{1}{2}$-inch movement, and the rods are attached either directly, or by using a $\frac{1}{2}$-inch spacing collar, which is supplied with the instrument. Inside micrometer heads for measurement of bore diameters larger than 8 inches have a full inch of screw travel.

The smallest bore that can be measured with this type of micrometer is 2 inches in diameter (exceptionally $1\frac{1}{2}$ inches), and the maximum diameter depends on the available rods, a practical upper limit being about 32 inches. A handle can be attached to the head to permit measurement at greater depths inside a bore (Fig. 2-26).

Bore diameter measurements by transfer. Small bores cannot be measured directly because of space limitations. It is possible, however, to transfer the inside dimension to be measured by using an appropriate means that will represent the reverse replica of the inside length. The resulting physical outside length can then be measured with a standard outside micrometer. Examples of such transfer devices are the following:

The *small hole gage* (Fig. 2-27), consisting of a split ball that can be expanded to the size of the diameter of the bore to be measured. Rotating the

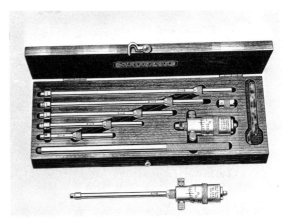

The L. S. Starrett Co.

Fig. 2-26. Inside micrometer set with interchangeable measuring rods and handle for inserting the instrument into deeper holes.

knurled knob of the handle advances the inside screw whose conical end causes the ball halves to separate to the required extent. The actual spread of the balls is small, on the order of $\frac{1}{8}$ inch, and therefore these gages are supplied in sets encompassing a wider range of dimensions.

The *telescoping gage* (Fig. 2-28) has a tubular member to which a handle is attached at right angle position. The tubular member has either one or two plungers, which are under spring pressure and telescope into the fixed tube. The free ends of the plungers or, in the case of a single plunger, the closed end of the fixed member also has spherical forms and

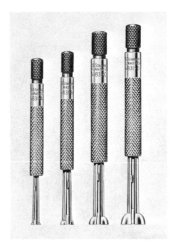

The L. S. Starrett Co.

Fig. 2-27. Small hole gage for the transfer of actual bore dimension, which is then measurable with a regular outside micrometer.

The L. S. Starrett Co.

Fig. 2-28. The use of a telescoping gage in taking the size of a bore for ultimate measurement by an outside micrometer.

are hardened to serve as contact elements. When introduced into the hole to be measured, the previously retracted plungers are released to extend to a length equal to the diameter of the object. In this position the plungers can be locked again by turning a knurled screw in the end of the handle. Subsequently, the thus fixed length of the telescoping rod can be measured with a regular outside micrometer.

Three-point contact internal micrometer. Several of the difficulties connected with bore diameter measurements by a micrometer can be reduced when using the three-point internal micrometer, shown in Fig. 2-29. The self-aligning property of this instru-

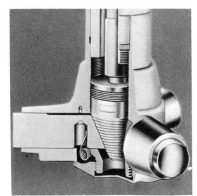

Brown & Sharpe Mfg. Co.

Fig. 2-29. Three-point contact, internal micrometer with self-centering properties.
(Left) General view.
(Right) Partial cross-section showing the mechanism for the controlled expansion of the contact plungers.

ment is particularly useful when measuring deep bores, for which purpose an extension part can be attached to the basic tool.

A precisely ground spiral on a cone surface functions as a seat for the shafts of the three self-aligning measuring points. These are contained in the measuring head of the instrument and are spaced 120 degrees apart. Advancing the cone along its axis spreads the engaged measuring points radially, resulting in a larger envelope circle, and the cone movement in the inverse sense causes the spring loaded contact members to retract. The cone is attached to a spindle whose axial position is shown on the micrometer sleeve and thimble. The smallest graduation of the micrometer varies from 0.0001 inch to 0.0005 inch, depending on the size and range of the instrument.

Three-point internal micrometers are available in sizes from 0.275-inch to 8-inch bore diameters. The individual tools have measuring ranges varying from 0.075 inch for the smallest size, to a full inch for larger sizes. Usually these micrometers are procured in sets to cover a more extended range of measurable bore diameters.

It is advisable to recalibrate these instruments periodically, using a standard ring gage and applying the regular micrometer adjustment procedures (see Figs. 2-21 and 2-22).

Special Micrometer Instruments

The following examples of measuring instruments, based on the application of the micrometer screw as the measuring member, should illustrate the wide uses of the micrometer principle.

Indicator micrometers (Fig. 2-30). The narrow range indicator, whose mechanism is coupled with the movable anvil of the micrometer frame, has graduations in 100-microinch or 50-microinch increments. The smallest graduation of the thimble is in thousandths and the sleeve of the illustrated type carries no vernier graduations. There are other models that have verniers on the sleeve; in these the indicator acts as a fiducial gage.

When making measurements with the instrument without vernier, the micrometer screw is advanced to the thousandth mark nearest to the final size as signaled by the movement of the indicator pointer from its rest position. At this point, the measured size is read by combining the size shown on the thimble and the position of the indicator pointer.

The built-in indicator improves the repeat accuracy of the micrometer measurements and adds to

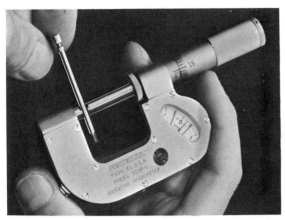

Federal Products Corp.

Fig. 2-30. Indicator micrometer with adjustable tolerance marks. Usable as a regular micrometer and also as an indicating snap gage.

the versatility of the instrument applications because of the following:

a. The constant measuring force, as checked by the pointer movement, is particularly valuable for comparative and repetitive measurements;

b. Measuring errors caused by mistakes in evaluating the vernier position are eliminated;

c. The micrometer can be used as an adjustable snap gage with added indicating potential, when the range of size variations does not exceed the measuring spread of the indicator. A lever permits retracting the anvil while the object is introduced so as not to mar the work surface. The adjustable indexes on the indicator sector can be set to the limit sizes of the object; and

d. False diameter measurements due to a slanted position of the micrometer, instead of it being strictly normal to the work axis, are avoided. (The minimum reading is observed on the indicator, as the true diameter value when measuring in the axial plane of a cylindrical object.)

Depth micrometers are used to measure the distance of an object feature from a flat reference surface (Fig. 2-31). Examples of application are the measurement of steps, the depth of flanges or the bottom surface in a bore, and the height of an object feature in relation to a reference surface, when accessibility permits it to be simultaneously contacted by the base member and the spindle face.

Bench micrometers (Fig. 2-32) can substantially improve the precision of micrometer measurements, particularly when the objects to be measured are

The L. S. Starrett Co.

Fig. 2-31. Depth micrometer for measuring the distance of protruding features from a flat reference surface.

small. The stable position of the instrument during the measurements permits a more precise locating of the work and the heavy base adds to the rigidity of the instrument. Bench micrometers are usually equipped with both large-diameter spindles and thimbles, permitting a finer pitch thread for higher sensitivity, and direct reading in ten-thousandths, or smaller.

Micrometer heads without a frame as an integral part have wide applications in the instrument

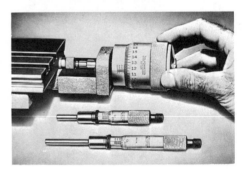

The L. S. Starrett Co.

Fig. 2-32. Bench micrometer for the sensitive measurement of small parts. The indicator serves auxiliary functions such as the assurance of uniform gaging force and the option of also using the device for comparator measurements.

Fig. 2-33. Micrometer heads as independent units and mounted on an instrument slide.

building field. One of the more frequent uses is the controlled displacement of slides, either single or cross slide (Fig. 2-33). Commonly, the micrometer will move the slide against a moderate spring pressure, which is applied to assure a positive contact between the micrometer spindle face and the contact point on the slide face. Springs exerting a uniform force along the entire slide traverse are preferred.

Micrometer Advancements

The technological advancements of the dial indicator and digital electronics previously discussed (see caliper advancements) have also been separately incorporated into the micrometer. The *dial comparator micrometer* (Fig. 2-34) is equipped with an indicator that takes the place of the vernier scale typically found on the sleeve and thimble. The indicator facilitates measurement reading speed and accuracy.

The *digital micrometer* (Fig. 2-35), which has an electronically generated digital display, is now without a line graduated measurement scale. The digital display is easy to read and may be reset to zero from any measuring position. Also, the display can be set to read in either inches or millimeters. Finally, due to the electronic nature of the digital micrometer, data collection and statistical evaluation via computer are possible (see Fig. 2-36).

Mahr Gage Co. Inc.

Fig. 2-35. Digital micrometer.

Mahr Gage Co. Inc.

Fig. 2-34. Dial comparator micrometer.

Mahr Gage Co. Inc.

Fig. 2-36. Digital micrometer with minicomputer.

3.
Fixed Gages

Fixed gages embody the direct or reverse physical replica of the object dimension to be measured. The gage may represent the part dimension either in its nominal condition—*the master gages*—or in one of its limit conditions resulting from the specified tolerances of the dimension—*the limit gages.*

The master gages, which constitute a direct replica, are primarily used as *setting gages* for setting up comparator type measuring instruments or as *reference standards* for calibrating measuring tools that require periodic readjustment. Master gages are dimensioned to represent the dimension to be gaged either in its basic size or the median size of the designed tolerance zone.

The limit gages are made to contain the reverse replica of the dimension to be checked, and are intended for the purpose of *inspection gages*. Because manufactured parts are made, as a rule, to specific tolerances, the acceptable size of the part is determined by its limit dimensions. Accordingly, for the inspection by fixed gages of any critical dimension on a manufactured part, two gages are usually required that are made to constitute the *design limits* of that dimension.

The gaging portion of the fixed gage is manufactured to consistently retain a specific size. This goal is usually accomplished by means of *rigid gage elements*, although gages with adjustable members are also used as fixed gages, particularly in limit gage applications. For the purpose of gage classification, however, the functional service of the gage should be considered the essential characteristic; gages that physically represent fixed sizes during their intended use, belong to the broad category of fixed gages.

Limit gages can be designed to check a *single distance*, length or diameter, or may serve the inspection of *multiple dimensions*. This latter may comprise contours, a combination of contours and length dimensions, geometric forms individually or combined with length, such as tapers and cylinders, or the interrelation of several forms, sizes and their location on a single body or within an assembly.

CLASSIFICATION OF FIXED GAGES

Table 3-1 lists the more common types of fixed gages and assigns them to somewhat arbitrary categories. These categories are not based on any standard classification, but have been selected to facilitate a review of that multiplicity of measuring tools that fall under the general term of fixed gages.

ADVANTAGES OF FIXED GAGES

Fixed gages constitute one of the pillars on which interchangeable manufacturing has rested since its early origins. The nearly exclusive position in inspection that the fixed gage enjoyed in the original stages of interchangeable manufacturing has been widely contested by the indicating types of comparator instruments. These comparator instruments have two advantages: (1) a sensitivity greater and more dependable than the skill and judgment—the "feel"—of the user, on which the effectiveness of fixed gages rests, and (2) the capability of indicating in positive values the amount of deviation from a nominal dimension.

The comparator type gages, however, are only replacing the fixed gages in some of their applications

TABLE 3-1.　CATEGORIES OF COMMONLY USED FIXED GAGES — 1

CATEGORY	DESIGNATION AND SKETCH OF TYPICAL GAGE	DIMENSIONS AND CONDITION GAGED	APPLICATION EXAMPLES
Master Gages	Master setting disc	Nominal dimension or limits of gaging range	Setting comparator type measuring instruments, calibrating air gages, adjusting micrometers, checking limit gages.
Limit Length Gages Rigid or Adjustable	Adjustable limit plug gage　　Adjustable limit snap gage	Single length dimension or diameter	Checking the diameter of round external or internal features, or the distance between parallel boundary surfaces. The gaging is carried out by using sequentially the GO and the NOT GO gage of the toleranced dimension.
Cylindrical Limit Gages	Cylindrical limit plug gage　　Cylindrical limit ring gage	Maximum and minimum material condition of basically round parts. The gaging extends over an annular tolerance zone.	Inspecting the combined effect of dimensional variations and geometric irregularities of nominally round inside and outside surfaces.
Geometric Form Gages	Taper plug gage　　Taper ring gage	Regular geometric forms individually or in combination with size.	Checking straightness, perpendicularity, etc. with straight edges, squares. Geometric bodies, like a cone with taper gages.

TABLE 3-1.   CATEGORIES OF COMMONLY USED FIXED GAGES — 2

CATEGORY	DESIGNATION AND SKETCH OF TYPICAL GAGE	DIMENSIONS AND CONDITION GAGED	APPLICATION EXAMPLES
Multiple Dimension Gages	Limit thread gage plug Limit thread gage ring	Features whose adequacy is determined by the combined effect of several parameters.	Threads, serrations, splines, etc. whose feature form, the sizes of the critical dimensions and their relative locations must be checked in their cumulative effect on the effective size.
Contour Gages	Male Female Radius gages (templates)	The conformance of the object contour with the gage contour.	Checking radii, fillets, screw thread pitch, gear contours, etc. by mating gage with object and then observing the condition of coincidence either directly or by means of optical magnification.
Assembly Gages	Assembly gage for coaxial bores	Size and geometric form of individual surfaces, as well as their interrelations such as coaxiality, alignment, etc.	Housings with several coaxial bores; size, alignment and location of keyways; indexing positions of grooves, etc.

as *working* or *inspection* gages. Even the most sensitive types of comparator gages must still rely for their setting on masters, which are usually fixed type gages.

Fixed gages provide many advantages, which account for their important role in engineering production. Some of these advantages, selected at random, are listed in the following:

1. *Consistent in form and dimension.* Fixed gages are essentially free from errors due to drift of the original adjustment, nonlinear response, effect of power variations and other extraneous factors that necessitate regular calibration and occasional correction on many comparator type measuring instruments.

2. *Positive dimensional information.* Limit gages supply unambiguous "yes" or "no" decisions regarding the acceptability of the inspected part. Beyond the reasonable care in handling the gage, the human factor barely enters into the judgment.

3. *Portable and independent from power supply.* Fixed gages can be used in any part of the plant and in field service as well and can be carried or mailed without the need of auxiliary equipment and setups.

4. *Checks multiple forms and interrelations.* Fixed gages can be designed to check combinations of several dimensions comprising lengths, diameters and angles. Other types of fixed gages serve to inspect interrelated features for size, location, form, alignment, and so forth. In this latter case the fixed gage is checking the combined effect of these parameters with regard to the functional adequacy of the inspected features, by determining a condition that is also known as the virtual size of a member.

5. *Assemblywise functional inspection.* Fixed gages are particularly useful in the checking of part members whose meaningful geometric irregularities cannot be readily detected by gages that do not provide complete reverse replicas of the critical part portion.

6. *Uniform reference standards.* For simple and complex forms, even with multiple toleranced dimensions, fixed gages are considered of particular value for correlating measurements between manufacturing and inspection, supplier and receiver, parallel or complementary production of interchangeable parts in different plants and so forth.

7. *Economics.* The relatively moderate cost of fixed gages frequently makes this type of inspection the most economical in wide areas of repetitive manufacturing.

MASTER GAGES

The designation master gage, in this context, refers to the application of the gage and not to its quality level, although fixed gages in master gage applications are usually required in the highest accuracy grades. The specified accuracy of these gages, however, may be regarded as a consequence of the particular function that the master types of fixed gages are expected to perform.

Master gages are used in many diverse applications, but none of these comprises the direct measurement of production parts. Master gages are reference standards, used for setting, checking, adjusting or calibrating different types of measuring instruments, including the limit types of fixed gages.

For each nominal dimension one master gage is commonly used. There are a few exceptions to this general practice, such as in the case of certain types of air gages with adjustable amplification. In these latter uses, the master gages are required to be in pairs, or to be comprised of two dimensions, representing the setting range of a particular comparator instrument.

Table 3-2 lists a few typical applications for master gages. For each application category, the table also indicates examples of gage dimensioning principles, although these must not be regarded as rigid rules. Finally, the table also mentions actual types of gages that are characteristic for a particular master gage category and these are shown in Figs. 3-1 through 3-4.

THE PRINCIPLES OF LIMIT GAGE MEASUREMENTS

Limit gages are made to sizes that are essentially identical to the design limit sizes of the dimension to be inspected. When the actual part size or form differs from the corresponding dimensions of the gage, either of the following two cases will occur, depending on the dimensional conditions of the part:

a. The part feature being gaged does not fill the permissible boundaries specified in the design. In this case the mating of the part and the gage is possible, although the physical contact will be only partial, the rest of the corresponding surfaces being separated by an air gap; and

TABLE 3-2.   APPLICATION EXAMPLES FOR MASTER TYPE FIXED GAGES

APPLICATION	EXAMPLES OF DIMENSIONING PRINCIPLES	TYPICAL GAGE	
		DESIGNATION	FIGURE
Setting of Comparator Type Measuring Instrument	The design size of the object to be gaged, or the median dimension of the expected size range, e.g. in the case of gaging for segregation.	Master setting disc	3-1
Checking of Limit Gages	The design size of the master gage will be equal to the gage limit size in the direction the max. material condition of the limit gage to be checked.	Taper master ring for checking a taper plug gage.	3-2
Establishing Setting Range for Airgage Adjustment	Size limits constituting a spread which (a) comprises the expected scatter of the actual part size, and (b) fits into the range of instrument indications at the selected amplification.	Master setting rings for calibrating air gages.	3-3
Standards for Measuring Instrument Calibration	The nominal size of the reference dimension.	A one inch standard for setting a 1 inch to 2 inch range micrometer.	3-4

b. The part feature to be gaged exceeds the specified boundaries. Excess material on the part surface will prevent the part and the gage from being assembled due to material interference.

Whenever a specific limit gage can enter or can be entered by the part, the assembly of the inspected part with another member, whose mating physical boundaries do not exceed those of its gage, can be accomplished. Consequently the part will be acceptable for assembly or, in current shop language, it can "go." Hence, limit gages, which by their mating with the part determine this condition of acceptance, are named *GO gages.*

Definition: The GO limit gage contains the reverse physical replica of the dimension to be measured at its maximum permissible material condition. The limit for the maximum material condition is (a) the

minimum size for an inside feature, and (b) the maximum size for an outside feature.

The maximum material condition affects the assembly by producing the minimum clearance in loose fits and the maximum interference in press fits.

In addition to permitting assembly in manufacturing, the dimensional requirements for an engineering part must also assure the functional correctness of the assembled components. Whereas excessive tightness can prevent the process of assembly, excessive looseness will endanger the proper operation of the mated parts. Too much looseness in running fits will cause excessive play, and looseness in press fits will defeat the objective of a sustained solid connection.

It follows that for assuring the functional adequacy of the part, its minimum material condition must also be kept under control. The part feature, at

Giddings & Lewis Measurement Systems

Fig. 3-1. (Left) Master setting disks in three basic styles. The single diameter designs are used for comparators and the double section type for setting limit snap gages.

Fig. 3-2. (Right) Master thread gage plug.

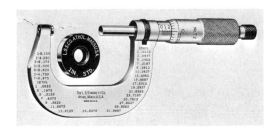

The L. S. Starrett Co.

Fig. 3-4. Plain master disk of one-inch diameter. It is used as a setting standard for a 1- to 2-inch range micrometer.

its minimum material condition, must at least fill the space whose dimensions are specified, as the minimum material limit, on the product drawing. When a fixed gage is made exactly to this limit size, the part will not readily assemble with the gage, unless the actual material condition of the part feature is less than the specified minimum value. Under these latter circumstances, however, the part becomes unacceptable for functional reasons. The gage that detects this defective condition by entering or being entered by the inadequate part is known as the *NOT GO* gage.

Definition: The NOT GO limit gage contains the reverse physical replica of the dimension to be measured at its minimum acceptable material condition. If the NOT GO gage assembles with the part being inspected, this indicates that the dimension to be measured is incorrect and is considered a basis for rejection.

The basic principles of inspection by means of limit gages are illustrated schematically in Fig. 3-5, using cylindrical features for examples.

Giddings & Lewis Measurement Systems

Fig. 3-3. Master setting rings for calibrating air gages with air plugs as probe members.
(Top) A pair of spread gages for small diameters.
(Bottom) Ring gage in the style used for large diameters.

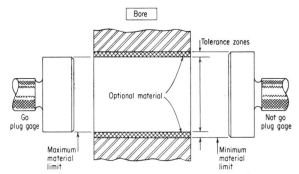

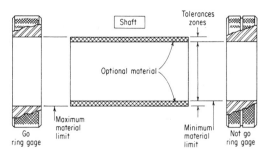

The design sizes of the gage are associated with one of the limit material conditions of the object feature:
Go gages with the maximum material condition
Not go gages with the minimum material condition

Fig. 3-5. Schematic illustration of the principles of dimensioning for limit gages.

LIMIT GAGES FOR LENGTH DIMENSIONS

The function of the length limit gages is to check whether the inspected critical dimension of the object is within the specified size limits. To ascertain its acceptability, the object must be checked for both size limits of the dimension, using subsequently the GO and the NOT GO gage.

The inspection for the limit sizes of the dimension can be accomplished with two separate gages, but more frequently gage designs are used that permit the functions of the GO and of the NOT GO gages to be carried out with the appropriate members of a single gage. These members can be located at opposite ends of the instrument, or both sizes incorporated into a single gage member, located in line, the GO size in front and the NOT GO size behind it. This latter variety of combined GO–NOT GO gages is also known as *progressive gages*.

The length dimension to be checked can be the distance between two parallel faces, or the diameter of a cylindrical surface. The length or diameter to be inspected can be bounded either by walls inside the distance (for external dimensions) or by walls outside the distance (for internal dimensions). External lengths are checked with snap type gages and internal lengths with end type gages.

Limit gages for a single length dimension can be made as *rigid gages or as adjustable gages,* the latter permitting adjustment of the contact members to specific limit sizes.

Originally the plain length dimension gages were made in the rigid type of design only. Typical examples are the rigid snap gages, with single or double jaws, and the end measuring rods. Later, adjusting mechanisms were developed that provided dependable guidance and solid locking of the contact members, combined with means of sensitive adjustment (Fig. 3-6). The range of adjustment for the contact members, also known as buttons, varies according to design. Commonly, the range of adjustment is on the order of $\frac{1}{8}$ inch for the internal gages, and $\frac{1}{4}$ inch for the external gages, but frequently may be wider, particularly for the larger gage sizes.

The advantages of adjustable limit gages are obvious, when considering the great variety of limit sizes resulting from different purpose fits. Further benefits result from the ability to compensate for wear by readjusting the position of the contact buttons. Consequently, the adjustable types of limit gages have, in most applications, replaced the rigid types of gages for the measurement of single length dimensions. The rigid types of limit gages are, how-

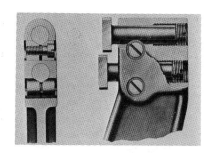

Pratt & Whitney Cutting Tool and Gage Div., Colt Industries

Fig. 3-6. The adjusting and locking device for adjustable snap gages. The fine-pitch adjusting screw permits the sensitive positioning of the gaging buttons, which are then effectively locked between the bushing and the nut.

ever, still preferred for certain specific purposes, for economical reasons and because they are tamperproof.

External Measurements—Adjustable Limit Snap Gages

An example of an adjustable limit snap gage is shown in Fig. 3-7. These gages are made in different styles and sizes, but most designs have the following common characteristics. The gage consists of a C-form frame with adjustable gaging members in the jaw of the frame. These members can be adjusted within a specific range to provide two gage sizes corresponding to the limit sizes of the dimension to be gaged. The gage distances created by the adjustable members are located one behind the other, for progressive gaging. This arrangement permits the gaging of both limit sizes of the object to be carried out in a single movement. The gaging movement may consist in swinging the gage, as shown in Fig. 3-8, or the gage may be designed with an extended anvil (Fig.

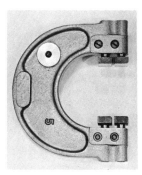

Giddings & Lewis Measurement Systems

Fig. 3-7. Limit snap gage with four individually adjustable buttons. Square-head gaging buttons are shown, but round-head buttons are also available.

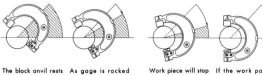

The block anvil rests on top of work, supporting the weight of the gage.

As gage is rocked downward, work should pass the first ("GO") button.

Work piece will stop at second ("NOT GO") button if it is within limits.

If the work passes both buttons, it is undersize.

Pratt & Whitney Cutting Tool and Gage Div., Colt Industries

Fig. 3-8. The process of inspecting the diameter of a cylindrical part with the aid of a GO–NOT GO limit snap gage. The common anvil of the gage is resting on the object, and the two buttons are engaged subsequently.

3-9), serving as a guide for the object when advanced to the subsequent gaging positions.

Depending on the style, adjustable limit snap gages are available for sizes up to 12 inches. The range of adjustment varies as a function of the frame size; the smallest gages can usually be adjusted over a range close to 0.200 inch, while the range of adjustment for the larger gages may reach a full inch.

The form of the contact members may be selected to be best suited to the conditions required by particular object configurations. Individually adjustable buttons (Fig. 3-7) can have different forms and contact areas, depending on how they are to be used as, for example, on small gaging surfaces or close to a shoulder. For most applications, snap gages with only two adjustable buttons and a single solid anvil are preferred. When nominally identical parts are to be inspected continuously, *extended anvils* (Fig. 3-9) provide significant advantages by reducing the gaging time and giving long wear life to the gage. *Snap gage stands* (Fig. 3-10) can even further shorten the gaging time for small parts by requiring less physical effort from the inspector than the use of hand gages.

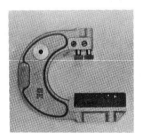

Giddings & Lewis Measurement Systems

Fig. 3-9. (Left) Extended anvil snap gage.

Fig. 3-10. (Right) Snap gage stand to hold snap gages with a bushing-type clamp.

Giddings & Lewis Measurement Systems

Fig. 3-11. Adjustable limit length gage of the progressive design with individually adjustable gaging buttons.

Adjustable Length Gages

The basic design of adjustable length gages, shown in Fig. 3-11, is similar to the concept found in adjustable snap gages, permitting easy setting and convenient handling of the gage. Although there are no specific limitations regarding the smallest size of adjustable length gages, the maximum size to which adjustable snap and plug gages are made is up to about 40 inches.

With appropriate design varieties of these gages, the gaging of either external or internal dimensions can be accomplished. External dimensions will include outside diameters, and the internal lengths mostly distances between parallel walls, such as flanges on a shaft. The GO and NOT GO sizes comprised in the same gage are attained with either double-sided gage heads or progressive design. For the latter purpose, the gaging buttons can be located in line on the same gage arm, or in fork-shaped arms holding the contact buttons for limit size gaging at close distance behind each other.

Internal Measurements—Adjustable Limit Plug Gages

Adjustable limit plug gages are illustrated in Fig. 3-12. These gages serve the inspection of distances between parallel surfaces or of diameters of larger size bores. The contact surfaces of the buttons are flat or spherical, the spherical being mandatory for bore measurements. The gages have a pair of contact buttons on each end of the handle; the buttons are adjusted respectively to the GO and NOT GO limits of the dimension to be inspected.

The adjustable limit plug gages are available in two styles that have different ranges of adjustment. One style is made for shorter distances, varying from $2\frac{1}{2}$ inches to 4 or $4\frac{1}{2}$ inches, in increments of $\frac{1}{4}$ inch, which are covered by the range of adjustment of the individual gages. The other style, designed for larger distances, up to 12 inches, provides adjustment ranges of $\frac{1}{2}$ inch for each gage.

Giddings & Lewis Measurement Systems

Fig. 3-12. Adjustable limit plug gages for inspecting bore diameters up to about 12 inches.

CYLINDRICAL LIMIT GAGES

By definition, the diameter of a cylindrical object is a single length dimension. Experience indicates, however, that mating cylindrical objects with nominal diameters that should provide appropriate fit conditions frequently cannot be assembled, or do not produce the required interference. This may be the case even when the functionally unsuitable parts have been inspected with a limit gage intended for checking single length dimensions and have been found to have the specified size.

Such conditions of noncorrespondence between the measured diameter and the effective size of the part are generally due to form irregularities. A single diameter defines the size of a cylindrical object only when its cross-sectional form is perfectly round. The manner in which roundness irregularities can cause the effective size of the part to differ from the measured diameter is discussed in detail in Chapter 14.

The effective size of a basically cylindrical object can be expressed by the diameter of an imaginary envelope cylinder containing all the surface elements that determine the maximum material condition of the feature. In actual application, this condition is satisfied with the minimum size envelope cylinder for an external feature, and the maximum size envelope cylinder for an internal feature.

The cylindrical fixed gage, whether a plug or a ring gage, physically constitutes that envelope cylinder whose relations to the material limits of the object can be dependably checked by means of effected (GO) or attempted (NOT GO) assembly with the part.

In addition to their important property of providing a realistic determination of the inspected feature size, cylindrical limit gages have various other advantages as well. Noncontinuous surfaces, resulting from interruption by grooves or consisting of ribs only, particularly those having uneven—"odd"—numbers, have only virtual diameters and should be checked with full cylindrical limit gages.

When small bores are to be inspected for size with limit gages, space limitations will exclude the use of other than cylindrical gages, in this case plug gages. The rigid design of cylindrical gages will be considered a guarantee for dimensional consistency and a desirable safeguard against deliberate or inadvertent changes in the gage size.

On the other hand, cylindrical gages have certain limitations that can affect both the gaging time and the tooling cost of the inspection process.

Limit sizes are only exceptionally combined in a single cylindrical gage member; usually two gages or two independent gaging members are required for the function of the GO and the NOT GO gage. The use of two gage members sequentially requires more time for the gaging process than needed for progressively arranged gaging members.

For each limit size, a separate cylindrical gage body is needed, and when excessive wear occurs on the surface of the gage, either an expensive salvaging process must be applied, or the gage must be discarded. This is an economic disadvantage compared with the adjustable limit bore gages that can be used for different parts with closely similar dimensions.

Cylindrical Limit Plug Gages

Cylindrical limit plug gages are used for gaging the limit sizes of round holes and, occasionally, of grooves with parallel walls. Plug gages are commonly preferred in *double-ended* design when used for limit gage applications (Fig. 3-13). For particular inspection conditions, *single ended* plug gages can offer certain advantages. While the gaging member of a plug gage is commonly made for a single size, *progressive plug gages*, (Fig. 3-14) that combine both limit sizes in a single member may also be selected.

Originally, plug gages were made of a single piece of steel. That design was superseded by the assembled type of plug gages, consisting of separate gaging members and a handle, complemented with fastener elements. The advantages of the assembled design are obvious considering the free choice of appropriate materials for these different members, the interchangeability of parts and the reversibility of the gaging member for prolonged service life.

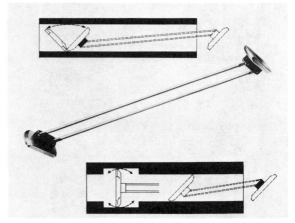

Pratt & Whitney Cutting Tool and Gage Div., Colt Industries

Fig. 3-13. (Left) Double-ended plug limit gage with interchangeable gaging members in taper lock design.

Fig. 3-14. (Right) Progressive type limit plug gage with GO section in front, followed by the NOT GO section. Interchangeable gaging member of trilock design.

Because cylindrical limit plug gages are used for a rather wide range of sizes, a single basic design would not be practical. As an example, for sizes above an 8-inch diameter the weight of a plug gage of the commonly used type would be excessive and a single handle could not provide the sensitivity of handling needed for dependable limit gaging. *Annular type gages* (similar to that shown in Fig. 3-18) were designed for sizes from 8-inch to 12-inch diameters; these gages have two ball handles near the periphery of the gage body, 180 degrees apart.

Table 3-3 shows the different basic designs of cylindrical plug gages that are widely used throughout the industry.

Spherical Diameter Plug Gages

The spherical diameter plug gage (Fig. 3-15) is a special and unique type of cylindrical limit plug gage. The spherical form of the gaging member eliminates many of the problems confronted with the use of conventional cylindrical plug gages. The spherical diameter plug is simply inserted into the hole to be checked at an angle other than 90 degrees. Once the spherical plug is inside the hole, it is rocked through its own axis. The GO gage will not meet resistance and the NOT GO gage will, provided the hole being checked is within tolerance. The spherical diameter

Flexbar Machine Corporation

Fig. 3-15. Double-ended spherical diameter plug gage.

Flexbar Machine Corporation

Fig. 3-16. Double-ended spherical diameter plug gage with flexbar.

plug permits the detection of taper, barrel and hourglass shapes and ovality. Also, when equipped with the FLEXBAR® system (Fig. 3-16) the spherical diameter plug gage can be used to check the full length of a deep bore, as well as dimetral recesses, chambers and undercuts within a bore.

Cylindrical Limit Ring Gages

Functionally critical dimensions of basically round external surfaces can only be assured for assembly or operation when tight tolerance are maintained for both the size and the geometry. Frequently it is permissible and practical to combine these parameters into a single dimension for the purpose of inspection. The limit sizes for the effective diameter of a basically round shaft can be checked in a convenient and dependable manner with cylindrical limit ring gages.

The actual use of limit ring gages is commonly confined to applications where the *effective* part diameter must be inspected, or where solid ring gages are preferred because of the consistent calibration they provide. Otherwise, the gaging of shaft diameters with adjustable limit snap gages is more economical in tooling cost and in gaging time.

Cylindrical limit ring gages are available in sizes ranging from about 0.060-inch to over 12-inch diameters. The design of the gage bodies varies according to the size range for which they are used. Table 3-4 is a survey of the more common designs of cylindrical ring gages. Because ring gages with identical outside dimensions are frequently used for both the GO and the NOT GO sizes, an annular groove is machined as a distinctive mark on the periphery of the NOT GO gages.

TABLE 3-3.   BASIC DESIGNS OF CYLINDRICAL PLUG GAGES

APPLICABLE FOR SIZE RANGE	DESIGNATION	SKETCH	DESCRIPTION OF THE DESIGN
0.030 to and including 0.760 inch	Wire type plug gage design		The wire type members of the gage are held in a collet equipped handle which will firmly retain the gaging wire in a more or less extended, or reversed position.
Above 0.059 inch to and including 1.510 inch	Taper lock plug gage design		The gaging members have taper shank which is forced into the taper hole of the handle, resulting in a dependably rigid gage. Drift hole permits the removal of the gaging member from the handle.
Above 0.760 inch to and including 2.510 inches	Reversible or Trilock plug gage design		The end faces of the handle have three radially arranged wedge shaped locking prongs which engage the corresponding grooves on the face of the gaging member. Both faces of the gaging members have similar grooves, permitting the plug to be reversed when worn. A single through screw secures the meshing wedge elements, causing them to stay in a rigidly locked position.
Above 2.510 inches to and including 8.010 inches			
Above 8.010 inches to and including 12.010 inches	Annular plug gage design		Weight reduction of the large size gages to the limit consistent with stability is the leading objective of this design, consisting of a rim as gaging member and an integral web for rigidity. The center of the web is bored out to reduce weight; threaded holes in the retained section receive ball handles.

TABLE 3-4.   BASIC DESIGNS OF CYLINDRICAL RING GAGES

APPLICABLE FOR SIZE RANGE	DESIGNATION	SKETCH	DESCRIPTION OF THE DESIGN
Above 0.059 inch to and including 0.510 inch	Ring body with bushing as gaging member	Go Not go	Using an inserted bushing of hardened steel permits making the gage body of soft steel, thereby facilitating the manufacture.
Above 0.510 inch to and including 1.510 inch	Solid ring gage of plain cylindrical form	Go Not go	Ring gages made of a single piece of hardened steel are considered mandatory for sizes larger than 0.510 inch, and are optional for smaller sizes.
Above 1.510 inches to and including 5.510 inches	Solid ring gage with flanged periphery	Go Not go	Basic design for gages above 1.510 inches. Some manufacturers offer as an alternate style master gage rings to 4-inch diam. in thinner than standard design — for handling ease — with a neoprene ring substituting for the steel flange.
Above 5.510 inches to and including 12.260 inches	Solid ring gage with ball handles in the flange		To facilitate handling during the gaging process, ring gages above 5.510-inch diameter have two ball handles.

The measurement of qualified plain internal diameters for use as master rings and ring gages is formally specified in the American National Standard B89.1.6M-1984 (R1988).

FIXED GAGES FOR MULTIPLE DIMENSIONS

Purpose and Operation

For parts or features having a basic geometric form, such as round or bounded by two parallel planes, the measurement of a single dimension will usually suffice to determine the dimensional adequacy of the part. Other types of bodies, although in current use or even standardized, have forms that require several dimensions to establish their actual size. The multiple dimensions that determine the size and configuration of the part may consist of lengths, angles, contour forms or a combination of these parameters.

Deviations of the actual size from the nominal values for these dimensions will interact, either causing cumulative error or partial canceling out, finally resulting in the *effective size* of the part.

Although the product drawing will usually specify the tolerances of the individual dimensions, the *functional* adequacy of the part can frequently be evaluated by its effective size. Based on the individual tolerances, the limit dimensions of the feature may be visualized to create a boundary layer, or shell, built around the nominal size of the part's surface. For the part to be dimensionally acceptable, all portions of the surface that constitute the effective size must be contained in that shell; in a cross-sectional view, that imaginary shell will appear as the tolerance zone of the illustrated dimensions.

The multiple dimension limit gages constitute the reverse replicas of the external and internal boundary surfaces of said tolerance shell. Limit gages that satisfy the described conditions will check the effective size of the part, obviating the measurements of the individual dimensions.

Multiple dimension limit gages are currently used for the inspection of various standard configurations, and for special measuring tasks, as well. In the category of standard forms whose effective sizes are frequently inspected with limit gages are threads, tapers and splines. A brief discussion of limit gages for these types of configurations will follow. It should be pointed out, however, that because limit gages check the effective size of parts, the inspection of multiple dimension features with limit gages is also

being applied, as a sound practice, to many other configurations or interrelated surfaces.

Thread Limit Gages

Thread limit gages are made for the gaging of internal (hole) or external (shaft) threaded surfaces, such as the bores of nuts and the outside thread on screws or bolts. In accordance with the concept of limit gaging, these gage members are available in GO and NOT GO sizes, conforming respectively to the maximum or minimum material limits of the inspected feature.

The limit sizes for the significant thread dimensions are standardized for different "classes," a designation for the degree of specified accuracy. The design sizes on the gaging surface of thread gages are based on these limit dimensions. In designing the gage dimensions, the possible interferences that can occur when inspecting several dimensions simultaneously with a single limit gage must also be taken into account.

Thread limit plug gages (Fig. 3-17) are essentially similar in design to the blanks of the plain limit plug gages. Thread limit plug gages are also made in wire, taper lock, trilock, and annular design (Fig. 3-18) as listed in Table 3-3, with comparable size ranges for the selection of the applicable styles.

A special type of thread limit plug gage is the *setting plug gage*. Although similar in appearance to the regular thread plug gage, it is not intended for the gaging of threaded bores in manufactured parts. The function of the setting plug gage is to serve as a reference standard for setting adjustable thread ring gages. The thread form of setting gages is truncated, with the thread crest and root relieved to avoid interference with the corresponding elements of the ring gage thread and to assure the actual contact of the functionally important thread flanks.

Giddings & Lewis Measurement Systems

Fig. 3-17. (Left) Thread limit plug gages, double ended design, with taper lock gaging members.

Fig. 3-18. (Right) Annular plug thread gage for bore diameters above 8 inches.

Giddings & Lewis Measurement Systems

Fig. 3-19. Thread limit gages adjustable model.
(Left) GO type.
(Right) NOT GO type.

Thread limit ring gages (Fig. 3-19) are made in adjustable design and differ in this respect from the plain cylindrical ring gages that represent a solid type construction. To make the gages adjustable, the blanks of the thread ring gages have three radial slots, two of these ending in terminal holes, the third being a through slot. The through slot is bridged by the adjusting device (Fig. 3-20) consisting of a split hollow adjusting screw, acting through a sleeve on a shoulder on the opposite side of the slotted section. Driving in the adjusting screw will spread the adjusting slot and thereby enlarge the diameter of the gage bore. A locking screw driven from the opposite side into the threaded bore of the split adjusting screw will cause the latter to expand, locking it into position without exercising any eccentric force that could alter the adjustment or distort the gage.

Taper Plug and Ring Gages

Taper body forms are widely used for accomplishing precisely aligned, yet detachable, connections between mechanical members, more particularly in machine tools. Frequently, it is also required that these connections assure in a positive manner, the mutual locations of the assembled members in the axial direction as well.

The critical dimensions of machine tapers are the included angle and the diameter at a specific reference level. The latter dimension is frequently indicated on the taper gage by an appropriate reference mark. For reasons related to their application requirements, machine tapers are frequently used with tangs; the size and location of this auxiliary member on machine tapers are additional dimensions that need to be inspected during the gaging process.

Taper limit gages for machine tapers are made for both internal and external tapers in the form of plug and ring gages, respectively. Depending on whether machine tapers with or without tangs are to be inspected, taper limit gages are made for either of these application purposes (Fig. 3-21).

Machine tapers differ widely in size, angle and other characteristics due to the function of their intended application. This may range from a small size twist drill shank to a heavy milling machine spindle nose. Design variations also exist between tapers originating from different manufacturers and those adopted by industry as standard for specific categories of products.

The large variety of machine taper designs and sizes may be considered a drawback from the viewpoint of fixed gage applications. However, the many inherent advantages of inspecting multiple dimensions with the aid of fixed limit gages usually outweigh the disadvantages. Taper limit gages are, therefore, in general use as convenient and dependable means for the dimensional inspection of machine tapers.

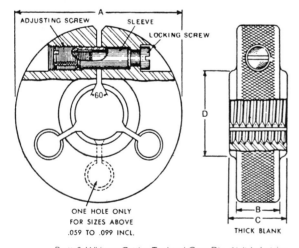

Pratt & Whitney Cutting Tool and Gage Div., Colt Industries

Fig. 3-20. Setting and locking device for adjustable thread ring gages. Details of construction.

Morse Twist Drill & Machine Co.

Fig. 3-21. Taper plug and ring gage designs with tang.

Spline Plug and Ring Gages

Splines are essentially multiple keys. Their function is to restrain the relative rotation of a shaft and a mating bore member, and also to transmit torque. The fit between the mating members can be chosen to produce a solid joint, or it may be a loose fit, permitting a relative axial displacement, for example, shifting of gears on a splined shaft. The different fits will, of course, require specific sizes for the corresponding dimensions of the participating members (see Fig. 3-22).

Depending on design specifications, governed by such factors as overall dimensions, the torque to be transmitted, part configuration and manufacturing considerations, one of the different basic forms—involute, serrated or straight-sided—of splines will be selected. The fit conditions resulting between the mating splined parts will be affected by the actual sizes and forms of the spline elements, including diameters, tooth thicknesses, spacing and so forth. The many participating dimensions of a splined assembly would make it a rather tedious task to determine the resulting fit conditions by measuring the individual dimensions. Consequently, inspection by gaging the effective size of the entire part with fixed gages is particularly warranted for both internal and external splines.

Splined plug gages are made from blanks whose design varies according to the size range they are intended to cover. Plug gages with integral handles are used up to 2 inches, although from 1.5 inches upward, gages with detachable spline gaging members can also be made. Up to an 8-inch diameter, either pilot type members or tapered tooth plug gages can

Vinco Corporation

Fig. 3-22. (Left) Fixed spline ring gage of nominal size. Such fixed gages may be used for the direct inspection of manufactured parts, functioning as limit gages.
(Right) Or they can serve as setting gages for indicator type spline gaging instruments such as the proprietary design shown.

be selected, the latter type being made either with integral or detachable handles.

Splined ring gages are made in the plain type or in the pilot type varieties, in sizes up to an 8-inch effective diameter. Other designs of spline ring gages, usually in the pronged type, are intended for checking spline relation, spline alignment or spline thickness.

CONSTRUCTION AND DIMENSIONING OF FIXED GAGES

The Design of Fixed Gages

The expected function of the fixed gages and the dimensions to be measured are variables that necessarily require a wide diversity of gage types. Gages satisfying the same functional requirements could also be designed in different ways, with solid or detachable gaging members, varying in form, external dimensions and in the design of the connecting elements. This complexity of variables requires a systematic and widely accepted approach to prevent chaotic conditions in the design and manufacture of these basic tools of dimensional inspection.

It is fortunate that a basic system of gage design has been developed by the American Gage Design Committee (ADG), as a cooperative effort on the part of manufacturers, users, and nonindustrial organizations. The work of this committee has resulted in a comprehensive standard of gage design,[1] which specifies the construction and the external dimensions of fixed gages to be used for various applications and size ranges. The standard also supplies detailed information on the design and dimensioning of the connecting elements for gages with detachable members. The data supplied in the preceding sections of this chapter on gage styles for different size groups are based on this standard. Specifications on the functional dimensioning of the gaging members and on the applicable size tolerances are not part of the AGD standard's objectives.

The Material of Fixed Gages

The most important material properties in fixed gage applications are dimensional stability and wear resistance. Modern heat treating methods, including deep-freezing, when judiciously applied to properly selected steel grades, can provide a high degree of stability. In addition to advances in heat treating processes, alloying elements in steel grades cur-

---

[1] American National Standard B47.1-1988.

rently selected for the manufacture of fixed gages substantially improve the wear resistance of the gaging members thereby extending the useful life of the gage. Realizing the harmful effect of surface roughness on the capability of resisting frictional wear, leading gage manufacturers are using special lapping processes to produce an exceedingly fine surface finish and also to improve the geometry of the gaging members.

Although for the majority of applications fixed gage members made of alloy steels prove entirely satisfactory, there are specific applications that warrant the use of even more wear-resistant gage materials. Such conditions may exist for gages in extended use to check very tightly toleranced dimensions, or for gaging parts made of materials having unusually abrasive properties.

For applications requiring a particularly high degree of wear resistance, fixed gages are also made from materials or by using special processes, examples of which are listed below:

1. *Chromium plating* increases wear resistance and by reducing friction is also beneficial to the gaging operation. Another application for chromium plating is for restoring worn gages to the original sizes in a salvaging process.

2. *Flame-plated tungsten carbide*, although producing a coating of only a few thousandths of an inch thick, it can very substantially increase the wear life of fixed gages.

3. *Tungsten carbide* materials for solid plug gage members provide very great stability and wear resistance. Their relatively high price and sensitivity to chipping are limiting; however, the actual use of carbide gages to applications where extended usage or the presence of extremely abrasive work surfaces are providing the economic justification for the selection of carbides as gage material.

Because the coefficient of thermal expansion of tungsten carbide differs from that of steel, a controlled temperature environment may be required when gaging tightly toleranced dimensions on parts made of steel, with the aid of tungsten carbide gage members.

4. *Ceramic gage members* are considered to provide the greatest degree of wear resistance. However, some of the drawbacks mentioned for the tungsten carbide materials are even more pronounced for ceramics, particularly the brittleness and the differential coefficient of thermal expansion.

The Gage-Maker Tolerances

The limit sizes of a part dimension, as specified on the product drawing, allow a certain tolerance range for variations in the manufacturing process. The fixed gages used for checking the limit conditions on manufactured parts, necessarily have a certain degree of size variations also, for which appropriate tolerances must be set.

Adhering to the principle that no part must be accepted in inspection, whose actual size is outside the design limits, the gage tolerances will have to be allocated within the part's design limits, to the detriment of manufacturing tolerances. This condition will reduce the range of permissible manufacturing variations and could affect the cost of producing the part. Therefore, it is desirable to retain the largest possible portion of the design tolerances for manufacturing purposes, which can be achieved by tightening the gage-maker tolerances. Because this latter approach will necessarily reflect on the gage cost, conflicting goals arise that require a compromise solution resulting in conditions closest to an optimum cost balance.

Similarly to the basic part tolerances, the currently used gage-maker tolerances have also several classes, and are stepped for different size groups. Table 3-5 shows the widely accepted gage-maker tolerances for plain cylindrical gages for which only a single dimension is toleranced. Thread gage tolerances, on the other hand, are specified for several significant dimensions, in line with the basic principles of multiple dimension gaging.

There are no rigid rules established for the selection of the preferable class of gage-maker tolerances. However, in view of the economic aspects just discussed, certain principles are usually applied, which can be summed up as follows. The gage-maker tolerance selected for either the GO or the NOT GO gage member, may absorb 5%, but must not exceed 10% of the total design tolerance for the part dimension to be gaged. For part dimensions with very liberal design tolerances that were established in view of the functional purpose of that particular product, even less than 5% of the total tolerance zone would be absorbed by the actually selected gage-maker tolerances on either end of the part tolerance range.

Wear Allowances for Limit Gages

When a GO gage is made exactly to the maximum material limit size of the dimension to be gaged, the slightest wear of the gaging member will cause the

TABLE 3-5.   GAGE-MAKER TOLERANCES FOR CYLINDRICAL PLUG AND RING GAGES

RANGE AND TOLERANCES					
Size Range		Class XX	Class X	Class Y	Class Z
Above	To and Including				
.029	.825	.00002	.00004	.00007	.00010
.825	1.510	.00003	.00005	.00009	.00012
1.510	2.510	.00004	.00008	.00012	.00016
2.510	4.510	.00005	.00010	.00015	.00020
4.510	6.510	.000065	.00013	.00019	.00025
6.510	9.010	.00008	.00016	.00024	.00032
9.010	12.010	.00010	.00020	.00030	.00040
Direction of Tolerance					
GO work plugs +		GO check plugs −	NO GO work plugs −		NO GO check plugs +

NOTE: Master setting rings for calibrating high amplification air gages are also made in Class XXX, having tolerances one half of those for Class XX. For special applications master rings, in sizes to 1.000 inch maximum, are available in Class XXXX, made to tolerances exactly one half of those for Class XXX.

gage size to pass the design limit of the object. This condition will then require the gage to be retired, either for salvaging or for being discarded.

For reasons of gage economy, it is customary to provide for a certain amount of wear when dimensioning the gage, unless particular manufacturing conditions of the product to be gaged make the application of this safeguard against early gage retirement unadvisable. Providing for gage wear is accomplished by shifting the design size of the gage toward an increased gage material condition, that is a larger outside diameter for a plug gage, and a smaller bore diameter for a ring gage.

The change in the design size of the gage for the purpose of compensating for wear will apply only to GO gages. No wear allowances are needed for NOT GO gages, whose wear develops in the direction of safety, that is, in a sense, where the gage will tend to reject parts with increased severity, automatically reducing the useful range of the manufacturing tolerances.

As a general rule, the wear allowance used in the design of GO limit gages is equal to 5% of the total tolerance range assigned to the part dimension to be gaged. Although the wear allowance, by shifting the design size of the gage, reduces the useful tolerance range of the manufactured part in addition to the similar potential effect of the gage-maker tolerances, the application of wear allowance is warranted in many cases, particularly when substantial gage wear is expected.

Allocation of Gage-Maker Tolerances

Deviations from the design dimension cannot be avoided completely, even when applying the great care and excellent manufacturing techniques common in gage-making practices. However, it is possible to limit the amount of these deviations by establishing tolerances, and to reduce or even entirely eliminate the detrimental effect of these tolerances on the functional objective of the gage. This latter goal can be accomplished by the proper allocation of the gage-maker tolerances.

For *limit gages*, the gage-maker tolerances should be allocated in a manner that does not detract from the reliability of the inspection process. Therefore,

**TABLE 3-6.   THE ALLOCATION OF WEAR ALLOWANCES
AND GAGE-MAKER TOLERANCES IN THE DIMENSIONING OF FIXED GAGES**

GAGE CATEGORY	DIMENSIONING PRINCIPLES	SCHEMATIC ILLUSTRATION*
Master Gage (Setting gage for comparator instruments)	Gage maker tolerances bilaterally distributed, using for basic gages size: (a) the design size of the object, or (b) the median size of the permissible object size limits.	
"GO" Plug Limit Gage	Basic gage size: Min. limit of feature size. Wear allowance (optional) added to the gage size. Gage maker tolerances are added unilaterally to the basic gage size.	
"NOT GO" Plug Limit Gage	Basic gage size: Max. limit of feature size. Gage maker tolerances are: (a) unilaterally subtracted from basic gage size; (b) bilaterally spreaded, straddling basic gage size; (c) (exceptional) unilaterally added to basic gage size.	
"GO" Ring Limit Gage	Basic gage size: Max. limit of feature size. Wear allowances (optional) subtracted from the gage size. Gage maker tolerances subtracted unilaterally from the basic gage size.	
"NOT GO" Ring Limit Gage	Basic gage size: Min. limit of feature size. Gage maker tolerances are: (a) unilaterally added to the basic gage size; (b) bilaterally spreaded, straddling basic gage size; (c) (exceptional) unilaterally subtracted from gage size.	

*The proportions are distorted to illustrate the relative location of the tolerance zones. As a general rule, the width of the gage maker tolerance zone must not exceed one-tenth part of the object's tolerance zone.

the gage-maker tolerances for limit gages are generally allocated unilaterally, based on the limit size of the dimension to be gaged, and directed toward the tightening of the manufacturing tolerances. This direction of tolerance allocation will result in the following senses of permissible variations:

| For GO gages toward increased material condition in the gage | Plug diameter might be larger Ring bore diameter might be smaller |
| For NOT GO gages toward the lesser material condition in the gage | Plug diameter might be smaller Ring bore diameter might be larger |

Some manufacturers apply bilateral tolerances in the dimensioning of NOT GO gages, spreading the gage tolerance zone symmetrically on both sides of the NOT GO gage design limit. This practice is supported by the consideration that the wear of the NOT GO gage will tend to *reduce* that side of the gage tolerance zone that constitutes a liberalization of the specified part tolerance. The adoption of bilateral tolerancing for NOT GO gages may be contingent on customer's approval when critical assemblies are involved.

In the case of *setting gages* the bilateral allocation of the gage-maker tolerances is commonly applied, with the purpose of reducing the potential absolute value of the gage error. By assigning the selected tolerance zone to symmetrically straddle the nominal size, deviations, when they occur within the permissible limits, will never exceed one half of the total tolerance range.

The principles of allocating wear allowances and gage-maker tolerances in the dimensioning of fixed gages are illustrated diagrammatically in Table 3-6.

4.
Gage
Blocks

GENERAL ASPECTS

The Need for Dimensional Reference Standards

All length dimensions in the industry and the sciences are expressed in magnitudes based on the standard unit of dimensional measurement, such as the international inch or the meter.

The basic reference unit is the international meter bar. It, along with authoritative duplicates, is available as the primary standard of length to a limited number of national standards laboratories. Today the use of a specific light wavelength as the international basic standard of length has increased the number of sites in possession of an absolute standard. However, the high cost of equipment necessary in making measurements based on light wavelength has made this method prohibitive for most manufacturing settings.

Although theoretically it is conceivable that all gages used in manufacturing and inspection should be checked by using basic measuring standards, considering the millions of gages in continuous use at all levels of industry, such a procedure would obviously be impractical.

Therefore, these primary standards must be used to calibrate high quality secondary standards that are close replicas of the former. These secondary standards form the basis for the creation of a dependable link between the basic standard and the instruments actually used for the measurement of industrial products.

Considering the inherently narrow gaging range of high magnification comparator instruments by means of which precise comparison can be made between the lengths on physical objects and the ref-

erence pieces, the latter would have to be available in very small increments, such as 1/10,000 inch, or even less. Because the range of precisely measurable dimensions in any plant will frequently extend over several inches, said system—when rigidly adhered to—could necessitate a stock of several times ten thousand basic reference pieces.

Fortunately, this problem has been effectively solved by a concept, developed and implemented at the end of the past century by a Swedish engineer, Carl E. Johansson, whose name has since that time been closely associated with his important invention, the gage block or, in shop language, the "Jo-block."

The Development of Gage Blocks

Historically, as some sources report, while studying the methods of interchangeable manufacturing at the Mauser weapon factory in Germany, in 1895, a member of the Swedish delegation, C.E. Johansson, saw in practice the enormous number of special gages used in that plant. He conceived of the idea of substituting built-up gages, consisting of individual blocks, for any of the large number of special length gages. Johansson realized that by selecting the sizes of individual blocks properly, a set of these could be used to build up, by combination, a wide range of different length dimensions in very fine increments. The original sets of blocks that Johansson devised, and also made by a specially developed lapping process, consisted of 102 blocks, arranged in three series.

Although the original lapping process on which the inventor had to rely consisted of a cast iron disc adapted to a sewing machine frame, the first gage blocks—as named by Johansson—found good use in the manufacture of rifles. This initial success started a rapid expansion in the use of gage blocks. In 1911, Johansson founded his own gage factory in Sweden, providing expanded facilities for the manufacture of gage blocks.

For some time, gage blocks were only rarely used in manufacturing. This is understandable in view of the extreme skill and extensive hand work that the manufacture of accurate gage blocks required. It was with the development of certain manufacturing processes, particularly by the application of mechanized lapping, that the production of precise gage blocks became less dependent on individual skill, and its efficiency rose to a level permitting wider applications of these important tools. The work of Major William E. Hoke, then head of the Gage Division of the U.S. Bureau of Standards, pioneered this important manufacturing development, which finally led to the present-day general use of gage blocks on most levels of engineering production.

Major Requirements for Gage Blocks

The essential purpose of gage blocks is to make available end standards of specific lengths by temporarily combining several individual elements—each representing a standard dimension—into a single gage bar. The combination of single blocks must result in a bar of reasonable cohesion, whose actual dimension truly represents, within specific limits, the nominal dimension sought for a particular application

To accomplish this purpose the following major requirements must be satisfied:

a. The individual elements must be available in dimensions needed to achieve any combination of sizes within the designed range and graduation of the set;

b. The accuracy of the individual blocks, in every significant respect, must be within accepted tolerance limits;

c. In the built-up combination, the individual blocks must be attached so closely to each other that the resulting bar will have a length that, for most practical purposes, is equal to the added sizes of the individual blocks of the assembly; and

d. The attachment of the elements to each other must be firm enough to permit a reasonable amount of handling as a unit, yet when taken apart, the individual pieces should be reusable without any harm to their original size or other essential properties.

To satisfy the above listed general requirements, the gage block sets consist of elements in arithmetically determined sizes and manufactured in adherence to rigorous dimensional and geometric specifications, which will be discussed in subsequent sections. The transient, yet reasonably solid and practically gapless, attachment of the individual blocks to each other is accomplished by "wringing"; the preconditions, the effect and the technique of this process of temporary attachment, are described in a later section of this chapter.

DESCRIPTION AND TECHNICAL REQUIREMENTS

Boundary Dimensions

The American National Standard for gage blocks, ANSI B89.1.9 M-1984 (R1989), distinguishes three basic forms—rectangular, square and round; the two commonly used types are shown in Figs. 4-1 and 4-2. The Standard specifies the boundary dimensions, the applicable size tolerances and the design of the through-holes for tie rods of those gage block types that contain this feature.

Table 4-1 lists the cross-sectional dimensions and tolerances of gage blocks, as specified by the standard; all domestic and most foreign manufacturers of gage blocks comply with these specifications.

The rectangular form is the more widely used in this country and, particularly, abroad. However, there are several domestic manufacturers who supply square gage blocks in addition to their full line of rectangular blocks. In certain applications the square blocks, although more expensive, are preferred; it is claimed that, because of their large surface area, square blocks wear longer, and adhere better to each other when wrung to high stacks. Some manufacturers make square blocks with center holes, whose purpose is to permit the use of tie rods as an added assurance against the wrung stacks falling apart while handling, particularly in shop use.

Pratt & Whitney Cutting Tool and Gage Div., Colt Industries

Fig. 4-1. (Left) Rectangular gage blocks.

Fig. 4-2. (Right) Square gage blocks with center hole.

TABLE 4-:.   CROSS-SECTIONAL DIMENSIONS AND TOLERANCES OF GAGE BLOCKS
Dimensions in inches

SIZE	STYLE	CROSS-SECTIONAL DIMENSIONS	
		DEPTH	WIDTH
Up to 0.01	Rectangular	$0.355 \begin{array}{l}+0.005\\-0.010\end{array}$	$0.787 \begin{array}{l}+0.450\\-0.229\end{array}$
0.01 to 0.20	Rectangular	$0.355 \begin{array}{l}+0.020\\-0.010\end{array}$	$1.181 \begin{array}{l}+0.074\\-0.084\end{array}$
0.20 through 20	Rectangular	$0.355 \begin{array}{l}+0.010\\-0.010\end{array}$	$1.378 \begin{array}{l}+0.010\\-0.207\end{array}$
0.05 through 20	Rectangular[1]	$0.531 \begin{array}{l}+0.005\\-0.005\end{array}$	$1.500 \begin{array}{l}+0.010\\-0.010\end{array}$
0.01 through 20	Square with 0.265 ± 0.005 hole (70° to 84° countersink)[2]	$0.950 \begin{array}{l}+0.010\\-0.010\end{array}$	$0.950 \begin{array}{l}+0.010\\-0.010\end{array}$
Up to 2	Round	$0.875 \begin{array}{l}+0.01\\-0.01\end{array}$ in diameter	
3, 4, 5 and 6	Round	$1.25 \begin{array}{l}+0.01\\-0.01\end{array}$ in diameter	

Dimensions in millimeters

SIZE	STYLE	CROSS-SECTIONAL DIMENSIONS	
		DEPTH	WIDTH
Up to 0.3	Rectangular	9 +0.13, −0.26	20 +11.4, −5.8
0.3 through 10	Rectangular	9 +0.5, −0.26	30 + 1.9, −2.1
10 through 500	Rectangular	9 +0.5, −0.26	35 + 0.26, −5.9
Up to 500	Square[3]	24.1 +0.2, −0.2	24.1 + 0.2, −0.2

[1] This style of gage block has one 0.250-inch hole through the sides of the 1-inch block, and two 0.250-inch holes through the sides of each block longer than 1 inch.

[2] Square blocks shall be countersunk on both sides for blocks 0.2 inch long and longer; blocks under 0.2 inch are not countersunk; 0.1-inch carbide wear blocks are countersunk on one side only.

[3] This style gage block has 6.7 millimeters ±0.1-millimeter center hole. The hole is countersunk on both sides 70 degrees to 84 degrees for blocks 5 millimeters long and longer. Blocks under 5 millimeters are not countersunk. Carbide wear blocks 2.5 millimeters long are countersunk on one side only.

The rectangular blocks are less expensive to manufacture, and adapt themselves better to applications where space is restricted, or excess weight is to be avoided.

Although gage blocks, in their entirety, are precision products made to rigid specifications, the greatest emphasis is laid on the two gaging surfaces on opposite sides of the blocks.

Gage Block Length—The Functional Dimension

For end standards, such as the gage blocks, the distance between the gaging surfaces represents the nominal dimension. In gage terminology, this distance is known as the *length* of the gage block. It is not necessarily the largest boundary dimension of the block, whose nominal size might be only 0.100 inch or even less.

The length is the most important single dimension of the gage block. It is the dependable closeness of its actual size to the nominal dimension of the gage block that primarily determines the accuracy of this end standard.

Although some manufacturers have long since established their own values for permissible deviations from the nominal length size, those are identical or closely related to the tolerances of length specified by the American National Standard ANSI B89.1.9M-1984 (R1989) "Precision Gage Blocks for Length Measurement (Through 20 in. and 500 mm)."

Metric gage blocks, generally made in rectangular form, are finding growing application in the domestic industry in conjunction with its current transition to the metric system; however, the dimensional tolerances and various other specifications of metric gage blocks very closely agree with those of inch gage blocks of comparable nominal size.

The American National Standard B89.1.9M-1984 (R1989) as well as the Federal Specification GGG-G-15C establish four accuracy grades for gage blocks; they are the following, the corresponding former grade designations being quoted in parentheses:

Tolerance Grade 0.5 (formerly Grade AAA)	Used for basic reference standard in calibration laboratories; also referred to as "grand masters."
Tolerance Grade 1 (formerly Grade AA)	For the purpose of gage room reference blocks.
Tolerance Grade 2 (formerly Grade A+)	Used generally by quality control for setting (mastering) and checking indicating and limit gages serving inspection and process control.
Tolerance Grade 3 (compromise between former Grades A and B)	Working gage blocks for use in tool room and production areas.

Table 4-2 lists the length tolerances for gage blocks as established by the Standard, with the maximum values of additional deviations for measurement error shown in parentheses following the last tolerance value of the applicable size group.

Geometry of Form

The length tolerance of gage blocks can only be meaningful when the gaging surfaces, whose distance from each other constitutes the measurable length, are substantially flat and parallel.

As a rule, the total value of flatness and parallelism errors must not be in excess of the maximum range of length tolerances permitted for any particular grade of gage block. These considerations are reflected in the flatness and parallelism tolerances of the American Standard, as listed in Table 4-3.

Errors in flatness, in addition to the potential effect on the length accuracy, must be kept at a minimum level for other reasons as well:

a. Deficient flatness interferes with the exact measurement of the length;

b. Lack of flatness can impede the "wringability"—the capacity of the gage blocks for being attached to each other by wringing (see "Handling of Gage Blocks" in this chapter); and

c. Deviations from flatness, by reducing the contacting area of the surface, can also shorten the wear life of the gage blocks.

Although the flatness tolerances apply to the area of the gaging surface omitting a border zone of 0.02 in. from the side faces, the plane of the border zone shall not be above the plane of the gaging surface.

While the sides of the gage block do not directly affect the length dimension, a reasonable level of accuracy also must be maintained in this respect. This requirement refers primarily to the squareness in relation to the gaging surfaces as well as to each other. Errors in the squareness of the sides must not exceed 5 minutes of arc, for gage blocks 0.400 inch long or longer.

Surface Condition

There are several reasons why great importance is attributed to a high degree of smoothness of the gaging surfaces:

a. The dependable measurement of the gage block length by means of contact instruments (electronic comparators) could be adversely affected by a surface roughness condition where crests of significant magnitude are protruding over the mean physical surface;

b. A high degree of smoothness is needed to provide a good reflecting surface for the very important interferometric length measurements (Fig. 4-3);

TABLE 4-2.   LENGTH TOLERANCES OF GAGE BLOCKS – 1

Tolerance dimensions in microinches

NOMINAL SIZE THROUGH (INCHES)	TOLERANCE GRADE			
	0.5	1	2	3
1	+1 / −1	+2 / −2	+4 / −2	+8 / −4
2	+2 / −2	+4 / −4	+8 / −4	+16 / −8
3	+3 / −3	+5 / −5	+10 / −5	+20 / −10
4	+4 / −4	+6 / −6	+12 / −6	+24 / −12
5		+7 / −7	+14 / −7	+28 / −14
6		+8 / −8	+16 / −8	+32 / −16
7		+9 / −9	+18 / −9	+36 / −18
8		+10 / −10	+20 / −10	+40 / −20
10		+12 / −12	+24 / −12	+48 / −24
12		+14 / −14	+28 / −14	+56 / −28
16		+18 / −18	+36 / −18	+72 / −36
20		+20 / −20	+40 / −20	+80 / −40

Tolerance dimensions in micrometers

NOMINAL SIZE THROUGH (MILLIMETERS)	TOLERANCE GRADE			
	0.5	1	2	3
10	+0.03 / −0.03	+0.05 / −0.05	+0.10 / −0.05	+0.20 / −0.10
25	+0.03 / −0.03	+0.05 / −0.05	+0.10 / −0.05	+0.30 / −0.15
50	+0.05 / −0.05	+0.10 / −0.10	+0.20 / −0.10	+0.40 / −0.20

TABLE 4-2.   LENGTH TOLERANCES OF GAGE BLOCKS – 2
Tolerance dimensions in microinches μm

NOMINAL SIZE THROUGH (MILLIMETERS)	TOLERANCE GRADE			
	0.5	1	2	3
75	+0.08 / −0.08	+0.13 / −0.13	+0.25 / −0.13	+0.45 / −0.23
100	+0.10 / −0.10	+0.15 / −0.15	+0.30 / −0.15	+0.60 / −0.30
125		+0.18 / −0.18	+0.36 / −0.18	+0.70 / −0.35
150		+0.20 / −0.20	+0.41 / −0.20	+0.80 / −0.40
175		+0.23 / −0.23	+0.46 / −0.23	+0.90 / −0.45
200		+0.25 / −0.25	+0.51 / −0.25	+1.00 / −0.50
250		+0.30 / −0.30	+0.60 / −0.30	+1.20 / −0.60
300		+0.35 / −0.35	+0.70 / −0.35	+1.40 / −0.70
400		+0.45 / −0.45	+0.90 / −0.45	+1.80 / −0.90
500		+0.50 / −0.50	+1.00 / −0.50	+2.00 / −1.00

Pratt & Whitney Cutting Tool and Gage Div., Colt Industries

Fig. 4-3. The surface of a gage block viewed through an optical flat. The straight pattern of the interference fringes proves the excellent flatness of the gaging surface.

c. Surface roughness in excess of values that are considered permissible for specific grades of gage blocks can impede the dependability of dimension transfer when gage blocks are used for comparator setting; and

d. Surface roughness also reduces the wringability and the resistance of the gage block surfaces against wear and corrosion.

Although a few random scratches of minimum depth might not affect the functional accuracy of the gage block, provided that there are no protruding burrs present, scratches are objectionable in appearance. Adverse functional effects may also be expected when size, spacing and number of scratches or streaks exceed a minimum level. Therefore, the Federal Specification establishes the maximum limit for acceptability of these marks on the gaging surface, usually caused by coarse abrasive grains during the lapping operation. The maximum permissible limits for these surface imperfections are listed in Table 4-3.

TABLE 4-3.   FLATNESS AND PARALLELISM TOLERANCES OF GAGE BLOCKS

Tolerance dimensions in microinches

NOMINAL SIZE THROUGH (INCHES)	TOLERANCE GRADE			
	0.5	I	2	3
I	I	2	4	5
2	I	2	4	5
3	I	3	4	5
4	I	3	4	5
5		3	4	5
6		3	4	5
7		3	4	5
8		3	4	5
10		4	5	6
12		4	5	6
16		4	5	6
20		4	5	6

Tolerance dimensions in micrometers

NOMINAL SIZE THROUGH (MILLIMETERS)	TOLERANCE GRADE			
	0.5	I	2	3
10	0.03	0.05	0.10	0.13
25	0.03	0.05	0.10	0.13
50	0.03	0.05	0.10	0.13
75	0.03	0.08	0.10	0.13
100	0.03	0.08	0.10	0.13
125		0.08	0.10	0.13
150		0.08	0.10	0.13
175		0.08	0.10	0.13
200		0.08	0.10	0.13
250		0.10	0.13	0.15
300		0.10	0.13	0.15
400		0.10	0.13	0.15
500		0.10	0.13	0.15

Recent progress in the analysis and measurement of surface roughness reveals the limitations of average values as a means to indicate the significant characteristics of very smooth surfaces. This aspect is more thoroughly discussed in Chapter 15 on surface texture. It will suffice to point out that the specified average surface roughness values are considered minimum requirements, and the more discriminating methods of surface analysis, particularly the microinterferometry, are widely applied for the inspection of the surface condition of higher grade gage blocks.

Nondimensional Characteristics

There are various parameters and conditions of the gage blocks that deserve attention and care, although they do not directly affect the dimension represented by the blocks. Some of these conditions have a bearing on the appearance only; others can interfere with the proper use or with prerequisites important for the maintenance of a dependable gaging system.

a. *Appearance*. Observable surface defects are, as a rule, undesirable. Definitely, no burrs of any kind are tolerated on the gaging surfaces. To avoid injuries of the wringing surfaces in service and handling, no sharp corners are tolerated; the edges of the wringing surfaces must be smoothly rounded.

b. *Homogeneity of the gaging surfaces*. Carbide gage blocks, made by powder metallurgical processes, may have less uniformly homogeneous surfaces than steel blocks. Very slight porosity would, in general, hardly affect the length of the gage block measured as the distance between the predominant gaging planes. However, pits on the surface can, in rare cases, permit the rounded tips of the indicators, while in contact with the gage block, to penetrate deeper than the predominant gaging surface. To reduce the incidence of false measurements due to this condition, the diameter, length or width of pits on the wringing surfaces must not exceed 0.001 inch.

c. *Identification marking*. Each gage block must be marked with its nominal size, a mark identifying the manufacturer, and a serial number that may be common to all blocks in a particular set. The marking must be legible, applied in a location where it does not interfere with the serviceability of the gage block and must not wear off easily.

Markings for similar identification purposes are also required for the wooden or plastic cases in which the gage block sets are supplied and stored,

Pratt & Whitney Cutting Tool and Gage Div., Colt Industries

Fig. 4-4. Large gage block set in a case with individual marked compartments for each block.

with individual compartments for each single block and accessory piece (Fig. 4-4).

THE MATERIAL OF GAGE BLOCKS

For several decades following their invention, gage blocks were made exclusively of steel. Generally, a *high carbon, chromium alloyed steel* grade was used, similar to the composition selected for the manufacture of antifriction bearing components. It was found that this quality of steel possesses several properties beneficial to gage blocks, such as good hardenability, minimum quenching distortions and reasonable wear resistance. Because of these favorable properties, to which the relatively low material prices, ease of manufacture and, particularly, of lapping should be added, essentially the same grade of steel is still used to manufacture the majority of gage blocks for a wide range of industrial applications.

By improving the cleanliness of the basic steel material, closely controlling its composition and applying special heat treating processes, such important properties as hardness, homogeneity and stability have been further improved.

Stability refers to the property of gage blocks of retaining their original size with a minimum of change (generally growth, less frequently, shrinkage) resulting from phase transformation in the metallic material. When not controlled properly, phase transformation changes can seriously affect the original dimensional accuracy of the gage blocks during a period of several months or years.

A few manufacturers of gage blocks prolong the wear life of steel gage blocks by applying *chromium plating* to the surface, thereby substantially improving the stain resistance as well. In some rare instances, chromium plating is used to salvage gage blocks that lose size due to wear, although the economic and technical aspects of such a procedure may be regarded as questionable. When applying chromium plating on gage blocks, great care must be exercised to avoid peeling while in use.

The valuable properties of chromium as an alloying element of steel, combined with those of nickel, are used with great advantage for the manufacture of gage blocks of stainless steel. The use of *stainless steel* as a gage block material assures favorable properties comparable to those of chromium plating with the added advantages of great hardness of the entire block and excellent dimensional stability. Certain compositions of stainless steel also permit the obtaining of exceptional surface smoothness, a condition of particular value for gage blocks.

Because advances in powder metallurgy technology contributed to wider applications of *metallic carbides,* efforts were made to utilize the excellent wear resistance of some sintered carbides for the manufacture of gage blocks. There are two types of carbides that are primarily used for gage blocks and accessories, particularly wear blocks (described later in this chapter): *tungsten carbides* and *chromium carbides.* Although substantially more expensive than steel, in applications where wear is a troublesome factor, such as in some grinding areas, the highly superior wear resistance of carbide gage blocks frequently justifies their use. When selecting carbide gage blocks, due regard must be given to the differential in the coefficient of thermal expansion as compared to steel (see Table 4-4), the proneness to chipping and the not always homogeneous surface texture of sintered carbide blocks. It is a question of evaluation to determine the areas where the advantages of carbide gage blocks surpass their known limitations.

TABLE 4-4. AVERAGE COEFFICIENT OF THERMAL EXPANSION OF MATERIALS COMMONLY USED FOR THE MANUFACTURE OF GAGE BLOCKS AND FOR COMPONENT PARTS IN ENGINEERING PRODUCTION

MATERIAL	COEFFICIENT ($\times 10^{-6}$ inch/inch/°F)
Tool steel, hardened	6.4
Stainless steel (410)	5.5
Tungsten carbide	3.0
Chromium carbide	4.5
Fused quartz	0.3
Copper	9.4
Aluminum	12.8
Zirconia (stabilized)	3.1

FORMULA FOR CALCULATING CORRECTION VALUES WHEN MEASURING DISSIMILAR MATERIALS AT TEMPERATURES OTHER THAN 68°F:

$$MD = L(\Delta k)(\Delta t)$$

where

MD = measuring difference in microinches (plus or minus);

L = length;

Δk = difference in the coefficient of thermal expansion of the gage block and of the object materials;

Δt = difference of the temperature from 68°F (plus or minus).

Most recently, stabilized zirconia (zirconium oxide) has been used as a material in the manufacture of gage blocks. Zirconia, a white crystalline compound, is a refractory ceramic with many physical properties that are ideally suited for this new use. Gage blocks made from zirconia are lightweight, more wear-resistant than carbide, completely corrosion-resistant, have long-term stability and a coefficient of thermal expansion close to that of steel (see Table 4-4). However, zirconia blocks reach changes in ambient temperature more slowly than steel blocks and as such should be allowed to soak at the anticipated operating temperature prior to use. This slow adaption to temperature change may be viewed as a benefit, as the ceramic block tends to be less open than steel to the dimensional inaccuracies caused by the introduction of body heat through normal handling. Ceramic blocks wring well together and in combination with steel blocks. Burrs raised on zirconia blocks, as a result of ordinary use, may be stoned in the same manner as steel blocks (see Figure 4-17).

Other materials are sometimes selected for the manufacture of gage blocks, but their actual use is very limited. Gage blocks made of *fused quartz* are resistant to wear and can be polished by methods used for optical lenses to the highest degree of smoothness. The transparency of quartz is of some advantage for checking the proper wringing of the blocks. Another distinguishing feature of quartz gage blocks is their property of producing no protruding burrs when scratched in use. On the other hand, the greatly different coefficient of thermal expansion of fused quartz (see Table 4-4) may be a source of potential error when used at other than normal temperatures.

The Coefficients of Thermal Expansion of Gage Block Materials

As indicated above, there are various important properties of gage blocks that may vary widely, depending on the material used for their manufacture. Some of these properties are purely technical in their nature, others—such as initial cost and wear life—are judged in their economic aspects as well. However, there is one important factor that must never be left unconsidered because of its direct bearing on the dependability of measurements by gage blocks: *the coefficient of thermal expansion.*

All gage blocks are calibrated at the standard temperature of 68°F. When the blocks are used for reference standards under conditions where all elements of the gaging, the instrument, the standard and the specimen are precisely kept at this standard temperature, no error due to heat expansion should be expected. However, if any of the elements has a temperature different from the standard, its length will change by expanding or contracting.

Gage blocks may be used as reference pieces for comparator measurements on parts that are substantially similar in mass and form to the gage blocks, and were made from the same kind of material, treated to comparable hardness. Under such conditions, exactly equal departures from the standard temperature of both the gage blocks and the specimen will not affect the accuracy of the comparator gaging. In practical applications, this may be the case when hardened steel parts are measured with steel gage blocks, in a constant temperature environment.

Considering the variety of gage block materials, and also of the materials of component parts that must be precisely measured, deviations from the standard temperature will cause differential rates of dimensional changes in the reference piece and the object. For this reason it is mandatory for critical length measurements to compensate for the difference in length variations due to the nonuniform rates of thermal expansion. Table 4-4 shows the average values of the coefficients of thermal expansion for the gage block materials discussed in this section, and for some frequently used component part materials. A simple formula is also presented, which permits compensating, by computation, for the errors in the measurements caused by differential dimensional changes at temperatures other than the standard.

THE MANUFACTURE OF GAGE BLOCKS

The manufacture of gage blocks is particularly critical in the last stages of the process. The preceding operations, considering the simple shapes of these standards, are carried out by conventional metalworking methods. However, heat treatment, starting from the first annealing, through the successive stages of hardening, tempering and related stabilizing, is of greatest importance for the ultimate precision of the gage blocks. The heat treatment processes, of course, must be adapted to the particular type and grade of steel of which the blocks are made, and prominent manufacturers of gage blocks have developed their proprietary techniques to better accomplish the objectives of wear resistance and stability.

It is particularly in this latter respect that substantial improvements are accomplished by scientifically developed stabilizing processes that reduce the warpage of even the thinnest regular size blocks, and avoid dimensional changes due to phase transformations in the hardened gage block material. The dimensional changes in improperly stabilized gage blocks can amount to a few millionths of an inch per inch of length, during a period of one or several years, thus seriously jeopardizing the original accuracy obtained during the manufacturing stage.

The final metalworking operation in the manufacture of gage blocks is the lapping. This finishing step must develop the following critical parameters of the gage blocks:

a. Geometry of form, particularly the flatness and the parallelism of the measuring surfaces;

b. The surface condition (texture) of the operating sides; and

c. The distance between the parallel measuring surfaces, the "length" of the gage blocks.

Prior to the lapping operation the blocks are precisely ground, closely approaching the final form and size, with a dimensional allowance of about 0.0005 inch. Most manufacturers apply three lapping operations, rough, semifinish and finish. The stock removal is distributed between these subsequent operations on the order of about 60%, 36% and 4%, thus leaving for the last finishing step only about 20 millionths of an inch of material to be removed.

Special lapping machines are used, with fine-grain cast iron (Meehanite) lapping discs usually in a horizontal position. The lower disc, called the lap, supports the blocks and the upper lap floats on top of the blocks. A special holder, known as the spider, with appropriate openings to retain the individual blocks, is placed between the laps. By means of planetary gears a combination of eccentric-rotary motions, at varying speeds and in changing directions, is imparted to the blocks through the spider. The variations of the resulting pattern, and the frequent in-process interchange of the individual gage block positions, contribute to the desired end results. A further factor of major importance is the selection and care of the lapping compound, which is made up of very fine abrasive grains of rigorously controlled quality, suspended in special soluble oils.

Reclamation of Worn Gage Blocks

The short survey of the manufacture of gage blocks, particularly the reference to the meticulously established and controlled stock allowances for the finishing steps, explains why it is seldom practical to rework worn blocks to a next smaller size, or to reclaim them by chromium plating. The most critical operations, and also the most expensive ones, are those by which the final dimensions are imparted to the blocks; the value of the semifinished blocks, including material that might be saved, hardly covers the extra costs caused by semifinished parts in varying sizes.

GAGE BLOCK SETS AND ACCESSORIES—THE HANDLING OF GAGE BLOCKS

The Composition of Gage Block Sets

Gage blocks are most frequently procured and used in sets comprising a specific number of blocks in different sizes. The nominal sizes of the individual blocks within a set are determined mathematically with the purpose of permitting the assembly of any length combination from the size of the smallest block over the entire range of the set, in increments of, say, 0.0001 inch, or even smaller.

Gage block sets are available in the trade in smaller or larger combinations, depending on the desired range and increments. Table 4-5 lists as examples the compositions of a few typical gage block sets; variations from these examples are quite common.

The larger sets also have the advantage of permitting the assembly of several combinations, simultaneously, to equal or different total lengths.

The Handling of Gage Blocks

For selecting the individual blocks needed to be assembled, by combination, into a stack of specific length, it is practical to proceed in the manner indicated in Table 4-6. After the individual blocks that go into the combination have been selected, they are assembled into a unit by a technique known as *wringing*. By wringing, the blocks are attached to each other with considerable force, capable of overcoming the gravitational pull that acts upon the stack (see Fig. 4-5). The properly wrung blocks are so closely attached to each other that, for all practical purposes, the space between the component members can be disregarded and the combined length may be considered equal to the added lengths of the individual blocks. The phenomenon of the attachment by wringing is attributed to the combined effect of molecular attraction and of adhesive action, the

TABLE 4-5.   THE COMPOSITION OF TYPICAL GAGE BLOCK SETS

These are the recommended sets, but gage block sets in many other compositions are also available

The individual blocks contained in the sets are listed in series

RECOMMENDED SET DESIGNATION	PIECES IN THE SET	NUMBER OF BLOCKS PER SET	SIZE RANGE, inches	SERIES INCREMENTS (STEPS), inch
Standard 81 block set	81	9 49 19 4	0.1001–0.1009 0.101 –0.149 0.050 –0.950 1.000 –4.000	0.0001 0.0010 0.050 1.000
Limited 36 block set	36	9 9 9 1 5 2 1	0.1001–0.0009 0.101 –0.109 0.110 –0.190 0.050 0.100 –0.500 1.000 –2.000 4.000	0.0001 0.001 0.010 0.050 0.100 1.000 4.000
Thin 28 block set	28	1 9 9 9	0.02005 0.0201 –0.0209 0.021 –0.029 0.010 –0.090	0.00005 0.0001 0.001 0.010
Thin 21 block set	21	1 9 11	0.01005 0.0101 –0.0109 0.010 –0.020	0.00005 0.0001 0.001
Long 8 block set	8	4 2 2	5– 8 10–12 16–20	1.00 2.00 4.00

NOTE: When gaging surfaces are exposed to excessive use, a pair of wear blocks is recommended. The sets' cases usually provide extra compartments for two wear blocks, 0.050 or 0.100 in. thick; these are generally procured together with the basic set.

Pratt & Whitney Cutting Tool and Gage Div., Colt Industries

Fig. 4-5. When gage blocks in perfect condition are wrung together the resulting clinging power is capable of resisting the action of gravitational force upon a long stack.

CEJ Gage Company

Fig. 4-6. The sequential steps in wringing together rectangular gage blocks.
(Left) Holding the blocks crosswise and exerting some pressure, while carrying out a slight swiveling motion will cause the blocks to adhere to each other.
(Center) The adhering blocks are then slipped into parallel position with a slight pressure.
(Right) The wrung blocks form a single unit for the purpose of instrument setting and measurement.

latter due to a very thin film of moisture or oil. The technique of gage block wringing is shown in Fig. 4-6.

Gage Block Accessories

For the purpose of transferring the exact length dimension represented by a specific combination of gage blocks onto a body where that dimension must be established, two methods may be considered:

a. Direct transfer, using individual blocks or a few blocks wrung together; and

b. Transfer with the aid of auxiliary elements.

TABLE 4-6. ASSEMBLY OF GAGE BLOCK STACKS

SUGGESTED PROCEDURE FOR SELECTING THE INDIVIDUAL BLOCKS TO BE COMBINED INTO THE REQUIRED TOTAL LENGTH

RULES FOR GAGE BLOCK COMBINATION

a. Adhere to a system of procedure and avoid try-and-error methods.

b. Write the desired dimension on paper, and continue with putting down in writing the selected gage block sizes and the dimension still to be covered.

c. Start with eliminating the last decimals of the desired dimension, by selecting blocks having the same number in the equivalent decimal places.

d. Proceed with this method of reduction until the combined value of the selected block sizes is equal to the desired dimension.

In general, it is preferable to cover the desired dimension with the minimum number of gage blocks, unless more than a single combination is to be assembled simultaneously. In this case a "balanced condition", with approximately equal number of blocks in each combination of comparable nominal length, should be approached.

EXAMPLE: For the combination of gage blocks into a specific total length. Available set of 81 blocks. Desired dimension: 1.7865 inch.

FOR A SINGLE COMBINATION		FOR TWO COMBINATIONS OF EQUAL LENGTHS			
		COMBINATION "A"		COMBINATION "B"	
DIMENSION TO COVER	GAGE BLOCKS SELECTED	DIMENSION TO COVER	GAGE BLOCKS SELECTED	DIMENSION TO COVER	GAGE BLOCKS SELECTED
1.7865 − .1005 1.686	0.1005	1.7865 − .1005 1.686	0.1005	1.7865 − .1003 1.6862	0.1003
− .136 1.550	0.136	− .146 1.540	0.146	− .1002 1.586	0.1002
− .550 1.000	0.550	− .240 1.300	0.240	− .136 1.450	0.136
− 1.000 0.000	1.000	− .300 1.000	0.300	− 0.650 0.800	0.650
		− 1.000 0.000	1.000	− .800 0.000	0.800

The auxiliary elements used for length transfer must satisfy requirements:

1. Protect the gage block stack from falling apart due to being hit inadvertently while in use;

2. Represent as closely as possible the original length of the gage block stack when transferring its dimension; and

3. Provide convenient means for the length transfer for diverse applications under varying conditions of available space and geometric configuration of the object.

Figure 4-7 illustrates a set of frequently used gage block accessories. The contents of accessory sets are selected with the purpose of handling the majority of applications where an indirect transfer of the gage block lengths is needed. A typical set may contain the following elements:

1. *Jaws* of different sizes and overall lengths. Jaws are usually supplied in nominal sizes of 0.100, 0.200, 0.250, 0.500 and 0.625 inch, with overall lengths varying accordingly from about $1\frac{5}{8}$ inches to 4 inches. The measuring ends of these jaws are flat on one end and convex on the other, the convex form serving measurements on the inside of bores.

2. *Beams* of I cross-section, nominal size 0.750 inch and overall lengths of 4 inches and 6 inches.

Beams are useful for setting up reference levels in height gage applications, and also for establishing reference distances for internal indicator gage adjustment.

3. *Scribers* are used for layout work.

4. *Center points* are useful for checking the lead of threaded parts; they may also be used for checking the axis location of small bores with respect to a reference plane.

5. *Foot blocks* are indispensable for applications where the gage block stack is used as a height gage to establish a very accurate distance normal to the top of the surface plate.

6. *Adjustable holders* are used in conjunction with any of the listed accessories to assemble them with the gage block stack into a single unit. The holders are not a substitute for the wringing of the blocks, because intimate contact between individual blocks is best established by wringing. After a stack of blocks has been wrung together, it is placed into a holder of appropriate size and the selected accessory elements are added.

The adjustable holders are supplied in different lengths to accommodate stacks of various nominal dimensions. The useful length of the holder is determined by the size of the opening and the travel of the adjusting screw. With the exception of the smallest sizes, the holders are commonly equipped with split nuts to shorten the adjustment time.

7. *Tie rods* are substituted for the holders in the case of square gage blocks with center holes. Figure 4-8 shows the use of tie rods whose rough adjustment is accomplished by the closely spaced pin holes, and the final adjustment is made with the

CEJ Gage Company

Fig. 4-7. Frequently used types of gage block accessories assembled into a set consisting of jaws, scriber, center point, tram points, straight edges and foot blocks, together with gage block holders in different lengths.

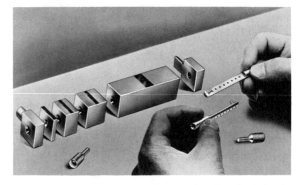

Pratt & Whitney Cutting Tool and Gage Div., Colt Industries

Fig. 4-8. Square gage blocks with center holes, assembled into a stack for internal measurement with the aid of adjustable tie rods.

screw. The possibility of avoiding outside clamps by the use of tie rods proves to be a valuable asset of square blocks when the operating space is limited.

8. *Straight edges.* Although not an accessory that can be assembled with the blocks, such as is the case with the items listed above, straight edges, particularly the triangular type, are frequently provided in gage block accessory sets. Commonly used edge lengths are $1\frac{1}{2}$, 3 and $5\frac{1}{2}$ inches, and the edges are slightly rounded to prolong the wear life of these tools. Straight edges, whose applications are by no means limited to gage block measurements, are useful in checking the level coincidence between the end face of the gage block stacks and the workpiece. This kind of coincidence check is, of course, limited to applications where there is no need for more sensitive measurements or, because of space limitations, where indicator gages cannot be accommodated.

The Accuracy of Gage Block Accessories

The transfer members in gage block based measurements must have accuracies that are comparable to those of the basic length elements. However, this relationship in accuracy does not necessarily mean equal orders of magnitude. Actually, the required accuracies for gage block accessories are defined in that order of magnitude that conforms to the needs of the particular gaging operation for which the accessories are generally used. For example, while the actual thickness, representing the nominal length dimension, of measuring jaws must not deviate from the nominal size by more than 0.000030 inch, center points and tram points are required to be accurate only within ±0.000300 inch. The permissible height of foot blocks are ±0.0001 inch.

Wear Blocks

Although used as an accessory, these are regular gage blocks of round number nominal size (0.100 or 0.050 inch length) made of steel or carbide. Wear blocks are placed on both ends of gage block stacks in applications where substantial frictional wear can be expected. When worn, the wear blocks can easily be replaced, and the original accuracy of the regular elements of the set can be maintained much longer.

APPLICATIONS FOR GAGE BLOCKS

The uses to which gage blocks may be put, where such end standards are the most appropriate, or even the only means of precise measurements, are too numerous to permit listing.

Therefore, it is only possible to refer here to a few applications that are considered typical. These examples of use are presented in groups, comprising measurements commonly accomplished with a specific grade of gage blocks. However, it should be remembered that the grade of gage block to be actually selected for a specific service must always be governed by the accuracy requirements of the operation. Consequently, the following association of specific grades and applications serves the purpose of providing typical examples only, and must not be regarded as fixed rules.

1. Applications for Grade "0.5" and Grade "1" Gage Blocks

a. *Basic reference standards.* Gage blocks of the highest grade of accuracy are the ideal and generally used means to maintain a link with the national primary standards of length. In this application a specific set of gage blocks is considered the *grand master* of a particular plant or laboratory. The length dimensions of this plant reference standard must be directly traceable to the masters of the National Bureau of Standards. (See the following section on gage block calibration.)

b. *Reference standards for gage calibration.* For the calibration of working gage blocks and other types of reference gages, comparator measurements are used, with the purpose of comparing the actual size of the plant secondary standards to the reference standards of equal nominal length. Comparator measurements with modern electronic gages can be carried out to a very high degree of sensitivity and repeat accuracy. Consequently, the precision of the end results are primarily a function of the inherent accuracy of the reference standards on which the measurements are based.

c. *Calibration of sensitive measuring instruments.* The repeat accuracy and indicating linearity of sensitive comparator type measuring instruments must be checked with reference elements whose accuracy of length and geometric form is compatible with the expected precision of the calibration. Gage blocks of different sizes, in length increments distributed over the indicating range of the instrument, are an excellent, and also conveniently available, means for checking the accuracy of response of comparator gages.

2. Applications for Grade "2" Gage Blocks

a. *Checking limit gages.* Both external and internal limit gages can be checked for size to a high degree of accuracy by the aid of gage blocks. The check is directed at determining whether the measuring dimension of the limit gage is equal to its designed dimension.

The operation requires a certain amount of skill to maintain the blocks in a properly aligned position in relation to the measuring surfaces of the gage to be checked and also to detect the presence of dimensional deviations. Gap gages can be checked directly with a gage block stack, although the mounting of wear blocks is advisable when the gage blocks are frequently used for such inspections. For the checking of plug gages and gage rings, accessories must be used, consisting of jaws or other types of appropriate contact tips, assembled in a holder with the required combination of gage blocks.

b. *Setting adjustable limit gages.* The process is similar to the checking of rigid limit gages, yet requires the additional operation of adjusting the contact members of the limit gage into proper agreement with the design dimension as represented by the gage block assembly (Fig. 4-9).

c. *Measuring setting gages.* Setting gages, such as those used for adjusting all types of comparator instruments on which repetitive gaging is carried out, are commonly inspected for size on gage room comparators, with gage block assemblies as masters.

d. *Use as comparator setting gages* (Fig. 4-10). Setting gages may not be available at the shop for a nonregular type of comparator measurement, or the required accuracy may exceed the limits to which standard setting gages are usually made. The convenience of using gage blocks for the purpose of set-

CEJ Gage Company

Fig. 4-10. (Left) Setting a comparator gage with the aid of gage blocks.

Fig. 4-11. (Right) Setting a special gage—the CEJ Sine Protractor—to the desired angle with the aid of appropriate gage blocks.

ting gages is obvious with respect to both, the ready availability in most of the required length dimensions and the high level of accuracy. However, conventional setting gages are preferred for regular use, due to economic considerations, and also for reasons of process control.

3. Applications for Grade "3" Gage Blocks

The use of working gage blocks, whether of the Grade B. or of an intermediate grade exceeding the accuracy requirements of Grade B. is the most varied. With the continuous tightening of product tolerances, and also as the result of the wide application of highly sensitive comparator instruments, the use of gage blocks in toolmaking, and also in actual production, is becoming more and more general.

a. *Direct measurement of distances* between parallel surfaces, for example, widths of grooves, slots, keyways, and so on. The direct use of gage blocks for checking the conformity of gap dimensions to specifications is particularly convenient for narrow grooves, or where space limitations—such as on internal surfaces—are making the use of bulkier types of measuring tools difficult. Another example of the direct use of gage blocks, in this case several blocks wrung together, is shown in Fig. 4-11.

b. *Checking and adjusting mechanics' measuring tools.* Micrometers can be checked for linearity (which might be affected by screw thread wear) with gage blocks of different sizes that are staggered within the range of the measuring tool. Gage blocks are also useful for the zeroing of micrometers with measuring ranges in excess of the screw travel, which is usually one inch. In such micrometers, the measuring spindle does not contact the anvil even in a fully closed position; inserting the corresponding size gage block into the gap is a convenient method of adjustment.

Pratt & Whitney Cutting Tool and Gage Div., Colt Industries

Fig. 4-9. Adjustable snap gage is set to the exact jaw opening with gage blocks.

c. *Limit gages assembled from gage blocks and accessories.* Assembled limit gages, consisting of gage blocks as distance members and of added contact elements, all built into a single unit with the aid of an appropriate holder, provide a great versatility of quickly assembled, yet very accurate limit gages (Fig. 4-12). It is even possible to combine the GO and NOT GO dimensions of the required limit gage into a single unit, by using two pairs of jaws with the appropriate gage blocks in a common holder. However, in actual gage room or shop practice, the use of assembled limit gages is reserved for temporary applications only. It would be uneconomical to tie up, during an extended period, elements of an entire gage block set for the repetitive checking of a single dimension.

d. *Establishing reference distances for the transfer of dimensions.* Where a common datum plane is available for the object and the reference distance, precise gaging by comparison becomes feasible and is frequently applied. The common datum plane may simultaneously accommodate all three essential elements of the gaging process: the object, the reference piece and the transfer tool. The most common application for this kind of comparative gaging is found in surface plate work. The reference pieces generally used here are gage blocks, either in the form of plain stacks (Fig. 4-13), or assembled in a holder, usually comprising a foot plate as well (Fig. 4-14).

When medium accuracy of comparison is the only purpose, indicating transfer instruments can be obviated, and the transfer carried out with a jaw attached to the gage block assembly. Some manufacturers even supply indicators that can be directly at-

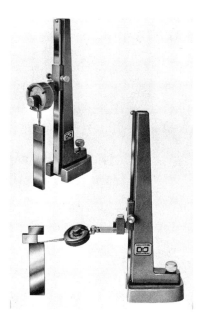

CEJ Gage Company

Fig. 4-13. Establishing reference lengths for height transfer gages with gage block stacks.

tached to the gage block assembly for uses where the sensitivity of mechanical dial indicators is considered adequate (Fig. 4-15).

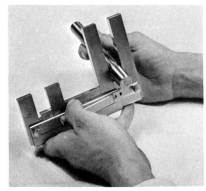

Do ALL Company

Fig. 4-12. Gage blocks assembled with jaws into a limit gage with GO and NOT GO spans.

(Left) The L. S. Starrett Co.
(Right) Do ALL Company

Fig. 4-14. (Left) Gage blocks assembled with foot plate and scriber jaw for surface plate layout work.

Fig. 4-15. (Right) Rapidly assembled comparator gage consisting of a gage block stack combined with a dial indicator.

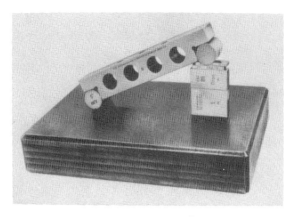

Brown & Sharpe Mfg. Co.

Fig. 4-16. Sine bar supported by a gage block stack produces an accurately controlled incline.

e. *Angle measurements with sine bar.* The implementation of the sine bar method of precise angle measurement (detailed in Chapter 10) is predicated on establishing a side of very accurate length and normal to the base. The right angle thus formed is subtended by a hypotenuse of known length, represented by the sine bar. The double requirement of accurate length and perpendicularity to the supporting base can best be accomplished with gage blocks, whose combined length can be easily selected to result in the desired slope angle of the sine bar (Fig. 4-16).

CALIBRATION OF GAGE BLOCKS

Impairment of Gage Block Accuracy

The original accuracy of gage blocks is subject to deterioration from three major causes:

1. The dimensional instability of the gage block material (steel or metallic carbides);

2. Wear due to physical contacts with other metallic surfaces, such as gage contact points, or friction during the wringing of the blocks. Service wear is an unavoidable, although controllable, consequence of the regular use of gage blocks; and

3. Damages inflicted to the blocks during storage and handling (corrosion, nicking, scratching, indentations and so forth).

Any of these factors can subtract from the original accuracy of the gage blocks, eventually resulting in dimensional and other kinds of changes beyond the tolerance limits of the grade to which the particular block nominally belongs. Except for injuries caused by careless handling, these dimensional changes do

not occur suddenly. Some of the blocks within a particular set may not be affected to a significant extent at all; for some others, it might take several months or years of continuous service before the gradual deterioration of the original accuracy progresses to an objectionable magnitude.

The circumstances that affect the maintenance of the original accuracy of the gage blocks are many and are varied. Examples of these are the inherent dimensional stability of the block material, its wear resistance, the frequency of usage, the kind of members with which the blocks are brought in contact while in use, for example, carbide tips of comparator gages, the environmental conditions (dust, humidity, corrosive atmosphere and so forth) and, perhaps the most important single factor, the competence and the care applied in the handling of the gage blocks.

Periodic Calibration of Gage Blocks

Gage blocks, because of their function as basic reference standards, must be dependably accurate within the established limits of their classification. Considering the prevalence of the comparator type of measurements whenever high sensitivity gaging is required, the accuracy level maintained by particular production areas, or even of an entire plant, may be dependent on the reliability of their basic reference standards.

In view of the diverse factors that can harmfully affect the consistent accuracy of gage blocks, these reference standards must be inspected at regular intervals. Such regular inspection—generally referred to as calibration—should be part of a gage surveillance program, which every manufacturing plant where precision measurements are required must establish and rigorously maintain.

The required frequency of gage block calibration usually depends on two sets of factors:

a. The accuracy grade to which the gage blocks are classified; and

b. The extent of dimensional deterioration that can be anticipated in view of the usage circumstances of the blocks.

Tentatively, a six-month period for recalibration may be considered. Actual experience, based on the results of repeated inspection or contractual commitments, for example, Mil. Std. specifications, may indicate a change of this schedule. In some cases the recalibration of the gage blocks at three-month intervals will be judged desirable, whereas in other applications a once-a-year frequency will prove sufficient. Whatever schedule is finally adopted, strict adherence to it must be observed.

Inspection of Gage Blocks for Calibration

Considering the great importance of the reliable accuracy of gage blocks, their calibration must follow a strict pattern. Standard practices for gage block calibration comprise the following inspection steps that must precede, but not necessarily in the listed order, the measurement of the significant length dimension:

1. *Preparation*, which usually consists of

    a. Cleaning, now generally carried out in ultrasonically assisted solvent baths; and

    b. Temperature equalization and stabilization by keeping the blocks on a soaking plate (known also as a heat sink), for an appropriate length of time. Its purpose is to practically eliminate temperature differentials between the masters and the blocks to be checked. This preparatory step should immediately precede the dimensional gaging, and it is customary to keep the soaking plate in the close vicinity of the comparator gage.

2. *Visual inspection* is primarily directed at detecting scratches, nicks and similar flaws caused during the use and handling of the blocks in service. Inspection is usually done under a magnifying glass, but when doubt exists regarding the acceptable condition of the gaging surfaces, the roughness or scratch depth may be checked by means of a highly sensitive tracer instrument or by microinterferometry. The visual check should also comprise the condition of the corners. Protruding burrs are particularly harmful to the functional adequacy of the blocks and can cause damage to other blocks when being wrung together. Minor burrs can be removed by applying a fine-grain dressing stone, made of some natural rock, such as ''Arkansas'' stone or black granite (Fig. 4-17).

3. *Inspection of gage block geometry.* Two major parameters require inspection during recalibration: flatness and parallelism.

    a. *Flatness* can best be checked with an optical flat under monochromatic light. Departures of the light interference bands from the straight are indicative of defective flatness. The extent of deflection, evaluated in fringe width or fractions thereof, should be compared to the tolerances for flatness errors.

Giddings & Lewis Measurement Systems

Fig. 4-17. The use of a dressing stone to remove minute burrs from the wringing surface of gage blocks.

    b. *Parallelism* can also be checked with the use of a pair of optical flats and with the aid of a master block having a nominal size equal to that of the block to be checked. The parallelism of the master is assumed to be well within tolerance requirements. The inspection takes several steps and requires specialized skill. Parallelism can also be checked in conjunction with the dimensional inspection of a comparator; when taking several length measurements across the gaging surface of the block, gradual differences in the measured dimensions are indicative of parallelism error.

4. *Functional test—checking the wringing quality.* The capacity of the gage blocks to be wrung together without any significant gap, and to form an integral assembly of known size, is one of the basic properties on which many uses of these end standards rest. Because the wringing capacity can be adversely affected in service, the periodic inspection of this property is desirable. Experienced operators can carry out this test by wringing the blocks to be tested to a master of known quality. A more reliable test consists in wringing down the cleaned block on an optical flat and inspecting the wrung surface for the presence of coloration. Color indicates an air gap; its area must not exceed 5% of the entire wrung surface of the gage block.

Comparator Measurement of Gage Block Lengths

The method consists of comparing the length of the gage block to be calibrated to the length of a master of identical nominal length. The master must be accurate to a very high degree (grade 1 or better). Even with this high inherent accuracy of the master, its known deviation from the absolute nominal value must be considered when evaluating the comparator measurement. Other parameters of the master that could affect the precision of the length dimension must be at least equal to grade 1 specifications.

The measured length should be the perpendicular distance between the two gaging surfaces of the block. When a single measuring spot is used, it should be located essentially in the middle of the block face. For square blocks with center holes, it is customary to take two readings, on opposite sides of the hole, halfway between the edges, and then to average the measured values.

The instrument used for gage block calibration should be an electronic comparator of the highest degree of sensitivity and stability. The least graduation of the meter must be one microinch, or less. The comparator stand must be of rigid design, made of well stabilized material in excellent workmanship, with particular emphasis on the consistent perpendicularity of the gage spindle in relation to the staging surface of the specimen table (Fig. 4-18).

The potential harmful effect on the precision of the gaging due to minor flatness errors in the instrument table, or the accumulation of minute dust particles, can be substantially reduced by limiting the actual contact of the specimen to points protruding over the general level of the table. Some manufacturers of comparator instruments prefer three points for establishing a measuring plane; others advocate a single locating point of about $\frac{1}{32}$-inch square. The

Brown & Sharpe Mfg. Co.

Fig. 4-18. Electronic gage block comparator with one-millionth-inch least graduation, supplemented with digital readout and interfaced to a personal computer and dot-matrix printer.

accurate alignment of the gaging spindle with the support point requires particular attention.

A preferred way to avoid the possible detrimental effect on the gaging precision of errors in squareness or flatness originating from the supporting table consists in excluding the role of the latter from the actual gaging process. This can be accomplished by means of differential gaging, using two sensing heads on the comparator, perfectly aligned but opposed to each other. More details on differential gaging are discussed in Chapter 7.

By using comparator instruments of the differential system, the measured dimension will be the distance between the opposed gaging tips, as established by the physical body of the gage block being measured.

Light Interference Measurement of Gage Blocks

Whereas gaging by a comparator instrument supplies the numerically expressed values in microinches of the deviations from the master's length, measurements by light interference must be interpreted. Although that interpretation may be facilitated and increased in its accuracy by means of sophisticated accessory devices (fringe counters) attached to the basic optical instrument, the direct readout in microinches is not an inherent characteristic of light interference measurements. On the other hand, light interference measurements, when applied to the inspection of gage blocks, provide distinctive advantages:

1. *Dependable accuracy* is assured by the process being based on natural physical values, the wavelengths of known sources of monochromatic light.

2. *Sensitivity*, on the order of one microinch or finer, can be accomplished with proper equipment capable of producing distinct fringes and equipped with means of precise fringe fraction evaluation.

3. *Versatility*. The light interference method and much of the available equipment permit the exploration of important characteristics of the gage blocks, other than the length dimension. The following pertinent parameters can be inspected and measured with instruments utilizing the light interference phenomenon: (a) flatness; (b) parallelism; and (c) surface texture, with regard to waviness, by sweeping the fringes across the gage surface.

Special equipment, also operating on the principles of interference fringes of light waves, is available for the primary purpose of detailed surface tex-

Pratt & Whitney Cutting Tool and Gage Div., Colt Industries
Carl Zeiss, Inc.

Fig. 4-19 (Left) Gage blocks of equal nominal size, when wrung to a common base and observed under an optical flat, can be checked for identical length by the pattern of interference fringes.

Fig. 4-20. (Right) The Zeiss Interferometer for gage block measurements.

Brown & Sharpe Mfg. Co.

Fig. 4-21. Automatic laser interferometer built by TESA metrology.

ture analysis (microinterferometry). However, these instruments are not intended for carrying out length measurements.

Table 4-7 presents a survey of the light interference methods and equipment used for the measurement and inspection of gage blocks (Figs. 4-19, 4-20 and 4-21).

Certification of Calibration

a. *The concept of traceability.* The calibration of gage blocks—unless based on accepted values of natural constants—should result from comparison with reference standards whose calibration is traceable to the National Bureau of Standards. Traceable means, in this context, the availability of documentary proof to the effect that the reference standards used for calibration, were inspected, approved and certified by the NBS. Calibration of the masters by the NBS must have taken place not more than six months prior to their use as gage block calibration reference standards.

b. *Certificates of calibration.* A certificate of calibration should be issued by the laboratory that carried out this systematic inspection of the gage blocks. The certificate should attest to the accuracy and the conditions under which the reported results were obtained, and also confirm the traceability of the reference standards used for calibration. The certificate lists the individual gage blocks, their sizes and serial numbers, with the amount and direction (plus or minus) of the measured deviation from the nominal size.

It is a standard practice in gage block calibration to mark on the certificate the individual blocks whose length or other pertinent dimensions, or their condition, are not in compliance with the existing specifications. The certificate will also indicate the accuracy grade of the inspected blocks; the marking of the deficient blocks is based on the tolerances defined for that particular grade.

Utilization of the Calibration Results

Individual gage blocks that, as the result of the calibration process, were found to lack compliance with the specification requirements, must be replaced. It is not permissible to continue using worn or damaged gage blocks for their original application. Not even by taking recourse to a system of compensation in the evaluation of gaging results should the adherence to the rule of mandatory replacement be relaxed.

The assignment of worn blocks to a lower grade of accuracy should the measured dimensions still comply with the requirements of that lower grade is permissible in some cases. However, great caution must be applied if confusion is to be avoided as to the actual quality of the reassigned gage blocks. Change of the serial number of the demoted block, and the addition of an appropriate marking are advisable safeguards.

SPECIAL-PURPOSE END STANDARDS

The function of an end standard whose gaging surfaces represent the boundaries of a specific dis-

TABLE 4-7. LENGTH MEASUREMENT OF GAGE BLOCKS BY LIGHT INTERFERENCE

BASIC SYSTEM	CATEGORY OF INSTRUMENTS	OPERATING PRINCIPLES
Comparative Length Measurements	Optical Flats	The simplest, yet rather effective application of the light interference phenomenon. A master of known accuracy and the gage block to be checked are observed simultaneously under a common optical flat. Deviation of the fringes from the uniformly spaced and straight pattern are indicative of height differences and geometric errors. The character and the magnitudes of these errors can be evaluated with an approximation whose accuracy is dependent on the skill of the inspector (Fig. 4—19).
	Light Interference Gage Block Comparator	The instrument consists of an optical system producing light interference fringes on the reflected image of the specimen. As a reference piece, a master of known accuracy and having the same nominal size as the block to be checked is used. For measurement by optically assisted comparison, both the master and the gage block to be checked are wrung onto the optical flat stage of the instrument. The instrument presents the fringed images of both blocks in an overlaid image. Fringe displacements to be detected in the image patterns are the reflection of lack of geometric congruence (Fig. 4—20).
Absolute Length Measurements	Optical Fringe Displacement Evaluation	Absolute length measuring interferometers, such as the Automatic Laser Interferometer built by TESA Metrology (Fig. 4-21), are used for determining the lengths of solid bodies in relation to a known physical unit. The wavelength of light originates from a specific source (i.e., a laser). Because of the critical conditions under which such measurements must be made, absolute length determination by light interferometry is only exceptionally used for measuring gage blocks. Example for application is the measurement of the basic reference standards of the National Bureau of Standards.
	Electronic Fringe Count with Digital Readout	Absolute length measurements are conveniently accomplished by an instrument known as fringe count micrometer. The instrument automatically counts the light interference fringes over the distance the machine spindle must travel to contact the top surface of an object placed on the measuring table. The spacing of the individual fringes under the actual ambient conditions of the gaging is related to the known dimension of a certified master. The number of counted fringes as announced by the readout device of the instrument is the factor which will be multiplied by the fringe spacing to obtain the total length dimension. The accuracy of this measurement is considered to be better than ± 2 microinches.

tance or geometric form (for example, parallelism, an angle) is not necessarily limited to gage blocks of conventional design. Although gage blocks are the most versatile end standards, there are applications where other forms of dimensional standards are better suited for the particular use. A few of these special-purpose end standards are discussed in the following.

1. *Measuring rods* (round gage blocks) are particularly useful when datum planes must be transferred beyond the measuring range of micrometer screws or similar fine-increment measuring devices. One example of this application is the case of measuring tables of optical instruments (toolmaker microscope, optical projector) whose micrometer screws cover a measuring range of one inch only, whereas the useful table movement is a multiple of this dimension. Intermediate lengths are easily measured by inserting measuring rods, which are available in full inch sizes, between the micrometer spindle and the anvil, thus transferring the datum plane for the micrometer spindle by the required multiples of one inch (Fig. 4-22).

2. *Stepped blocks* consist either of a series of gage blocks arranged in a staggered manner to provide steps at fixed intervals, or are made of a single bar of steel. Stepped bars are used in some types of measuring machines, and also as reference elements for checking screw thread pitch accuracy. A typical application for stepped blocks is found in surface plate work, where the desired height levels for dimension transfer can quickly and precisely be established with instruments having a stepped bar for a master (Fig. 4-23). Each step of the gage block stack represents a one-inch distance from the base surface when the indicating instrument is at the zero position. Any intermediate dimension, in increments of 1/10,000 of an inch, can be produced by raising the entire stepped block with the aid of a built-in micrometer screw. Then the nominal value of the selected step, plus the amount by which the stepped block is

Jones & Lamson / Textron
The L. S. Starrett Co.

Fig. 4-22. (Left) Measuring rods inserted between the anvil and the micrometer screw spindle can extend the measuring capacity of an optical projector table by several inches.

Fig. 4-23. (Right) Optical height gage using a series of gage blocks for the step master. The optical part of the instrument is used for the subdivision of the one-inch long distances that are derived from the individual elements of the basic gage block stack.

raised, produces the required height level for transfer by means of an appropriate height gage.

3. *Angle gage blocks.* There are various conditions where the angular position of a surface in relation to a reference plane can best be checked by adding the compensating angle to the surface to be measured. Angle gage blocks, having known angular values, are convenient means for providing, in the form of a physical body, the desired compensating angle. Thereby a plane will result whose position can precisely be measured in relation to the selected reference plane. Angle gage blocks are supplied in sets that permit setting up any combination of angles from 0 degrees to 90 or 100 degrees, in increments of one minute, or one second of arc, depending on the number of individual blocks contained in the set. Angle gage blocks are discussed in greater detail in Chapter 10.

5.
Comparative Length Measurements with Mechanical and Electronic Indicators

GENERAL EVALUATION OF THE MEASURING SYSTEM AND INSTRUMENTS

The Principles of Comparative Length Measurement

Indicators are displacement sensing, amplifying and indicating or displaying instruments. Their purpose is (a) to detect variations in a specific distance, as determined by a reference plane established at a fixed position relative to the instrument and by a selected gaging point on the surface of the object, and (b) to display on a dial, graduated scale or digital display the amplified version of the sensed dimensional variations.

When the measurement of length is the objective of the process, the indicator—whether mechanical or electronic—will typically be used to compare the actual length of the specimen to the length of another physical body. As such, indicators are typically thought of and used in connection with the process of comparative gaging. However, recent advances in electronic indicators, as discussed later in this chapter, make absolute length measurements possible. Yet, when an indicator is used in comparative measurement, a master of known size is substituted for the nominal length of the dimension to be gaged. Thus, the indicator instrument only measures the amount and shows the deviation that may exist in relation to the nominal size.

When devising or evaluating indicator measurements of length, it must be remembered that the indicator in itself is not a complete measuring instrument. As shown diagrammatically in Fig. 5-1, for the purpose of length measurements the indicator must be complemented with devices serving the following functions:

a. Staging the object on a reference plane that is coincident with one end of the distance to be measured (the datum of the measurement); and

b. Holding the indicator instrument in a positive position from that reference plane, with the effective movement of the gage spindle in alignment with the distance to be measured.

The use of indicator instruments is not limited to plain length measurements. Many other conditions of object geometry can be observed and measured using indicators for sensing and reporting elements. Thus, the indicator can be used to measure the deviations from the nominal size of a distance taken from a datum that is not necessarily the opposite ob-

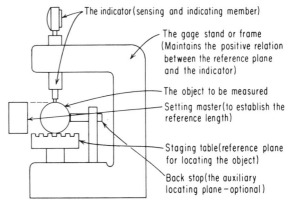

Fig. 5-1. Diagram of the principal elements in the comparative measurement of length.

ject boundary—as used in basic length measurements. Other convenient datums may be selected, such as the axis (between centers) or another surface portion of the object. In other applications, even an extraneous surface to which the object is related when in a specific position may be chosen for the datum, e.g., in surface plate based measurements.

Neither are indicator measurements limited to a stationary condition. When the object geometry is such that the distance to be measured has the same nominal size for any position in a specific direction from the datum, indicators are used for continuous measurements on moving objects, for example, run-out measurement of rotating machine elements.

Major Applications of Indicator Instruments

Table 5-1 surveys some characteristic applications of indicator instruments. Although illustrated with dial indicators and presented in conjunction with the discussion on mechanical indicators, these measurement categories also apply to electronic instruments with digital displays.

Advantages of Mechanical Indicators

The broad category of instruments known as mechanical indicators comprises some basic types that belong to the most widely used tools of dimensional measurements in metalworking production. Although there are various other systems of displacement amplification, as will be discussed later in this chapter and in Chapters 6 and 7, mechanical amplification provides many advantages in indicator applications.

A few of these advantages are given in Table 5-2, with the main purpose of pointing out certain aspects that deserve consideration when selecting the best-suited type of instrument for a particular set of measuring conditions. It must be realized, however, that not all the advantages listed in the table necessarily refer to all models of indicators. Nor are these advantages the exclusive properties of mechanical indicators, although the mechanical system of amplification lends itself best in utilizing the potentials of the described beneficial characteristics.

SYSTEMS OF MECHANICAL AMPLIFICATION

Mechanical indicators operate by different systems of displacement amplification, which may consist of gear trains, levers, cams, torsion strips, reeds, or a combination of these or other systems.

The direction of the original displacement that can be sensed by the indicator may be in line with the instrument spindle's axis, or normal to the contact lever swinging over a very small angle, where the lengths of the arc and of the chord are considered practically equal. Attachments, or actually integral members, to redirect the sense of the original displacement while transmitting it to the contact element of the indicator instrument may also be used. These can serve either to reverse the direction of the original motion by means of an equal arm lever, or to translate it at right angle, using a bell crank type device.

The pointers of mechanical indicators generally pivot around a suspension axis. The pivoting movement may extend over a complete circle, even comprising several turns, which requires a dial form of scale. This is the most commonly used form of display in mechanical indicators, and is characteristically present in all types of dial indicators. Dial form scales are also used for several types of precision (sweep movement) indicators with a relatively long measuring range, requiring extended scales. Such movement can most conveniently be displayed in a circular arrangement, even when the effective sweep of the pointer does not comprise a complete turn.

Other types of precision indicators have a short measuring range only, in some cases extending over not more than 20 or 25 graduations in either direction. Short-range indicators have sector type display faces, using a relatively long pointer for increasing the ultimate rate of amplification. Another type of display, with application limited to stationary (bench type) comparators, consists of a scale applied on the periphery of a sector with an extra large radius, again as a means of additional amplification.

Although the large and ever-increasing variety of mechanical indicator systems makes a complete listing impractical, Table 5-3 surveys the basic systems of displacement amplification most frequently used in mechanical or essentially mechanical indicators.

DIAL INDICATORS

Definition of the Category

Dial indicators are mechanical instruments for sensing and measuring distance variations. The mechanism of the indicator converts the axial displacement of the measuring spindle into rotational movement, which is then amplified by mechanical means, and finally displayed by a pointer rotating over the face of a graduated dial.

TABLE 5-1. CHARACTERISTIC APPLICATIONS OF INDICATOR INSTRUMENTS — 1

CATEGORY OF MEASUREMENTS	EXAMPLE FOR THE PROCESS		DISCUSSION
	DIAGRAMMATIC SKETCH	INSTRUMENT	
Comparative measurement of length		Comparator gage for external measurements	Basic application is for comparative measurements in a stationary instrument which is set to the nominal part size with fixed gage type masters.
Comparative measurement of bore size		Indicator plug gage	Sectors of a split cylinder are expanded to contact the bore walls, the indicator signalling the position of the sectors. The instrument is first set with a ring gage and will then indicate the bore size relative to the master.
Size comparison by transferred distance.		Test type indicator in surface plate work.	Using a common reference plane, regularly a surface plate, the size of the master, such as gage blocks, can be compared with the object size.
Distance from fixed reference plane		Depth indicator gage	Compares the distance of a surface from an offset face located in a plane normal to the dimension to be measured.

TABLE 5-1.   CHARACTERISTICS APPLICATIONS OF INDICATOR INSTRUMENTS – 2

CATEGORY OF MEASUREMENTS	EXAMPLE FOR THE PROCESS		DISCUSSION
	DIAGRAMMATIC SKETCH	INSTRUMENT	
Limit size gaging		Indicator snap gage	For the repetitive inspection of length dimensions, particularly diameters, on parts whose gaging can more conveniently be carried out with a hand held instrument.
Bore size comparison and gaging of diameter variations		Bore indicator gage	For the comparative measurement of bore diameters and for the concurrent exploration of diameter variations along and around the bore surface.
Consistent distance of a surface from a datum plane (Interrelated surfaces)		Squareness measuring indicator assembly	Parts located to permit one degree of freedom for rotation (see example) or sliding in relation to positive datum planes, are checked with indicator.
Runout of rotating bodies		Bench centers with indicator in adjustable holder	Part rotated around its actual or virtual axis, will reveal the amount of runout (radial variations) by the pointer movement of an indicator in contact with the part's surface.

TABLE 5-1.   CHARACTERISTIC APPLICATIONS OF INDICATOR INSTRUMENTS — 3

CATEGORY OF MEASUREMENT	EXAMPLE FOR THE PROCESS		DISCUSSION
	DIAGRAMMATIC SKETCH	INSTRUMENT	
Multiple dimension gaging		Multiple dimension indicator gage assembled of standard components	Several dimensions, either referenced from fixed stops, or in their interrelations, are measured simultaneously in special gages using indicators for measuring members.
Distances of traversing motions		Cross slide on precision machine tools fitted with indicator	Long range dial indicators in contact with the face of a machine tool or instrument slide, is a simple, yet effective method of measure traversing distances from fixed reference positions.
Mutual positions of interrelated features		Concentricity indicator gage with locating and guide fixture	The position and the geometry of a bore in relation to other features of the same object are effectively measured with special indicator gages.
In-process gaging		"Arnold" type grinding gage	The measurement of critical dimensions of an object while still in a process setup is feasible with the aid of special type of indicator gages.

TABLE 5-2.   CHARACTERISTIC ADVANTAGES OF MECHANICAL INDICATORS

ASPECTS	DISCUSSION
Long measuring range	Mechanical indicators, particularly those operating by the rack and pinion system, have measuring ranges which extend over several turns — in certain models up to 20 or even more — of the hand over the dial. This design properly permits measurements in small increments over quite substantial distances, not even approached by any other system of indicators.
Small overall size	The small overall size of the whole indicator instrument, inclusive of sensing, amplifying and indicating elements, is of advantage for measurements where space is confined, or when using several indicators mounted at close distances to each other.
Positive contact and controlled measuring force	The possibility of selecting instruments whose contact force is best suited for a particular application can be of critical importance. In some models of mechanical indicators the measuring force can be varied and adjusted over a rather substantial range.
Ruggedness	Particularly when used under shop conditions or on operating machines where substantial vibrations are present, the relative shock resistance of many types of mechanical indicators can be a deciding factor for the instrument selection. Most types of mechanical indicators are less sensitive to inadvertently caused over-travel, than comparable instruments operating by other systems.
Independence of power supply	The operation is not limited to any particular location; therefore measuring instruments comprising mechanical indicators can be set up anywhere in the plant or used in the field. Power lines are not interfering with the mounting of the instrument in the optimum measuring position and fluctuations in the power supply never affect the inherent indicating accuracy of mechanical indicators.
Stability and repeatability of indications	Drift which could occur in instruments whose operation is based on transducing the linear displacements of the probe into electrical or fluid pressure variations, is not present in mechanical indicators. Instrument errors, when due to mechanical causes, e.g. binding, wear, play, etc. usually occur with consistency, permitting prompt detection and correction or some degree of compensation.
Economic advantages	The initial cost of mechanical indicators of comparable sensitivity is usually less than that of other systems, besides the availability of a wide range of less sensitive models at fractions of the cost of precision type instruments. The plain models of mechanical indicators primarily the dial types, can be maintained and repaired in the plant of the user at reasonable cost, standard replacement components being readily available. Several precision types of mechanical indicators are practically free of wear and loss of accuracy during many years of sustained usage.

TABLE 5-3. SYSTEMS OF DISPLACEMENT AMPLIFICATION USED IN MECHANICAL INDICATORS — 1

| DESIGNATION | OPERATING PRINCIPLES | | FORM OF DISPLAY | COMMONLY AVAILABLE | |
	SCHEMATIC SKETCH	DESCRIPTION		SMALLEST GRADUATION	MEASURING RANGE
Rack and pinion		Measuring spindle integral with a rack, engaging a pinion which is part of an amplifying gear train	Dial	From 0.0001 to 0.001 inch	0.025 to 0.250 inch Specials to several inches
Cam and gear train		Measuring spindle acts on a cam which transmits the motion to the amplifying gear train.	Dial	From 0.00005 to 0.0001 inch	From 0.010 to 0.025 inch
Lever with toothed sector		Lever with a toothed sector at its end engages a pinion in the hub of a crown gear sector meshing the final pinion.	Dial	From 0.0001 to 0.001 inch	From 0.010 to 0.100 inch
Compound levers		Levers forming a couple with compound action are connected through segments and pinions	Sector	From 0.00002 to 0.0005 inch	From 0.002 to 0.020 inch

TABLE 5-3.   SYSTEMS OF DISPLACEMENT AMPLIFICATION USED IN MECHANICAL INDICATORS — 2

| DESIGNATION | OPERATING PRINCIPLES | | FORM OF DISPLAY | COMMONLY AVAILABLE | |
	SCHEMATIC SKETCH	DESCRIPTION		SMALLEST GRADUATION	MEASURING RANGE
Twisted taut strip		Straining caused by the tilting of the knee connected to the gage spindle causes the twisted taut band to rotate proportionally, this motion being displayed by the attached pointer.	Sector	From 0.00001 to 0.0005 inch	From 0.0012 to 0.040 inch
Lever combined with band wound around drum		Movement of the gage spindle tilts hinged block; this causes swing of the fork which induces rotation of the drum.	Large sector	From 0.00001 to 0.0002 inch	From 0.0012 to 0.020 inch
Reeds combined with optical display		Parallelogram reed transfers spindle movement to a deflecting reed with extension carrying a target which partially obstructs the optical path of a light beam.	Shadow on the face of a sector	From 0.00001 to 0.0001 inch	From 0.0005 to 0.008 inch
Tilting mirror projecting light spots		Spindle movement transferred by lever action tilts intermediate mirror thereby changing position of the projected light spot on the graduated screen.	Stepped light spot on the face of a sector	From 0.000025 to 0.00005 inch	From 0.0028 to 0.0055 inch

Federal Products Corp.

Fig. 5-2. Dial indicator with back cover removed, showing the essential elements of its mechanism.

In the commonly used dial indicators, the measuring spindle carries, usually as an integral part, a rack whose teeth are meshing with a pinion, the pinion being part of a gear train. The indicator's mechanism is thus accomplishing the double function of converting the linear displacement into rotary motion, and then amplifying the rotary motion by means of several sets of driving gears meshing with substantially smaller pinions. Figure 5-2 shows the internal mechanism of a generally used type of dial indicator.

The category of dial indicators may need a definition to better distinguish it from other types of mechanical indicator instruments. The display of the amplified spindle motion by a pointer, also called a hand, on a graduated dial is not an exclusive characteristic; there are various other types of mechanical indicators with graduated dials, the rack-and-pinion mechanism, although the most common design, being only one system of movement conversion found in dial indicators.

Essentially, a displacement indicating mechanical instrument may be considered a dial indicator when its design is basically in compliance with the American Gage Design (AGD) specifications, which will be discussed in a following section. Distinguishing characteristics derived from these specifications are the dial form of display, the essential boundary dimensions and the measuring range of the instrument, which must be not less than $2\frac{1}{2}$ turns of the hand on the dial.

Advantages and Limitations of Dial Indicators

Dial indicators represent the most widely used basic type of mechanical displacement indicating in-

struments. There are numerous reasons for that preference and a few characteristic ones are mentioned:

a. Amplification mechanisms operating by gear trains permit a sturdy construction that is less sensitive to shocks and vibrations than most other systems of design, making the dial indicator ideal for shop uses.

b. A long measuring range may be provided, when needed, limited only by the practicable length of the effective rack motion. Dial indicators also usually permit a reasonable amount of overtravel (spindle displacement beyond the effective indicating range of the instrument) without injuring the mechanism.

c. The dial form of display, with several turns of the hand, can encompass an extended indicating range within compact outside dimensions, yet retain a distinct resolution by an adequate spacing of the graduation lines.

d. The favorable economic aspects, in initial as well as in maintenance cost is also an advantage. Better grades of dial indicators are designed to facilitate repairs and to permit the use of interchangeable replacement parts.

Several limitations of dial indicators will now be pointed out for the sake of completeness, which is needed for judicious instrument selection. It must be remembered, however, that various improvements in design and execution, some of which are also mentioned below, can substantially reduce the adverse effect of these potential limitations.

a. The gear teeth mesh with some clearance, causing backlash and lagging response, particularly when wear progresses. Accuracy is substantially improved by effective backlash eliminating devices (commonly, hairsprings) and the consequences of tooth wear are reduced by precisely machined and heat treated gears.

b. The effect of imperfect gear form, or play, results in cumulative error, which is usually proportional to the length of the gaging travel. For this reason, dial indicators with extended measuring range are not recommended for precision applications.

c. A clearance of the measuring spindle in its guides is needed to prevent binding and to provide an unimpeded operation movement; this clearance can result in positional variations of the rack in relation to the meshing pinion. Precision dial indicators, in which a cam is substituted for the rack-and-pinion system, are not affected by said looseness in the measuring spindle guides.

d. Play in the pivot bearings due to initial inaccuracies or wear affects the precise meshing of the gear teeth, causing measuring errors. Jeweled bearings substantially improve the consistent accuracy of gear locations.

The American Gage Design [1]—Specifications for Dial Indicators

The American Gage Design (AGD) Standard states minimum requirements for essential characteristics of dial indicators serving linear measurements. Although originating in 1945, the specifications of the standard are still valid as now documented in the American National Standard ASME/ANSI B89.1.10M-1987 and are regarded as guidelines for the design and selection of dial type linear displacement indicators.

In view of the general acceptance of the AGD standards, it is appropriate to review their essential specifications.

Figure 5-3 shows the standard nomenclature for dial indicator components. The dial, which is attached to the bezel, should be rotatable, and a clamp should be provided for locking the bezel in the desired position. The standard covers four size groups whose main dimensions are shown in Table 5-4. It is required that the effective spindle travel produce a minimum of $2\frac{1}{2}$ turns of the indicator hand; for mechanical reasons, the effective travel usually starts

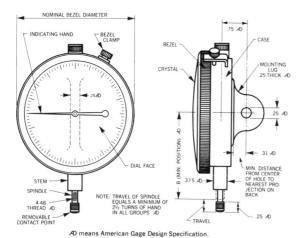

Fig. 5-3. Standard nomenclature for dial indicator components as specified in the Commercial Standard (CS9E) 119–45.

---

[1] Commercial Standard (CS9E) 119–45.

from $\frac{1}{12}$ to $\frac{1}{4}$ turn beyond the idle position of the pointer. This is a minimum requirement and does not exclude the use of longer measuring ranges when warranted by application conditions.

The dial markings (graduation lines) represent decimal values and the numbering of the dial must always state thousandths of an inch, regardless of the class of the markings. The designation class, in this context, refers to the smallest value of the dial markings (graduations). Four classes are specified in the standard, representing different rates of amplification and resulting in dial marking values of 0.001, 0.0005, 0.0001 and 0.00005 inch, respectively. It is evident that the smaller the value of the individual markings, the shorter the range that the indicator can cover by the specified $2\frac{1}{2}$ turns of the hand. Most dials have 100 markings around the entire periphery of the dial; consequently, the useful measuring range is generally 250 times the minimum graduation value. However, in view of the interrelation between dial diameter and the number of applicable graduations, certain types of dial indicators have more or less than 100 markings for a full circle.

The AGD standard also contains specifications regarding the mounting dimensions of dial indicators (see Fig. 5-3). By adhering to these specifications, dial indicators of different makes become interchangeable in their fixture mounts. This condition simplifies the design of measuring devices using dial indicators for displacement indicating members.

As a further development toward interchangeability in the field of industrial measuring instruments, the AGD mounting dimensions are also frequently adopted for many other types of indicators, both the mechanical and those operating by another system of amplification.

The Accuracy of Dial Indicators

The American Gage Design (AGD) Standard distinguishes two categories of potential errors to be considered for the assessment of dial indicator accuracy:

a. *Repetition of readings* using the following wording: "Reading at any point shall be reproducible through successive measurements of the spindle to plus or minus one fifth graduation."

b. *Accuracy.* This refers to calibration accuracy, also known as linearity of indications. The standard's wording is the following: "The dial indication shall be accurate to within one graduation plus or minus, at any point from the approximate ten o'clock position to the final two o'clock position ($2\frac{1}{3}$ turns)."

**TABLE 5-4. NOMINAL DIAL (BEZEL) DIAMETERS AND CLASSES OF MARKING
FOR AGD DIAL INDICATORS**

(According to Commercial Standard (CS9E) 119–45)

AGD GROUP	NOMINAL BEZEL DIAMETER (inches)		DISTANCE (inches) BETWEEN THE CENTER OF DIAL AND SPINDLE END (MIN. POSITION)	VALUE OF SMALLEST GRADUATIONS	
	Above	To & Inc.		(in.) ENGLISH	(mm.) METRIC
1	1-3/8	2	1-5/8	.0001 .0005 .001	.005 .01
2	2	2-3/8	2	.00005 .0001 .0005 .001	.001 .002 .005 .01
3	2-3/8	3	2-1/8	.0001 .0005 .001	.001 .002 .005 .01
4	3	3-3/4	2-9/16	.00005 .0001 .0005 .001	.001 .002 .005 .01

This specification requires that the variations of the actual object size in relation to the setting master must be displayed by the pointer position on the dial with a maximum error of plus or minus one graduation, at any position within the 2½-turn effective measuring range. The calibration accuracy determines the confidence that can be placed in the measurement of size variations by the use of the instrument.

The standard does not specify how this spread of permissible indication errors of the gage may occur; whether it is a gradual gain or loss, with magnitudes in straight proportion to the gage spindle travel (as it is frequently assumed), or whether it might have a periodic, or even random, appearance within the boundaries of the calibration tolerance zone. Figure 5-4 illustrates the interpretation of these specifications for arbitrarily selected yet closely typ-

ical cases, where the deviations of the pointer indications in relation to the actual size occur in a single cycle along the entire measuring range (Fig. 5-4(A)), or show a periodic pattern repeating for each turn of the pointer (Fig. 5-4(B)). The actual accuracy of an indicator can best be established by means of a complete calibration process, as explained in conjunction with the calibration of micrometers (see Chapter 2).

Assuming the conditions shown in Fig. 5-4 to be typical, the following general directions can be established for the purpose of reducing the effect of instrument calibration errors on the result of indicator measurements:

a. It is advisable to set the indicator to a dimension that is approximately in the center of the spread over which the actual object size is expected to vary.

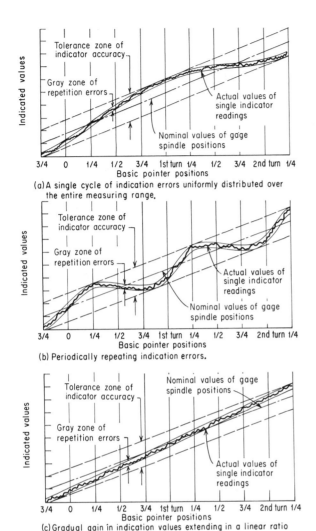

(a) A single cycle of indication errors uniformly distributed over the entire measuring range.

(b) Periodically repeating indication errors.

(c) Gradual gain in indication values extending in a linear ratio over the full range of pointer travel.

Fig. 5-4. Interpretation of the concepts "repetition of reading" and "calibration accuracy," based on imaginary, yet typical, examples of dial indicators. (*Note:* The indication errors are shown in an exaggerated scale to illustrate the concept.)

b. Use a dial indicator of that amplification class (represented by the value of dial graduations) in which the expected size variations to be measured will spread over a limited section of the total indicator range only (about ⅓rd turn or less). From the typical error curves of Fig. 5-4, it appears obvious that by this approach the effect of calibration errors can be kept between narrower limits than the total calibration tolerance range of the instrument.

c. When an actual calibration chart of the indicator, similar to that shown on Fig. 2-19 for microm-

eters, is available, then the area of best linearity within the entire indicator range can be selected for the actual measurement, thereby reducing even further the gaging inaccuracy that result from instrument calibration errors.

However, the above directions need not be followed as a general rule. In most cases, with the possible exception of particularly critical measurements, the required gaging accuracy will be less than the systematic error of the judiciously selected indicator instrument, thereby obviating any particular restrictions regarding the usable portion of the indicator's measuring range. Nevertheless, concentrating the measurements on a relatively narrow section of the total indicator range, distant from the range boundaries, will generally prove beneficial to accomplishing more consistently accurate instrument indications.

Dial Indicator Components, Accessories and Attachments

The diversity of dial indicator applications and the variety of requirements resulting from different uses and purposes has led to a wide range of alternative designs. While several of the major indicator dimensions are specified in the recommendations of the AGD standard, flexibility of design has been retained for many other component details.

A few of the more important component categories will be discussed in the following review of the designs that are made available by major manufacturers of dial indicators.

Dials. Although the dial design with 100 graduations around the complete periphery is the most common, other types of dial scales with 200, 80, 60, 50, 40 or 20 graduations are also used for different types of dial indicators. The greater number of graduations reflects a larger measuring range for the same graduation value (based on the 2½-turn minimum movement), while the wider spacing of graduations provides a better resolution. The size of the dial is an important factor to be considered when selecting the type of graduation best suited for the intended use and application conditions. In this latter respect, due consideration should be given to the circumstances of dial observation, comprising the distance of the dial from the regular working position of the operator, and the purpose of the measurement, for example, whether for single layout work or for continuous inspection operation.

The numbering is normally applied to every tenth graduation, with the exception of dials whose individual graduation lines do not represent full decimal fractions of the inch; as an example in the case of the 0.0005-inch graduation values, every twentieth marking line is numbered. According to the standard, the numbering will always express in thousandths of an inch the travel of the hand on the face of the dial. The numbers are in ascending order, starting from zero, and there are two basic types of arrangement for the sequence of the numbering:

1. *Balanced dials* (Fig. 5-5) have the numbering symmetrically arranged in both directions from the starting zero, the largest number, common to both sides in most cases, being opposite the initial position. One side is usually marked plus (+) and the other minus (−), signifying the sense of the measured deviation from the base position. This type of dial numbering is preferred for comparative measurements, where the indicator is usually set to zero for the size corresponding to the nominal dimension, the purpose of the indicator gaging being to determine the direction and the amount of any deviation from that base size.

2. *Continuous dials* (Fig. 5-6) carry the numbering around the dial in sequential order, usually in a clockwise direction. This numbering arrangement is favorable for the indicator measurement of linear displacements, such as result from the travel of machine or instrument slides. Continuous dial numbering is particularly preferred for indicators with extra long measuring range, where the main dial is equipped, as a supplementary device, with a *revolution counter*.

Mountings. There are two basic ways to mount a dial indicator: (a) by the stem, and (b) by the back.

Federal Products Corp.

Fig. 5-5. (Left) Dial indicator with balanced dial, graduated in 5/10,000ths of an inch.

Fig. 5-6. (Right) Long-range type dial indicators with continuous dials. Left, with graduations in 0.001-inch and revolution counter; and right, with graduations in 0.010 inch.

The stems of dial indicators are normally round and of ⅜-inch diameter, although the standard does not exclude the use of smaller stem diameters. Dial indicators, manufactured in Europe in metric dimensions, usually have 8 millimeter stem diameters. Stem mounting is more common than back mounting, and provides the advantages of permitting simple adjustment in the axial direction, as well as the rotation of the indicator around its axis for the convenient observation of the dial.

Mounting by the back is preferred when the means of adjustment are assured by the design of the fixture mount. In some applications where space is limited, back mounting may be the only choice, with the added advantage of more positive clamping of the indicator than provided by stem mounting. While the vertical central bracket, or mounting lug, is the commonly used and normally furnished design, other types of mounting backs are also currently available to suit varying application conditions. Figure 5-7 shows the usually available alternate designs for dial indicator back mounts.

Contact points. The spindle end of standard dial indicators has a No. 4-48 thread bore and the interchangeable contact points are made with a mating screw shank. Figure 5-8 illustrates a selection of optional contact points that differ with regard to two major characteristics: the length of the contact member and the form of the contact face.

Although contact members of different lengths are currently available, the ¼-inch long contact points are normally used. Longer contact members are required when space limitations prevent the mounting of the indicator close to the surface to be measured. The lesser stiffness of the longer contact members is the main reason for avoiding their use unless warranted by the application conditions.

The general form of the contact face is that of a spherical segment with ³⁄₁₆-inch radius. The spherical form of the face provides essentially single point engagement with most object surface forms and makes the measuring contact less sensitive to minor misalignment in the mutual position of the work and of the indicator.

Flat contact points are preferred for diameter measurements on convexly curved surfaces, such as balls, cylinders and cones, to make the precise centering of the object less critical for obtaining the correct indicator reading. However, flat contact surfaces are sensitive to alignment errors in a plane perpendicular to the direction of measurement. Therefore, it is advisable to abstain from the use of flat contact points, unless their face can be located precisely in relation to the indicator spindle (normal to the spin-

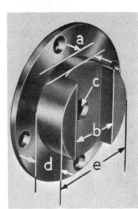
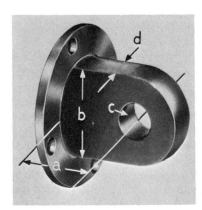

Federal Products Corp.

Fig. 5-7. Various types of standardized indicator backs with corresponding dimensions: (a), (b), etc. (Top Left) Post bracket. (a) $\frac{5}{16}$ in.; (b) $\frac{7}{32}$ in. (Top Right) Screw bracket. (a) $\frac{1}{2}$ in.; (b) $\frac{11}{16}$ in.; (c) $\frac{1}{4}$ in.-28. (Bottom Left) Flat back. (Bottom Right) Adjustable bracket. (a) $\frac{1}{8}$ in.; (b) $\frac{3}{8}$ in.; (c) 8-32; (d) $\frac{1}{4}$ in.; (e) $\frac{11}{16}$ in. (Center Bottom) Offset bracket or vertical lug, centered. (a) $\frac{1}{2}$ in.; (b) $\frac{11}{16}$ in.; (c) $\frac{1}{4}$ in.

dle axis) and the staging table (in parallel planes). When proper locating cannot be assured, an intermediate form of point having a larger than normal radius, may be the best choice.

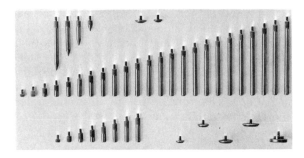

Federal Products Corp.

Fig. 5-8. Samples of interchangeable contact points for dial indicators showing different sizes, with spherical (standard), tapered, flat end, wide face, and button forms.

Taper points permit measurements in the bottom of grooves, at the root of gear teeth and so forth, where the regular contact point could not penetrate.

Special contact points may be selected for particular applications, with regard to the geometry (form and size) and the material of the point. With respect to material, the tips of the contact points may be hard chromium plated, made of tungsten carbide, sapphire, or even of diamond to assure a high degree of wear resistance for the continuous gaging of abrasive types of materials or of surfaces in motion.

Dial indicator accessories is the collective designation for various optional elements, which then become integral parts of the instrument. The following are examples of frequently selected dial indicator accessories:

a. Adjustable tolerance limit hands to aid the dimensional inspection of lots consisting of nominally identical parts;

b. Lifting lever to raise the rack spindle while inserting the part into the measuring position, thereby avoiding undue wear on the indicator points, or the marring of the object surface; and

c. Steam adaptors, guards and bushings to protect the instrument spindle, when the indicator has to withstand adverse conditions of use or of clamping.

Attachments for dial indicators. Currently available attachments serve to adapt standard dial indicators to application conditions that could not be handled by using the basic instrument only.

a. Hole attachments for indicator measurements on internal surfaces that are not accessible to regular dial indicators, or where the indicator would be difficult to observe when mounted to reach inside the bore (Fig. 5-9). This attachment consists of an equal arm lever, pivoting around a fulcrum pin mounted

Federal Products Corp.

Fig. 5-9. (Left) Hole attachment mounted on a dial indicator, enabling the inspection of runout on internal surfaces.

Fig. 5-10. (Right) Right-angle offset attachment for dial indicators. Design with flat-spring hinge.

in a bracket, which can be attached to the stem of the dial indicator.

b. Right-angle attachment (Fig. 5-10) for use when the indicator dial must be observed at right angles to the direction of measurement, or space limitations prevent the normal mounting of an indicator. The attachment works essentially without distortion, and will offset at right angles the direction of the sensed displacement before transferring it to the indicator spindle. This is accomplished by a bell crank having arms of equal effective length. In the design shown in Fig. 5-10, a flat spring hinge serves as fulcrum and this is attached to a bracket that can be mounted on a standard indicator stem. Other designs use a screwed pin for fulcrum, which might be less accurate than the solidly anchored flat spring hinge but permits, by means of interchangeable levers, the right-angle attachment to be used as a hole attachment as well.

Federal Products Corp.

Fig. 5-11. Mounting bracket to accept dial indicators, with adjustable-type back that permits the setting of the indicator's position to accommodate parts in different sizes.

c. Adjustable mounting brackets (Fig. 5-11) are dimensioned to accept dial indicators with rail-type backs (see adjustable type in Fig. 5-7). By sliding the indicator in the guides of the mounting bracket and locking it in the desired position, a substantial range of adjustment can be accomplished by relatively simple and inexpensive means.

ELECTRONIC INDICATORS

Definition of the Category

Electronic indicators are the digital equivalent of mechanical dial indicators. The size and physical appearance of the digital electronic indicator are close matches to those of the traditional dial indicator (see Figs. 5-12 and 5-13). This resemblance is not coincidental; the digital electronic indicator is now used in place of the traditional dial indicator in many inspection situations.

As shown in Fig. 5-12, the digital electronic indicator can, through the use of both digital and analog displays, provide a numeric readout and a graduated dial. The digital display gives an exact and easy-to-read numeric measurement, whereas the analog display simulates the sweeping of a hand on a

Federal Products Corp.

Fig. 5-12. Maxum digital electronic indicator.

Ono Sokki Technology Inc.

Fig. 5-13. Electronic digital measuring indicator.

Ono Sokki Technology Inc.

Fig. 5-14. Long range digital measuring indicator.

traditional mechanical dial indicator. Where an analog display is deemed unnecessary, an electronic digital measuring indicator with digital display can be specified (see Fig. 5-13).

The axial displacement of the electronic indicator's measuring spindle converts linear movement into a useable electronic signal through the use of a *linear variable differential transformer* (LVDT). The electronic signal is then amplified and passed through an analog-to-digital conversion for use in a *liquid crystal display* (LCD). Both the LVDT and the LCD are more fully discussed in Chapter 7.

Advantages and Limitations of Electronic Indicators

The electronic indicator provides the user with several advantages when compared with the traditional mechanical dial indicator:

a. The digital display is easy to read as it instantly displays an exact reading, thereby eliminating the need for dial position interpretation.

b. With the touch of a button, the electronic indicator can be switched from an English unit (inch) to a metric display.

c. Indicators are typically used for comparative length measurement, however, the electronic indicator can also give absolute length measurements (see Fig. 5-14).

d. When equipped with a remote transducer (Fig. 5-15), the electronic indicator can be positioned for ease of reading and several transducers can be clustered in a small space that would prohibit the use of mechanical indicators.

e. An electronic indicator can be tied via an RS232 interface to a personal computer for purposes of data collection. These data are then used to automatically compute statistical process control information such as X-bar and R charts.

There are two primary limitations to the use of electronic indicators:

1. An electronic indicator, unlike a mechanical dial indicator, requires electrical power via either a battery or a 110-volt AC power source. Therefore, without a new or charged battery, or an AC adapter and power source, the electronic indicator will not function. The typical life of a battery used in an electronic indicator is 200 hours of continuous use.

2. The electronic indicator is somewhat temperature sensitive. Most electronic indicators will operate properly at temperatures between 50°F and

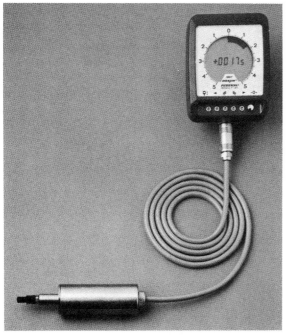

Federal Products Corp.

Fig. 5-15. Maxum electronic indicating unit with remote transducer.

110°F, however, this temperature range is sometimes violated in shop settings where indicator gages are used.

The Accuracy of Electronic Indicators

As previously discussed, there are two types of electronic indicators: those that display both digital and analog information (see Fig. 5-12), and those that display only digital information. Each of these major electronic indicator types has a level of accuracy that is different and is expressed in somewhat different terms.

An electronic indicator that displays both digital and analog information is necessarily limited in the range of movement possible. This range of possible movement is called the *digital range* and would typically not exceed 0.080 inch total. The accuracy of this style of measuring instrument is expressed as a percentage of the digital range (for example, 1.0%). Therefore, an indicator with a digital range of 0.080 inch would be accurate within 0.0008 inch. As the digital range is reduced to a 0.040 inch total, the accuracy is 0.5%.

The electronic indicator style that displays only digital information (Fig. 5-13) has its level of accuracy expressed in terms of resolution. Here *resolu-tion* is defined as the smallest amount of contact member deflection that will produce proportional signals (see Table 7-1, Chapter 7). This style of indicator has a resolution of 0.0001 inch and an accuracy of plus or minus one resolution from a known standard at 68°F.

Electronic Indicator Components, Accessories and Attachments

Because electronic indicators were designed to take the place of American Gage Design (AGD) type mechanical dial indicators, items such as electronic indicator backs and threaded contact points conform to ANSI/AGD specifications. However, there are several items that have been designed solely for use with this newer electronic indicator technology, including the following.

a. *Dials.* For digital electronic indicators (Fig. 5-12) that also have analog displays, a number of dial faces are available in a wide variety of graduations and in both inch and metric.

b. *Mounting Racks.* A special mounting rack (Fig. 5-16) is available to cluster indicators, equipped with remote transducers, for ease of viewing by the operator/inspector.

c. *RS232 Interface.* Electronic indicators can be linked to a personal computer of light-emitting diode (LED) gage display (Fig. 5-17) via multiwire cables and the industry standard RS232 interface for statistical process control. Also, the linear gage sensor

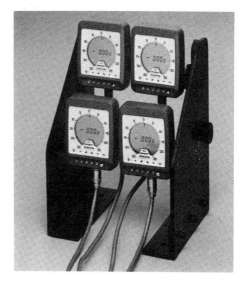

Federal Products Corp.

Fig. 5-16. Mounting racks for electronic indicating units.

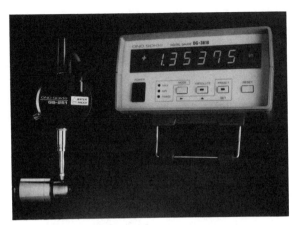

Ono Sokki Technology Inc.

Fig. 5-17. Electronic digital measuring instrument with RS232 interface for direct data collection via digital display or microcomputer.

(Fig. 5-18) can be used to receive and send remote readings of measurements.

d. *Contact Tips.* Contact tips for electronic indicators come in a variety of familiar styles in addition to the new roller tip (Fig. 5-19). The roller tip is designed to permit continuous profile readings as well as allow thin stock (for example, sheet metal) to be

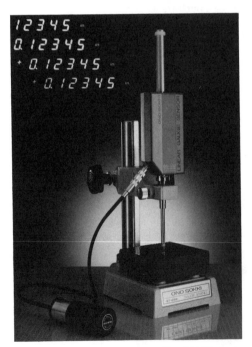

Ono Sokki Technology Inc.

Fig. 5-18. Linear gage sensor with stand.

Ono Sokki Technology Inc.

Fig. 5-19. Roller contact tip on a linear gage sensor.

eased in below the contact tip without the benefit of a lift lever.

TEST TYPE INDICATORS

The Basic Design of Test Type Indicators

The term test type indicators designates a special type of indicator characterized by the direction of the sensed displacement. While for most types of indicators this is parallel to the axis of the instrument spindle, test indicators are designed to sense and to measure displacements that occur in a direction essentially perpendicular to the shaft of the contact point or, using another designation, to the probe arm.

The displacement, which is acting on the test indicator's contact point in a direction normal to the axis of the probe, causes the swing motion of a lever, the shorter arm of which is the shaft of the contact point (see Fig. 5-20). The longer arm of that instrument lever ends in a toothed segment that engages a pinion. The pinion is in the hub of a sector with geared periphery and functions like a crown gear. This crown gear meshes with a final pinion that is attached to the spindle of the indicator hand.

The displacements sensed by the probe of the instrument are thus amplified by a transmission train comprising the main lever of unequal arm length, the sector and two pinions. A flat spring acting on the

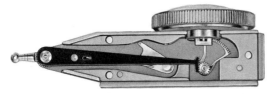

Federal Products Corp.

Fig. 5-20. Cross-sectional view of a test type indicator, showing the long lever with toothed-segment end and the crown-gear type sector as characteristic elements of the amplifying mechanism.

lever provides the gaging force, whose direction can be reversed by a lever switch.

With regard to the general form of the dial and the rotatable bezel, test indicators resemble the common dial indicators. However, due to the special design of the amplification system, the effective gaging range of test type indicators seldom exceeds a single turn of the hand, and in many models it is even less. This condition, as well as the basically less sturdy amplifying mechanism, limits the use of test indicators to applications where the special adaptability of the test type is needed or it presents other advantages in comparison to the standard dial type of indicators.

Test indicators are available with two different dial arrangements, which are designated as the parallel and the perpendicular style (Figs. 5-21 and 5-22). As seen from the illustrations, these designations refer to the position of the dial in relation to the main lever of the gage. The choice of the style should be guided by the most convenient observation of the dial during the measuring process.

The probe member of the test indicators, which functions as the shorter arm of the instrument lever, is not an integral part of but is positively attached to the hub section of the lever. The connection between

Federal Products Corp.

Fig. 5-21. (Left) Test type indicator with parallel dial arrangement.

Fig. 5-22. (Right) Test type indicator with perpendicular dial position.

the probe is not rigid; the probe can be swiveled and then locked in the selected position solidly enough to securely withstand the applied gaging force. Most types of test indicators permit a swivel motion of the probe over an arc of about 180 degrees, this being one of the features assuring the flexible adaptability of the test type indicators. Varying the angular position of the probe, without affecting the result of the measurement, is made possible by the geometry of motion conversion on which the design of test indicators is based.

Because in most types of test indicators the operational swing motion of the lever for the entire gaging range seldom exceeds one or two degrees in either direction, the linear displacement to be measured is considered to produce an essentially proportional angular motion. The angular value of that motion is unaffected by the swivel position of the probe arm, when the coincidence of the swivel axis with the fulcrum of the instrument lever is assured.

Cosine Error in Test Indicator Measurements

For the basic types of indicators, whether of the dial type or having a sector form, the mutual alignment of the instrument and of the work is not subjected to intentional variations. The surface on which the size of the displacements have to be measured will be located in a plane essentially perpendicular to the axis of the indicator spindle.

The test type indicator, on the other hand, permits a greater flexibility in the location of the instrument in relation to the work surface. Actually, in this flexibility of setup lies one of the major advantages of the test indicators for certain types of measuring applications. While, in principle, the size variations to be measured should occur in a direction perpendicular to the probe arm of the indicator, small deviations from this theoretically correct position have only negligible effect on the accuracy of the measurement.

The terms *small variation* and *negligible effect* are, of course, vague and were chosen to point out the presence of a judgment area that must be evaluated in the perspective of the actual accuracy requirements of a specific measuring task.

The measuring errors that result as a consequence of deviations from the theoretically correct mutual position of object and instrument are explained in Fig. 5-23. As illustrated, these errors are the function of the θ angle of mutual offset, and will increase in an inverse cosine ratio. Hence, this type of measuring discrepancy is commonly known as the *cosine error*. Figure 5-23 also shows, in percentages of the

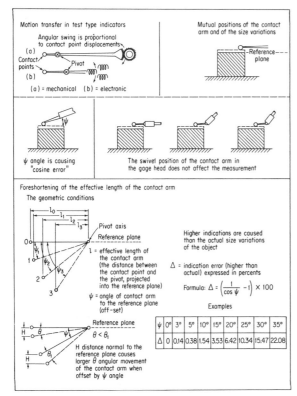

Federal Products Corp.

Fig. 5-23. Geometric conditions in test indicator measurements and their potential effect on the resulting indications—the cosine error.

indicated dimension, the value of the interfering cosine error, calculated for different offset angles.

Test Indicator Applications

Because of the wide variety of possible uses, a survey of test indicator applications can be of informative value only, without even approaching completeness. Nevertheless, it is believed that examples of characteristic applications, as shown in Table 5-5, will be helpful in recognizing those gaging positions under which comparative measurements can best be carried out with the test type indicators.

For appreciating the advantages offered by test indicators it is worthwhile to visualize the use of regular dial type indicators under application conditions similar to those illustrated in Table 5-5. It becomes obvious that without additional attachments to regular dial indicators very few of these measurements could be carried out at all, or only with the dials oriented in a direction that is awkward

for observation. Indicator attachments, however, by introducing an additional linkage into the chain of motion translation, may possibly detract from the original accuracy of the instrument.

The test type of indicator design is not limited to mechanical instruments. It is very frequently used in electronic gages and, exceptionally, in air gages also. The discussed basic principles of application, inclusive of the potential introduction of the "cosine error," are valid for test indicators in general, independently of their system of amplification.

INDICATOR GAGES

The General Function of Indicator Gages

It has been discussed previously that for accomplishing length measurements, the indicator instrument must be complemented with auxiliary devices. Only by establishing the reference plane of the measurement and locating both the object and the indicator in a definite relation to that common reference plane will the indicator instrument be capable of carrying out measurements of length variations.

Frequently these complementary functions of the gaging process are secured by means of appropriate gaging setups, which also incorporate the indicator instrument. In many other cases it is found to be more convenient, dependable and/or economical to unite the indicator instrument with members that serve the needed complementary functions, resulting in a single tool, generally known as an indicator gage.

In the following examples, indicator gages that are predominantly equipped with dial indicators are shown. Although the dial type indicator with its wide gaging range is the best suited instrument category for the majority of indicator gage applications, the design principles of indicator gages also permit the use of other types of indicators. When the attainable sensitivity requires, and the needed measuring range permits, indicator gages are equipped with sector type (sweep movement) precision indicator instruments, which will be discussed in a following section of this chapter.

A common property of all indicator gages must be remembered. Indicator gages are always comparator type measuring instruments and require the use of a setting gage for establishing the basic measuring position. The function of the gage is to indicate coincidence, or the amount and direction of deviation from the basic size, when the occurring variations are within the measuring range of the instrument.

TABLE 5-5.   CHARACTERISTIC APPLICATIONS FOR TEST TYPE INDICATORS — 1

DESIGNATION	DIAGRAMMATIC SKETCH	DESCRIPTION
Contacting hard-to-reach surfaces		The basic design of test type indicators permits measuring contacts to be established on surfaces with narrow clearance or behind protruding features of the object.
Measurements on internal surfaces		Measurements on internal surfaces, such as inside bores of stationary or rotating objects, can be carried out with test type indicators in a manner permitting the unobstructed observation of the dial.
Transfer of height dimensions		For transferring height dimensions from a reference piece to the object surface, a basic operation in surface plate based measurements, test type indicators are generally preferred, and actually needed when using step blocks for referencing.
Measuring side surfaces inside grooves		In exploring the geometry of the vertical side surfaces inside grooves, to measure straightness, parallelism or perpendicularity, the flexible adjustment of the test type indicators, particularly the swivel movement of the contact arm, are necessary gage properties.

TABLE 5-5.   CHARACTERISTIC APPLICATIONS FOR TEST TYPE INDICATORS – 2

DESIGNATION	DIAGRAMMATIC SKETCH	DESCRIPTION
Measurements along vertical surfaces		Indicator measurements along vertical surfaces for determining the inherent and the relative geometry, require optimum observational position for the dial. The vertical type test indicators are the preferred instruments for this type of measurements.
Simultaneous gaging of several closely located surfaces		When the geometric characteristics of several closely located surfaces have to be observed in relation to a common datum, such as in run-out measurements referenced from an axis, test indicators permit the orienting of all dials essentially in the same direction.
Exploring the geometry of deep lying surfaces		Deep lying surfaces, where space is restricted, can frequently be reached with test type indicators only. For this kind of measurement, the vertical type indicator is generally preferred because of the easy observation of the dial.
Centering of a rotating member relative to a bore surface		For centralizing rotatable machine tool spindles in relation to a bore, test type indicators facilitate contacts even in hard-to-reach areas and, at the same time, permit a convenient observation of the dial.

Adjustable Indicator Snap Gages

The adaptability of the adjustable snap gage (see Chapter 3) is combined with the advantages of indicator gaging in the adjustable indicator snap gages. The indicating property of this gage permits combining the functions of both the GO and NOT-GO gage into a single instrument, thus reducing the required inspection steps from two to one. In addition to enabling a plain decision as to acceptance or rejection, which characterizes the fixed gages, the use of indicator snap gages permits an evaluation of the actual size of the gaged dimension. This property is of particular value in production gaging, because it can supply guidance for the proper adjustment of machine tools.

Indicator snap gages are made in different designs with regard to the indicator's measuring range, its accuracy and the smallest graduation of the dial, to name a few readily visible characteristics. Other features may be the appropriate accuracy of the frame and its adjusting, locking and motion translation mechanism. However, various other characteristics, a few of which are listed below, may be considered common to most types of indicator snap gages:

a. Adjustment over an extended range, making the gage adaptable for measuring part features of widely different nominal sizes;

b. Provision for locking the adjustable member in the selected setting position;

c. Retracting device for relieving the gaging force while bringing the part or the gage into the proper gaging position; and

d. Adjustable back stop to aid in positioning the object for repetitive measurements, particularly for gaging the diameters of round parts.

Regarding their general construction, three basic types of indicator snap gages may be distinguished: (a) plain indicator snap gages, (b) translated motion indicating snap gages, and (c) indicating precision snap gages, equipped with sector type indicators for measuring members.

The *plain indicator snap gage* (Fig. 5-24) consists basically of a frame, which in its general form resembles that of a micrometer. One end of the bow carries the adjustable anvil; the other end is equipped with the indicator, with its spindle in line with the anvil shaft. Either the contact point of the indicator is used directly to sense the object surface, or a separate contact piece, frequently a steel ball is attached to the indicator point. Some limitations of application flex-

Federal Products Corp.

Fig. 5-24. (Left) Plain indicator snap gage with the indicator spindle aligned to the dimension to be gaged.

Fig. 5-25. (Right) Translated motion, indicator snap gage has an internal, motion transfer mechanism connecting the spring-loaded lower contact to the sensing point of an encased indicator.

ibility have to be accepted with this less expensive type of design. The extending indicator instrument can interfere with the use of the gage where space is limited; the gage is also exposed, and thereby more easily damaged. Furthermore, the dial may be inconveniently oriented for precise observation, and forces acting directly on, but not in line with, the indicator spindle could cause inaccuracies in the measurement.

In the *translated motion indicating snap gages* (Fig. 5-25) the gage frame has an integral lever device that is connected with the stem of the movable anvil and that transmits the sensed displacement to the built-in indicator instrument. The rigid anvil of the gage is adjustable for setting the gage and can then be locked securely in position. The compactness of design permitting use in areas where space is limited, the protected position and the convenient location for observation of the indicator are a few of the favorable features that distinguish this type of indicator gage.

The indicator snap gage, regardless of type, may also be equipped with an electronic indicator (see Fig. 5-26). Again, when applied to the snap gage, the electronic indicator provides several advantages, as previously discussed in this chapter.

Depth Indicator Gages

Similar in its general design to depth micrometers, the depth indicator gage also consists of a beam (or base), whose face functions as the reference plane of the gaging, and of the measuring member with its

Federal Products Corp.

Fig. 5-26. Snap gage with electronic indicator and bench stand.

The L. S. Starrett Co.

Fig. 5-27. Depth indicator gage in an application where the perpendicular distance between offset parallel surfaces is gaged.

sensing tip reaching beyond the reference plane. Substituting the indicator, commonly the dial type, for the screw micrometer assures a more convenient and rapid reading of the sensed distance, this being a characteristic or definite value for repetitive measurements. However, when selecting depth gages, due consideration must be given to limitations that are common to all indicating type measuring instruments: the need for setting masters and the limited range of action. The limited range of action may be effectively compensated for by using long-range dial indicators and interchangeable measuring rods of different lengths.

Figure 5-27 shows the use of a plain depth indicator gage for measuring the height of a shoulder in relation to a flat surface. Actually the gage is located on the flat surface of the shoulder, and the measured dimension is the depth of the plate from that datum level. This example was selected to illustrate the essential conditions of depth gage application: the measurement of the distance between two parallel, yet offset, planes.

More common uses are the measuring of the depths of blind holes, recesses and grooves. In these applications the beam of the depth indicator gage is usually resting on supporting points whose positions straddle the location of the contact pin. Fairly accurate measurements can be made this way even from a cylindrical surface, when aligning the beam parallel to the axis of the object. When adapted to the depth indicator gage, the electronic digital in-

dicator can provide either comparative or absolute measurements as desired (see Figs. 5-13 and 5-14).

Shallow Diameter Gages

For the comparative measurement of diameters on certain types of outside or inside features, a special category of indicator gages, commonly known as shallow diameter gages, may be considered the best suited measuring instrument.

As illustrated in Fig. 5-28, which shows a basic type of shallow diameter gage, this category of indicator gages comprises instruments having the following essential members:

1. The *frame*, which is a rail type structure providing guidance and retainment for the adjustable contact members and the rest feet.

2. The *anvil post* carrying the reference contact member; it is adjustable along the frame in conformance with the distance to be gaged. After adjustment, the anvil post is locked into the selected position and functions as a fixed reference member during the gaging.

3. The *gaging head* may also permit adjustment along the frame prior to being locked in a fixed position. This adjustment allows an essentially symmetrical arrangement of the contacts along the frame. The gaging head carries the sensing contact for the gaging, as well as the transfer mechanism for transmitting the sensed size variations, as an undistorted displacement movement, to the indicator spindle.

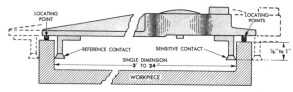

Federal Products Corp.

Fig. 5-28. Shallow diameter gages representing a general form of execution.
(Top) Shallow diameter gage for outside measurements.
(Center) Shallow diameter gage for inside measurements.
(Bottom) Diagram for alternate arrangements of shallow gage elements for inside (solid line) and for outside (dotted line) measurements.

4. The *indicator instrument* may be mounted, depending on the model of the gage, directly on the gaging head, or on a separate base on the frame, which can also be adjusted to secure contact with the motion transfer device of the sensing head.

5. The *rest feet,* equipped with the contact points, must have equal length and are also adjustable along the frame. The rest feet are interchangeable and can be arranged either inside or outside the gaging contacts. The two rest feet have the function of locating and supporting the gage in a reference plane, which is parallel to and at a predetermined distance from the actual measuring plane on the object.

The use of the shallow diameter indicator gages is necessarily limited by two requirements:

a. The diameter to be gaged (or more precisely stated its acceptable measuring plane) must be lo-

cated reasonably close to a face surface on the object; and

b. The face surface should be flat and also square to the feature axis to a degree adequate to represent the datum plane for the measurement.

When these conditions are satisfied, the use of shallow diameter indicator gages can provide several distinct advantages some of which are discussed in the following:

a. Flexibility of application due to the "building block" principle, on which the design of most types of shallow diameter gages rests, making the instrument adjustable for a great variety of measuring tasks, for example, plain lengths or diameters, on external or internal features;

b. Wide range of adjustment, covering sizes that, for other types of indicator gages, may require several instruments;

c. Adaptability to the majority of gaging surfaces, whether cylindrical or taper, freely accessible, or confined behind shoulders, inside grooves, in recesses and so forth, by virtue of the many types of gaging tips that can be used by insertion into the gaging posts. Figure 5-29 shows, as random examples, a few unusual dimensions that can be measured successfully with the aid of interchangeable contact tips for locating, referencing and sensing; and

d. The positive establishment of the gaging plane in relation to a datum face on the object. The maintenance of a fixed depth of the gaging plane in relation to the datum is particularly important in measuring length or diameter on taper or curved surfaces, as well as behind shoulders. By positively locating the shallow diameter gage on its rest feet, the selected measuring plane will consistently be maintained, thereby avoiding one of the persistent sources of potential error in diameter measurement.

The position of the indicator instrument can be selected to be parallel to the gaging plane, or perpendicular to it. In addition to the convenience of dial observation, considerations arising from the applicable *motion transfer mechanism* will guide the choice.

When a parallel indicator position is selected, the generally used transfer mechanism consists of a parallelogram linkage, commonly known as the reed type (Fig. 5-30(A)). Although an excellent system for transferring short displacements and in applications requiring only a moderate length for the spring leafs, displacement translation through reeds can display

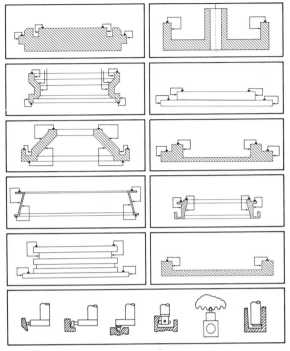

Standard Gage Co.

Fig. 5-29. Characteristic examples of unusually located diameter dimensions. These can be conveniently measured, in proper relation to the datum surfaces, with the aid of shallow diameter gages using appropriate contact elements such as are shown at the bottom.

meaningful system lags (up to several percentage points), when transferring movements that are sensed at a substantial distance from the indicator instrument.

When the position of the indicator is perpendicular to the plane of measurement, motion transfer devices with direction changing properties are needed. Lever type mechanisms are applied in one of the several available varieties. The more commonly used types are shown in Fig. 5-30. Type (b) is the plain bell crank, which is selected to contact difficult-to-reach surfaces. However, unprotected bell cranks are sensitive to shocks, which could affect the accuracy of the delicate cone pivots generally used to function as the positively located fulcrum of that lever system. The plunger type transfer devices are preferred for the majority of uses, because they are less vulnerable than the exposed crank type. Furthermore, in the plunger type the crank is protected and has no direct contact with the work surface, so that it is less subject to wear that could affect the geometric balance of the two lever arms. Two basic models of plunger type transfer mechanisms are commonly used, one having a spring-loaded plain pin for the plunger (type (c)), whereas in the other type the plunger is precisely guided at both ends, the contact with the crank being located between the guided sections (type (d)).

Bench Comparator Gages

Indicators—whether mechanical or operating by another system—when used on a stand that provides a locating surface for the object and a positive retainment for the indicator instrument, will constitute a type of gage that is commonly known as a bench comparator. Figure 5-31 shows the typical example of a plain bench comparator equipped with a

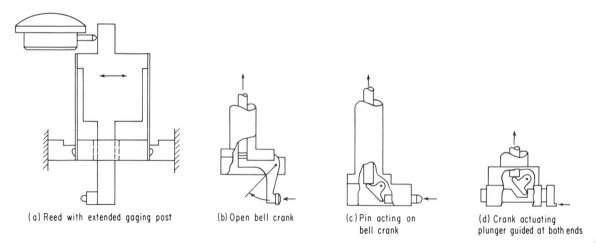

(a) Reed with extended gaging post   (b) Open bell crank   (c) Pin acting on bell crank   (d) Crank actuating plunger guided at both ends

Fig. 5-30. Principles of motion transfer mechanisms used in shallow diameter gages.

Federal Products Corp.

Fig. 5-31. Basic type comparator stand with dial indicator, adjustable support arm and work-staging platen mounted on the base.

regular dial indicator. Figure 5-32 shows the plain bench comparator equipped with an electronic indicator.

Important requirements for any type of bench comparator are the following:

a. The adjustability of the indicator position relative to the object stage. The adjustment is usually accomplished in two steps: the rough positioning of the gage arm along the column of the comparator, and the fine adjustment of the indicator position in

its immediate holding member. In other types of comparators, the fine adjustment is accomplished by raising or lowering an auxiliary staging table.

b. Locating the object in a plane perpendicular to the gaging movement of the indicator spindle. The staging surface may simply be the top of the comparator base, a table attached to the column, interchangeable anvils or platens fitting the comparator base, but also any appropriate type of special fixture.

For repetitive measurements, which represent the predominant application for comparator gaging, it is practical to complement the basic equipment of bench comparators with auxiliary work-staging devices. The positive location of the object for gaging in a comparator may be assisted, and the staging time substantially shortened, by properly selected attachments, examples of which are named in the following:

a. Serrated table tops to reduce the contact area and to avoid dirt penetrating between the stage and the work;

b. Narrow anvils to permit a better grip on the work and/or to locate the object on a specific area of its surface;

c. Center attachment to hold work between centers; also for checking runout (Fig. 5-33, Left);

d. Back stop of adjustable design to locate a cylindrical workpiece in a position where the axes of the indicator spindle and of the part are contained in a common vertical plane; and

e. Vee-blocks, either the straight wall or the twin roller type (Fig. 5-33, Right), for positive locating and

Federal Products Corp.

Fig. 5-32. Bench comparator stand with electronic indicator.

Federal Products Corp.

Fig. 5-33. Auxiliary work-holder accessories mounted on a comparator base.
(Left) Center attachment.
(Right) Twin rollers functioning as a Vee-block.

aligning of cylindrical parts. Although convenient in several respects, this locating system has drawbacks similar to those of the Vee-type arrangement of locating stops, which are discussed in the following section, in connection with the Y-type plate gages.

Additional types of mechanical bench comparator gages will be reviewed in the section on precision indicators.

Plate Gages

Figure 5-34 illustrates the general design of plate type indicator gages, a designation that is frequently used for a type of bench comparator equipped with a slanting staging plate to rest and to locate the part to be gaged. This bench comparator design is convenient for the rapid gaging of flat and relatively thin parts, most frequently disks or rings—for example, ball and roller bearing rings—whose measurement must be made in a plane parallel to at least one of the object faces.

The plate has a flat surface, on which one may find wear strips producing the contact plane for the object to be gaged. Inserted strips or buttons, when made of hardened steel or carbide, will increase the wear life of the gage and also reduce the incidence of dirt or chips affecting the correct positioning of the part.

The gage plates have slots arranged either in the form of a T in horizontal position, or in the form of a Y in erect or inverted position. Inside the slots are the adjustable locating stops and the sensitive con-

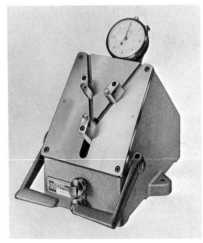

Federal Products Corp.

Fig. 5-34. Plate gage type comparator with Y groove arrangement for inside diameter measurements. The three adjustable contact buttons serve for supporting, centering and gaging the object.

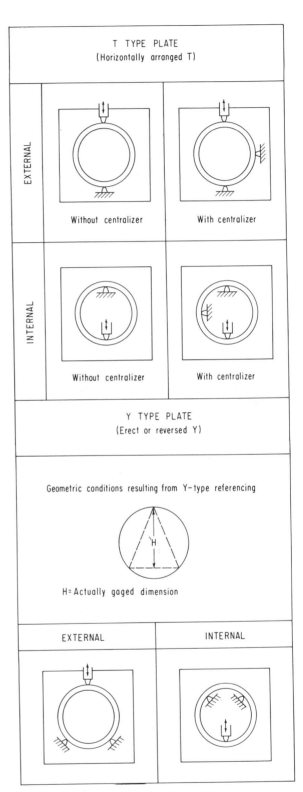

Fig. 5-35. Principles of object location in plate gage type measurements.

tact point. Setting gages are used for adjusting the stops and the contact point in compliance with the nominal size of the dimension to be gaged. The contact member is connected to the indicator, and a lever of generous size in the front of the gage permits the rapid retraction of the contact member for inserting and removing the object.

The principles of part locating on plate gages are shown in Fig. 5-35. Plates with T form slots have one stop opposite the contact member for actual diameter measurements, whereas the second stop in the horizontal slot is used optionally as a centralizer (see Fig. 5-36). This additional stop reduces the locating time, and is also preferred with less experienced inspectors who might have difficulties in consistently establishing the "high point reading" in diameter measurements.

The Y form slots have stops in each of the converging legs and these stops must be adjusted symmetrically to the center plane of the Y, which contains the contact point. The double stop, with locating effect comparable to the Vee-block, permits a positive and precise location of the part on the gage plate, but has the drawback of presenting a dimension for gaging that is not actually the diameter of the part, but the height of an inscribed isosceles triangle. When the angle included between the stop slots is small, the effect of these geometric conditions on the end result of the comparative gaging is frequently considered insignificant, and is neglected in shop practice in view of the advantages of the positive object location that can be assured by the twin stops.

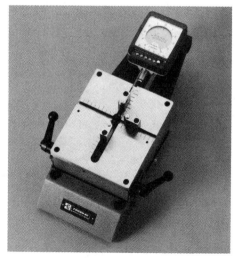

Federal Products Corp.
Fig. 5-36. I.D./O.D. plate gage with electronic indicator.

INTERNAL INDICATOR GAGES

Comparative Measurement of Internal Dimensions

Comparative measurements of distances between internal surfaces are generally carried out with instruments resembling the external types of indicator gages. However, there are a few essential differences in the operational requirements that reflect on both the gaging process and the design of the gages.

In internal measurements there is a minimum practicable gaging distance, this being determined by the space into which the instrument or its sensing member can penetrate. In comparison, indicator gages serving external measurements have capacity ranges that commonly start from zero.

The locating and aligning of the gage on inside features is usually more difficult than on a corresponding outside surface, the latter providing unobstructed observation of the relative positions and of the engagement of the contact members. The problems of correct positioning and alignment inside deep bores are particularly pronounced and require effective countermeasures to reduce their disturbing effect on the gaging of the true diameter.

Measuring the bore diameter away from its axial plane will show smaller than actual size, whereas using a gaging plane that is not perpendicular to the bore axis tends to increase the observed size. To reduce the effects of these potential sources of error, indicator bore gages are generally equipped with centering devices that assist in locating the gage in the axial plane of the bore. Then, by rocking the shaft of the gage (Fig. 5-37), the perpendicular plane position can be found as evidenced by the minimum indication on the dial.

To produce in manufacturing a bore of precisely cylindrical form and within close size limits is generally a more difficult proposition than the production of a corresponding outside feature of basically cylindrical form. Problems connected with the precise measurement of bore dimensions may be partly responsible for that relative difficulty. These conditions are also recognized in the International Standards Organization (ISO) recommendations for fits, where a grade representing more rigorous tolerances is usually specified for the shaft than for the mating bore, for example, American Standard and ISO tolerance limits H7/g6.

Inaccuracies of the bore size that are determined by measuring the diameter can result from plain dimensional variations, but can also be caused by irregularities of form that affect the diameter of the bore. While plain dimensional errors in a bore of es-

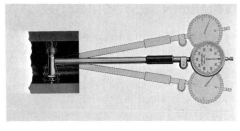

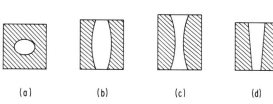

(a)   (b)   (c)   (d)

Mahr Gage Co.

Fig. 5-37. (Top) Rocking a bore indicator gage in the plane of the gaging contacts to establish the proper measuring positions, as evidenced by the point of inversion in the change of indications.

Fig. 5-38. (Bottom) Typical form irregularities of basically cylindrical holes which can be detected and measured with bore indicator gages: (a) oval; (b) barrel-shaped; (c) hour-glass form or bell-mouthed; and (d) tapered.

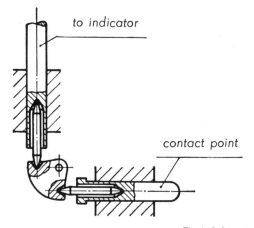

to indicator

contact point

The L. S. Starrett Co.
Federal Products Corp.
Mahr Gage Co.

Fig. 5-39. (Top Left) Cylinder gage of earlier design; an instrument widely used in motor repair work for exploring the form irregularities of worn cylinder bores.

Fig. 5-40. (Top Right) Bore indicator gage with centralizing plunger and adjustable reference point.

Fig. 5-41. (Bottom) Bell crank type, motion transfer mechanism of a high sensitivity, bore indicator gage. Precisely nested struts permit an almost frictionless transfer of the contact point motions to the indicator.

sentially regular form can be detected by measuring a single diameter, irregular forms require a more thorough exploration of the bore dimensions.

Bore indicator gages are effective instruments for detecting and measuring those types of form irregularities that affect the diameter of the basically round bore. Figure 5-38 illustrates a few of the more commonly occurring bore irregularities that belong to this category of form errors. Indicator bore gages permit the observation and measurement of these conditions when the internal surface is explored by moving the gage in the direction of the bore axis and—for some types of irregularities—repeating this process in several axial planes. Actually, one of the first applications of indicator bore gages, the cylinder gages used in motor service operations (Fig. 5-39), originated from the need of detecting and precisely measuring bore diameter variations by means of a portable instrument.

Bore form irregularities causing diametrical variations will be signaled by the indicator gage as the spread of successive pointer positions on the dial. In addition to these variations, the instrument will also supply information regarding the actual size of the bore diameter, this in the form of deviations from the basic size. The basic size of the measurement must be established in a preceding setting process, which requires appropriate setting gages.

System of Internal Indicator Measurement

The various benefits that can be derived from using indicating type gages for internal measurements led to a rather extended adoption of this gaging method. Although similar in their basic principles, internal indicator measurements can widely differ in the actual application requirements. Differences in the part size and form, in the desired accuracy, and also in the objectives of the measurement have resulted in internal indicator gages of widely varying design.

Without striving toward completeness, a survey is presented in Table 5-6, showing diverse indicator gaging systems that are commonly used for internal measurements. The sketches are diagrammatic only, having the sole purpose of illustrating the essential principles in which the various systems differ.

Examples of Internal Indicator Gages

The gages described in this section were selected as examples to illustrate the actual design of gages having certain operating principles listed in Table 5-6. Completeness—a goal that would hardly be practicable in view of the many available makes and models—has not been attempted, nor has the selection been guided by the relative value of the instruments. Although proved and widely used gages are described, for most types there are various other makes of comparable value, and also other models of unquestionable usefulness.

While bench type bore gages with mechanical indicators are still used and provide good service, particularly under manufacturing shop conditions, no examples for this category of gages will be presented. The reason for that choice is that, whereas in the case of hand gages designed for bore measurements the mechanical indicator is in many respects superior to other amplification systems, in the stationary type of bore gages the air and electronic amplification systems are preferred, particularly for higher levels of accuracy.

Bore indicator gages with centralizing devices. One of the more commonly used types of bore indicator gages is shown in Fig. 5-40. With the exception of the smallest sizes that—because of space limitations—have only a leaf spring member to help the centralizing, the sensing head of these gages has two spring-loaded plungers equally spaced on both sides of the contact point. The spring action of these plungers effectively assists in quickly finding the central plane of the hole and then only a slight rocking motion of the gage is needed to accomplish a fairly accurate bore measurement.

The four contact points, those of the centralizing plungers, the reference and the sensing contacts, being contained in a common plane, and the gage sensitively detect all bore form irregularities that are reflected by diameter variations. An essential requirement of bore indicator gages, the sensitive and backlash-free motion transfer is accomplished in this design with the aid of a pantograph spring. Other designs have bell cranks, actuated by a plunger to which the sensing point is attached (Fig. 5-41).

There are models of bore indicator gages whose centralizing devices differ from the principles dis-

cussed above. This variety may be characterized as having a single centralizer plunger either with the contact surfaces resembling a split spherical segment (these are less sensitive to a slightly cocked position of the gage), or the contact surfaces have the form of circular segments whose virtual centers are in a common plane with the reference and sensing contacts (this arrangement facilitates the gage alignment in a normal plane of the bore).

These alternate designs of the centralizer, while making the gage alignment easier and quicker, may tend to conceal certain types of bore form irregularities, to which the point contact centralizers are more sensitive. The different properties of various designs serving essentially the same purpose may be considered a typical example for circumstances where the choice of the best suited model from several comparable measuring instruments must be guided by the prime objectives of the measurements.

Indicator plug gages with expanding segments. These indicators bore gages (Fig. 5-42) unite the centering and the gaging functions into the same set of members, two ring segments with a radius somewhat smaller than that of the bore to be measured. The contact members also have a small back taper, resulting in a reduced area contact at two diametrically opposed sides of the part's bore. For special purposes, individual types of segments can be used such as, for example, measuring the pitch circle diameters of internal gears using segments equipped with contact pins or balls.

The actual contact with the bore wall is accomplished by expanding the segments through the action of an internal taper plug member of the gage. As soon as the retracting lever is released, the taper is advanced axially by spring pressure and produces an expansion of the segments proportional to the axial

The Comtor Company

Fig. 5-42. Indicator bore gage with expanding segments. Pressing the button retracts an internal, tapered plug that, under spring pressure, tends to spread the segments for contacting the opposite walls of the bore being measured. The figures show the consecutive steps of the gaging process.
(Left) Setting the instrument with a gage ring.
(Center) Introducing the segmented probe into the bore of the part.
(Right) The dial pointer shows the deviation from the setting size.

TABLE 5-6.   SYSTEMS OF INDICATOR GAGES FOR INTERNAL MEASUREMENTS — 1

CATEGORY	DESIGNATION	DIAGRAMMATIC SKETCH	DESCRIPTION
Portable Indicator Gages for Bore Diameter Measurements	Shallow bore gages		Simple instrument, commonly of modular design, with extended measuring range and great flexibility of adaptation for diverse object configurations. The face of the part is generally used for gage locating surface. Application limited to shallow depths.
	Plain bore indicator gage with two-point contact		Lack of centralizer makes the alignment of the gage with the axial plane of the bore difficult. Application generally limited to small bore sizes where space limitations prevent the use of centralizing devices.
	Expanding segment bore indicator gage		Effective centralizing action results from the segments expending under spring action. To assist centralizing, the segments have diameters only slightly less than that of the bore. Slender back taper of segments reduces the contact areas.
	Bore diameter indicator gage with centralizer slide		The most generally used basic type of indicator bore gages with a spring loaded slide whose two points or rails straddle the contact plunger of the gage and effectively assist in centralizing the diametrally arranged contact points.

TABLE 5-6. SYSTEMS OF INDICATOR GAGES FOR INTERNAL MEASUREMENTS — 2

CATEGORY	DESIGNATION	DIAGRAMMATIC SKETCH	DESCRIPTION
Portable Indicator Gages for Bore Diameter Measurements	Bore indicator gage with three-point contact		The simultaneous and uniform expansion of the three contact plungers provides rapid centralizing action. The dial indications do not reflect true diameter variations. Effective for detecting three-point out-of-round condition of the bore.
	Three-blade bore indicator gage		The two external blades are rigid and serve for centralizing, while the centrally located movable blade acts on the indicator which is adjustable in its retaining sleeve. Good centralizing and aligning permits rapid gaging. For the setting of the indicator a ring gage is needed.
Stationary Bore Indicator Gages	Bench type bore indicator gage		Operating principles comparable to those of the shallow bore gages, using a plate to support and locate on one of its faces the part to be gaged. Centralizing by two adjustable, yet functionally rigid stops which also serve for gage reference points.

TABLE 5-6.   SYSTEMS OF INDICATOR GAGES FOR INTERNAL MEASUREMENTS — 3

CATEGORY	DESIGNATION	DIAGRAMMATIC SKETCH	DESCRIPTION
Bore Location Gages	Bore concentricity and location gage		The body of the gage is rotatable in a sleeve locating the gage referenced from the selected datum surfaces. The instrument indicates the radial variations of the bore surface from the reference axis.
Internal Groove Diameter and Location Measurements	Swing arm groove diameter indicator gage		Rapid adjustment over a wide range, long retraction to penetrate into deep grooves and light weight are advantages. Swing movement of the measuring arms produces indications closely proportional to the measured linear distance.
	Adjustable slide groove diameter indicator gage		The adjustment to the nominal size is accomplished by setting the slide of the fixed contact along the rail to the desired position. Retraction and gaging motions are either swing or linear displacement of the sensing arm acting on the indicator.
	Adjustable slide groove location indicator gage		To measure axial location, groove width or separation on internal surfaces. Adjustment is similar to the preceding type and the measuring stroke of the sensing contact is usually linear.

travel of the taper plug. This expansion actuates the indicator graduated in ten-thousandths of an inch, which then displays the amount and direction of the bore size variation relative to the setting size. Although the actual measuring range of the gage is only ±0.006 inch, the ready interchangeability of the expansion plugs permits the use of the same indicator head (called the "amplifier") for a wide range of bore sizes. For checking the setting of this type of indicator gage, a ring gage is needed for each nominal size, and only minor adjustment can be made by rotating the indicator bezel.

Indicator plug gages with expanding blades (Fig. 5-43) compare the object bore diameter to the setting dimension of the gage by means of a three-line contact established by the mutually expanding blades. The contact faces of these blades have the form of cylindrical segments with radii less than the radius of the bore to be gaged. The two outer blades are rigid and integral with the gage body, while the central blade, with contact face opposite to those of the former, has a well guided but virtually frictionless movement. During the insertion of the gage into the bore, the central member is retracted by pressing a thumb lever to avoid wear on the contact faces. When released, spring pressure causes the central blade to expand and all three segments to thereby contact the bore wall. The displacement of the central member is sensed by an indicator, which displays the value and the direction of any deviation from the setting diameter. Interchangeable blade

heads provide a rather extended measuring range specified as capacity steps, whereas the intermediate dimensions are covered by the rough and fine adjustments of the indicator instrument.

The three-line measuring contact, as compared to the purely diametrical two-point contact, provides the advantages of facilitating the precise alignment and centralizing in the operation of the instrument. By using special blades with face forms adapted to the job, internal size variations of threaded bores and the bores with odd numbers of grooves, the inside diameters over rollers of roller bearing assemblies and many other unusual forms can be measured. Because this type of gage is practically insensitive to certain forms of irregularities, such as bell mouth, barrel form, and taper, its use should be limited to features where such irregularities of form are not expected, or are not considered significant.

Small bore indicator gages with expanding plug (Fig. 5-44) operate on principles similar to those of the expanding segment type bore gages. The flexible contact members are radially expanded by an internal taper plug, whose axial displacement actuates the indicator, as shown diagrammatically in the cross-sectional drawing. In most types of small bore gages the spring pressure urging the gage spindle is sufficient to exert the force required for expanding the contact members. The contact elements commonly have a spherical form, resembling a split ball, for effective centering. The alignment of the gage with the bore axis is accomplished by slightly rock-

Mahr Gage Co.

Fig. 5-43. Expanding blade type indicator bore gage with interchangeable blade heads and high-sensitivity indicator. Coarse and fine adjustment of the indicator position extends the effective range of the heads.

Foster Supplies Co.

Fig. 5-44. Small bore indicator gage with expanding plug actuated by the axial movement of a conical pin.
(Left) Gage with trapezoid plug for blind holes.
(Right) Cross-sectional diagram.

ing the gage and observing the reversal point of the indications. For diameter measurements in blind holes close to the bottom, plug members having trapezoid cross-sectional form are also available. Readily interchangeable plugs, supplied in sets with the mating setting rings, provide a rather wide effective measuring range for these instruments.

Internal Groove Indicator Gage with Parallel Jaw Movement

The precise measurement of the diameters of internal grooves, such as used for seal rings, retainer rings and similar purposes, can be quickly carried out with specially designed indicator gages. These gages must possess two particular properties that are not required for the common types of bore gages:

1. An extra long retraction movement to permit the contacts to penetrate behind shoulders; and

2. Interchangeable contact tips, including those of sturdier dimensions for general measurements, and also thin ones for gaging inside narrow grooves.

Figure 5-45 illustrates a general type of parallel jaw internal indicator gage for grooves or similar measurements behind shoulders. This gage has a pistol grip handle with coarse and fine adjustment, and a thumb-actuated jaw retractor. During the mea-

suring process the gage is supported by the fixed lower reference jaws that are resting on the work surface. The motion of the sensitive jaw is transmitted directly to the indicator through a "friction-free" transfer mechanism. The retraction range of this particular type of gage is about $\frac{1}{2}$ inch.

Internal Groove Indicator Gage in Swing Arm Design

A particularly wide flexibility of application distinguishes the internal groove indicator gage (Fig. 5-46), which is intended for use where the dimension to be gaged is specified in thousandths of an inch, or coarser. The setting of the gage within its wide effective range of $\frac{3}{8}$ of an inch to 6 inches is accomplished by simply turning a knurled knob to obtain zero indication for the setting size and then locking the adjustment. Jointed end sections of the main arms also can be adjusted to align the contact tips with the plane of the measurement. Interchangeable tips and centering arms permit carrying out measurements with either two- or three-point contact, the former with or without the use of a centering attachment.

Although the long-swing arms reduce the rigidity of the instrument with potential detracting effects on the accuracy of the gaging, the rapid setting, the wide gaging range, the convenient observation of the integral indicator dial, and the availability of contact tips in many different forms and sizes are considered useful characteristics of this gage design.

The L. S. Starrett Co.

Fig. 5-45. Indicator gage for internal grooves, with parallel jaw movement. Interchangeable jaws and contact tips (inserts) accommodate different size ranges and groove dimensions.

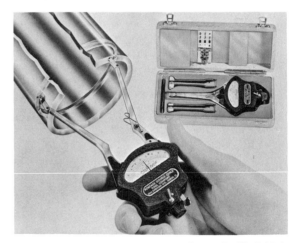

Portage Double-Quick, Inc.

Fig. 5-46. Internal groove indicator gage with swing movement of the contact arms. The wide adjustment range of the gage permits measurement of bores from $\frac{3}{8}$-inch to 6-inch diameter.

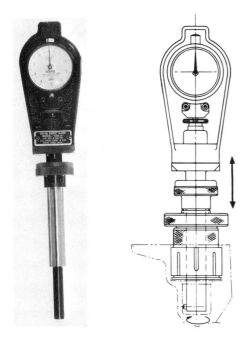

A. G. Davis Gage & Eng. Co.

Fig. 5-47. (Left) Bore concentricity gage in one of the basic designs, with long locating body for checking a deep-lying bore.

Fig. 5-48. (Right) Gage checks bore concentricity when rotated, and the parallelism of the bore wall when moved in an axial direction.

Concentricity and Bore Location Indicator Gages

For bore relationship measurements a particular type of indicator gage is available. This is commonly designated as a concentricity gage (Fig. 5-47), which takes its name from a category of measurements representing its predominant application.

Round holes in engineering components are frequently specified with tight tolerances for concentricity and location. Concentricity, in this context, will usually require the accurate alignment of the axes of two or more in-line holes of equal or different diameters. One of the holes, which serves as the datum for the concentricity measurement, will locate the instrument by guiding the cylindrical gage body. Differences in the diameter of the datum hole and of the smaller gage body are conveniently compensated by inserting a bushing of appropriate dimensions into the guide hole. The shaft of the gage, actually a cut-down extension of the body, carries the adjustable radial contact member that is con-

nected to the indicator. The contact point will travel over the surface of the bore to be explored, when the gage is rotated while being guided by the datum hole (Fig. 5-48). Variations in the radial distance of the explored bore surface from the axis of rotation of the gage will be detected by the contact and signaled by the indicator as the amount of bore eccentricity. By moving the gage up and down in the guide bore, while the sensing point stays in contact with the wall of the bore to be measured, the indicator will show the presence and the amount of deficient parallelism of the related bore surfaces.

Concentricity indicator gages also can be used very effectively for measuring hole location in relation to physical datum surfaces on the part. Indications resulting from this gaging will satisfy the concept of true position dimensioning. To carry out this kind of measurement, auxiliary fixturing is required, whereby the basic position of the hole axis relative to the datum surface can be established physically (Fig. 5-49). This kind of fixture consists essentially of a plate equipped with contact elements to locate it on the datum surface of the object, and with a precise bore to guide the gage.

The sensing point of the gage will usually contact the bore wall directly; however, in cases where hole location is the sole objective of the gaging, a transferred hole position, implemented by a gage pin snugly fitted into the bore, may also be used (see Fig. 5-49).

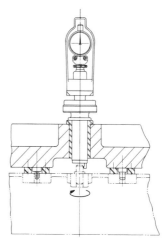

A. G. Davis Gage & Eng. Co.

Fig. 5-49. Concentricity of a bore can be checked with respect to a virtual axis whose location is related to external datum surfaces. Dowel is used to "transfer" the bore position for gaging.

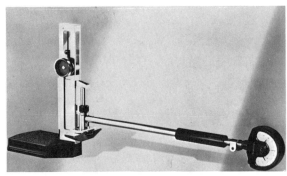

The L. S. Starrett Co.
Mahr Gage Co.
Ex-Cell-O Corporation

Fig. 5-50. (Left) Setting a bore indicator gage with a ring gage; both held in mutually parallel position by resting on a surface plate.

Fig. 5-51. (Center) Using a gage block stack with beams, in a foot-stand type holder, for setting a bore indicator gage.

Fig. 5-52. (Right) Adjustable setting gage for bore indicator gages.

Setting of Bore Indicator Gages

Bore indicator gages, being comparator type measuring instruments, must be set to the required nominal size, to which the actual size of the bore to be measured will then be compared. The indicator's function, in this application too, is to detect and to show the amount as well as the direction of any deviation from that nominal size. The setting of the bore indicator gage consists of adjusting a gage member, usually the reference contact point, to produce zero indication on the dial when the gage is in measuring contact with the selected nominal length.

The manner in which this nominal length is made physically available for the purpose of indicator gage setting can vary, the different methods having their particular advantages and limitations. The more frequently used methods of internal indicator gage setting will be reviewed in the following.

The use of an *outside micrometer* is the quickest and most readily applicable method even under regular shop conditions, but it provides limited accuracy only. The inherent inaccuracies of larger size outside micrometers, and the difficulty of establishing dependable location and alignment of the spherical measuring points of the gage on the small area of the micrometer contact faces, make this method of gage setting inadvisable for measurements where better than 0.001-inch repeat accuracy is required.

Certain types of bore indicator gages, particularly those intended for repair shop uses, provide means for reversing the process of setting bore indicator gages with the aid of a micrometer. Such bore indicator, or cylinder, gages (see Figure 5-39) are first

coarsely adjusted by feel to fit the bore to be explored for form variations. Thereafter the contact plunger of the gage can be locked while still inside the bore, using the knurled knob on a stem protruding from the gage body, and after having retracted the gage from the bore, the distance over the locked gaging contacts can be measured with a regular micrometer.

The ring gage method used for setting masters is the method least exposed to errors in adjusting the bore indicator gage, and also the only practical method for precisely setting three-point contact gages. The setting of bore indicator gages with the aid of gage rings is facilitated when the indicator gage has a flat face parallel to the axes of the contact plungers. In this case, the gage can be placed on a surface plate that also supports the ring gage, the instrument and the master thus having a common reference plane for the setting operation (Fig. 5-50). The major drawback of using ring gages for setting results from the need for providing separate ring gages for each setting dimension. Exceptionally, ring gages having a size close to the required nominal dimension can be used, when the difference in nominal dimensions is substantially less than the measuring range of the indicator. Ring gages are the preferred setting masters for the continuous or repetitive measurement of identical bore sizes.

The gage block assembly in a clamp with beams at both ends (Fig. 5-51) provides a highly accurate setting master for internal indicator gages. This may be considered the most frequently used setting method where dependable gage setting accuracy is required and no special setting masters are available. Gage blocks belong to basic gage room equipment;

consequently, they constitute the means of a readily applicable setting method. A possible drawback arises from the need for assembling a gage block stack and accessories for each new setting operation, this being a somewhat time-consuming process.

Setting gages for bore indicators (Fig. 5-52) are convenient tools for the rapid setting of internal indicator gages to any desired size within the extensive range of this auxiliary instrument. The essential element of the device is a step bar of gage block accuracy, representing one-inch intervals; the intermediate positions are set with the aid of a large drum micrometer with direct reading in ten-thousandths of an inch. The overall accuracy of the instrument is on the 0.0001-inch order, which may be considered adequate for the majority of measurements carried out with the aid of indicator bore gages.

HIGH SENSITIVITY MECHANICAL INDICATORS

General Characteristics of High Sensitivity Mechanical Indicators

This category comprises mechanical indicators with amplification systems designed for accurately measuring small displacements under gaging conditions that require lesser contact forces than are common in general shop applications for indicator measurements. These mechanical indicators, usually operating by a lever-based amplification system, are designed to sense displacements in the direction of the gage spindle axis, and are distinct from the test type indicators that also use the lever as the primary amplifying member. Most models use a rather long pointer for final amplification, with indicator faces of sector form, and the pointer describes a sweep motion only, instead of a complete rotation. Because of these characteristics, the designations "sector type indicator" or "sweep movement indicator" are also frequently used to describe a mechanical indicator of the high sensitivity category.

While the lever system permits a high degree of amplification as mechanical systems go, requiring fewer members than a gear train of comparable amplifying characteristics, the measuring range is substantially smaller. Considering, however, that high sensitivity indicators are commonly selected for inspecting parts with close tolerances, the short measuring ranges are generally found sufficient to encompass the entire spread of the part's tolerance zone. Practical experience has shown that the rate of amplification of the indicator selected should permit the part to be gaged in increments equal to about a

tenth part of its tolerance range. Selecting a poorer indicator sensitivity would detract from the desirable accuracy of the gaging, while substantially higher sensitivity could cause the limits of the part tolerance zone to extend beyond the measuring range of the instrument.

Sector Type Precision Indicators

Figure 5.53, in a cross-sectional view, illustrates the major mechanical elements of a particular design that was selected as a characteristic example of the sector type mechanical indicators. A special form of lever, which avoids the unreliable knife edge suspension, carries a toothed segment in engagement with a small pinion, a coil spring acting as backlash eliminator. The displacements of the measuring spindle are transmitted to the lever by means of a precision ball rolling over a finely lapped sapphire surface. The gaging pressure is supplied by a tension spring acting on a lever arm. The gaging pressure can be temporarily relieved by a plunger actuated through a flexible cable, which lifts the spring-urged lever arm. The remote control of the lifting mechanism protects the gage from inadvertently disturbing the setting and from the effect of body heat originating from the inspector who operates the instrument.

Figure 5-54 shows the same type of indicator mounted on a comparator stand, this being one of the

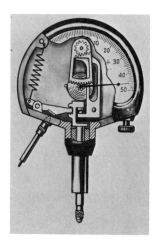

Mahr Gage Co.

Fig. 5-53. (Left) Cross-sectional view of a sector type, high sensitivity, mechanical indicator showing the principal elements of the amplifying mechanism.

Fig. 5-54. (Right) Sector type, sensitive indicator mounted on a comparator stand. The flexible cable permits the retraction of the indicator spindle by remote control.

characteristic, although not exclusive, applications for high sensitivity indicators. The fine adjustment of the pointer for zeroing while the gage spindle is under measuring pressure is carried out with the knurled head screw at the right side of the instrument. The tolerance hands are adjustable to indicate the limit sizes of the part being gaged, thus providing a convenient guide for the inspection of lots consisting of nominally identical parts.

This type of precision indicator is available in three different models with amplifications of about 180×, 900× and 2250×, displaying scale graduation values (smallest increments) of 0.0005 inch, 0.000050 inch and 0.000020 inch, respectively.

High Sensitivity Indicator Gages

The sector type indicator mechanism also has applications in indicator gages when object dimensions with relatively narrow tolerance range must be inspected by using a portable type of instrument. A typical example for this category of gages is the Zeiss Passameter, whose cross-section is shown diagrammatically in Fig. 5-55. As illustrated, a spring-activated plunger, which carries the contact point, is acting on a special lever whose longer arm ends in a toothed segment in engagement with a pinion. The knob-actuated relieving lever acts on the plunger. Figure 5-56 shows the general view of the one-inch capacity model of this instrument. Other models, with 2-, 3- or 4-inch measuring capacity, all with one-inch adjustment range, are also available.

These instruments are actually adjustable indicating snap gages with the distinguishing characteristics of a very accurate and sensitive amplifying mechanism operating over a relatively long measuring range. The sector type dial of the Passameter has distinctly readable graduations about 0.045 inch apart, each space representing 0.0001 inch contact plunger displacement. The total indicator range is ±0.0035 inch. Adjustable tolerance limit hands,

pressure relieving knob and interchangeable back stops are part of the standard equipment.

Precision Indicators with Twisted Strip Type Mechanism

An original system of mechanical amplification is used in the Swedish Mikrokator type indicator instruments. It is a known phenomenon that when a string of flexible metal strip is twisted, and then a tensile force is applied along the axis of the twisted member, the resulting elongation will cause a change in the pitch. This change of the geometry, actually a rotation of the stretched strip, stays proportional to variations in the elongation, as long as it occurs within a specific range, the length of which depends on the design characteristics of the twisted strip.

The amount of rotation induced by the stretching can be measured by attaching a pointer to the strip, and observing the pointer deflections in front of a graduated segment that is located in a plane normal to the strip axis. In the instrumental application of this phenomenon, a perforated bronze band of rectangular cross-section is used, and the stretching movement, which originates from the displacement of the gage spindle, is transmitted through a special lever, called the knee. The operational principles are shown diagrammatically in Fig. 5-57; Fig. 5-58 illustrates the essential functional elements of the Mikrokator indicator.

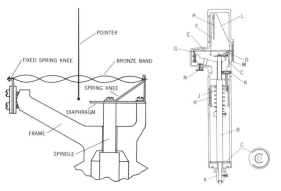

CEF Gage Co.

Fig. 5-57. (Left) Diagram illustrating the operating principles of twisted strip type, mechanical indicators.

Fig. 5-58. (Right) The essential functional elements of a Mikrokator mechanical indicator: *(A)* replaceable measuring tip; *(B)* rigid actuating spindle; *(C)* steel support diaphragms (two); *(D)* spring knee; *(E)* twisted bronze band; *(F)* tapered, drawn-glass pointer; *(G)* fixed spring knee; *(H)* spring retainer; *(J)* measuring pressure spring; *(K)* measuring pressure adjustment; *(L)* tolerance indicator; *(M)* tolerance indicator setting tabs; *(N)* zero adjustment screw; *(P)* Mikrokator scale.

Carl Zeiss, Inc.

Fig. 5-55. (Left) Cross-sectional view of a "Passameter," a type of adjustable snap gage with sector system indicator mechanism.

Fig. 5-56. (Right) General view of the Passameter, an indicating snap gage.

The twisted band principle of displacement amplification permits a good flexibility of design, which is reflected in the extended range of Mikrokator indicators currently available, all operating on the same basic principles. Amplifications can be selected to indicate for each scale graduation a size variation of 0.0001 inch for the lowest, and 0.0000005 (one half of a microinch) for the highest sensitivity. Although this is an impressive choice of amplifications, the Mikrokators are precision instruments that are not intended for measuring wide size variations.

The actual measuring range of the instrument depends, of course, on the rate of amplification and on the extent of the scale. For the scale extent, two basic types of Mikrokator instruments are available, one with a 100-degree sector, and another covering a pointer rotation of over 250 degrees.

The gaging pressure of Mikrokators is regularly 250 grams (about 8.8 ounces). Other models of this type of indicator are available with adjustable gaging pressures, ranging from 1 to 40 ounces for general measurements, or from 0 to about 1.05 ounces (30 grams) for the gaging of soft materials or components with very thin cross-sections. These low gaging pressures are feasible because of the practically frictionless operation of the twisted band gages.

A very high degree of accuracy characterizes this system of indicator instruments. Most models are accurate to ±1%, and even for the models with the highest degree of amplification the error range does not exceed ±1½% of the pointer indication. Because there are practically no wearing elements in the gage, the original accuracy can be maintained during several years of continued service.

Although the design variables discussed above provide a wide range of potential uses, the inherent form and the boundary dimensions of indicator instruments operating on the twisted band principle could prove to be a limiting factor in applications where the gage must operate in a restricted area.

Mikrokator instruments are used for external and internal indicator hand gages, and also as test indicators. However, the most common application is in bench comparator gages (Fig. 5-59), where the full potentials of consistent accuracy at high amplification can be utilized with a minimum of interference from extraneous conditions that are unavoidable in free-hand gaging operations.

Floating Lever Indicators—The Sigmatic Gages

The floating level indicator, or sigmatic gage, refers to a purely mechanical gage (see Fig. 5-60) with

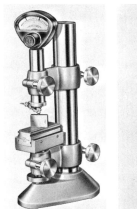

CEJ Gage Co.
Pratt & Whitney Cutting Tool and Gage Co. Div., Colt Industries

Fig. 5-59. (Left) Mikrokator indicator mounted on a comparator stand equipped with elevating nut for the fine adjustment of the staging table.

Fig. 5-60. (Right) General view of the "Sigmatic" mechanical comparator.

high amplification potentials, which utilizes the parallel displacement properties of the reed suspension[1] and the positive fulcrum position obtained by employing steel strip hinges. The linear displacements of the reed suspended gaging member are transformed into swing movements of a hinged assembly. The resulting angular displacement is then amplified by a double leverage system, permitting amplification rates that can be selected from 300× to 5000×.

Figure 5-61 illustrates the operating principles of the gage with the reed suspended floating block (1, 2 and 3) carrying the gage spindle (5). The displacements of the suspended block are transmitted through the adjustable wedge (6) to the hinged console (4), which constitutes the shorter arm of a lever system. By adjusting the distance (14), which separates the fulcrum line from the line of action of the wedge, the ratio between the linear displacement and the angular swing can be varied. The other arm of the hinged lever is a forked member (7) whose extremity describes an arc having several times the length of the original displacement. This movement is further amplified by a transmission system consisting of a ribbon (8) attached to the fork ends and wound around a small drum that is set into rotation

---

[1] A description of reed flexure linkages is given later under the heading of "Optically Assisted Mechanical Comparators."

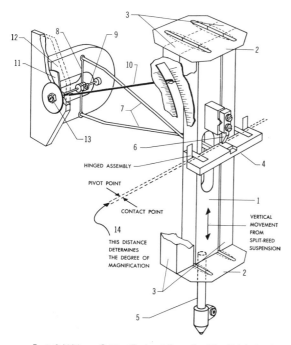

HINGED ASSEMBLY

PIVOT POINT

CONTACT POINT

14

THIS DISTANCE
DETERMINES
THE DEGREE OF
MAGNIFICATION

VERTICAL
MOVEMENT
FROM
SPLIT-REED
SUSPENSION

Pratt & Whitney Cutting Tool and Gage Co. Div., Colt Industries

Fig. 5-61. Diagram illustrating the functional elements and the operational principles of the "Sigmatic" comparator.

MECHANICAL INDICATORS WITH ELECTRICAL LIMIT-POSITION SIGNALING

This category of gages covers measuring instruments that, although operating essentially as mechanical displacement amplifying devices, differ from the conventional indicators by having a lever with electrical contact points that are either

(a) Substituted for the regular pointer and scale (Fig. 5-62); or

(b) Equipped with the electrical contact lever as an additional element, still retaining the regular sector scale and pointer (Fig. 5-63).

While the plain electrical contact type sensing heads produce signals only at the end positions of a size range; which must be established in setting the gage, the indicator type signaling instruments also permit observation and measurement of the intermediate sizes of the object.

The mating contacts of the contact lever device are adjustable in order to close the respective circuits when one of the tolerance limits of the object has been reached or passed during the measuring process. Some manufacturers claim a discriminating sensitivity of 10 microinches for the mechanical sensors operating as limit switches and producing the

Mahr Gage Co.

Fig. 5-62. (Left) Mechanical sensing heat with adjustable electrical contacts for limit-position signaling.

Fig. 5-63. (Right) Sector type, mechanical indicator complemented with electrical limit-position signaling contacts acting in coincidence with the adjustable tolerance markers. The light box indicates the dimensional zone into which the actual size of the inspected part fits.

by the swing movement of the fork. The drum spindle *(9)* carries the long pointer *(10)*, which indicates along a sector scale the highly amplified version of the sensed displacement. The pointer spindle bearing *(13)* and the members of the magnetic rotation stabilizer *(11, 12)* complement the list of essential elements in the Sigmatic gage.

The amplifying mechanism of the Sigmatic gage is adaptable for gaging multiple dimensions by mounting several basic mechanisms into a common assembly arranged to have contacts with the critical dimensions of the object. In this application the actual size of the different dimensions is not displayed, the inspection having the sole purpose of determining the actual conditions of the gaged dimensions in relation to the tolerance limits. The gaged size conditions of each individual dimension can be announced on a single panel by lights indicating "OK," "oversize" or "undersize." A special type of Sigmatic multiple dimension gage uses air columns for displaying the conditions of several dimensions on the panel, but even in this version of the instrument, the actual gaging, as well as the basic amplification are purely mechanical functions.

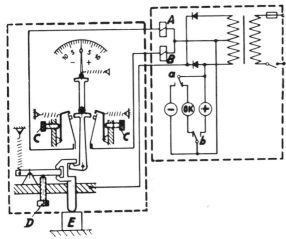

Mahr Gage Co.

Fig. 5-64. Schematic circuit of an electrically signaling mechanical indicator. *(A and B)* relays; *(C)* adjusting screw for the electric contacts; *(D)* lifting screw; *(E)* gage block for zero setting.

positive light signals. These characteristics of the gages result in an inspection tool whose performance can hardly be duplicated by plain mechanical comparators even of the high sensitivity category.

When using sensing heads without indicator scale, the two limit positions of the gage must be set with the aid of two masters or gage block stacks, representing the limit sizes. Indicator instruments equipped with *complementary* electrical contacts, on the other hand, require only a single master for setting, usually, but not necessarily, corresponding to the nominal size of the part. Following this initial setting, the tolerance hands of the indicator are brought to the desired limit positions guided by the indicator's scale graduations. This adjustment will also set the mating contacts of the electrical switch lever, and signals will be produced when the pointer of the indicator traverses the tolerance marks.

Figure 5-64 shows a simplified circuit diagram of a mechanical indicator with electrical components, which combines the potentials of scale indication with electrical limit-position signaling. The electrical signals originating from the gage head at the set limit positions can also be utilized for several additional mechanical functions, such as recording, counting and sorting.

OPTICALLY ASSISTED MECHANICAL COMPARATORS

Instruments assigned to this category utilize the leverage produced by a light beam projected on a

distant screen to complement the measuring displacements that were detected and then initially amplified by mechanical members. It may be a question of interpretation whether this type of measuring instrument should be classified as an optical device, or as a special category of mechanical comparator.

Classification as a mechanical comparator may be justified by the consideration that the light beam is used in these instruments to perform a function essentially identical with that of a mechanical lever. On the other hand, in the air and electronic gages, classified as independent categories and discussed in separate chapters, physical phenomena, such as the flow or pressure of air or an electric current, are subjected to variations as a function of the size differences in the measured object. These influenced physical phenomena then produce analogous signals that are proportional to the size conditions of the object being measured. Finally, the modulated signals are reconverted into a geometric form, such as the travel of a pointer along a scale, or the movements of a pen on the ruled strip of a recorder, displaying the proportional, but highly amplified, versions of the original displacements.

The Sheffield Visual Gage

The Sheffield Visual Gage is a measuring instrument that uses the combined amplification potential of mechanical linkages and of a light beam. Although not a recent design, the "visual gage" is a well proven comparator instrument that is still widely preferred because of its valuable properties. Functionally, the gage is not dependent on power whose fluctuations could affect the measuring accuracy, and it uses electricity only for the lamp that produces the light beam. Figure 5-65 shows the general view of one form of execution, and Figure 5-66 illustrates diagrammatically the mechanical amplification system of the gage.

The operation of the essential mechanical elements of the gage—the optical part serving as a supplementary amplification step only—rests on certain basic characteristics of steel strip flexure linkages, commonly known as *reeds*. The distinctive properties of reeds, which also are used extensively in other types of measuring instruments are shown diagrammatically in Fig. 5-67, and can be summarized as follows:

a. Two parallel blocks connected by two, or any even number of steel strips of equal size and stiffness, form a reed type flexure linkage. When a force normal to the connecting strips is acting on one of

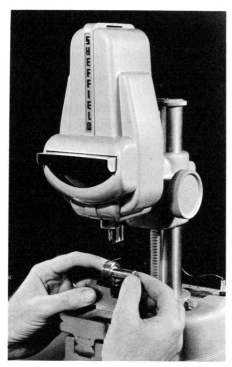

Giddings & Lewis Measurement Systems

Fig. 5-65. General view of the "Sheffield Visual Gage" in use for measuring the pitch diameter of a thread gage over wires.

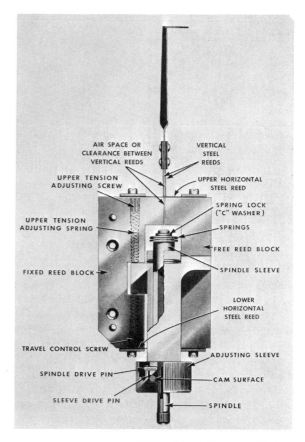

Giddings & Lewis Measurement Systems

Fig. 5-66. Cross-sectional diagram of the Sheffield Visual Gage, listing the major functional members.

the blocks, while the other block remains firmly anchored, the connecting strips will flex, resulting in the displacement of the movable block. However, the mutually parallel position of both blocks will remain unchanged, in a manner comparable to two opposite sides of a parallelogram (Fig. 5-67A).

b. Two steel strips, parallel but kept apart to maintain an air gap, are joined at one end, while the separated ends are attached to two individual blocks (Fig. 5-67). When the blocks are mutually displaced in the direction of the strip axes, the parallel strips will try to slip past each other. The joint at the farther end prevents this natural movement, and the whole strip assembly bends in a direction that depends on whether the movable block exerts a push or a pull on its attached strip. For small magnitudes of displacement, which is a rule for fine measurements, the bending of the strip assembly may be considered proportional to the distance over which the blocks were mutually displaced.

When an extension arm is attached to the joint of the strips, the deflection can result in a swing over an arc of rather extended length, which is a multiple

of the original displacement of the base blocks. This relationship is the essence of the mechanical amplification produced by the visual gage.

In addition to this initial amplification, which is produced by mechanical means, a further amplification is provided by the optical system of the gage. This latter consists of a light source with lenses that produce a collimated light beam. The extension of the flexing parallel strips carries a target that falls into the optical path of the light beam. The shadow of the target will appear on the graduated glass scale, located at a distance in order to result in an optical lever of the required ratio.

The combined amplification of the mechanical reed system and of the superimposed optical lever varies from 500:1 to 10,000:1, depending on the gage model.

Sturdy construction and great flexibility of adaptation distinguish the visual gage. The almost fric-

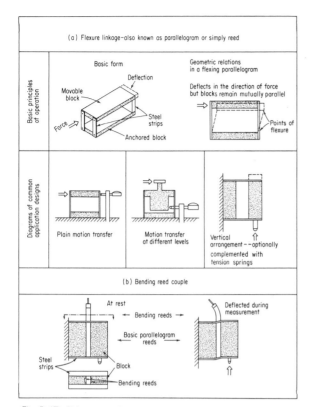

Fig. 5-67. Principles of reed type, motion transfer devices.

tionless reed mechanism is not subjected to wear even after years of continued service and the original accuracy is not affected by long usage. While certain types of manufactured parts may require a lesser gaging pressure than found in the visual gage, the positive action of the reed spring urged gage spindle is of advantage for the majority of length measurements that are needed in general production.

Tilting Mirror Type Optical Comparators

Figure 5-68 illustrates a length measuring comparator in which only the initial stage of amplification is carried out by mechanical means; the major amplification step is accomplished by an optical system.

The initial stage consists of a short lever carrying a mirror, which is tilted when the measuring plunger is acting on the lever. The amplification rate of this stage is only 4:1.

An additional 250× (or 500×) amplification is obtained by purely optical means through a front surface mirror reflecting onto the graduated screen the illuminated target mark, which is produced on

the tilting mirror by a light projection system. Tolerance limit marks on the screen can be adjusted by knurled head screws.

A heavy base, interchangeable anvils, adjustable back stop and height adjustment on the threaded column by a knurled collar complement the equipment of the comparator gage.

The predominant use of optical means in the amplification system has several distinct advantages. High total amplification, in this example 1000× and 2000×, can be obtained by relatively few elements. Other instruments, operating on similar basic principles, can provide even higher rates of amplification, although usually at the price of a comparable reduction of the measuring range. The working accuracy of the tilting mirror type optical comparator is usually better than $\pm\frac{1}{2}\%$ of the total measuring range. The only moving member of the mechanism being the plunger with the tilting mirror lever, very little wear may be expected and the original accuracy of the instrument is retained through protracted use.

Figure 5-69 shows the actual instrument selected to illustrate the principle. This particular type of tilting mirror mechano-optical comparator is available in two models with measuring ranges of ± 0.002750 or ± 0.001400 inch, displayed on a graduated screen of over 5 inches in length. The minimum graduations of the screen represent 50 or 25 microinches, respectively, depending on the particular model's rate of amplification.

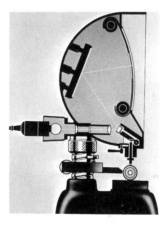
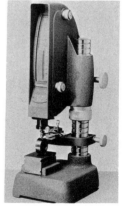

Leitz/Opto-Metric Tools, Inc.

Fig. 5-68. (Left) Diagrammatic cross-sectional view of a tilting mirror type, mechano-optical comparator, illustrating the operating principles.

Fig. 5-69. (Right) The Leitz "Tolerator," a mechanical comparator with optical amplification and display.

6.
Pneumatic Gaging

THE OPERATING PRINCIPLES AND GAGE SYSTEMS

General Definition

The operation of air gages (the tools of pneumatic gaging) is based on phenomena that occur when pressurized air, while escaping through an orifice of controlled size, is impeded in its free flow by impinging against a solid surface. The distance of that surface from the air escape opening will affect the air flow velocity and create back pressure upstream of the orifice. By maintaining, at a constant level, the pressure and the volume of the air that is introduced into the system, variations in the upstream air will be a function of the obstruction to the air escape. To create an effective obstruction, the solid surface facing the orifice must be located close enough to actually reduce the area of air escape.

Within a specific dimensional range, which depends on the design characteristics of the gage, variations in the air flow and pressure will be almost proportional to the changes in the distance between the office and the obstruction.

In actual gage applications the orifices are the bores of nozzles mounted into appropriate gage bodies, which keep the air escape openings at a positive distance from the object wall, when the dimension being measured is at nominal size. Variations in the actual size of the part will modify the distance between the orifice and the obstruction, changing the velocity and the pressure of the air in the system. The changes caused by differences in the escape rate of the air can be displayed on the scale of an appropriate indicator instrument. The display usually represents a substantial magnification of the actual variations in the distance between the orifice and the object surface.

The Operational Systems of Commonly Used Types of Air Gages

Although the operation of all air gages is based on common principles regarding the conversion of distance variations into changes of gaging air characteristics, the gage systems that have been developed differ in many respects.

Table 6-1 compares the essential operational characteristics of frequently used systems of air gages. The listing follows a classification by the physical phenomena on which the operation of particular types of air gages is based, with a brief description of the applied conversion and amplification process complementing the selected system designations.

The actual operation of the different systems of air gages, as well as a few of their functional characteristics, will be discussed later in this chapter in a more detailed manner.

The Applicational Advantages of Pneumatic Gaging

The potential advantages of pneumatic gaging are clearly indicated by the fact that, although the majority of air gage systems in current use were developed only during the past three or four decades, the gaging of dimensional variations with the aid of pressurized air is now widely accepted throughout the entire metalworking industry.

Air gages are particularly preferred for measurements that are repetitive or are carried out under conditions that, in general, are not favorable to sensitive measurements. Typical examples for the latter are the in-process gaging applications. However, the use of air gages is not limited to said fields of preferential choice. Actually, there are few general areas

TABLE 6-1.   SYSTEMS OF AIR GAGES — OPERATIONAL PRINCIPLES OF DIFFERENT CIRCUITS

UTILIZED PROPERTY OF PRESSURIZED AIR	MEASURING INSTRUMENT SYSTEM AND SIMPLIFIED SCHEMATIC SKETCH	BASIC PRINCIPLES OF OPERATION
Flow (Velocity)	Air flow gage with rotameter tube	A glass tube with tapered bore, mounted over a graduated scale, serves as indicator. Inside the bore a float is lifted by the air flow. Unrestricted flow keeps the float in uppermost position, restricted flow causes it to sink lower along the tube bore.
	Velocity type air gage with Venturi chamber	Velocity change in the air flow when passing through the venturi chamber generates a basic pressure differential. Obstruction to the escape of the air through the gage head causes velocity variations which are indicated by the gage.
Pressure (Back Pressure)	Water column back pressure air gage	Uses the self-balancing properties of communicating vessels, the constant pressure being maintained by the height of a water column. The level of the liquid in the graduated gage tube indicates the amount of obstruction which is facing the escape jets in the gage head.
	Directly indicating Bourdon tube type pressure gage	Obstruction to the air escape through the gage head results in back pressure which will deflect the Bourdon tube of the indicator. The gage is calibrated to indicate length variations with the selected rate of amplification.
	Air gage with variable amplification	This system contains two control valves in the circuit and permits adjustment of the rate of amplification over a wide range. The second control device which operates as an adjustable bleed valve supplements the function of the basic restriction adjustment.
	Differential type air gage with fixed amplification	Non-adjustable matching metering restrictions of parallel channels provide a reference pressure which permits the use of a single setting master. Different amplification can be accomplished by exchanging the restrictions.

of comparative dimensional measurements where air gages would not be applicable, or even definitely warranted.

Air gaging, of course, is limited to a particular field of comparative measurements where the range of dimensional variations is relatively small. Considering, however, that the largest number of dimensional measurements are required in continuous or mass production and these are, with very few exceptions, of the comparative type, the potential field of air gage applications can easily be visualized.

Table 6-2 compares the various measuring instrument characteristics that are commonly considered when selecting the gaging system best suited for a particular application. The listing is limited, for topical purposes, to requirement categories that can be satisfied successfully with air gages of appropriate design, whereas aspects in which other systems are, generally, superior have not been included. A rigid delineation of suitability is, of course, not feasible because exceptions may be readily found depending on application circumstances that could not be evaluated in a general comparison.

A few other pertinent properties of air gages need to be discussed to complement the information contained in Table 6-2. These are mainly characteristics of air gages that could limit their applicability under specific circumstances.

The range of size variations that can be measured with air gages is functionally limited. The order of magnitude of that range, when using open air jets, is about 0.003 inch, although the actual size spread may vary depending on the design of the particular gage. Substantially wider ranges, about 0.020 inch or even larger, can be covered when using contact type sensing members.

Air gages of the basic open jet type operate by referencing the dimension to be measured from an area on the object surface, as distinct from the theoretically single point referencing that characterizes most of the contact type gages. Consequently, air gage indications are sensitive to the geometry of the reference area, and will react differently, for example, on a curved than on a flat surface. Because of this property of air gaging, the surface geometry of the air gage setting masters must be essentially identical with the surface of the object to be gaged. As an example, when measuring the diameters of internal or external cylindrical surfaces with open jet air gages, gage block stacks cannot be used as substitutes for ring or plug setting gages, although use of the latter is considered a standard procedure for contact type comparator instruments.

Direct air gaging with open jets—as contrasted to indirect air gaging with mechanical contact elements—is generally not adaptable to surfaces with closely spaced discontinuities. Examples of these discontinuities are grooved, serrated or threaded surface areas or parts made of porous materials.

While gages that operate by establishing mechanical contact with the object surface will sense the highest points on a rough surface (the maximum material condition), the response of the air gage will reflect the obstruction that the surface presents to the free flow of the air escaping through the gage nozzles. A rough surface will permit some air to escape through the valleys between the ridges of its surface. Consequently, the resulting air gage indications will reflect a mean level of the surface and not its envelope level, which is being measured when mechanical contact gages are used. Considering that for most assembly purposes it is the maximum material condition that determines the fit relation between mating parts, this averaging property of open jet air gages could prevent their use for parts with relatively rough surfaces that must be inspected for closely toleranced fits. As a general rule, based on an approximate evaluation of the resulting conditions, the following limit values may be considered for open jet gage applications:

For a Size Tolerance Range of (inch)	Maximum Permissible Roughness of Surface, AA (microinches)
0.0002 to 0.0003	
>0.0003 to <0.0010	10 to 15
0.001	32
	50

There are certain applications where the property of air gages to indicate the mean surface level, instead of the envelope, is considered desirable. In this context, it may be of interest to note that there are pneumatic instruments for surface roughness measurement whose operation is actually based on that sensitivity of air gages to the roughness conditions of the gaged surface.

Later in this chapter the indirectly sensing or contact type air gage will be discussed. Because air gages of this particular type operate by establishing mechanical contact with the surface to be measured, the limitations of application due to conditions of surface geometry and texture in open jet air gaging do not have to be considered. However, for the majority of air gage uses, the open jet type is preferred. Thus, some of the advantages as listed in Table 6-2 apply only to air gages of the open jet type.

TABLE 6-2.   APPLICATION ADVANTAGES OF PNEUMATIC GAGING — 1

	DESCRIPTION	EXAMPLES OF THE BENEFICIAL EFFECT ON THE GAGING PROCESS
Noncontact	The basic open jet air gages operate with a clearance between the object surface and the sensing member of the gage.	Air gages will not mar the object surface. They cause no penetration into the surface even when gaging soft materials. The contact surface of the gage head is not subjected to wear. Air gages can be used to measure the sizes of objects with abrasive surfaces, even while in movement.
No sliding members	The operation of air gages does not involve sliding surfaces at all in the case of the direct sensing type, and to a limited extent in some if the indirectly sensing types. Indications result from air supported floats or from the flexing of indicator members.	Frictional forces will not interfere with the conversion of the sensed displacements into proportional changes in the air flow. Wear will not develop and cannot affect the useful life or the original accuracy of the gage.
Self-cleaning of the target area on the object surface	The pressure of the open air jet impinging on the target area of the size sensing clears momentarily that portion of the object surface from loose foreign substances.	Direct air gaging, because of the cleaning action of the open air jet, is much less sensitive to dust and dirt particles, grit and grinding fluid, which occur under shop conditions, than any of the contact type gaging systems.
Small overall size of the sensing members	Air gage heads which contain the air nozzles are available commercially in sizes which are commonly much smaller than those of the sensing members of most other systems of gages.	The small outside dimensions of the essential sensing element, the air nozzle, permits measuring the size variations on hard-to-reach surfaces, or to use gages with several sensing points located close to each other.
Self-aligning and -centering inside bores	The small clearance between the gage head (air plug) and the bore wall, combined with the balancing effect of multiple jets, practically eliminate the effect of alignment and centralizing errors.	Air gages can be operated by unskilled personnel, or in automatic processes, without affecting the inherent accuracy of the system. This property of air gages is particularly valuable in the otherwise critical measurement of bore sizes.

TABLE 6-2.   APPLICATION ADVANTAGES OF PNEUMATIC GAGING – 2

CHARACTERISTIC	DESCRIPTION	EXAMPLES OF THE BENEFICIAL EFFECT ON THE GAGING PROCESS
Wide range of amplifications	Air gages can operate with amplifications ranging from about 250 times, up to about 20,000 times, or even higher.	Air gages can be adapted to a wide range of different work tolerance zones, frequently using the same basic amplifier unit and exchanging only the restriction elements and the indicator scales.
Remote observation	The gage heads which house the air nozzles are self-contained elements, connected by a thin flexible hose to the amplifier-indicator unit.	While the gage head must be placed in a position adapted to the configuration of the object, the remote location of the amplifier and of the indicator protects the more sensitive elements of the system, and permits a convenient, as well as precise observation of the gage indications.
Continuous indication	Air gages will indicate, for most designs with very little delay, the momentary position of the object surface relative to the sensing member.	The response speed of air gages to the variations of the dimension to be measured can be increased to permit measurements on moving objects (rotational or linear) or on parts which can dwell for a very short time only in the gaging position, a condition which is common in automated gaging processes.
Averaging and combining capabilities	The air channels which terminate in the air jet nozzles can be combined into a common passageway which will indicate on the dial a single size resulting from the aggregate amount of the air escaping through several nozzles.	Technical bodies can be inspected for size and for regularity of form by checking simultaneously several points on their surface. Air gages can also be used to indicate the computed value of size variations which were detected by interrelated air circuits.
Adaptability to multiple dimension gaging	The compactness of the indicator element of air gages permits the mounting of several indicators adjacent to each other, to display for quick scanning the size conditions of multiple dimensions.	The inspection process of parts with several toleranced dimensions takes substantially less time when a number of dimensions can be checked simultaneously by means of a synoptical display on adjacently mounted indicator scales, using identically adjusted zero points for the nominal size conditions, and equipped with individual tolerance limit markers.

TABLE 6-2.　APPLICATION ADVANTAGES OF PNEUMATIC GAGING – 3

CHARACTERISTIC	DESCRIPTION	EXAMPLES OF THE BENEFICIAL EFFECT ON THE GAGING PROCESS
Initiating potentials	The analog variations of the air pressure inside the instrument can be used either directly, or through a relay, to initiate and to sustain desired processes.	Recording the measured sizes, calibrating the preceding gages in a cascade type gage arrangement, signaling by lights specific size conditions, feeding back information on dimensional variations to actuate size control adjustments on the machine tool element, are examples of actions and functions which can be initiated by properly equipped air gages.
Economic advantages	Air gages have a relatively low initial cost, are inexpensive to maintain and are virtually free of wear. They also have a flexibility of adaptation that permits a high rate of instrument utilization.	The basically simple design of air gages and the possibility of flexible adjustments for size and calibration result in lower initial cost than for most other gage systems of comparable sensitivity. The absence of friction in the instrument and of the contact members during the gaging process eliminates wear and prolongs the useful life of the gage.

THE PRINCIPAL ELEMENTS AND OPERATION OF PNEUMATIC GAGING SYSTEMS

For the comparative measurement of dimensional variations by pneumatic gaging, a system consisting of several basic elements is needed. The design and manner of operation of the actual instrument may vary, but no pneumatic gaging system can operate without comprising each of the following basic elements, which are also indicated on the schematic drawing in Fig. 6-1: pressurized air; pressure reducing and regulating valve; metering device (restriction); indicator; gage head or sensing member; and setting master.

The above-mentioned elements are now considered in greater detail. The operation of any pneumatic gaging system must rely on a continuous supply of pressurized air, with a minimum pressure of about 30 psi (pounds per square inch). The actual minimum pressure and the volume of air required will depend on the general design of the instrument, as well as the number of and orifice area of the air escape nozzles. As a guide for approximate estimates, it may be considered that a single nozzle with

about a 0.040-inch bore diameter will consume a minimum of about 10 cubic feet of air per hour.

It is mandatory that the air received from the shop line for the purpose of pneumatic gaging should be reasonably clean, with a minimum of humidity and free of oil. Unless these conditions are assured, it is to be expected that solid impurities and excessive liquid will soon clog the fine filters of the air gage unit, cutting down their effectiveness.

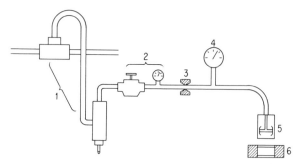

Fig. 6-1. Principal elements of pneumatic gaging systems: *(1)* continuous supply of pressurized air; *(2)* pressure reducing valve; *(3)* metering device (controlled obstruction); *(4)* indicator (pressure or flow rate); *(5)* gage head; and *(6)* setting master.

The actual gage unit should be preceded by an individual moisture separation and filtering device, to extract from the air the last remnants of solid particles, and to separate effectively the moisture and oil that may have passed the general cleaning system of the shop line. Obstruction of the narrow passages in the metering elements of the air gage unit by condensed moisture or entrapped solids is the most frequent source of defective, erratic or inaccurate operation of air gages.

The pressure regulator is a precision valve that may be independently installed in the system as a part of the gage control unit. Most systems comprise a pressure gage immediately following or attached to the pressure regulator, as a means for the dependable control of the system's air pressure.

The actual design of the control unit will vary for different makes and models and will be governed by the type of pneumatic gaging system in which it operates (see Table 6-1). In addition to the basic functional elements, the control units of air gages contain complementary devices such as adjusting and bleed valves, as required by the particular system. Because the various air gaging systems that are in current use differ in several essential aspects of operation and performance, the air gage control unit will be discussed in detail in the following section.

The sensing part of the pneumatic gaging system has for its essential element the air escape orifices, commonly provided by the controlled size bores of the nozzles (some manufacturers designate these elements as "jets"). For protection and for locating during gaging in the proper reference position, the nozzles are mounted either into portable gage heads or on stationary gages. The nozzles in the gage head can eject the air directly against the object surface, operating as an open jet, or against an intermediate gage element, which will be in mechanical contact with the object surface.

Functionally complementary to the sensing members are the setting masters, although they are not actually operating elements of the pneumatic gaging system.

Both the design aspects of the sensing members and the great variety of types and sizes in which these air gage elements must be made to accommodate the diverse measuring problems warrant a more detailed discussion in subsequent sections of this chapter.

The Speed of Response in Pneumatic Gaging

In the majority of air gage applications the gage head is located at some distance from the control unit. Due to this distance, and also because of the compressibility of the air, the effect of the variations in the gaging position does not reach the control unit instantaneously. While size variations of the object will promptly affect the air flow at the nozzles, further upstream these changes will manifest themselves only after some interval. The time gap between the sensing and the indications will depend on various factors, such as the length of the air line between the nozzles and the indicator, the type of gage system and the design of the control unit.

Flow type air gages have a relatively quick response, but their use is essentially limited to applications where only visual observation of the indications is required. Because these indications result from the position of the float that is enclosed in a tube, no actuating performance can be derived from it. This property practically excludes the use of flow type air gages when the gage must initiate certain functions related to the object size, such as the emission of light or sound signals, the actuation of sorting machine elements, the recording of the measured sizes, the control of machine tool movements and so forth.

In air gages that operate by indicating back pressure, the response is slower because the compressibility of the air also contributes to the delayed transmission of the variations that originated at the object surface. This delay is further increased by intermittent gaging when, due to temporarily unrestricted orifices, the air pressure in the system is permitted to fall much below the indicating range.

In commonly used air gages with pressure type indicators, the time that elapses from the sensing of the size variations to the appearance of the full indication is usually on the order of about one second and may be quite acceptable. Actually there are some, although rare, examples of continuous type measurements where the short-term variations are not meaningful and the smoothing out of the higher frequency "back-hash" of the indications is considered useful.

In various other gaging applications, however, the lagging indication may prove objectionable. Examples of these types of applications are: in-process gages operating on grinding machines; segregating machines and other automatic gages with short dwell of the part in the gaging position; and detection of size variations of fast moving objects.

Various devices have been developed and are commercially available to increase the effective response speed of back-pressure type air gages. A few of these devices are illustrated by indicating their functions:

a. Filling the bourdon tube of the pressure gage with a virtually incompressible liquid to reduce the volume of air whose back pressure must develop to obtain corresponding gage indications;

6. Restricting the unimpeded escape of the air through the orifices when the gaging head is not in operation, by using a spring-urged cover sleeve around the gage head;

c. Counteracting the unrestricted air escape by an auxiliary air supply relay whose operation automatically discontinues as soon as a specific back pressure develops during the actual gaging process; and

d. Using a high speed relay to compensate for additions to the volume of the instrument system; such increase of the overall volume will result when air-actuated supplementary devices, for example, electro-pneumatic switches, are added to the circuit.

Zero Setting and Amplification Adjustment

Most types of air gage control units are equipped with means for effecting various adjustments. Two of these adjustments deserve special mention: the zero setting and the amplification adjustment, frequently termed calibration.

The zero setting is commonly accomplished by means of a bleed valve. It is common practice in checking toleranced dimensions to establish on the gage a scale range bounded by two limit positions. The zero setting consists of adjusting the indicating element of the gage (the pointer, the float or the water level) to that marking on the scale that was selected to signal coincidence with the nominal limit size represented by a setting master. For one of these limit points the zero marking of the scale is frequently selected, hence the designation "zero setting" for this kind of gage adjustment.

The amplification adjustment for the purpose of calibration is carried out with another precision valve of the control unit. The purpose of this adjustment is to obtain a range of gage indications that, in terms of the scale graduations, are in correspondence with the size span of the setting gages. The spread of the setting positions is usually, but not necessarily, identical with the tolerance range of the dimension being inspected.

Once set, the two adjustments can be expected to stay constant for a period of time, the length of which will depend on the design and condition of the instrument. Evaluation of the setting stability is based on that deviation in the indications that is considered still acceptable for the measurement

being carried out. It is a good practice to have the setting masters of air gages readily available for making periodic checks during the inspection process. Such checks may be limited to the zeroing but, if necessary, should also comprise the amplification adjustment. The frequency of these checks is usually established by experience with the purpose of providing a reasonable safeguard against gaging errors due to false indications.

Adjustable, Variable and Constant Amplification

While means for amplification adjustment are generally part of most air gage control units, there are systems that obviate the need for but also exclude the potential for adjusting the amplification of the instrument.

A contrasting objective is accomplished by a system of air gage control units that permits amplification variations to a degree of actually selecting widely different rates. The designation *variable amplification* will be used to distinguish the capacity of actually changing the amplification from the adjustment for calibration purposes.

The manner in which the objectives of fixed or flexible amplification are accomplished and their effect on the *design* of various gage systems will be discussed in the following section dealing with the control units of air gages. Here, the different capabilities will be evaluated from the viewpoint of gage *application*.

Variable amplification can be made available by two essentially different design approaches: by using exchangeable metering elements in the control unit, or by providing a very extended range of adjustment, substantially in excess of that needed for calibration. Whichever method is used, provision must be made for the convenient exchange of the indicator scales to make them correspond to the selected amplification.

The capacity of variable amplification is useful in plants where the same gage must serve, in turn, various measuring tasks that require different measuring ranges. A greater flexibility of instrument usage can be derived if there is a selection of amplification, or of its corollary, the gaging range. However, when an air gage is intended to continuously serve a single type of measurement only, no benefits result from the additional capacity of variable amplification.

The primary objective of designs that assure constant amplification is to dispense with the use of two setting masters for the same nominal size. Air gages having constant amplification can be set with a single master gage by zeroing on a specific mark on the scale. Once set, the indications of the gage will represent size variations that correspond to the values of the scale graduation, without the need for gage calibration. Because of the fixed amplification, the setting remains constant, whereas on the adjustable types of control units a periodic recalibration is needed to compensate for drift or adjustment instability.

In comparison, proponents of variable amplification systems claim that not even an excellently engineered, constant amplification system could be equivalent in accuracy and reliability to a precisely conducted and periodically repeated calibration process.

THE PNEUMATIC CONTROL UNITS

While the basic operating principles of all pneumatic gages are essentially common—converting distance variations into proportional changes of air pressure and of flow velocity—the methods of effecting this conversion and of indicating the resulting changes in the physical conditions of the gaging air, will widely differ, depending on the system of any particular type of air gage.

As indicated in Table 6-1, there are several currently used air gage systems, each possessing its own characteristic properties. Most of these different systems are associated with the distinctive air gage products of prominent manufacturers. While general superiority may not be claimed by any of these systems, the suitability of each will differ depending on the conditions and on the objectives of any particular gaging problem.

The difference in suitability as a function of the gaging condition warrants a thorough evaluation by the prospective user when making his choice of any particular system of pneumatic gages. Although economic factors, such as initial and operating costs, as well as several other, not primarily technical, conditions will weigh heavily when making the selection, the understanding of the essential functional differences between the various systems can prove to be a useful asset.

For this reason descriptions of the more widely used systems of air gages, as incorporated in the respective control units, are presented in this section.

Flow Type Pneumatic Gaging Systems

Pneumatic gages of the flow type system operate by sensing and indicating on the scale of the control

unit the momentary *rate of air flow*. Air gages utilizing the *changes in the flow characteristics of air* are also classified as belonging to the velocity system of pneumatic gages.

The actual design of air gages that operate by measuring the rate of air flow can differ; one of the best known designs will be discussed in detail.

In the system shown in Fig. 6-2, air from the plant supply line, after passing through a filter and a pressure regulating valve, is introduced into the bottom of the vertically arranged indicator through an adjustable restriction, by which the instrument can be calibrated. The flow indicator proper consists of a glass tube with tapered bore (also called a *developed tube* or a *rotameter tube*) that contains a freely moving float. The air passing through the rotameter tube is channeled, by means of plastic tubing, into the sensing member of the gage whose orifices permit

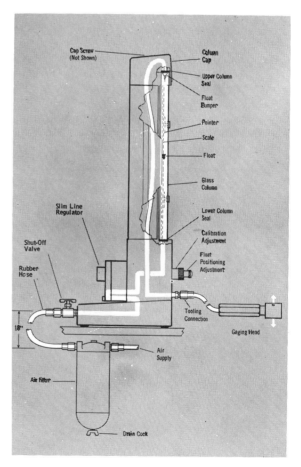

Giddings & Lewis Measurement Systems

Fig. 6-2. Operating elements of a flow type pneumatic gaging system.

the pressurized air to exit into the atmosphere. An adjusting screw at the top of the rotameter tube serves for positioning, or zeroing, the float during the setting of the instrument by means of a setting master.

Figure 6-3 shows an actual instrument that operates by the system of "rate of flow" measurement. In addition to the functional elements of the amplifying and indicating circuits with their auxiliary adjusting devices, the illustrated instrument also contains a special pressure regulator valve, to assure the delivery of air at a constant pressure for the operation of the gage. The rate of amplification by which the gage displays the sensed size variations of the object is primarily governed by the taper inside the glass tube. By exchanging the rotameter tube and the scale, the rate of amplification can be changed, making the instrument adaptable for several different tolerance ranges. Commonly used amplifications for flow type air gages range from 1,000 times to 20,000 times, and for special uses, even higher.

When a flow type air gage is to be used under conditions that require the amplification to be changed frequently, a special amplifier unit can be attached to the illustrated type of instrument. Gages equipped with this accessory obviate the exchange of tubes when changing over to a different amplification; only the scale must be replaced to correspond to the selected amplification.

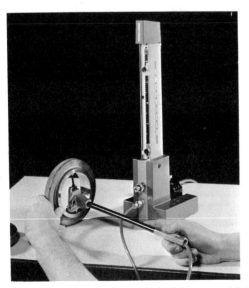

Giddings & Lewis Measurement Systems

Fig. 6-3. Flow type air gage control unit containing, in addition to the indicator column with the float, all the essential regulating and adjusting elements. The use of the central unit in conjunction with a bore gage head (air plug) is shown.

The operational process of the flow type pneumatic gage is explained in the following: when no obstruction hinders the flow of air through the nozzles in the gage head, the velocity of the air flow will hold the float inside the tube at its upmost position, mechanically restrained only by a bumper or stop. A partial restriction of the air escape through the sensing head's orifices will diminish the velocity of the air flowing through the developed tube. The air that is now flowing at a reduced velocity will be capable of supporting the float at a correspondingly lower level only and the calibrated scale behind the tube will indicate in inches the dimensional variation of the object that caused the changes in the rate of air flow.

The obstruction to the air escape is related to the material condition of the surface being gaged: the greater the material condition, the more the free flow of air will be restricted, resulting in a lower sustaining level for the indicator float. This response is expressed in the sense of the dimensional variation of the object as follows: greater material condition will reduce the actual size of an internal surface (for example, of a bore), but will increase the measured size of an external surface. When flow type air gages are used alternately for inside and outside measurements, it is advisable to use scales with markings to indicate the sense of object size variations that correspond to the changes in the float positions.

The essential simplicity of the system reflects on the initial cost of flow type air gages, and permits carrying out the maintenance of the instrument in the user's plant, without relying exclusively on the repair facilities of the gage manufacturer. Further advantages of this system are the quick indicating response, even in the case of extensively remote locations of the gage head and the indicator; the clear display of the indications on a long scale, easily observable from a distance of several feet; and the compact design that permits concentrating the complete air circuitry behind a slender base. By mounting several flow type control units in a common housing and installing the indicator tubes side by side at close intervals, measurements originating from several gaging points can be observed at a single glance (see Fig. 6-4).

While the distant viewing and the simultaneous observation of several gage columns are characteristic advantages of flow type air gages, they do not possess the capability of actuating auxiliary devices such as recorders and switches.

Flow type pneumatic gaging systems have also been touched by the electronic revolution in dimen-

Giddings & Lewis Measurement Systems

Fig. 6-4. Multiple-circuit control units of flow type air gages. The adjacent mounting of several columns on a common base permits the simultaneous observation of indications originating from different yet functionally interrelated gaging positions.

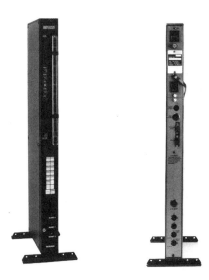

Giddings & Lewis Measurement Systems

Fig. 6-5. (Left) Air electronic column, front view.

Fig. 6-6. (Right) Air electronic column, back view.

sional measurement. An air/electronic control unit is now available (Fig. 6-5). The electronic column can be used in place of the traditional float type indicator column. The easily read liquid crystal display (LCD) digital and light-emitting dioxe (LED) analog displays are generated via a linear variable differential transformer (LVDT) and piezoelectric crystal transducers. Limit lights also aid the inspector and operator in both the speed and accuracy of reading indicated information. The electronic column accepts either flow or back-pressure type air gaging systems.

Figure 6-6 shows the back panel of an electronic column and reveals two important features. First, the 115 volt AC receptacle permits the stacking of columns into multiple-circuit control units for simultaneous observation. Second, the serial RS232 port permits interfacing of this unit to a personal computer for the purpose of data collection.

Velocity Differential Type Air Gages

Figure 6-7 shows schematically the arrangement of an air gaging system whose operation is based on measuring the changes in the velocity of air caused by varying the obstruction of the air escape. A venturi, a tube with two adjacent sections having different diameters, is used to convert the air velocity

changes within the system into minute pressure differentials. A sensitive flexible diaphragm will react instantaneously to the pressure variations and actuate, through an appropriate linkage, the pointer of a gage. The diaphragm, or a bellows, has an inside chamber that is connected to the entry section of the venturi, whereas the reduced diameter section of the venturi tube communicates with the enclosed space surrounding the bellows.

When the volume of the escaping air is reduced due to the proximity of the object surface to the nozzle orifice, the velocity of the air in the downstream section of the venturi will decrease, upsetting the balance of forces acting on the two opposite sides of the diaphragm. Depending on the sense of the resulting pressure variations, the bellows will expand or contract in a degree that is proportional to the size variations of the object causing that pressure differential. The amplified value of these size variations is then indicated by the pointer position over the calibrated scale of the gage dial.

Means of adjustment for the gage calibration and for the positioning of the pointer during the setting stage of the operation are integral parts of the system. The shifting of the pointer's position will not alter the amplification of the instrument, it being governed by the design dimensions of the venturi. When a different amplification is required, the venturi block of the unit must be exchanged, and the corresponding scale mounted on the indicator dial. This

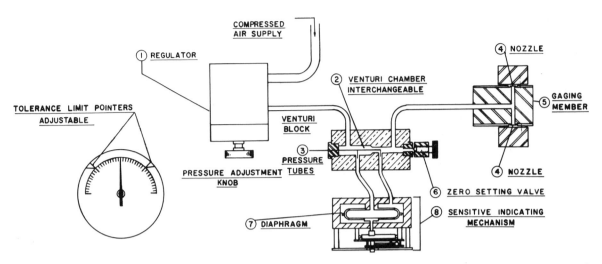

The Taft-Peirce Mfg. Co.

Fig. 6-7. Schematic illustration of the principal operating elements of a velocity type air gage using a venturi to produce proportional pressure variations.

is a relatively quick operation that can be carried out with the use of simple tools. For the setting and calibration of the instrument, two masters, representing maximum and minimum sizes, are required.

Figure 6-8 illustrates an air gage control unit that operates on the basis of the velocity differential principle, using a venturi tube. For gaging multiple dimensions, several of these units can be mounted into a common housing and the positions of the pointers observed simultaneously.

Air gaging systems that operate by sensing the velocity differential, although intended mainly for applications requiring visual observation only, also can be used in special control units that produce electrical signals or originate electrical impulses for control purposes. Air gages operating by the velocity differential system have several valuable properties,

The Taft-Peirce Mfg. Co.

Fig. 6-8. Control unit of a velocity type air gage.

such as a relatively quick response; operation with a rather large clearance between nozzle and object surface, thereby reducing the wear of the gaging members; and low air consumption.

The velocity differential type air gages are commonly used for measurements requiring magnifications from about 500 times to 5,000 times.

Pneumatic Gages with Air Pressure Controlled by Water Chamber

The water-chamber type pressure indicating air gage system represents the original concept of pneumatic gaging using a water chamber for balancing out the effects of line pressure variations. The system was originally developed by the manufacturers of Solex carburetors for inspecting the effective bore diameters of carburetor nozzles by direct air flow. The third illustration in Table 6-1, diagrammatically showing the operating principles of air gages, is based on the original version of the water-chamber type, air-pressure regulating system. Discussion of this system, however, is warranted, not only because of the historical background, but because the system is still regarded as one of the most sensitive and versatile systems with the added advantage of permitting the use of extended and distinctly readable scales.

The large water chamber used in the earlier designs made the instrument rather cumbersome and inconvenient to move; these drawbacks were eliminated and the basic advantages of the system retained. Figure 6-9 illustrates the operating principles

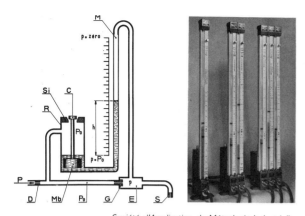

Société d'Application de Métrologie Industrielle

Fig. 6-9. (Left) Operating principles of a water-chamber type pressure indicating air gage.

Fig. 6-10. (Right) Liquid column air gages of the water-chamber system; single, double and triple unit types.

of an up-to-date design of the low pressure, water-chamber type air gage that differs only slightly from the high pressure versions of the same system.

Compressed air P is supplied to the air controller and is maintained rigorously at a constant pressure P_0. Some of that air is admitted into the pressure chamber E, and leaves it through the orifice S, which takes the form of one or more open jet nozzles, or of one of the numerous types of mechanical contact units. The cross-sectional area of G is constant, and the cross-sectional area of S varies according to the dimension of the component being measured. This causes variations in the pressure p in the pressure chamber E, and these variations are reflected by a manometer M, equipped with a scale for translating the variations in S into linear units.

The operating air pressure is maintained at a constant level by means of the pressure regulator, consisting of a chamber R, which contains a floating body Mb, whose upper section forms the valve C. Compressed air P, which passes through the restriction jet D, to reach the capacity chamber E, also exerts a pressure on the valve of the floating body, which compensates for the apparent weight of the body. The pressure P_0 thus obtained remains constant, independent of the variations in P, which only cause a corresponding displacement of the floating body and, subsequently, variations in the air flow between the valve C and its seat Si.

The preferred type of indicator-member of this system of pneumatic gages is the liquid column, which provides excellent visibility and high resolution. The scale length of the standard models is 20 inches, yet special models with scale lengths up to 48 inches are also available. Figure 6-10 shows typical representations of the air gages in the water-chamber system (Solex). These instruments are available in sets of one, two or three units—permitting any combination of the sets on a common base—when a larger number of gaging points must be inspected in a concurrently observable manner.

Pressure Type Air Gages with Fixed Amplification

The operating circuit of the control units used in basic models of pressure type air gages is shown diagrammatically in Fig. 6-11. When the free flow of the air through the orifices of the gage head is impeded, air pressure develops in the channel section connecting the gage head and the restriction. Because the pressure develops upstream of the gage head, it is frequently referred to as back pressure.

A pressure gage, usually of the Bourdon tube type, communicating with that section of the air channel where the back pressure develops, will indicate the variations in the air pressure. In the range that is used for the gaging, the changes in the air pressure are proportional to the variations of the obstruction to the air flow and, consequently, they indicate the relative distances of the object surface from the gage head nozzle.

The pointer of the gage, although actuated by the changes in the air pressure, is calibrated to indicate

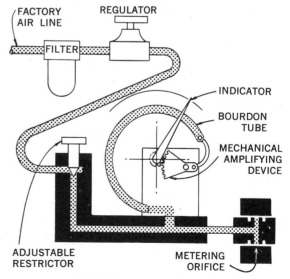

Sheffield Measurement, A Cross & Trecker Company

6-11. Operating principles of a basic back-pressure type air gage system.

the cause of the pressure variations, that is, the dimensional differences between the size of the object being gaged and the size of the setting master. Instruments of this type are available to display the size variations of the object at different magnifications, such as 5,000×, 10,000×, 20,000× or 40,000×, listed as examples for different models of control units.

Gage indications appear only when the actual size of the object is within the range to which the instrument was set. Otherwise, the pointer of the instrument will stay at one of the extremities of the scale: at the low end when the air escapes with none or only insufficient obstruction at the gage head, and at the high end when the clearance for the air escape is completely obstructed or reduced beyond the set limit values of the gaging range.

The metering restriction of the gaging air at its entrance to the control unit can be adjusted for the purpose of instrument calibration, that is, corrections of the amplification. For the calibration and setting of the instrument, two masters of known sizes are used that usually represent the limit sizes either of the object's tolerance range, or of the total scale value. Figure 6-12 illustrates an air gage control unit operating by the system described.

Pressure Type Air Gages with Variable Amplification

Pressure type air gages with variable amplification are comparable to the generally used pressure type air gages, with the exception of the complementary elements by means of which the rate of amplification can be varied over a very extended range. This extensive adjustment is accomplished with the aid of an adjustable metering restriction and an air bleed consisting of sensitive regulating valve. By using two masters to represent the limits of the size range to be covered by the gage (either full scale or a selected part of it), the changeover to the desired amplification is carried out by adjustments on the metering and bleed valves. It is practical to choose one of the fixed rates of amplification within the range of adjustment of the instrument, to permit the use of one of the standard types of dial scales.

Figure 6-13 illustrates the control unit of an air gage that operates by the system discussed above. The amplification of the particular type of air gage can be adjusted up to 28,000×, due to its "high gain" type pneumatic circuit. When only smaller amplifications are needed, the same unit—but without the amplifying relay—can be selected (and in this case the maximum amplification will be

Sheffield Measurement, A Cross & Trecker Company
Standard Gage Co.

Fig. 6-12. (Left) Back-pressure type air gage control unit with directly mounted air plug and a pair of setting gages.

Fig. 6-13. (Right) Control unit of an adjustable magnification back-pressure type air gage. The clip-on dial is useful when changing over to a different rate of amplification.

4,000×). The illustration shows that the scale part of the dial is detachable; these "clip-on" dials permit the easy exchange of the scale to one whose graduation is in correspondence with the amplification for which the gage has been adjusted.

In addition to its capability of covering a very great variety of tolerance ranges, the adaptability of the variable amplification control unit also permits the use of sensing members (air plug, air rings or contact type gaging elements) that were originally designed to operate with other systems or types of air gages.

The particular model of control unit shown in Fig. 6-14 is a dual type, comprising in a single case

Standard Gage Co.

Fig. 6.14. Dual type control unit of an adjustable magnification type air gage; permits the observation of indications resulting from two different dimensions, which may be functionally interrelated, at a single glance.

two individual gages, supplied by a common air line and filter-regulator device. The dual gage is useful when two dimensions are to be measured simultaneously with the gage indications observable at a single glance. The measured dimensions can be independent, representing two different features of the same object, or they can be related to each other. In the latter case, the gage indications also will supply information, in addition to the sizes of the individual dimensions, on their mutual relation that can be indicative of meaningful geometric conditions.

Differentially Controlled Constant Amplification Air Gages

Differentially controlled constant amplification air gages are designed to implement a gage setting concept differing from that commonly used in pneumatic gaging. In the systems discussed in the preceding sections, the amplification produced by the control unit can be adjusted, either for the purpose of gage calibration or, in a particular design, to obtain variable amplifications with the same control unit without exchanging its metering elements. However, for adjusting and calibrating those control units of more conventional design, two masters representing a known spread within the setting range are needed.

As shown in the schematic drawing in Fig. 6-15, the constant amplification type air gages are equipped with two metering restrictions, frequently designated as jets. These restrictions are nonadjustable and control two parallel air circuits, known as the reference channel and the measuring channel, respectively. One of these circuits leads inside a bellows, and the other circuit into the enclosed space surrounding the bellows. During the operation of the gage, when the air pressure acting on the opposite sides of the diaphragm or bellows is balanced, no deflection results and the metering handlinked to the bellows indicates zero. Imbalance in the operating air pressure, which is caused by size variations at the sensing end of the system, will deflect the bellows, resulting in proportional gage indications.

The balanced state should reflect the nominal size of the dimension being measured, which is physically embodied by the setting master. In this setting position, the air pressure of the differential circuit must be balanced manually, by adjusting the zero setting valve of the reference channel. Because the restrictions at the entrance of the parallel circuits are nonadjustable, the zeroing will not change the original amplification, which remains constant and linear over the entire scale range. As a result of this unique characteristic, this type of air gage requires a single master only for the zero setting of the indicator, and no adjustment or calibration of the amplification is needed.

The control unit of a differential type air gage is shown in Fig. 6-16, displaying a single master, to symbolize the unique setting characteristic of the system. Control units of this basic type are available for magnifications ranging from 1,250× to 20,000×. The differential system provides an extended range

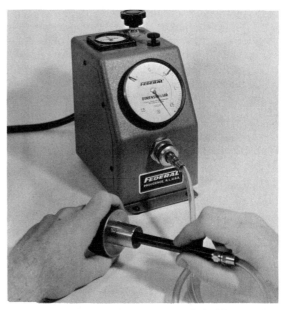

Federal Products Corp.

Fig. 6-16. Control unit of a constant amplification type air gage; the need for only one setting master is indicated by the single ring gage, to the side of the instrument.

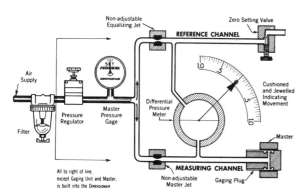

Federal Products Corp.

Fig. 6-15. Circuit diagram of the differentially controlled constant amplification air gaging system.

Federal Products Corp.

Fig. 6-17. Digital and analog air gage control unit.

of linear indications on the dial, permitting the coverage of a relatively wide dimensional spread during the measurement.

Figure 6-17 shows an electronic control unit for a differential type air gage. The electronic control unit provides both digital and analog displays and can be used in place of a traditional dial type indicator device, as shown in Fig. 6-16. The electronic unit provides the user with the advantages of reading accuracy, data output and collection as previously outlined in Chapter 5 and this chapter. Depending on the model, the digital air gage control unit has a measuring range of from 0.003 inch to 0.0003 inch total.

THE SENSING MEMBERS OF AIR GAGES

Operating Fundamentals of Sensing in Pneumatic Gaging

The function of the sensing member is to detect and to signal the size variations of the dimension being measured, by creating proportional changes in the upstream air flow. Such changes occur when a solid surface is placed in the proximity of the air discharge opening, reducing the cross-sectional area of the air escape to less than the orifice area.

In general, air gages operate by using rigid sensing members to insure the requirement of controlled clearance between the nozzle and the obstruction.

Exceptions are applications where the cross-sectional area of a very small bore is to be measured. This can be carried out by channeling the air flow from the nozzle directly through the object bore. However, the direct flow process is only applicable for bores whose diameter is smaller than that of the nozzle, thereby creating an effective restriction of the air flow.

The cross-sectional area that a partial obstruction leaves open for the air flow can be represented as a ring, when a flat obstructing surface placed normal to the nozzle axis is assumed. That imaginary ring will have a diameter equal to that of the nozzle bore, and a length that is determined by the clearance between the nozzle and the obstructing surface (Fig. 6-18). The clearance must be small enough in relation to the nozzle opening, in order to effectively reduce the air escape area.

The mathematical relationship of an equal passage, that is the theoretical limit of an effective restriction, may be expressed as

$$k \cdot d \cdot \pi = d^2 \frac{\pi}{4}$$

where

> k = clearance between the exit plane of the nozzle and the obstruction surface;
> d = diameter of the nozzle bore.

When solving for the clearance

$$k = \frac{d}{4}.$$

The effective restriction to the free escape of air from the nozzle, which will result from the partial obstruction of the air passage beyond the nozzle orifice, is not in an exact linear relation with the calculated value. As the distance between the nozzle and the obstruction approaches the equal passage

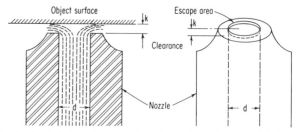

Fig. 6-18. Schematic illustration of the dimensions d and k that control the air escape area of an open jet nozzle during the gaging process.

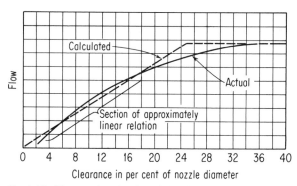

Fig. 6-19. Diagram showing the relationship between the calculated and actual volume of air escape in the essentially linear section of the air discharge curve.

level, the air flow diminishes at a higher rate than the calculated amount, mainly due to frictional drag. However, the relationship will stay very closely linear, deviating by not more than 1% or 2% for a substantial part of the effective restriction distance, which usually encompasses 15% to 20% of the nozzle bore diameter (see Fig 6-19).

Basic Types of Air Gage Sensing Members

The essential function of air gage sensing members—as discussed in the preceding section—is associated with impeding the flow of air through the orifices of the gage head at a rate proportional to the size differences of the dimension being measured.

The obstruction by which this restrictive effect is accomplished can be the surface of the object against which the jet of the escaping air impinges directly. A comparable effect can be achieved by using an intermediate member to create the obstruction; in this case the momentary position of the member in relation to the nozzle orifice is governed by the size of the object with which the intermediate element is in mechanical contact. Two basic methods of sensing can, consequently, be distinguished in air gaging: (a) the direct or open jet method; and (b) the indirect (close) or contact type method.

The actual air gage elements that operate by either of these two basic methods of sensing may differ in their form of execution and the purpose of their application. Table 6-3 reviews the commonly used general designs, indicating examples of their characteristic applications.

Design Principles of Direct Sensing Air Gage Heads

The sensing member of the open jet, or direct sensing system of air gages has the double function

of retaining and protecting the air escape nozzle, or several of them, and holding the nozzles in a position that is at a controlled distance from the surface of the nominal size object.

The securing of a positive distance between the air gage nozzle and the surface of the nominal object can be accomplished by either of two different methods.

One method uses an extraneous reference surface to serve both purposes, that of locating the object and representing the datum plane of the gaging. This is the basic method of referencing that serves the conventional comparator gaging, irrespective of the type of indicator being used. When applied in air gaging, a gage head with only a single nozzle is needed to sense the size variations of the positively located object (see Fig. 6-20A).

Air gage sensing members of appropriate design can also satisfy the requirement of holding the nozzles at a functionally controlled position from the surface of the nominal size object. This is accomplished with sensing members that contain more than a single nozzle (at least two, but frequently more), as shown in Fig. 6-20B. The term *functionally controlled* refers to the combined sensing effect of all participating nozzles.

An intermediate method is shown diagrammatically in Fig. 6-20C, with the workpiece resting on a stage that, however, does not have the role of a datum. The size of the object is gaged by the combined action of two opposed air nozzles.

The operation of the multiple-nozzle gaging heads is essentially based on the ability of air gages to sense the variations of the total air escape through several orifices, independently of whether or not all

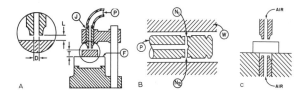

Standard Gage Co.

Fig. 6-20. Schematic illustration of open jet type air gage heads in different applications. (A) Single jet for measurements in relation to an extraneous reference surface, showing: jet cartridge *J*; comparator stage *F*; workpiece *T*; and air hose *P*. In the enlarged detail, the nozzle bore *D* and the distance *L* are factors determining the air-escape area. (B) Open jet, two-nozzle air plug for bore diameter measurements, showing: air plug *P*; workpiece *W*; and diametrically opposed nozzles N_1 and N_2. (C) Twin jets for the measurement of workpiece thickness, independent of the stage on which the part is resting.

TABLE 6-3.   AIR GAGE SENSING MEMBERS — THE PRINCIPLES OF THEIR OPERATION

SYS-TEM	DESIGN CATEGORY	SCHEMATIC SKETCH	APPLICATION CHARACTERISTICS
Directly Sensing	Single Open Jet		Requires positively located object staging. The applicable measuring process is essentially similar to the conventional comparator gaging.
	Multiple Open Jets		The compensating effect of multiple nozzle air gages, combined with the reduced clearance between gage head body and bore surface, imparts self-locating properties to the sensing member. Supplies average size information, but will not reveal the extreme size in the case of irregular object form.
Indirectly Sensing	Plunger Type		Substitutes a conical valve for the essentially flat or slightly curved obstruction which is presented by the object surface to an open jet. A substantial reduction of the amplification with a proportional extension of the gaging range can be achieved by selecting a suitable angle for the tapered valve.
	Leaf Spring Type		Two leafs of spring steel are holding the deflectable obstruction buttons in positions facing the orifices which are set back inside the gage body. The external surfaces of the buttons contact the object and transmit its size variations to the clearance for the air escape.
	Ball Contact Type		Essentially similar in operation to the leaf spring type, using spring loaded balls for contact and obstruction elements. Selected as a replacement for the direct measuring air plugs where the roughness of the surface or its interrupted design require positive contact sensing.

the nozzles are participating equally in the discharge of the gaging air.

In the majority of air gage applications, particularly on internal surfaces or for combined dimensions, gage heads with multiple nozzles are used. Such measurements offer the greatest benefits from the unique air gage characteristics that are assured by the combined sensing effect of judiciously distributed nozzles.

However, the functional property of air gages, which assures a constant total air flow as long as the sum of the clearances facing the individual nozzles remains the same, has only a qualified validity. It requires that the actual air escape through any of the individual nozzles be held within the limits of linear relationship, as explained in Fig. 6-19. This requirement must be satisfied by the design of the gage head, which establishes mechanical limits to the variations in the relative positions of nozzle orifices and object surface.

To illustrate the actual meaning of this requirement, assume an air gage system in which there is a linear relationship between nozzle clearance and air flow over a distance range of from 0.003 to 0.005 inch. Then to obtain reliable measurements, the gage body used must maintain a distance between the object surface and any of the nozzles of not less than 0.003 inch when the gage body is in actual contact with the object surface, and the lateral movement of the gage body must be restricted to a range of 0.002 inch.

Figure 6-21 illustrates, by the schematic drawing of an air plug, the design principles by which the nozzle positioning requirements described above are satisfied. The minimum distance is assured by the

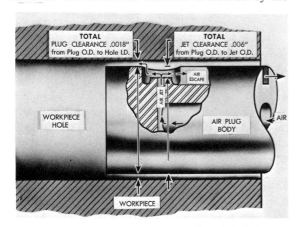

Federal Products Corp.

Fig. 6-21. Principles of design for an air plug, showing the recessed position of the nozzles. (Opposite nozzle not shown.)

recessed location of the nozzles inside the gage body, and the maximum distance is limited by the sideward movement of the plug body inside the object bore, that is, by the difference between the diameters of the bore and of the plug.

Referring again to Fig. 6-21, note that the recessed location of the nozzles is also helpful in protecting these sensitive gage elements from impacts that might distort the precisely calibrated orifices and thereby affect the accuracy of the gaging. Also, axial grooves in the plug surface, originating from the nozzle positions, permit the unrestricted discharge of the air after it has impinged on the object surface and fulfilled its gaging objectives.

The gage heads operating by the open jet or direct sensing system may be either standard gage elements or special members designed to meet a specific inspection problem.

Standard air gage heads are commonly available for the inspection of bores or of cylindrical outside features and are, accordingly, either air plugs or air rings. Although these gage members are considered standard elements, the classification refers to the general design and to overall dimensions only. The functional dimensioning and several details of the design must be adjusted to fit the requirements of the feature to be gaged, with regard to size tolerances, shape and gaging conditions. Commonly, the gaging heads of different systems of pneumatic gages are not interchangeable, although some degree of interchangeability can be assured with a few types of control units.

In the adaptation of the gage head to the size tolerances of the feature, it is a frequently followed practice to distinguish between a *design tolerance zone*, and *approach range* and an *overtravel range*, reserving for each of these zones adjacent sections on the indicator scale. It is also advisable to select gage heads that permit the adjustment of the indications to the middle of the range, that is, the zero mark, when a nominal-size master is being used.

It is necessary to establish whether the feature diameter should be gaged (a) along its entire length, (b) at any position, or (c) at a specific location. An example of the latter is the gaging of a blind hole at a specific distance from its bottom. Figure 6-22 illustrates standard gage plugs for inspecting (left) through holes and (center) blind holes. Similar considerations can arise when designing air rings; a standard model for through-going objects is shown in Fig. 6-23.

When inspecting a bore with an air gage, the objectives of the gaging can vary depending on the

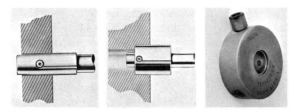

Federal Products Corp.

Fig. 6-22. Air gage sensing heads (air plugs) for round holes.
(Left) Through-hole plug.
(Center) Blind-hole plug.

Fig. 6-23. (Right) Basic type of air ring for measuring the diameters of straight shafts. One of the nozzles is visible in the bore of the ring.

meaningful design parameters of the feature to be gaged. One of the great advantages of pneumatic gaging lies in this flexibility of adapting the sensing member to the significant characteristics of the object's design geometry. Figure 6-24 illustrates schematically different forms of open jet air plugs with single and multiple nozzles.

Single-nozzle type air plugs are chosen when the straightness of a bore must be inspected in relation to a datum that is established by the plug body. Single jets are also used in some cases to inspect relative positions with regard to location or geometric relationship, relying again on a datum element that will result from the mechanical contact between gage and object.

Two-nozzle gage plugs are the most common in air gaging, because they check the diameter of the hole for size as well as for consistency, independently of other variables of form, that might be of lesser significance or should even be disregarded. The three-nozzle design is particularly sensitive to three-lobed conditions of the bore, a form irregularity that could pass undetected when relying only on

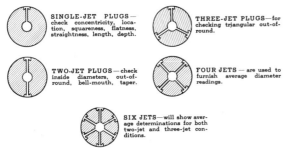

SINGLE-JET PLUGS— check concentricity, location, squareness, flatness, straightness, length, depth.

THREE-JET PLUGS—for checking triangular out-of-round.

TWO-JET PLUGS— check inside diameters, out-of-round, bell-mouth, taper.

FOUR JETS — are used to furnish average diameter readings.

SIX JETS—will show average determinations for both two-jet and three-jet conditions.

Federal Products Corp.

Fig. 6-24. Distribution of air channels and nozzles in single- and multiple-jet air plugs.

diametrical measurements. The four- and six-nozzle designs of air plugs, in addition to serving specific purposes, have excellent averaging capabilities and are useful for measuring bores whose mean sizes are considered to be the meaningful dimensions.

Not counting the air plugs and the air rings, most open jet air gage heads are of special design. It is a widely followed practice to submit the part drawings with size tolerances and other pertinent details to the manufacturer of the pneumatic control units, for the purpose of designing and making the appropriate sensing heads. These will usually contain air nozzles of the manufacturer's standard design.

Contact Type Sensing Members of Air Gages

Conditions that warrant or actually require the use of contact type sensing members in air gages are multiple and diverse. Table 6-4 lists examples of such conditions, any or several of which may be present in a particular gaging project.

The many conditions that favor contact type air gaging must not be construed as a proof of its superiority to open jet gaging. In the predominant majority of air gage applications, the open jet is the preferred system of sensing, whereas contact type sensing may be regarded as a means of extending the field of application for pneumatic gaging.

The probable reason for the lesser usage of contact type sensing members is the dominant application of air gaging in continuous inspection operations on nominally identical parts. The inspection of parts of the mass production category warrants the manufacture of special sensing devices; the flexibility of fixturing that is assured by contact sensing elements is only seldom needed.

The most frequent use of contact type sensing members is for the measurement on outside surfaces, although examples will be shown for inside gaging contact members as well. However, when small diameter bores must be measured, the limitations of contact type sensing members in comparison with the open jet gaging become evident.

There are two major groups of contact type sensing members, differing in the direction of displacement that the gage head is designed to detect: *axial* and *radial*.

Axial-contact sensing members, generally known as air cartridges, have for principal operating elements a plunger and a poppet type air valve (Fig. 6-25). The spring-urged plunger's movement is restricted to the axial direction and its outside end operates as the contact tip. The other end of the plunger works as the valve spindle whose axial displacement

**TABLE 6-4.   EXAMPLES OF CONDITIONS WARRANTING THE USE OF
CONTACT TYPE AIR GAGE SENSING MEMBERS — 1**

	Surface Roughness The contact type sensing member will measure the envelope size, which is the meaningful dimension for fits.	Gage Blocks	**Setting Air Plugs with Gage Blocks** The cost of setting rings can be saved in short run bore gaging by using air plugs with leaf spring type contact buttons.
	Interrupted Surfaces By observing the maximum reading of the indicator the proper location of the sensing point can be determined in a dependable manner.		**Extended Gaging Range** For covering an extended gaging range at low magnification, e.g. checking stock removal in lathe turning, contact members of the plunger type must be used.
	Narrow Lands Unless very carefully located on a land of ample width, open jets can cause false indications which easily pass undetected.		**Location Inaccessible by Open Jets** Example: Wall thickness variations gaged inside a bore and behind a shoulder with contact member combined with bell crank transfer element.
	At the Edge of a Hole Open jet cannot be used at the edge of a feature, because part of the surface against which the air stream must impinge is non-existent.		**Surface of Irregular Form Located from a Reference Plane** Example: Partly rounded beveled shoulder's distance from a positively located face.
	Porous Materials Porous substances, such as powder metallurgy products, can cause false readings with open jet even when pores are undetectable by unaided eye.		**Reversed Indications** Checking the matching conditions of a bore and of a shaft with a dual circuit indicator, using reversed action air plunger to contact the outside feature.

**TABLE 6-4.   EXAMPLES OF CONDITIONS WARRANTING THE USE OF
CONTACT TYPE AIR GAGE SENSING MEMBERS — 2**

	"Floating" Gaging of Outside Diameter Matched contact plungers with flat tips measure actual diameter unaffected by minor locating errors to which open jets are sensitive.		**Comparator Gage Stand** By substituting a contact type sensing member for a mechanical indicator, a standard gage stand can be promptly converted into an air comparator.
	Air Snap Gages A flexible substitute for the air ring, using contact type sensing member. Adjustable over wide range and indicates maximum diameter size.		**Test Indicator Air Gage** Gage heads which combine the positive contact and a right angle transfer motion make air gaging adaptable for surface plate based measurements.

in relation to the conical valve seat will alter the air escape area in the valve. By changing the taper angle of the valve design, the ratio between the axial displacement movement of the spindle and the resulting clearance inside the valve can be altered. This design feature of the plunger type cartridges permits extending the measuring range of the air gages to cover a spread ten times, or even more, over the range normally available with the direct sensing system.

Plunger type air gage cartridges normally operate with the valve orifice open, and the axial movement of the plunger, when in contact with the object surface, causes the valve to close partially. Depending on the application needs, the design can also be reversed, resulting in conditions where the valve is normally closed and the displacement of the plunger

results in the partial opening of the air escape area in the valve.

Radial-contact sensing members are available in different models, the most frequently used design being the leaf spring type, shown schematically in Fig. 6-26. This particular example represents a diametrically gaging air plug, which is one of the primary applications of this sensing member. As can be seen from the drawing, two buttons, supported by leaf springs, cover the deeply recessed air orifices in the gage body. The outside, spherical portions of the buttons project above the surface of the plug, and establish mechanical contact with the surface of the bore to be gaged. Depending on the bore size, the contact will press the buttons down into the plug body, proportionally reducing the air escape clearance in front of the orifices.

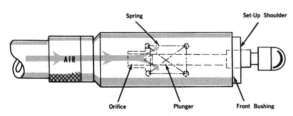

Giddings & Lewis Measurement Systems

Fig. 6-25. Functional diagram of a plunger type, mechanical contact, air gage sensing head.

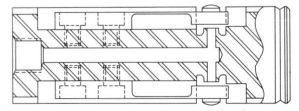

Standard Gage Co.

Fig. 6-26. Schematic drawing of a mechanical contact, air plug of the leaf spring type.

Depending on the surface of the bore being measured—which may be laminated, grooved, with keyways or slots or of through- or blind form—the design of the contact type air plugs can usually be adapted to the gaging conditions. Although air plugs constitute the prominent field of application for leaf spring type sensing heads, these members are also available with single leaves for measurement on outside surfaces, particularly when space limitations prevent the use of air cartridges.

APPLICATIONS FOR PNEUMATIC GAGING

The potential or preferential fields of application for pneumatic gaging are determined by either or both of the following two groups of factors:

1. The inherent merits of pneumatic sensing and amplification, as discussed in a preceding section; examples are listed in Table 6-2.

2. The adaptability of air gages for a wide range of different measuring applications, discussed in this section.

The objectives of pneumatic gaging, like those of any other system of dimensional measurement in engineering production, is the verification and assessment of dimensional and geometric conditions. It is, therefore, considered a functionally proper approach to base the review of potential areas of air gage applications on the different categories of dimensional and geometric conditions that must be explored and measured.

Table 6-5 follows this system of reviewing, complemented with examples for each category of conditions that are discussed in the survey. Although the table contains a wide variety of conditions to which pneumatic gaging can conveniently or even preferentially be adapted to carry out the measurement of components or features, the survey is not intended to represent a complete listing. Furthermore, the examples quoted were selected with the sole purpose of illustrating dimensional categories and are only representatives of particular groups.

Because pneumatic gaging operates by ejecting pressurized air at locations on the object surface that are precisely controlled by the gage and, when needed, by the workholding fixture, the selection of these points of ejection and the channeling of the air to the nozzles are essential variables for adapting pneumatic gaging to specific measuring conditions. For this reason the sketches of the table schematically indicate the path of the air channels and the location of the ejection nozzles.

Although sensing by open jet nozzle is the characteristic method used in pneumatic gaging, sensing with intermediate elements that operate by mechanical contact is also frequently applied. This alternative method of sensing was discussed in greater detail in the preceding section. In view of this choice, a few of the schematic drawings of the table indicate the use of contact type gage heads where their application is considered an acceptable variety, or even the preferred method of sensing.

Pneumatic Measurement of Interrelated Dimensions

It has been pointed out earlier that air gages permit the simultaneous measurement of several dimensions, even when closely spaced on the same object, with indicators conveniently located for rapid scanning. Although this unique capability of air gages is of great advantage when measuring interrelated dimensions, the scanning of the display by several indicators—even when the dials or columns are adjacently mounted—could become tiring to an inspector engaged in continuous measurements; the resulting fatigue can lead to observational errors.

Air gage indicators are usually equipped with adjustable tolerance markers that serve as guides for the inspector. However, in the case of interrelated dimensions, the tolerance range of the individual size is not necessarily identical with the tolerances specified for variations in the mutual relationships of the inspected dimensions. Actually, in many cases, the acceptable variations in the dimensional relationships—expressing regularity of form, proper location or orientation, and so forth—may represent only a fraction of the tolerances that have been set for the individual dimensions.

The application flexibility of air gages provides an excellent solution for these problems of nonidentical tolerances. Air gages of a particular type will permit the simultaneous observation and evaluation of indications representing both the size and the mutual relationship of two dimensions.

Examples of features and of components with nonidentically toleranced individual and interrelated dimensions are shown in Table 6-6. The listings in the table will assist in visualizing concrete applications where the relationship of two dimensions can be more critical than the absolute size of either dimension, individually.

The type of air gages that were designed for and are highly suitable for the simultaneous measure-

TABLE 6-5.   APPLICATIONS FOR AIR GAGES — DIVERSE CATEGORIES OF GEOMETRIC CONDITIONS WITH EXAMPLES OF FREQUENTLY MEASURED DIMENSIONS — 1

Diameters — Outside and Inside		The air gage will measure true diameter unaffected by minor variations in the location of the object, due to the compensating effect of two air jets which are opposed in position and acting on a common air channel.
Average Material Size of a Cylindrical Feature		The self-compensating properties of air gages permit sensing the cumulative effect of the obstructions facing each of the nozzles in the gage head. The resulting indications will reflect the average material size of the feature being measured.
Distance from a Reference Plane	A B	Air gage applications in the basic comparator type of measurements rely on a single jet held at a positive distance (A) from an external reference plane which also serves to locate the object, or (B) from a datum surface on the object itself.
Departures from a Nominal Plane	A B	Plain geometric forms, when referenced from a datum plane which is either parallel or normal, can be measured effectively with a single air jet or with an air cartridge, mounted in a functionally proper position on a staging fixture. Examples: (A) flatness; (B) squareness

TABLE 6-5.   APPLICATIONS FOR AIR GAGES — DIVERSE CATEGORIES OF GEOMETRIC CONDITIONS WITH EXAMPLES OF FREQUENTLY MEASURED DIMENSIONS — 2

Consistency of Size or Regularity of Form		Unintentional variations of size or form can be detected and measured with air gages using special fixtures or gage heads. The process may require a relative rotary or linear motion of the object and of the gage, e.g. to measure wall thickness variations, or a simple insertion of the gage head. Examples shown: (A) Out-of-squareness of a bore; (B) Bore chamber.
Interrelated Dimensions		The angular dimensions of a basically regular cone are determined by the difference of two diameters located at a specific axial distance apart. These combined dimensions can be measured accurately by air gaging, which will indicate both diameters, as well as their relative size. Indications may be displayed on a single dial.
Interrelated Positions or Geometric Conditions		Interrelated geometric conditions are frequently specified independently of size, like the parallelism of bore exclusive of diameter variations. Balancing air channels, which are sensitive to the critical dimension only, will show the required information on a double pointer indicator dial.
Size Matching of Mating Component Parts		For matching parts which are to be assembled with a specific clearance (like the inner and outer races of ball bearings) air gages will indicate the significant relative size as a single dimension. Circuits can be set up which indicate the individual sizes additionally, on separate dials.

**TABLE 6-5.   APPLICATIONS FOR AIR GAGES — DIVERSE CATEGORIES OF GEOMETRIC
CONDITIONS WITH EXAMPLES OF FREQUENTLY MEASURED DIMENSIONS — 3**

Multiple Dimensions		Multiple dimensions, either referenced from a common datum, or individually, are preferably checked simultaneously. Air cartridges with normal or reversed action, can be closely located in a workholding fixture and the measured dimensions may be observed at a single glance on adjacently mounted and remotely located indicators.
Contour — Deviations from a Nominal Form		The form of a body with curved contour, when defined by the separation of opposite points on the surface, can be inspected with air gages whose several sensing heads will detect and display on adjacently placed indicators the deviations of individual cross-sectional dimensions from their nominal values.

ment of the sizes of two individual dimensions, and of their relation as well, are frequently designated as *duplex gages* or *dual circuit comparators*. Essentially, the circuitry of this gaging system consists of two independent air channels, each feeding an appropriate number of nozzles for the sensing of a particular dimension. The principles for the layout of these channels are indicated on the features shown in Table 6-6. The amplifiers of these individual air channels are mounted in a common case that has a single indicator dial (see Fig. 6-27). The indicator differs, however, from the conventional type by having two pointers, one pivoted in the center of the dial and the other moving around the dial's circumference. Each is the indicator element of a separate air channel.

Regular type limit hands mounted on the crystal of the dial can be adjusted to indicate the tolerance range for the size. Another pair of hands, known as slave pointers, are adjustably mounted on the hub of the central pointer and travel together with that pointer. The gap between the slave pointers, whose position may vary but whose spread remains constant, represents the relationship tolerance range. In

order to meet the relationship tolerance requirements, the measured dimensions must cause the circumferential pointer to be in a position facing the gap between the slave pointers.

There are certain inspection conditions where the size of each individual dimension may or should be disregarded because it is considered functionally meaningless, or has already been the object of a preceding inspection process. In such cases, information regarding the individual sizes is unnecessary, or may even prove to be confusing. Examples of conditions belonging to this category of inspection cases are: the angle of a taper independently of the axial location of the gaging planes; the regularity of form when determined in the gaging process as a dimensional relationship (roundness, flatness, squareness); the relative size (clearance) of two part features that are to be assembled with a specific fit.

For this kind of measurement the pneumatic gaging offers the *computing type of circuits and control units*. The computing air circuit carries out the algebraic summation of the size variations that are sensed simultaneously by the gaging heads of two independent air circuits, and then present the com-

TABLE 6-6.   EXAMPLES OF GEOMETRIC FORMS WHOSE INTERRELATED DIMENSIONS ARE EFFECTIVELY MEASURED WITH DUAL CIRCUIT AIR GAGES

DESIGNATION OF THE PART OR FEATURE	INDIVIDUAL DIMENSIONS	SIGNIFICANCE OF THE DIMENSIONAL RELATIONSHIP
Tapered bore	Large and small diameters at a positive distance apart.	The included angle of the taper (indicated as a linear deviation from the nominal relation which was established with a setting master)
Ring	The thickness of the wall	Eccentricity of the bore in relation to the outside surface, observable as wall thickness variations
Straight bore	Diameters at different levels, measured in planes parallel to the face	Straightness and squareness of the bore, observable as the conformance of two axially separated diameters
Spacing of two parallel bores	Distances between opposite bore walls	Center distance between bores, determined by the conformance to the master of the combined separation of opposite bore walls
Round bore	Diameters in different orientations	The presence of ovality, revealed by the difference in the size of diameters measured in a common plane but normal to each other
Normally parallel sided flat part	Thickness at different points	Parallelism of the opposite surfaces measurable as the mutual similarity of thicknesses at different points of the surface

Moore Products Co.

Fig. 6-27. Duplex (dual circuit) type, air gage control unit with central and circumferential pointers. Tolerance markers of fixed and traveling (slave) types—both individually adjustable—are also visible.

puted amount as a single value on a plain indicator dial, or as a single line recording on a chart. The inspector's task is reduced to verifying whether or not the pointer position is between the limit markers on the dial, which were adjusted to embrace the tolerance spread of the relationship being inspected.

The computing type air circuits may also be used for the very precise measurement of a single length dimension with the aid of two opposed and matched gaging heads. In this process the observed variation is that of the gap between the sensing points, a distance that is independent of the object's staging conditions. This kind of application for computing circuits will be discussed more in detail under the heading of "Differential Gaging" in the next chapter, which covers electronic gages.

Diverse Nonbasic Applications for Pneumatic Gaging

In addition to the more common uses of air gages, which may be considered as basic applications, pneumatic gages are being used, to a growing extent, in measuring applications involving rather complex functions. Without attempting a thorough exploration of that large domain of actual and possible air gage applications, a few examples are given in order to indicate potentials of pneumatic gaging that frequently are not fully realized.

1. *Automatic sorting and segregating machines* are capable of checking several independent or interrelated dimensions on manufactured parts, at a rate of up to about 4500 pieces per hour. The machines will discard the out-of-size parts and segregate the acceptable ones into categories that can represent size increments as small as 0.0001 inch and, exceptionally, even less.

2. *Automatic post-process and in-process gages* attached to machine tools or integrated into production lines will indicate the dimensional conditions of the part just finished, or during the actual machining process. The gage will signal, e.g., by lights or audible warnings, the occurrence of unsatisfactory conditions. Although involving more complex installations, air gages can also be used for feeding back the measuring information through special relays (air-electric transducers) controlling the machine tool, thereby automatically initiating corrective actions for the observed size digressions.

3. *Pneumatic recorders.* Air gages operating by sensing variations in back pressure can, in principle, be used to supply actuating signals to pneumatic recorders of appropriate design that produce permanent records of the obtained measurements. Such recorded charts permit the checking of control trends or they may be used as quality control data.

The automation of inspection and in-process size control is greatly facilitated by the availability of component modules, such as pressure regulators, volume boosters, differential pressure transmitters, different types of relays, and many others. The layout and design of complex pneumatic gaging and actuating circuits that are intended to inspect specific types of parts and to initiate definite functions most always require the through experience of specialized gage manufacturers.

7.
Electronic Gages

BASIC CHARACTERISTICS AND APPLICATION ADVANTAGES

It is difficult to overestimate the impact electronic gaging has made on the field of dimensional measurement in recent years. Advances in the miniaturization of electronic components, the design and manufacture of integrated circuits, and the development of computer hardware and software have touched and changed virtually every facet of metrology now practiced in industry. Therefore, any discussion of modern electronic gages cannot be confined to a single chapter such as this. And as such, electronic gages have been introduced, pictured and discussed within the individual chapters of this handbook. Today, electronic gages are simply viewed as the natural evolution of specific gage types (that is, vernier calipers become digital calipers and mechanical dial indicators became digital indicators). Yet, certain basic electronic gaging principles can and should be discussed in a general way. This chapter will attempt to do so.

In metrology, the term electronic gages comprises a diverse category of measuring instruments capable of detecting and displaying dimensional variations through the combined use of mechanical and electronic components. Such variations are typically used to cause the displacement of a mechanical contacting sensing member in relation to a preset position, thereby originating proportional electrical signals, which are then amplified and indicated. Noncontact methods, covered later in this chapter, can also be used to generate electrical signals. The indicated variation generally appears on the face of either a meter or an LCD (liquid crystal display) analog whose scale is graduated in fractional values of the standard unit of length, the inch or the meter. Variations may also be displayed in numeric form utilizing light-emitting diodes (LEDs) or an LCD.

The basic principle of converting mechanical displacement into proportional electrical variations that then can be displayed was originally applied to electrical gages designed to make linear measurements in very fine increments on a comparative basis. More recently, however, great advances in the field of electronics have made available elements for use in both incremental and absolute length-measuring instruments. Such elements form the basis for what is now seen as an integrated system of quality control used in manufacturing (see Fig. 7-1).

The advantages of electronic gages that warrant the rapidly extending applications of this category of measuring instruments are multiple and diverse. Although most of these advantages originate from the basic concept of using electronically amplified electrical signals for measuring linear displacements, the concept's implementation is an essential factor in realizing the potential benefits of electronic gaging.

Because of the multiplicity of advantages that can be derived from the use of electronic gages in dimensional measurements, a tabulated form of listing and explanation should contribute to the clarity of a survey. Table 7-1 gives the essential advantages of electronic gages, classified into two groups, discussing the performance capabilities and dealing with the applicational aspects; it is intended to provide a guide for evaluating whether the electronic gage represents the most suitable category of instrument for a particular measuring task.

Since the attainable benefits are dependent on the manner in which the basic concept of electronic gaging is implemented, the discussion of characteristic

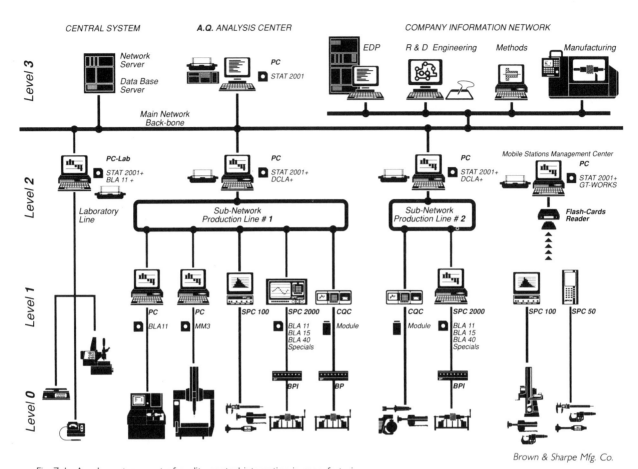

Fig. 7-1. A coherent concept of quality control integration in manufacturing.

Brown & Sharpe Mfg. Co.

properties must be limited to principles. Actually, the commercially available types of electronic gages differ widely in design and workmanship, the degrees of realization being governed by the objectives that the manufacturer intends to accomplish with a particular model of instrument. Electronic gages can and are being used for an extensive variety of purposes that require different capacities and accuracies.

OPERATIONAL SYSTEMS AND PRINCIPAL ELEMENTS

Functionally considered, electronic gages are length measuring instruments that operate by converting distance or displacement to proportional changes in electrical current or potential, with the aid of electromechanical transducers.

Various physical effects can be utilized in electromechanical transducers, such as the following:

a. The change of electrical capacitance between two metal plates caused by variations in the distance separating these plates. One of the plates, the stator, is fixed and the other plate, the reed, is free to be moved by the displacement of the gage contact tip;

b. Variations in the resistance of a wire due to mechanical strain (applications of the strain gage principle);

c. Changes in the electrical resistance caused by variations in the position of a movable contact;

d. Changes in self-inductance or mutual inductance resulting from the varying positions of a movable magnetic body; and

e. The voltage produced by bending a piezoelectric crystal.

TABLE 7-1.   FAVORABLE PROPERTIES OF ELECTRONIC GAGES — 1

PROPERTY	EXPLANATION	DISCUSSION
PERFORMANCE CAPABILITIES		
Sensitivity (Resolution) Expressed as the smallest input (contact member deflection) which will produce proportional signals	Will react to very small input produced by the displacement of the sensing contact member (either a spindle or a stylus).	In many designs the sensing member is integral with the instrument element whose displacement produces the electrical signals. The practically frictionless suspension of the sensing member assures the unimpeded transmittal of the distance variations to the electronic components whose resolution is considered infinite. The response is immediate and practically free from lag or hysteresis.
Repeatability Measured in linear units, computed on a 3σ basis	Repeating the exact positioning of the sensing member will produce the precise duplication of the original indication.	Due to the avoidance of mechanical linkages all backlash and play is eliminated. Most types of sensing member suspensions, while effectively restraining the contact deflections to a single path, permit its displacement without incurring other than molecular friction. The absence of mechanical or electrical hysteresis contributes to the excellent repeat accuracy of most types of electronic gages.
Calibration Accuracy Expressed in per cent of indicated scale value, representing the max. deviation from the exact size differential	The pointer over the meter scale should accurately indicate the distance over which the sensing member moved from its reference position.	The electronic components used in gages for dimensional measurements belong to systems selected to produce electrical signals precisely proportional to the distance of mechanical displacement which originated said electrical variations. The electronic gages are equipped with means to adjust the basically linear signals for producing pointer deflections which indicate with great accuracy the distance of the contact travel. Once set, the calibration is retained over an extended service period.

TABLE 7-1.   **FAVORABLE PROPERTIES OF ELECTRONIC GAGES – 2**

PROPERTY	EXPLANATION	DISCUSSION
PERFORMANCE CAPABILITIES		
Amplification Expressed as a ratio between pointer deflection and contact displacement	The ratio by which the deflection of the indicating member (pointer or pen) is a multiple of the gage contact's displacement.	The amplification potentials of electronics meet the maximum requirements of displacement measurements. Practical limitations are set by the technical need, the design of the mechanical gage components and by the distance range which must be covered (this latter being inversely proportional to the rate of amplification). A unique advantage of electronic gages is the simple switching to different amplifications. Currently the maximum ratio of amplification used in electronic displacement measuring instruments is 100,000 : 1 (for special purpose instruments, even higher).
Gaging Force Expressed in ounces or grams	The force required to establish dependable mechanical contact between the object surface and a gage sensing member. In most types of mechanical gages that force is increased for the purpose of secondary functions.	Electronic gages do not have to use a part of the gaging force for eliminating backlash and play between linkage elements such as found in mechanical gages. The light gaging force greatly reduces the amount of deflections on the contacting surfaces as well as the extent of strain in the participating mechanical elements; both of these can constitute major error factors in submicron measurements. Certain models of internal gages are equipped with means to adjust the contact force in accordance with the gaging conditions.
Speed of Response Expressed in hertz (cycles per second) for the pointer travel over the entire scale length	The inverse value of the time gap needed to transmit a sensed displacement into a full scale reading on a meter, display or recorder.	Many kinds of displacement measurements require gage indications which appear practically without delay. Examples are in-process gaging during machining operations, surface tracing, automatic sorting, etc. The instantaneous response of electronics, supported by the near frictionless suspension of the contact and indicating elements of electronic gages, provide response speeds unexcelled by any other measuring instrument system.

TABLE 7-1.   FAVORABLE PROPERTIES OF ELECTRONIC GAGES — 3

PROPERTY	EXPLANATION	DISCUSSION
APPLICATIONAL ASPECTS		
Versatility of Display	The gage indications can be displayed by various means, used alternately, or several of them simultaneously; these include the meter, the recorder, the oscilloscope, limit signals, digital volt-meters and digital displays.	Depending on the purpose and the conditions of measurement, one of several display methods could be the most suitable, or more than a single method operating concurrently might be desirable. The variety of indicating means available for displaying the electrical signals of the measuring instrument in linear values lends unequalled versatility to electronic gages. Also, with the push of a button, the display can be converted from inches to millimeters.
Flexibility of Adaptation	The transducer sensing members of electronic gages can be integrated into a great variety of measuring instruments.	The displacement sensing elements of electronic gages are compact in size, operate in different positions and, in special designs, in environments adverse to sensitive measuring instruments in general. Another important characteristic is the remote location of the control unit. These properties permit the use of electronic displacement measurement for many different applications, even where most other types of indicator instruments could not be mounted and/or observed.
Signal Utilization	In addition to indications, the electrical signals of the gage can also be used for initiating and controlling various associated functions.	Opening and closing chute gates in sorting, signalling out-of-tolerance conditions during in-process inspection, changing feed rate in automatic machining, compensating for machine maladjustments in auto-sizing, are examples of functions which can be initiated and maintained by the size-dependent signals of electronic gages.
Operational Convenience	The switching to different amplifications and the electrical adjustment of the zero position impart to electronic gages an operational convenience which, in these aspects, is unequalled by any other system of dimensional gages.	The setting of a sensitive comparator gage (also known as "mastering") to a nominal size represented by a setting gage, is a delicate and time consuming operation. The unique properties of electronic gages reduce the setting time and facilitate the adjustment to the exact reference position.

TABLE 7-1. FAVORABLE PROPERTIES OF ELECTRONIC GAGES — 4

PROPERTY	EXPLANATION	DISCUSSION
APPLICATIONAL ASPECTS		
Maintainability	Mechanical wear being practically nonexistent in electronic gages, these instruments can operate for years without the need for repair.	While most systems of mechanical contact gages are subjected to irreversible wear in the linkages and pivots of their amplification train, electronic gages need only occasional readjustment by a single screw to restore their original accuracy. Most types of electronic components possess a very long service life.
Readability	Digital electronic displays minimize the reading and interpretation errors often experienced with other gage types.	Digital displays are easier to read than meters and dials. And when supplemented with green (within specifications), yellow (caution), and red (outside of specifications) indicator lights, inspection errors are reduced.
Data Collection	Electronic dimensional data collection via microcomputer through the use of an RS232 interface eliminates the need for recording and calculating information by hand.	Today, many companies either wish or are required to keep detailed inspection data on parts produced and processes used within their plants. This may include documentation of process capability and control (X and R charts) among others. Such data collection and analysis is facilitated by the combination of electronic gages and computers.

Electronic Probes

In the majority of the currently available electronic gages, a particular type of electromechanical transducer has found application. That type of transducer is the linear variable differential transformer, frequently referred to by the abbreviated term LVDT.

Figure 7-2 shows the cross-sectional diagram of a linear displacement transducer of the LVDT type. This transducer has one primary and two secondary coils arranged in line and symmetrically. A small iron core, attached to a nonmetallic rod, can move axially within the hollow cylinder formed by the coils. The rod is integral with the measuring spindle and is precisely guided radially, its axial position being sustained by spring action. The spring action also produces the gaging force when the end of the measuring spindle is brought to bear against the surface of the work.

In the cross-sectional diagram of Fig. 7-2 the measuring spindle has a plain bearing guide and a helical spring for balancing the position of the core; however, other types of displacement transducers use a frictionless reed suspension for both these

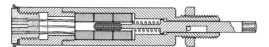

Daytronic Corp.

Fig. 7-2. Schematic cross-sectional drawing of a typical linear displacement transducer of the LVDT category.

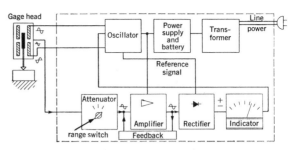

Brown & Sharpe Mfg. Co.

Fig. 7-4. Simplified block diagram of an electronic circuitry used in conjunction with an LVDT type displacement sensing transducer.

functions. This diagrammatic drawing illustrates the principles of the mechanical arrangement only; actual designs of gage heads with LVDT transducers vary widely in detail, to meet different applicational purposes, and also by the manufacturer's choice.

For operating the transducer, an excitation voltage of low potential and high frequency (for example, 6 volts and 2.4 kilohertz) is imparted to the primary coil and when the movable core is in a precisely centered position, the differential transformer induces equal and opposite voltage in the secondary windings. When in the gaging process the measuring spindle is moved, causing a change in the position of the integral core, the coupling between the primary and the secondary windings will be altered. This change in the core position produces a difference output voltage that is proportional to the displacement distance of the measuring spindle. The phase of this voltage will indicate whether the point being contacted on the work surface is under or over the reference position that was originally set to cor-

respond to the electrical null value of the LVDT signals (see Fig. 7-3).

Electronic length measuring instruments contain *several principal members* that are discussed briefly in the following section. These members are part of a circuitry such as shown, for the purpose of example, in the block diagram of Fig. 7-4.

Gage Heads

The gage head, with the gage spindle or stylus tip, is brought into mechanical contact with the selected gaging point or the work surface. The gage spindle is integrated or rigidly connected to the rod holding the core of the LVDT (or the functionally corresponding member when another transducer system is being used). The gage head also contains the transducer providing the necessary mechanical protection and magnetic shielding to this essential element.

The gage heads used in electronic displacement measuring instruments are made in three basic styles that are selected according to the application conditions.

a. The type of gage head shown in Fig. 7-5 has a true reed suspension, providing the most accurate guidance to the gage spindle with its integral core and can be operated with the least gaging force. The gage spindle has precisely axial displacement motion. This type of gage head is generally used in comparator gage stands and in similar applications where the overall dimensions resulting from the reed suspension are not objectionable and the highest degree of accuracy is sought.

b. The cartridge type head of cylindrical form usually has a ⅜-inch outside diameter. This is the standard diameter for dial indicator stems and is selected to permit interchangeability. Electronic gage cartridges may be used to replace mechanical indi-

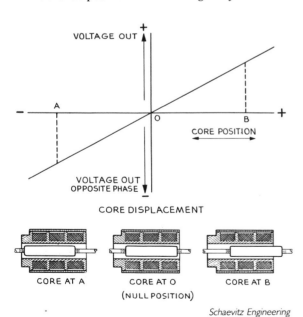

Schaevitz Engineering

Fig. 7-3. Interrelations between the core position and the resulting output voltage and phase in an LVDT type displacement measuring system.

Federal Products Corp.

Fig. 7-5. Reed spring gage head.

Federal Products Corp.

Fig. 7-7. Lever gage head.

Federal Products Corp.

Fig. 7-6. Cartridge gage head.

cators in gaging fixtures originally designed to operate with a dial indicator. The spindle of the cartridge has axial motion, like the preceding type, but because of space limitations the suspension is either by circumferentially slit spring steel disks, or by a helical spring assisted by a guide bushing. The cartridge type gage head is shown in Fig 7-6.

c. The test indicator gage head (lever type) is equipped with a spindle having a pivot suspension. This type of gage head is used in height gages or other applications where pivoting contact permits better access to the work surface than the axially displaced gage spindle. The lever type test indicator gage head is shown in Fig. 7-7.

Because of the comparative character of electronic length measurements, the gage head must be held positively in a fixed position in relation to a datum plane. Consequently the *gage stand* or a gag-

ing fixture are functionally associated elements of the gage head.

Connected with the gage head by a cable, and usually, remotely located from the actual gaging spot, is the *control unit* of the electronic gage. This instrument member, frequently designated as the amplifier, although that is only one of its operational tasks, comprises elements for the following functions (see Fig. 7-3):

a. Producing excitation current for the primary winding of the LVDT by means of an oscillator, to serve instruments that rely on frequencies in excess of 60 Hz:

b. Amplification of the relatively weak electrical signals emitted by the transducer in response to the sensed mechanical displacement;

c. Demodulating the amplified signals to make them available for display on the meter, or as signals for recorders and other types of output utilizing instruments;

d. Attenuation (change of amplification), pointer position shift (zeroing) and gain adjustment (calibration), with individual switches, knobs and similar controlling elements accessible on the outside of the control unit;

e. Signal display is accomplished using a meter with a graduated scale, an LED digital readout, an LCD analog an digital readout, or some combination thereof (see Figs. 7-8 and 7-9). Light-emitting diodes (LEDs) convert electrical energy into light energy using the phenomenon known as electroluminescence and come in three colors (red, yellow and green).

Mahr Gage Co.

Fig. 7-8. Universal measuring instrument with analog and digital display.

Federal Products Corp.

Fig. 7-9. Precision comparator stand with analog and digital LCD electronic amplifier.

Liquid crystal displays (LCDs) do not generate light energy; they use a phenomenon of dynamic scattering to alter existing light, making certain areas of the display appear bright or dark (black or light gray). This newest electronic display technology is very

popular in battery-operated equipment due to very low power dissipation.

f. As an operational power supply, a battery, required when self-contained operation of the instrument is preferred, making it independent of power line connections.

ELECTRONIC COMPARATOR GAGES

Comparator gages are the basic instruments for length comparison by electronically amplified displacement measurement. The measuring process is essentially similar to that applied in comparator gages equipped with mechanical indicators. However, there are differences in the applicable techniques due to the earlier discussed potentials, as well as limitations, of electronic gaging.

As with all comparator gaging processes, the gage first must be set to a reference dimension, either with the aid of gage blocks or of a setting master. The gage setting is greatly facilitated in electronic gages by the function known as zero shift. This device permits short distance adjustments of the pointer position to be made electrically, without actually changing the mechanical position of the gage spindle that carries the contact point. The zero shift is effected remotely from the gage stand and with great sensitivity, simply by turning a knob on the amplifier while observing the position of the pointer.

In comparators equipped with a sensitive mechanical indicator, yet without means for pointer position adjustment, the zeroing in the gage setting phase may be a very tedious process due to the unavoidable stick-and-slip of partially locked mechanical elements. The inconvenience resulting from the gage setting difficulties frequently leads to the practice of considering a random position of the pointer, as long as it is close to the scale zero, as the reference value and, in the subsequent gaging process, calculating the actual values of the observed deviations. The zero shift of electronic gages eliminates the need for improvised reference positions and avoids the possible calculating errors that that practice involves.

The range change, which all types of commercially available electronic comparators provide by means of the gain control, effectively compensates for the limitations that result from the relatively short measuring span found in electronic gages. In addition, this device also lends great application flexibility to electronic comparator gages. Using the same instrument for several types of measurements that extend over different tolerance ranges and require distinctive levels of sensitivity is of great advantage in applications where work assignments vary widely.

The very light gaging force that can be used in electronic comparators, where almost no mechanical friction needs to be overcome, is of great value when measuring workpieces with very fine finish that easily could be marred by heavier gage contact. Light gaging force is also required for parts made of soft material or with thin walls, which could suffer permanent deformation or elastic deflections when the force of the magnitude exerted by most types of mechanical indicators is applied.

The use of differential gaging and the utilization of the electrical signals for recording are further application advantages that frequently determine the choice of electronic gages as the most suitable type of comparator. These potentials of electronic displacement measurements will be discussed later in this chapter.

A characteristic example of electronic comparators is shown in Fig. 7-9. The gage stand is of sturdy construction designed to possess a stability commensurate with the unique sensitivity and high degree of accuracy assured by the electronic components of the instrument. The control unit is located separately and is connected by a cable only with the transducer sensing head mounted on the cross arm of the gage stand. That cross arm is precisely guided along the column of the stand, a hand-wheel actuated lead screw serving the vertical adjustment.

The control unit shown in Fig. 7-9 is of the analog/digital LCD display type. The amplifier is powered by a rechargeable nickel–cadmium battery that will operate for 10 hours under full load. This particular unit simultaneously computes the maximum, minimum, total indicator reading (TIR), nominal and actual gage head signal for dynamic measuring capability. The unit has three selectable ranges: (1) inch, (2) metric and (3) automatic, and selects the smallest range the signal allows for the best resolution. For communication with a data collection device (that is, computer or printer) the unit is equipped with RS232 output capability.

Differential Gaging of Linear Dimensions

Differential gaging designates a category of dimensional measurements where two sensing members, in simultaneous contact with the object surface, are mutually referencing their positions. Consequently, the measured dimension is the change in the relative positions of the two sensing members

during their contact with the object surface. When the two gage heads are in line and in an opposed position, the sensed dimension will be the change in the separation of the two gage tips, as brought about by the differences in length of the object with which said tips are in simultaneous contact.

These conditions differ from the conventional comparative gaging of external length dimensions, where the object and a single sensing member have a common reference plane functioning as one of the boundaries of the measured distance. These distinctive characteristics of the two systems of comparative dimensional measurement are shown diagrammatically in Fig. 7-10A using, for example, the gaging of a plain cylindrical object.

Various systems of instruments are available for implementing the principles of differential gaging. The operational principles of the more commonly used differential gaging systems are shown diagrammatically in Fig. 7-10B. Although length (diameter) dimensions were chosen to illustrate the principle, differential gaging is also used successfully for solving other measuring problems, such as the measurement of parallelism (observed as wall thickness variations), the conformity of contour in relation to a master and many others. The sensing members used in differential gaging with electronic comparator instruments can be either of the axially displaceable or of the pivoting (test indicator) type.

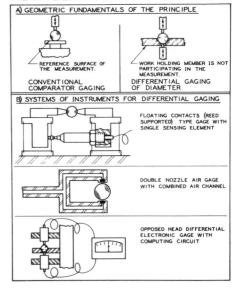

Fig. 7-10. Principles of instruments commonly used for the differential gaging of length.

The advantages of measuring length or diameter by differential gaging are not always realized. When the measurement is carried out under ideal conditions, that is, with the object sides perfectly flat and parallel, on an instrument staging table of impeccable flatness and exactly perpendicular to the axis of the gage head, then identical results could be expected, whether the gaging is carried out by conventional or by differential gaging methods.

Ideal conditions, however, cannot be expected as a general rule. Small irregularities in the flatness of the staging platform, a speck of dust, trapped air, or similar conditions, can affect the accuracy of measurements in the microinch order. Such uncontrolled, and often undetected, conditions will cause gaging errors that—when applying a geometric analysis—may be considered errors of the first order of magnitude (comparable to sine errors), errors of the second order of magnitude (comparable to cosine errors) or a combination of both.

The diagrammatic drawings of Fig. 7-11 compare the geometric conditions created by conventional as well as differential gaging in the assumed case when a minute burr or a speck of dust defects the object from its ideal position, which should be perpendicular to the axis of the gage head. While in the conventional method of comparative measurement both error sources, those of the first order and of the second order of magnitude, affect the result of the measurement, the differential gaging eliminates the interference of the more significant sine type error.

ELECTRONIC INSTRUMENTS IN HEIGHT GAGE APPLICATIONS

The test indicator type electronic gaging head attached to a pivoting, extendable, and tiltable cross bar of a gage stand, with means for controlled height adjustment and complemented with a fine adjustment device, results in a versatile gage, commonly designated as an *electronic height gage*. Although adaptable for varied applications, the term height gage seems warranted in view of the major use of this instrument: transferal of height dimensions and the comparison of the vertical separation between specific part surfaces, by referencing the measurement from a common datum plane. For the function of the datum plane, a surface plate is most commonly used.

Figure 7-12 illustrates a typical electronic height gage consisting of three essential members: (a) the gage stand; (b) the test indicator type sensing head; and (c) the amplifier unit with the meter.

The height gage stand is designed to be supported on a surface plate and for this purpose the bottom surface of the base has precisely machined contact pads, which must be flat and contained in a common plane within very close limits. The height gage stand for electronic measurements must be properly dimensioned for a high degree of rigidity and to pro-

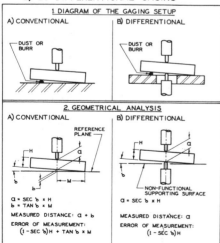

Fig. 7-11. Comparison of potential errors in thickness (diameter) measurement.
(A) By conventional gaging (sine error).
(B) By differential gaging (cosine error).

Giddings & Lewis Measurement Systems

Fig. 7-12. Electronic height gage assembly used for inspecting the length of a large tapered bearing roller by comparing its dimension to a gage block stack.

vide a stable support for the gage head even in the most extended position of the cross bar. The positive locking of the joints is also critical. The dimensioning of the gage stand, however, must not result in excessive weight, which could defeat one of the functional purposes of the stand, that is the transfer of height dimensions by the sensitive sliding of the stand on the surface plate.

For the vertical adjustment of the joint block that is holding the cross bar, the columns of height gage stands are often equipped with a rack and pinion device or with a friction roller guided in a groove. In the model shown in Fig. 7-12, the fine adjustment screw is part of the gaging head support. Other models, particularly those that instead of a cross bar, are equipped with a short horizontal arm only (see Fig. 7-13), achieve the fine adjustment by means of a flexure spring in the base of the stand that, when actuated by a thumb screw, imparts a tilt motion to the gage column. In other designs the turning of the thumb screw produces a sensitive vertical movement of the column.

Electronic height gages are generally used for the comparative measurement of linear distances on objects whose configuration satisfies the following requirements: (a) the surface being measured must lie in the horizontal plane; and (b) the distance to be determined must be referenced from a surface plate representing a plane parallel to the part surface on which the measurement is being carried out. The locating surface on the part is not required to rest directly on the surface plate, but may be supported on an intermediate parallel element of known thickness. Such an arrangement is needed for locating a surface that is in a setback position. When adequate support blocks are used, the indirect referencing will not distract from the resulting correctness of the height measuring process.

The linear dimension being measured in this process is designated a height because of the arrangement of the gaging process. The size of the dimension being measured is determined by comparing it to the height of a gage block stack, or to a level established by means of adjustable step blocks (Fig. 7-13).

Modern digital electronic technology permits absolute height measurement, as shown in Figure 7-14. With the push of a button, the digital display can be zeroed out at any position of the measuring probe (that is, the surface plate, parallels or a part feature). From the zero position, the probe is moved to measure a given height, with the difference in distance being displayed in absolute terms on the digital readout.

Another important application for electronic height gages in surface plate work is the verification of angles in conjunction with a sine plate. In this measurement, which will be discussed more in de-

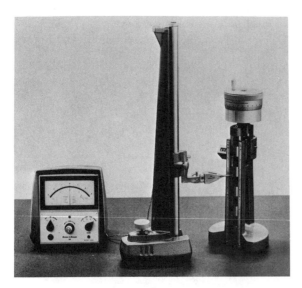

Brown & Sharpe Mfg. Co.

Fig. 7-13. Electronic height gage with sensitive position adjustment by a screw in the base. The adjustable step block, used in conjunction with the height gage, provides the required height levels for transferal and comparison.

Mahr Gage Co.

Fig. 7-14. Digital height measuring instrument.

tail in Chapter 10 on angle measurements, the setup of the part results in a condition where the top element of the part's surface is nominally in the horizontal plane. By referencing the electronic height gage at one point of that surface element, and then comparing the other points to that reference position, it is possible to detect and to measure sensitively any variations in the actual position of the explored surface element in relation to a horizontal plane.

It is possible, in principle, to use electronic height gages for comparator type measurements also. In such applications, particularly when small- or medium-size parts are measured, it is practical to retain the height gage stand in a fixed position, while the master and the parts are brought under the sensing tip of the gage. This operation could be carried out in any of several arrangements, such as: (a) sliding the part while it rests directly on the surface plate; (b) setting up a stage under the gage tip, similar to a comparator table; the stage may have guide rails and an end stop to reduce the gaging time in repetitive measurements; (c) holding the part in a locating support, such as a Vee-block, which is adaptable for sliding on the surface plate (in this case a master having the same form as the workpiece must be used for setting the reference size); and (d) holding the part on a precision slide with traverse movement actuated by a fine thread screw.

Diverse Applications for the Electronic Test Indicator

The test indicator type gage head, with which the electronic height gages are equipped (see Figs. 7-12 and 7-13), has many other uses, in addition to the surface plate based transferal, which was discussed in the preceding section.

These applications are essentially similar to those of the mechanical test indicators (see Table 5-5) and include the measurement of geometric interrelations such as runout, parallelism, flatness, wall thickness variations and various others.

The similarity of potential applications, however, does not result in equal practicality and suitability. The substantial difference of cost—or the order of about 10:1—requires a valid justification for the choice of an electronic test indicator as the replacement for a mechanical instrument. There are several technical aspects that indicate the use of one or the other basic category of instrument. It is well to consider the following distinctive characteristics of either system:

a. The sensitivity of an electronic gage can be substantially higher than that of a mechanical test indicator. Practical limits for the highest degree of sensitivity are on the order of 1/100,000 inch least graduation for the electronic system and 1/10,000 inch for the mechanical;

b. The indicating range of the mechanical indicators usually extend over 100 scale increments, whereas the range of most types of electronic gages is limited to 40 or 50 graduations of their scale. This limitation of the electronic gage is partially counterbalanced by its capacity to be switched over a lower amplification embracing a wider displacement range; and

c. Although there is no substantial difference between the overall size of a compact mechanical indicator and the transducer equipped sensing head of an electronic gage, the latter provides remote observation, as well as the possibility of signal utilization; these properties of the electronic gage can be of great value in specific applications.

ELECTRONIC COMPARATOR GAGES FOR INTERNAL MEASUREMENTS

The techniques required for the measurement of distances bounded by internal surfaces differs in several respects from those used in measuring the length or the diameter of external surfaces.

A few of the characteristic conditions that call for the application of distinctive measuring techniques are listed in Table 7-2, using for examples cylindrical features whose diameters are being measured.

While for external length or diameter measurements by means of electronic gages, a plain gage stand or even a height gage on a surface plate could prove sufficient, the carrying out of measurements on internal surfaces with a similar degree of accuracy would require a special internal comparator.

Two-Point Contact Internal Comparator Gages

Considering the conditions listed in Table 7-2 an internal comparator gage designed for measurements with better than 1/100,000th (10 microinches) sensitivity and repeat accuracy—an unusual but not extreme requirement for electronic gaging—should possess the following elements and capabilities:

a. A workholding plate or table to provide a staging plane.

**TABLE 7-2.   COMPARISON OF CONDITIONS AFFECTING GAGING TECHNIQUES
IN EXTERNAL AND INTERNAL LENGTH MEASUREMENTS
(DIAMETER MEASUREMENTS ON A PLUG AND ON A RING) — 1**

CONDITION	DIAGRAM		DESCRIPTION	EFFECT ON THE GAGING TECHNIQUE
	EXTERNAL	INTERNAL		
Referencing for gaging			Outside diameter is referenced from the locating plane; for internal measurements a separate reference point must be established.	When a stationary gage is used, which is essential for high precision measurements, the instrument must be designed to ensure the mutual referencing of the contact members.
Reliance on auxiliary surfaces			Outside measurements are made independently of other surfaces; for outside measurements the part rests on an auxiliary surface.	The part must possess an outside locating surface parallel to the gaging plane. The contact points of the gage must lie in the selected gaging plane.
Establishing the line of gaging			The outside diameter is the distance between two tangent parallel planes; the inside diameter is the distance between two points on the surface in the intersection of an axial and a radial plane.	Due to that difference in the basic geometric conditions, the locating and the aligning are much more critical for bore gaging than for outside diameters. The comparative measurement of bore diameters requires rather elaborate instruments.

TABLE 7-2. COMPARISON OF CONDITIONS AFFECTING GAGING TECHNIQUES IN EXTERNAL AND INTERNAL LENGTH MEASUREMENTS (DIAMETER MEASUREMENTS ON A PLUG AND ON A RING) — 2

CONDITION	DIAGRAM		DESCRIPTION	EFFECT ON THE GAGING TECHNIQUE
	EXTERNAL	INTERNAL		
Contact tip form	$r \to \infty$ R	$r < R$	The single contact tip for outside diameters may have a large radius approaching the flat; inside diameters require two tips with radii safely smaller than that of the bore.	The smaller contact tip radius reflects in several respects on the gaging conditions, requiring a reduced gaging force to limit the effect of elastic deformation, and are also more sensitive to alignment errors.
Stiffness of contact members			The gage spindle for outside measurements is not limited in size and the force acts in its axial direction. The usually weaker contact fingers for inside gaging are subjected to radially acting force.	Another reason why the gaging force must be kept light for inside measurements. The gaging force can have only a limited role in locating the part, in order to avoid harmful deflections.
Space limitations	$R \to \infty$	A_{min} $2R \geqq A_{min}$	Access to outside dimensions is usually unlimited, while for internal measurements adequate space is needed to accept the gage contacts.	After penetrating inside of the bore, the contact fingers must be located in the proper direction and in the selected reference planes.

b. Two contact fingers that can penetrate inside the feature, for example, a hole; the contact points on the fingers must be located in a common plane parallel to that of the staging. The distance between these contact members should be adjustable during the setting of a gaging process, for accommodating features of different dimensions within the design range of the instrument.

c. After completing the gage setting, the position of the fingers must be locked to assure that any change in the relative positions of the contact points will cause equal displacement of the transducer spindle. The precise parallelism of the gage contact plane in relation to the staging plane must be retained, unaffected by the setting and measuring movements of the contact fingers.

d. The contact force exerted by the two sensing points should be essentially equal. For this reason the positions of the contacts must be balanced in relation to the opposite feature walls that will be contacted in the gaging process. Table 7-3 shows diagrammatically three different design concepts directed at assuring the balanced contact conditions; these designs are applied individually by three prominent manufacturers of electronic internal gages.

e. In addition to the linear motions needed for the setting and for the positional balancing, the internal gage must have two additional movements along axes normal to the preceding ones: (a) a crosswise movement in the original measuring plane, but perpendicular to the line of contact, for finding the position of the maximum length, that is, the true diameter of a bore measured in its axial plane; (b) a vertical movement normal to the gaging plane, to transfer that plane to a different level. This vertical adjustment is primarily needed during the initial setting of the gage, but also may be required for exploring the geometry of the bore by making diameter measurements in different parallel planes. Geometric errors that reflect on the diameter, such as taper, barrel form and bell mouth, can be detected and measured with great accuracy by such a procedure.

This process of measuring the diameter in different parallel planes will produce a kind of mapping of the bore geometry. For this purpose it is of advantage when the table of the internal gage is equipped with an indicator showing the amount of vertical table adjustment. Unless the original design of the instrument provides a comparable purpose indicator, a long range dial indicator mounted on the

table and contacting a fixed anvil attached to the gage stand (see the last sketch in Table 7-3) will satisfy this requirement.

Figure 7-15 shows a widely used model of internal gage equipped with a standard transducer cartridge used in conjunction with a battery powered amplifier. The system of contact finger suspension used in this gage is shown in the second sketch of Table 7-3. The smallest increment of the indicating meter is 10 microinches, when using the highest amplification.

The internal gage illustrated in Fig. 7-16 is designated by its manufacturer as "master comparator" and incorporates design aspects aimed at measurements with the highest degree of accuracy with respect to both the repeat of the indications and the variations from the reference size. The contact finger suspension of this gage is shown diagrammatically in the last sketch of Table 7-3. The helical spring that is acting on one of the contact fingers, when not locked by a cam, has the purpose of establishing equal contact force during the setting of the gage (with gage blocks) and during the actual measurement. By assuring equal contact force, the accuracy of the comparative measurement will be less affected by the unavoidable deflections of the contact fingers and by the penetration of the contact points into the object surface due to the elastic deformation of the contacting surfaces. Although the combined effect of said variables might represent only a few millionths of an inch, these values are still in excess of negligible magnitudes in the case of an instrument that is

Federal Products Corp.

Fig. 7-15. (Left) Electronic comparator.

Fig. 7-16. (Right) Electronic comparator gage for calibration laboratory applications with 60,000 × highest amplification. Primary use is for internal measurements, but it is also applicable for outside dimensions.

TABLE 7-3. OPERATIONAL PRINCIPLES OF DIFFERENT TYPES OF INTERNAL COMPARATOR GAGES

DESIGNATION	DIAGRAM	DESCRIPTION
Single Reed Self-centering Table		One of the contact fingers is solidly anchored to a stationary gage member while the other rests on a reed. The instrument table is roller supported in order that the force exerted by the reed should be sufficient to move it into a position where the object is well centered, and in contact with the solid gage finger. Any change in the position of the reed block, which results from variations in the bore size, is picked up by the transducer sensing head.
Two Reeds in Tandem		Two reed blocks in line, each carrying a contact finger, will exert essentially equal contact force when the distance between the fingers is adjusted to the approximate bore size. For this purpose one of the blocks has guideways in which the contact finger can be moved by a screw to the required position. One of the blocks is holding the transducer cartridge in a bracket, while an extension of the second block carries the anvil against which the sensing point of the cartridge will bear. To assist the centering, the work is resting on two gage pins.
Two Reeds Superimposed		One of the two reeds functions like a cradle holding the entire sensing member of the gage in a self-adjusting suspension relative to the stationary table. A second reed mounted on the cradle and carrying one of the fingers assures equally distributed gaging force, the other finger being attached to the main reed. A practically frictionless floating contact results. Identical contact force is exerted for the setting and the gaging by a separate helical spring. Front-to-back motion of the table top permits centering in the axial plane of a bore without contacting the work.

designed to measure in increments as small as one microinch.

The gage shown in Fig. 7-16 can also be used for external measurements by reversing the direction of both the contact points and the spring action of the contact fingers. Measurements of external diameters on a horizontal comparator are indicated in various cases such as for parts that must be measured in a plane parallel to one of the face surfaces on which the part may be located, or on a part whose general form requires the measurement to be made while the part is resting in a horizontal position, thereby avoiding excessive deflection.

The Setting of Internal Comparator Gages

The basic geometry of internal surfaces does not permit using gage block stacks directly for setting up the gage to a specific nominal size in the manner currently applied for external comparator gages. To simulate the geometric conditions that characterize the internal surfaces, the contacted surface elements must straddle the gage sensing points. It is quite possible, and even ideally correct, to use gage rings as setting masters, yet in practice such a procedure is more often the exception than the rule. This condition is explained by the major field of application of electronic internal comparators, which is the measurement of ring gages and other individual high precision parts; therefore the ready availability of master rings of suitable size can only rarely be expected.

In view of the high degree of measuring accuracy for which this category of gages is designed and generally used, a commensurate level of accuracy must be assured in the setting phase of the process. Gage blocks of the appropriate grade, frequently AA or A+, are selected and their inherent accuracy is transferred to the sensing points of the gage with the least amount of additional error.

End beams having a flatness comparable to that of the selected grade of gage blocks should be used to assure the accuracy of the distance transferal. Standard type gage block holding frames to retain the stack of blocks together with the pair of end beams are used frequently for setting internal gages. However, it is difficult, and also of questionable reliability, to hold the gage block set on the staging table of the gage in a position where the parallel planes of the beam faces are exactly perpendicular to the line of action of the gage point displacement.

To reduce the effect of potential errors in the setting of internal gages, instrument manufacturers have devised and are supplying for that operation

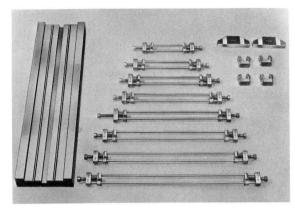

Federal Products Corp.

Fig. 7-17. Elements of a gage block holding clamp with transfer jaws and the auxiliary alignment rail for use during assembly.

functionally adapted gage block holding fixtures. Figure 7-17 illustrates a particular model of such devices, supplied as a complete set to accommodate a wide range of sizes. The set also contains an auxiliary element and an alignment rail for locating the gage blocks during their assembly into the holding device. Figure 7-18 shows the use of the fixture in setting an internal comparator gage. To assure the parallelism of the gage block stack to the table top, the former is resting on four precision balls that are held in brackets mounted at proper distance to straddle the central opening on the staging table.

When measurements with an expected repeat accuracy better than one millionth part of an inch are to be effected, the errors in the parallelism of the end beams that could occur to a very small degree even in the described type of special device are inadmissible. For the purpose of carrying out the gage setting with the highest degree of accuracy, the direct wringing of gage blocks to a master square block may

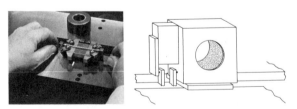

(Left) Federal Products Corp.

Fig. 7-18. (Left) The clamp device of Fig. 7-17 shown holding a gage block stack in application for setting the contact-point distance of an internal comparator gage.

Fig. 7-19. (Right) Diagram of a setting gage assembly using gage blocks wrung to a square master block to produce an accurately controlled internal span.

be considered. Such square blocks can be manufactured to equal the bet grade of gage block accuracy. The principles of a setting gage assembly, based on the aforesaid elements, can be seen from the diagram in Fig. 7-19. One of the advantages of the illustrated arrangement is the controllable parallelism of the functional surfaces. The mutually parallel position of the operational elements can be verified with very high sensitivity by using optical means, such as interferometry or an autocollimator.

DIVERSE APPLICATIONS OF ELECTRONIC GAGES

The electronic gages, by producing electrical signals, which can be amplified and are proportional to the distance of the sensed displacement, are adaptable to unique measuring applications and lend themselves to many secondary functions that are in positive relation to the displacement sensed by the gage.

The variety of application possibilities and the multitude of secondary functions exclude a complete listing. In addition, such an attempt would be utterly unsuccessful in view of continuous developments and progress in the application of electronics for dimensional measurements and controls.

For the discussion of these important aspects of electronic gages, the following procedure has been chosen:

a. Listing in a tabulated form with examples of utilization, the unique characteristics of electronic gages (see Table 7-4). Notations are added to indicate the sections or chapters where the pertinent instruments or processes are discussed in greater detail.

b. Discussing a few applications of electronic displacement gages and derived functions that differ from the basic types of length measuring uses reviewed in the preceding sections.

Digital Read-Out Systems for Electronic Gages

Reading a meter by associating the pointer position with a specific spot along the scale of the meter, and then expressing that position in numerical values by combining the printed numerals and the intermediate graduation lines, is a procedure burdened with potential errors. The errors may stem from different causes, the major culprits being: (a) parallax errors when observing the meter's face from an angle; and (b) errors in counting the graduation lines from the nearest numeral. Some degree of uncer-

tainty also originates from using judgment in assigning a value to pointer positions that are not exactly in coincidence with a scale graduation line.

The use of digital displays connected to the amplifier of an electronic displacement gage can eliminate the sources of readout errors that are associated with meter faces and pointers. A digital display, used in conjunction with a compatible transducer and amplified, will display in numerals and without ambiguity the corresponding value of the sensed displacement.

The displaced numerals are easy to read, even from a distance substantially greater than used for meter observation, and do not have to be read from a particular, precisely aligned position. Most digital displays permit four-digit presentation, providing a much higher resolution of the sensed displacements than a meter with 20 to 50 graduation lines on each side of the zero position. Various models of digital displays have graduation lines between the least significant digits to provide additional readout resolution (Fig. 7-20).

Range selection switching combined with decimal point shift are also available on certain models and add to the adaptation versatility of digital displays for the purpose of dimensional measurements. Most models of digital displays used for length measurements have adjustable reading retention, a feature of substantial value when the measured part dimensions must be manually recorded.

Mahr Gage Co.

Fig. 7-20. Digital universal measuring instrument.

**TABLE 7-4. SECONDARY AND AUXILIARY FUNCTIONS DERIVED FROM
ELECTRONIC DISPLACEMENT MEASURING GAGES**

UTILIZED CHARACTERISTIC	APPLICATION EXAMPLE	NOTE
Producing signals which are transferable by any appropriate electrical conductor	Transferring the electrical signals through a slip ring permits continuous rotation of the sensing head without affecting the accuracy of the indications displayed by a stationary amplifier.	Rotary indicators will be discussed subsequently.
Producing voltage variations proportional to the distance of the sensed displacement.	Indicating the numerical values of the displacement on a digital voltmeter; optionally, printing out the numerical values on a printer or writing to a disc for computer applications.	Digital voltmeter display and printing will be discussed subsequently.
The sensed displacements produce proportional variations in electric current or potential.	Signals may be transferred to a recorder which prepares a permanent record of the consecutive measurements.	Recorded length measurements will be discussed subsequently.
The lightweight and compact sensing members are adaptable to controlled translational and rotational motion and are operative in any position.	Contour tracing instruments which indicate and record with great sensitivity the actual form of a surface element in comparison to either a straight line or to a practically perfect circle.	Will be discussed in Chapter 13 on Profile Measurements and in Chapter 14 on Roundness Measurements.
Very light gaging force and extremely fast response to the sensed displacements.	Surface texture measurements by tracer instruments which either indicate integrated (averaged) values or supply chart tracings with selective magnification.	Will be discussed in Chapter 15 on Surface Texture Measurements.
The proportional electrical signals produced by linear displacement can initiate secondary functions which are precisely associated with limit positions of the displacement range.	Machine tool control functions, frequently termed auto-sizing.	
The range of the linear and continuous electrical variations resulting from the displacement of the sensing member can be subdivided electrically into adjacent sections producing stepped signals.	Automatic segregation of parts into different size groups.	

Numerical Recording Systems for Electronic Gages

The results of measurements effected on a series of parts often need to be recorded for reference purposes. Records are particularly important for calibration procedures, such as in the case of gage block verification.

Digital displays are sometimes used in conjunction with compatible storage discs or printers. These instruments may be simple, or they may be sophisticated, such as the elements used in the gage block calibration set shown in Fig. 7-21.

The essential measuring function, comparing the size of a test part to that of a master used for setting the instrument, is carried out by a standard electronic gage block comparator. The additional elements are a digital display indicator and interfaced to a personal computer with a dot matrix printer.

Following the necessary preparatory operations, which include the filling in of a calibration certification form so that it contains all the required information except the actual size of the gage blocks to be inspected, the semiautomated calibration process may be started. From this stage on, the operator's role is limited to the measurement of the individual gage blocks by comparing their size to that of the corresponding master. The measured size difference, as indicated by the meter of the amplifier can, at this stage, be transferred to the recording part of the setup. The printer will insert the value and the sign of the measured deviation into the appropriate space of the certificate form and then index to the position of the next size. A tolerance limit control is also a part of the instrumentation that, when set previously to a specific tolerance range, will cause the printer to add an asterisk to the deviation value should that exceed the preset tolerance limits.

In addition to saving time, which can be meaningful in the case of continuous gage-block calibration, the main advantage of the system lies in eliminating such sources of error associated with the manual recording and the subsequent transcription of the visually observed measurement results.

Recording Discrete Length Dimensions on a Strip Chart

When digital readout is not actually required but the consecutive measuring results obtained in inspecting a series of nominally identical parts should be recorded mechanically, the use of a strip-chart recorder may be considered.

Figure 7-22 shows an example of an electrical recorder with an in-process strip chart, the levels of the bar plateaus indicating the sizes of the measured individual dimensions. Such charts are produced with considerable savings in time by avoiding errors that frequently occur in manual recording. In addition to providing an easy-to-duplicate permanent record of a measurement series, the charts can also be used for a number of quality control purposes, such as:

a. Presenting a synoptic image of the sizes of a batch of parts in relation to the pertinent tolerance

Mahr Gage Co.

Fig. 7-22. Electrical recorder with strip chart.

Brown & Sharpe Mfg. Co.

Fig. 7-21. Electronic gage block comparator with digital readout and interfaced to a personal computer with a dot matrix printer.

limits represented on the chart by drawing two properly placed parallel lines;

b. Determining the span of size variations by drawing parallel lines tangent to the highest and the lowest bar, respectively; and

c. Displaying graphically the trend of machine performance, for example, that of a grinder, by connecting the adjacent bar plateaus with a single line; the parts are measured in the same sequential order as produced following the dressing of a wheel, or the adjustment of certain machine controls.

NONCONTACT GAGES

Another form of electronic gage is the noncontact thickness meter, which uses a capacitive measurement method to eliminate the surface damage to the material under inspection (see Fig. 7-23). The material thickness is calculated through the use of high precision capacitance type gap detectors. Materials that are conductors, semiconductors and insulators ranging from sheet steel to plate glass can be inspected for thickness to the submicron level using this type of instrument.

The principle behind noncontact thickness measurement is as follows. Gap detectors (sensors) measure displacement (gap) by calculating the capacitance between the sensor and the part being inspected. This capacitance is a function of the sensor's surface area, the opposing part material and the gap between them. Assuming the part and the face of the sensor are held parallel and given a known sensor surface area, the gap will be inversely pro-

portional to the capacitance. Then, through the use of processing circuits, the need for linearization is eliminated and a voltage directly proportional to the gap is determined. This voltage is converted to a digital signal and processed to permit the display of the material thickness.

Noncontact gaging provides many advantages over measurement methods requiring surface contact. Items such as compact recording discs and photographic films can be measured without the fear of scratching the surface. Continuous films can be measured automatically as they pass between the sensors. This measurement method also eliminates any influence that contact may have on soft or nonrigid materials.

Ultrasonic Gages

Ultrasonic gaging uses the principles of 1940s sonar. However, the electronic microprocessor advances of recent years has permitted the miniaturization of this technology into a useful form for the taking of thickness measurements. High frequency (1 to 20 MHz) short wavelength ultrasound is now used in industry to make accurate thickness measurements on most engineered materials (that is, metals, plastics, ceramics, composites, epoxies and glass). The measurement is made from only one side of the material as the ultrasound waves are sent through

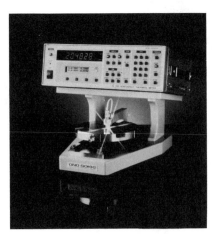
Ono Sokki Technology Inc.
Fig. 7-23. Noncontact thickness measuring system.

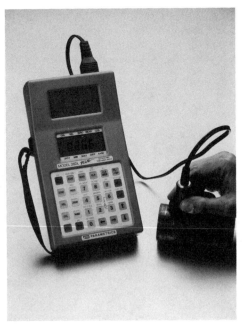
Panametrics Inc.
Fig. 7-24. Ultrasonic thickness gage.

the material to bounce *(echo)* off the opposite side and return to the sending/receiving transducer to be digitally displayed as a thickness reading (see Fig. 7-24).

The transducer converts sound energy into electrical energy, which is further processed electronically as the time interval between the initial pulse and back wall echo is recorded. Material thicknesses of from 0.020 inch to 10.00 inches can be measured to an accuracy of ±0.001 of an inch using ultrasonic gaging. However, different materials transmit sound at different velocities, making it essential to calibrate the ultrasonic gage using the material type to be measured.

Hall Effect Gages

Another commercially available method of thickness measurement utilizes the *Hall effect*, which is the production of a transverse voltage in a conductor in a magnetic field. Figure 7-25 shows a Hall effect thickness gage (a *magnetometer*), which places a magnetic probe on one side of the material being measured and a small target steel ball on the other

Panametrics Inc.

Fig. 7-25. Hall effect digital thickness gage.

side. The Hall effect sensor measures the gap between the probe and the target ball. All nonferrous materials up to 0.250 inch thick can be measured to an accuracy of from ±1% to ±3% of the material thickness with this type of magnetometer.

8.
Engineering Microscopes

The measurement of geometrically defined dimensions with the "light touch" of light is appealing, as well as positively advantageous for many metrological applications. Measuring conditions created by the particular geometric forms or the insufficiently defined boundaries of object features; the not readily accessible location of the dimension to be measured; the functional interrelation of critical dimensions, materials or finishes too sensitive for mechanical contacts; and the desirability of visual observation for alignment or for contour comparison,

are examples for the preferential application of optical measurements. The optical measuring instruments that provide either virtual or projected images of magnified portions of the test piece are the engineering microscopes and the optical projectors. A few models of engineering microscopes, representing advanced types, designed for versatile uses, are shown in Figs. 8-1 to 8-3.

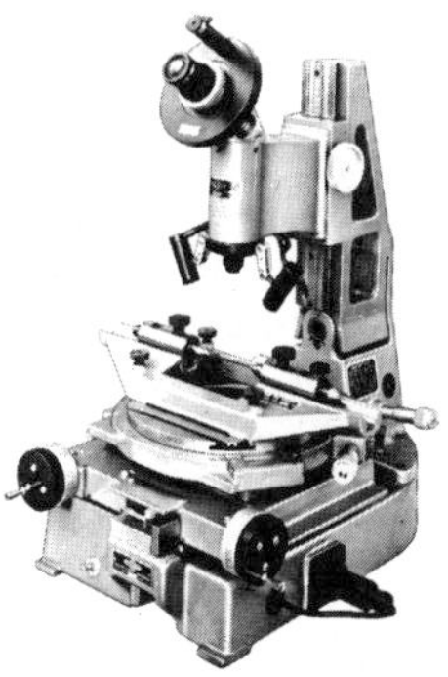

Carl Zeiss, Inc.

Fig. 8-3. Large, universal measuring microscope. Push-button control selectively brings the magnified scale sections of the cross-slides, the optical protractor, the rotary stage and the tilt-adjustment of the microscope head into the field of view of the microscope tube, for the convenient reading of the measuring values. A projection screen for contour comparison against transparent charts is mounted above the binocular tube.

(Left) The Gaertner Scientific Corp.
(Rigt) Leitz/Opto-Metric Tools, Inc.

Fig. 8-1. (Left) Toolmaker microscope for precision shop measurements. The protractor eyepiece indicates angles in one minute of arc increments; the rotary stage is mounted on the cross-slides, which have substage illumination; attached lamps are used for surface illumination. The cross-slides have micrometer screws for displacement and position measurement.

Fig. 8-2. (Right) Universal measuring microscope. The binocular observation tube contains an optical protractor and the cross-slides have individual master scales, which are observed in the display window of the optical readout system. The instrument accepts many types of optional accessories for part staging and measurement.

Functions of Engineering Microscopes

Engineering microscopes are optically assisted instruments for measuring the geometric dimensions and forms of small- and medium-size technical parts. These instruments provide the means for carrying out the following basic functions:

a. Magnification, the primary function of microscopes in general, used for presenting the enlarged view of the observed object area, either in its contours or as a surface image;

b. Referencing or aiming, accomplished by providing index lines on a transparent graticule inside the microscope tube, observed concurrently with the magnified image. The index lines guide in determining reference positions for specific surface elements or, when representing nominal contours, the alignment and comparison with the observed part section; and

c. Staging, which involves holding the object and displacing it along controlled tracks translationally or, as an option, rotationally, over measured distances.

Illumination, although indispensable for any optical process, is an auxiliary function. The source as well as the means of transmission for the light are generally integral elements of the engineering microscope, although illumination can be and occasionally is provided by members attached to the instrument, or even independently arranged.

General Operational Procedure

Dimensional measurements on the engineering microscope commonly involve the following operational process: The microscope part of the instrument permits the inspector to observe, at an appropriate magnification, the selected elements or features on the part's surface in a fixed position, which is coinciding with, or in a defined relation to the optical axis of the microscope. For establishing that position of coincidence, the part must be displaced by moving the stage on which it is supported. The direction of that displacement is precisely controlled and its distance is measured, thus supplying data for determining the separation between specific points on the part's surface that were brought sequentially into coincidence with the axis of the observation system.

The plane of staging is generally normal to the direction of observation and contains the two, mutually perpendicular, displacement axes. These may be considered to represent the X and Y axes of a system of rectangular coordinates. Consequently, any selected point on the observed part surface can be brought into the reference position of the optical system, as long as it is contained within the displacement range of the stage. This latter consists of a cross-slide system, moved by means of two lead screws, which also often function as the spindles of micrometers for the measurement of the displacement distance, unless the instrument is equipped with master scales for optical distance measurement.

Depending on the intended use with regard to the size range of the parts, the attainable sensitivity and accuracy, and the required versatility (for example, whether angles should also be measured), engineering microscopes are made in a large variety of designs, grades and capacities, with a more or less generous choice of optional accessories.

The many variables that result from the diverse types of instruments and from the extensive choice of accessories make it impractical to evaluate the design principles and the application potentials of individual models of instruments. For this reason the following discussion of functions and applications potentials is based on advanced models of these instruments. For users of engineering microscopes that represent less sophisticated models, or have for equipment only a limited number of accessories, it should be possible to visualize the applicable techniques and the attainable accuracies, when adapting the review to their actual equipment by a process of elimination.

To facilitate the evaluation of instrument capabilities, the last section of this chapter will discuss the different types of engineering microscopes. While not intended to provide an inventory, that survey, which is based on representative makes and models, may prove to be of assistance in selecting the type of engineering microscope best suited for a specific kind of measurement.

THE OPTICAL OBSERVATION SYSTEM OF ENGINEERING MICROSCOPES

The measurement of geometric conditions by a microscope is based on the observation of the object, or of one of its parts, at an appropriate magnification. This observation is carried out in a plane normal to the direction of viewing. It follows that, to be adaptable to dimensional measurements by microscope, the critical parameters of the object feature must be associated with a common plane of observation.

The ordinary method of producing a virtual plane of observation for optical measurements is by creating, with the use of collimated light rays, an illuminated field, a part of which will become darkened due to the presence of a physical body. The contour of the darkened area, as observed through the microscope, will represent the magnified replica of the object contour in the plane of observation. Although not actually a shadow because it is not projected on a surface, the darkened area (due to its appearance) is commonly designated a *shadow image* or, more correctly, a *silhouette*.

Features that cannot be brought into the path of light to intercept its rays can also be observed by the microscope as long as they are contained in an essentially common plane and are open for unobstructed viewing from at least one direction. In this case, the illuminating light must travel in a way similar to that of the viewing and the image is created by the reflection of the light from the object surface in the selected plane of observation.

While the unobstructed observation of the meaningful contour in an essentially single plane is commonly required for dimensional measurements by the microscope, exceptionally, contour elements that are partially obstructed can also be measured by using special accessories.

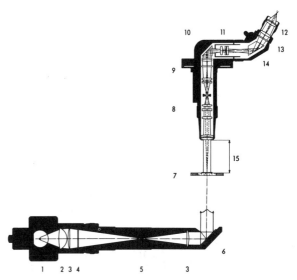

Leitz/Opto-Metric Tools, Inc.

Fig. 8-4. The optical system of an engineering microscope. The diagram illustrates a type that produces intermediate images and is equipped with substage illumination. *(1)* Point source of light; *(2 and 3)* collimating lenses; *(4)* filter disc; *(5)* aperture diaphragm; *(6)* deflecting mirror; *(7)* work stage; *(8)* optical system for the intermediate image; *(9)* graticule magazine; *(10)* deflecting prism; *(11)* microscope objective; *(12 and 13)* lens system of the adjustable eyepiece; *(14)* deflecting prism; and *(15)* focusing distance.

Optical System with Intermediate Image

The principles of design of a well proved system are shown diagrammatically in Fig. 8-4. In this optical system, a relay objective creates a one-to-one image of the staged object in a definite plane. Glass graticules, with index marks or contour lines of various commonly used, or even special, forms can be placed exactly in this optical plane. The intermediate image of the object may then be observed through the microscope that, for this particular model, contains in a rotatable turret three objectives to produce $10\times$, $20\times$, or $30\times$ magnifications.

Not all types of engineering microscopes have an optical system that relays the object image into an intermediate plane. However, the creation of an intermediate image has several distinct advantages that deserve attention. The plane of that image remains the same independently of the selected microscope magnifications; this property of the system permits variation of the magnification without having to change the distance of the object from the relay tube for refocusing. An adjustment of that distance is required only when the relay objective is exchanged to produce an initial magnification, such as

$5\times$, resulting in $50\times$, $100\times$ and $150\times$ total observed magnifications.

The graticules that serve as visual guides for contour comparison or that provide reference lines for establishing observational coincidence with specific elements of the object image are usually mounted into cartridges. The cartridges can be inserted into the microscope tube by means of a magazine changer, or may be contained in a turret, whose positions can be brought selectively into the field of observation. Since the graticules are located in the plane of the intermediate image, their observation can be carried out with any of the microscope's magnifications, without the need for exchanging the graticule for one prepared in another scale.

The full-scale transmission of the object image into the optical plane in which the graticule is placed avoids distortion due to errors in magnification. The observation of the image against a clear glass graticule provides a sharper definition than that attainable on a grained projector screen when identical magnifications are used.

The diagrammatic view of the optical system shown in Fig. 8-4 represents one specific form of execution, but variations in design are not unusual. The measuring microscopes illustrated in Figs. 8-1 to 8-3 have binocular tubes instead of the single eyepiece shown in the schematic cross-sectional drawing. The binocular tube can be tilted to the most convenient angle and may be adjusted for different interpupillary distances. This feature contributes to the ease of observation and greatly reduces the operator's eye fatigue during continuous work on the microscope.

Supplementary Optical Equipment

There are a few models of engineering microscopes that can be equipped, or are originally supplied with a projection screen to complement the viewing through the eyepiece of the microscope (see Fig. 8-3). Such screens are of distinct advantage when simultaneous observation by several people is desired, or when features of irregular form are being inspected by comparison to a transparent chart mounted on the screen.

While the understage illumination, such as shown in Fig. 8-4, is the most frequently used, not all object features lend themselves to the production of "shadow images" for precise contour observation. In such cases, illumination by light reflection is indicated. Light sources are available that illuminate the object surface with light rays essentially parallel with the line of observation, permitting optical measurements in the bottom of recesses or of blind holes. There are certain types of surfaces whose topography can be observed with better discrimination when the illumination strikes the surface at an angle: for such cases, adjustable light sources, attached to the microscope heads, are preferred. These are ordinarily used in pairs, emitting light of identical or different color, the latter selected to provide optimum definition of the observed surface details.

Referencing for Distance Measurements

One of the important functions of the measuring microscope's optical system is the referencing for distance measurements. The meaning of the term, in this context, is the establishment of boundaries for a dimension to be measured. While producing a visually observable coincidence of the boundary elements with the cross hair of the objective graticule is the most common method of optical referencing, it is not the only one. There are various alternate processes of referencing that are carried out with accessories other than the basic viewing system, although the operation of these accessories also relies on the use of optical devices.

The operation of said accessories is based on particular optical effects, such as the mechanically induced deflection of a reflecting mirror; the reflection of a grazing incidence light beam from the surface of the workpiece; or the optical reversal of images whose congruence indicates the optical alignment of the observed feature. Although operationally different when considering their functional purpose, these accessories also belong to the optical observational system of the microscope. While details of the design and operation of these accessories will be discussed later, in conjunction with the function of referencing, the existence of alternate systems is pointed out for the completeness of this general survey of the optical observation system.

THE STAGING AND DISPLACEMENT MEASURING MEMBERS

Part features, whether appearing as contours or as surface portions, are usually presented for microscope observation in a horizontal plane. For many types of objects, this condition can be satisfied most conveniently by actually supporting the part on a platform, usually a glass plate. Another frequently used method of workholding is between centers, in a plane parallel to that of the basic platform. The centers support the part directly when it is provided with center holes or by means of an arbor for parts with a bore. Further alternate workholding methods comprise Vee-blocks, magnetic blocks, vises and so forth.

Because the dominant plane of support is horizontal, workholding is often simplified by relying on gravitational force, perhaps assisted by a light clamping action. The basic method of workholding on the microscope and the versatility of alternate locating means make the staging particularly convenient and reliable for small- or medium-size parts.

The staging platform is a part of the microscope slides that provides precisely controlled translational motion in two mutually perpendicular directions, the X and Y axes of a system of rectangular coordinates. When equipped with a rotary table, the stage also has a rotational movement. Furthermore, most types of measuring microscope stages also have means for a motion that compensates for minor work orientation errors and corrects the alignment of the meaningful object element with the direction of the slide motion.

Although not physically part of the staging platform, the tilt motion of the upright is functionally associated with the staging operation. This tilt adjustment can be controlled to increments of 10 minutes, or smaller, and permits setting the line of view in a direction normal to the cross-sectional plane of a helix form. Alignment that satisfies these conditions is essential for the convenient observation of screw threads, worms and of similar parts with basically helical form.

In engineering microscopes, the displacement movement of the slides along the X and Y axes is measured either by means of micrometer screws or by comparing the slide position to the pertinent graduations of individual master scales. This comparison is effected by the optically guided and assisted observation of that section of the scale that coincides with the referencing target. The scale may be held in a fixed position, whereas the optics for observation travel with the slide; or the scale may be attached to the slide and the observation system rigidly connected to the instrument base. Either of these designs may be used. In addition, there are instrument models where one of the cross-slides has the master scale attached, and the other slide carries the optics to view the related scale, which is mounted to the stationary base.

The master scales used in measuring microscopes of the inch system have graduations commonly in 0.050-inch increments. Two different methods are known to be used for producing these graduations—photographic reproduction on glass and direct ruling on steel. The accuracy of such master scales is claimed not to exceed a maximum error of 0.000008 inch for any distance within the graduated range.

The optical readout device permits the observation of these graduations with a magnification on the order of 50×, resulting in a resolution that may be associated with a 20-microinch distance along the scale. This value corresponds to the least graduation of the most sensitive type of readout devices found on measuring microscopes.

The intermediate positions between the scale graduations are measured with the aid of adjustable graticules in the observation system. This may consist of an optical projector whose display window has a laterally adjustable vernier slide to bring the projected scale graduation line into a bifilarly centered position (Fig. 8-5), or of a micrometer eyepiece having a tiltable light deflector plate that is connected through a cam with a rotating graticule (Fig. 8-6).

In each of these systems, the position of the stage slides is determined by purely optical means. Any mechanical action involved in the readout process is limited to the displacement of graticules whose per-

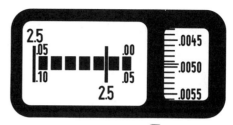

Leitz/Opto-Metric Tools, Inc.

Fig. 8-5. Projection window of a scale readout device with vernier graduations in 0.0001-inch increments. The screen with a series of blocks can be laterally displaced to bring the projected scale line (in this case the lower 2.5) into an accurately centered position between two blocks (in the example, the gap used represents 0.06 inch). The amount of screen displacement needed to create a centered position is shown on the right-hand scale (indicating 0.00495). The combined values of both scales represents 2.56495 inches.

manent rulings are not affected by this movement. The resolution of such measuring systems is 100 or 50 microinches, and on more recent models 20 microinches.

When the potential accuracy level of the micrometer-screw-actuated slide movement is being compared with that of the displacement position obser-

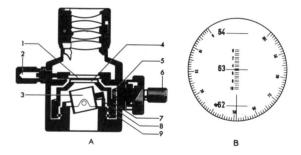

Leitz/Opto-Metric Tools, Inc.

Fig. 8-6. The micrometer eyepiece of a scale-reader operating on the principles of the optical micrometer. (A) Cross-sectional diagram showing: *(1)* vernier scale graduated in 0.1 millimeter or 0.005 inch; *(2)* adjusting knob for the 0.005-inch scale; *(3)* plane-parallel plate (optical glass); *(4)* rotating disc with circular divisions (one graduation equals one micron or 50 microinches); *(5)* bevel gear; *(6)* knurled knob for rotating the vernier scale and axial cam; *(7)* ring gear of this tubular frame for holding the vernier scale; *(8)* axial cam for tilting the plane-parallel plate; and *(9)* tangential coupling. (B) Field of view as observed through the eyepiece.

Tilting the plane-parallel plate produces a controlled, lateral displacement of the projected scale graduation, permitting its centering between the nearest bifilar reference lines on the stationary screen. The tilting is accomplished with an axial cam whose rotation is coupled with the graticule having circumferential graduations, also observable on the readout screen. The example indicates 63.551 millimeters, the least graduation lines representing 1 micrometer (about 40 microinches).

vation based on master scales, it must be remembered that the translational movement of most micrometers is limited to 1 inch. Consequently, when distances exceeding 1 inch must be measured on a microscope equipped with a micrometer slide, the full inch elements of the distance must be established with gage blocks, which are inserted between the anvil and the face of the micrometer spindle. Spring pressure is applied to the slide to make it advance toward the anvil to a position where the face of the micrometer spindle bears against the anvil directly, or through the inserted gage blocks.

In addition to the inconvenience of having to use gage blocks as complementary elements for accomplishing the desired length combination, the probability of gaging errors due to foreign particles and air pockets on the interfaces, or caused by distortion in the spring-loaded members, must also be considered.

How the difference in the basic design and operation of these two systems of displacement measurement affects the attainable measuring accuracy is illustrated in Fig. 8-7. The chart shows the general measuring accuracy levels of both systems, namely, the micrometer head in combination with gage blocks, and the master scale with optical position observation.

The plots, based on formulas shown in the illustration, indicate the magnitude of potential errors at a 2σ level, which are inherent in the analyzed system. However, the actually accomplished measuring accuracy, which is dependent on many factors with cumulative or compensating effects, can substantially deviate from the potential error level shown on the chart.

The earlier models of measuring microscopes had the limitation of requiring a controlled temperature environment for their accurate operation. The glass used for the master scales had a coefficient of thermal expansion much smaller than that of steel and, unless the 68°F (20°C) reference temperature was observed, measuring deviations had to be expected.

Although for the majority of microscope applications the effect of minor temperature variations may be meaningless, new glass materials have been developed for master scales that possess a thermal expansion comparable to that of steel.

LENGTH MEASUREMENTS WITH THE MICROSCOPE

The term *length*, in this context, designates the separation of two points that represent the boundaries of a straight-line distance. The two points also establish the direction in which that distance must be measured.

Straight-line distances are the predominant basic elements of geometric dimensioning—another important category of elements is the angular separations—with most features on engineering drawings being defined by several length dimensions. These may be related to external or internal surfaces, may be determined by actually observable points, or may be determined by virtual points only, virtual points being defined indirectly with the aid of physically present elements of the feature. Several length dimensions may be independent of each other, or interrelated. Interrelations may be locational (for example, a specific distance apart), orientation-defined (for example, parallel, perpendicular) or have a common datum (for example, coordinate dimensions).

This brief recapitulation of conceptual fundamentals should indicate the aspects in which the dimensions of technical parts are analyzed for evaluating the feasibility of microscope-based measurements.

The process of length measurement by which linear dimensions of technical parts are determined, always involves three basic functions:

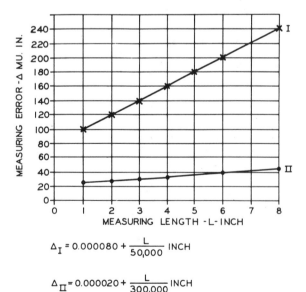

$$\Delta_I = 0.000080 + \frac{L}{50,000} \text{ INCH}$$

$$\Delta_{II} = 0.000020 + \frac{L}{300,000} \text{ INCH}$$

Fig. 8-7. Levels of potential errors in length measurements by microscope slides: (I) equipped with micrometer screw supplemented with gage blocks; (II) equipped with optically observed master scale.

a. Establishing on a physical body the location of an imaginary line along which the length dimension must be measured.

b. Determining on the surface of the part the two boundary points of the length section to be measured along that reference line, and finally,

c. Measuring the separation of the boundary points.

The first two functions may be designated the "referencing phase" of the process, while the third function involves the distance measurement. The measurement of linear distances with the microscope is always accomplished by the controlled displacement of the instrument stage in relation to the optical axis of the sighting. The operation of the microscope's displacement-measuring system has already been discussed in the preceding section dealing with the general description of the instrument. This operation, its functions and capabilities are essentially the same for all linear measurements on the microscope.

The referencing functions, however, are varied and will differ depending on the design characteristics of the length dimension to be measured, and are also affected by the physical location of the dimension on the actual part. Therefore, in discussing the microscope's applications for measuring diverse length dimensions, the emphasis will be on the referencing phase of the entire operation.

The diversity of physical locations for the dimension to be measured excludes the application of a single basic system of referencing for all types of length dimensions. By providing different systems of referencing, the scope of applicability of the engineering microscope is substantially extended. In the description of the referencing operations for microscopic measurements, the application of the different referencing systems will be discussed separately.

Referencing by Graticule Cross Hairs

Referencing by graticule cross hairs is the most common system of referencing where the object surface is observed under magnification through the microscopic tube. A graticule with hairlines is an element of the optical observation system, and is mounted into the plane of the virtual image. The orientation of two mutually perpendicular hairlines, when observed in the basic protractor positions, is parallel to the direction of the slide movements, and the intersection of the lines coincides with the op-

tical axis of the microscope tube. Most types of microscope graticules contain, in addition to the basic cross hairs, other reference lines as well. These are in precisely controlled relations to the basic cross hairs and facilitate the referencing of certain types of frequently measured features.

The hairlines of the graticule, besides providing the required reference point in the field of observation, are of valuable assistance in two more respects, both important for the correct measurement of length:

a. Targeting the reference point in the proper location on the object surface, in relationship to feature elements specified on the product drawing; and

b. Orienting the staged object in a manner where the required direction of length measurement is parallel with the translational movement of either of the instrument slides.

These additional potentials of referencing by sighting greatly contribute to extending the application range of the microscope with regard to the diversity of measurable dimensions.

As indicated earlier, two basic types of *illumination* may be used on the microscope.

1. The transmitted-light system with substage source, resulting in a silhouette image (diascopic illumination); and

2. The incident-light system, producing an image by light reflection (episcopic illumination).

Although the silhouette image provides a sharper definition of the contour, in many applications the reflected image is the only means by which the surface can be viewed for dimensional referencing. Examples of technical parts, mostly roller bearing components, referenced by both types of illumination, are shown in Tables 8-1 and 8-2.

When measuring parts of essentially cylindrical or similar form, particularly with highly reflective surfaces, attention must be given to a condition that can cause referencing errors if not properly controlled. Such errors occur in the diascopic (silhouette image) sighting when using large light source apertures (Fig. 8-8). The unobstructed light rays of the substage source are supposed to pass the object tangentially, thus producing a clear-cut definition of its surface boundaries. However, stray rays may glance from the object surface above its cross-sectional plane, thus transposing the observable silhouette line closer to the axis of the object. Due to this

TABLE 8-1.   REFERENCING FOR LENGTH MEASUREMENTS BY TRANSMITTED LIGHT — 1

DESIGNATION	DIAGRAM	DISCUSSION
Complete dimensional inspection of a machine shaft		Shafts used in testing machines and similar critical applications have several closely interrelated length and diameter dimensions. The shaft should be measured in a single setup to keep cumulative errors at a minimum.
Linear dimensions of a flanged ring in axial section		Several outside and inside dimensions of a body of revolution, such as a roller bearing ring, when observed in an axial plane, can be associated with a rectangular system of coordinates and measured directly with the microscope.
Total and effective length and diameter dimensions of a cylindrical roller		In addition to the maximum dimensions resulting from the total length and diameter of a cylindrical bearing roller, the functionally meaningful effective side lengths and face diameters need also to be measured. Optical inspection on the microscope, preferably with the aid of knife edges, supplies all the required dimensions in a single setup.
Coordinate dimensions of corner radii and chamfers on a bearing ring		Coordinate dimensions are frequently specified to assure noninterference of corner features on mating members; the microscope's capability for accurate measurements along the X and Y axes is in compliance with these requirements.

TABLE 8-1. REFERENCING FOR LENGTH MEASUREMENTS BY TRANSMITTED LIGHT — 2

DESIGNATION	DIAGRAM	DISCUSSION
Coordinate dimensions of shaft fillets		The optical observation of a silhouette representing the magnified contour of a body of revolution in an axial cross-sectional plane is the most convenient and reliable method for measuring the characteristic dimensions of corner features in set-back locations.
Outside and inside diameter dimensions of a stamped cup		Stamped and formed parts usually have edges which do not provide positive contact surfaces for mechanical measurements. The microscope, also offering means for measurements in several parallel planes, substitutes the optical viewing of tangent reference lines for the unreliable mechanical gaging contact.
Pocket dimensions in a punched sheet metal part		The rectangular pockets in a needle roller thrust bearing separator, particularly in the smaller sizes, are inaccessible to the sensing members of mechanical contact gages. Optical viewing on the microscope permits the direct measurement of both the length and width dimensions in mutually perpendicular directions.
Virtual dimensions of a taper plug gage		Taper plug gages which must have positively controlled diameters measured at the intersection of the side surface and the end face can be inspected correctly with the microscope, unless the specified gage tolerances require finer discrimination than feasible by optical length measuring instruments.

TABLE 8-1.   REFERENCING FOR LENGTH MEASUREMENTS BY TRANSMITTED LIGHT – 3

DESIGNATION	DIAGRAM	DISCUSSION
Virtual diameter of a tapered bore	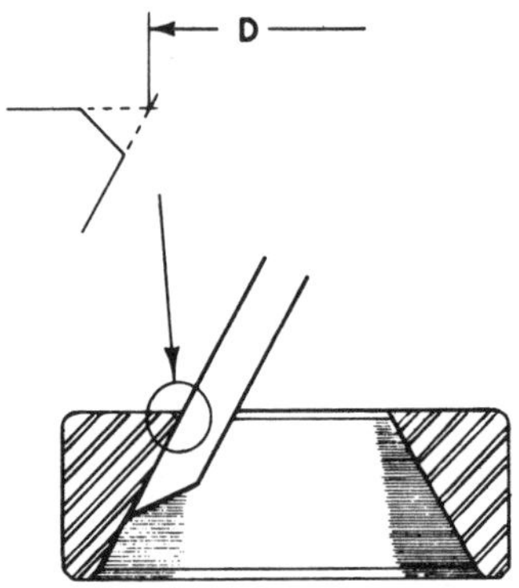 	The bore diameter of a tapered roller bearing outer ring in the plane of the part's face is not physically present because of the bore chamfer. This dimension can be measured correctly on the microscope using a knife edge as the mechanical extension of the bore surface element.

optical phenomenon, the observable diameter of the part may become smaller by as much as 0.002 inch, a potential error that is definitely intolerable for precision measurements.

To reduce the effect of this aberration, modern measuring microscopes are equipped with an adjustable diaphragm in front of the light source. Instructions are usually provided for its proper setting

in order to assure the best lighting for parts of different diameters. A further, very effective improvement consists in using knife-edge blades mounted in the cross-sectional plane of the object (the axial plane of the workholding centers) and adjusting the blades to produce a narrow light slit for observation.

Referencing by Optically Indicating Mechanical Contact

Mechanical contact type referencing can be of advantage in certain measuring applications, whereas in various others it is the only means by which dependable referencing can be accomplished on the microscope.

Examples of conditions where mechanical contacts in sensing a particular area of the surface will supply more positive guidance for the measurements than the direct optical observation are:

a. Test pieces whose sides are not exactly parallel to the line of sight and the measurements must be made in a specific horizontal plane; and

b. The culminating point of a large radius arc that must be found for referencing the line of measurement.

Other important uses for the mechanical contact result when surfaces are not directly observable through the microscope tube because of obstruction by other elements of the object. Examples of this type of mechanical contact measurements are internal features, such as threads, tapers and diameters behind shoulders (see Fig. 8-9).

Leitz/Opto-Metric Tools, Inc.

Fig. 8-8. (Left) Effect of an excessive light source aperture on the observational accuracy of cylindrical bodies. Diagram shows exaggerated proportions.

Fig. 8-9. (Right) Measuring microscope equipped with an optically indicating contact sensor ("zero checker") for measuring the internal diameter behind flanges of a roller bearing outer race mounted directly on the staging table.

TABLE 8-2.   REFERENCING FOR LENGTH MEASUREMENTS BY REFLECTED LIGHT

DESIGNATION	DIAGRAM	DISCUSSION
Radial width dimensions of corner radii and of the flat area on the face of a roller bearing ring		The effective area of the face is determined by subtracting from the total radial width the dimensions of the relieved edges. The pertinent dimensions can be measured on the microscope using transmitted and reflected light concurrently. The graticule cross hairs, being in precise alignment with the measuring directions of the microscope stage, assist in assuring the correct locations of the distances being measured.
Diameters of concentric annular zones on the face of a tapered bearing roller		The concentric ring zones on the face of a tapered bearing roller, namely those formed by the edge radius, the spherical zone and the indent, have specific functional roles requiring controlled dimensions which can be measured accurately when viewed under the reflected light of a microscope.
Ring zones of a stamped cup		The dimensions of stamped and drawn parts which are specified by the diameters of the zone boundaries, are generally very difficult to measure by mechanically contacting instruments. The reflected light permits the transition lines separating the adjacent zones to be defined with a precision adequate for most uses.
Internal flange width dimensions of a cylindrical roller bearing outer ring		The effective radial width of an internal flange, and the edge-radius coordinates in the plane of the flange back-face, are difficult to measure by any other method than optically, such as under the reflected light of the microscope.

Very sensitive and dependable contact elements are available for microscope applications. These operate on principles comparable to the systems of autocollimators. Deflections of the probe caused by contact with the workpiece affect the position of a tilting mirror that reflects into the ocular the image of the cross hair projected by the light beam (see Fig. 8-10). A graticule with bifilar reference lines and graduations (Fig. 8-11) is mounted into the eyepiece, and the reflected image of the original cross hair appears superimposed on the reference graduations. The precisely centered position of the cross hair between the bifilar reference lines indicates the zero (fiducial) position of the contacted surface in relation to the optical axis of the instrument. The graduations of the graticule indicate in 2-micron increments the displacement of the contacted surface from its fiducial location. The corresponding graduations on inch versions are in 0.0001-inch increments.

Precision balls with calibrated diameter are generally used for contact tips, and the sensing direction of the probe is reversible without affecting the location of the zero position. Due to the use of an optical lever for magnification, the 2σ repeat accuracy of the sensing is stated to be ± 0.2 micron (8 micro inches). 2σ is frequently used as a measure of reliability in appraising the indicating accuracy of measuring instruments, and expresses two standard deviations, comprising about 95% of all occurrences.)

Special microscope heads that operate with the optically indicating mechanical sensing probe are combined with an optical length-measuring device permitting the measurement of vertical displacements with a sensitivity of 0.00005 inch (Fig. 8-12). The combination in the microscope of these two functions, the mechanical sensing and the fine incremental vertical displacement measurement, provides a unique capacity for measuring various types of internal features at an accuracy level usually unattainable by any of the known substitute methods.

Examples of the application of the described sensing devices in measurements on the microscope are shown in Table 8.3.

When evaluating the application potential of mechanical contact probes, it must be remembered that the last graduation of the optically observed scale is 50 microinches, although the bifilar position of the cross hair can be established with greater sensitivity. Consequently, for the measurement of discrete length dimensions in submicron (less than approximately 40 microinches) increments, a microscope with a mechanical contact probe is not an equivalent substitute for an electronic comparator.

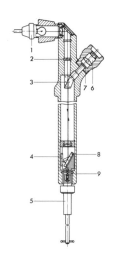 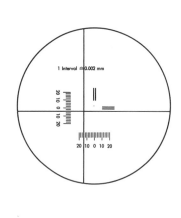 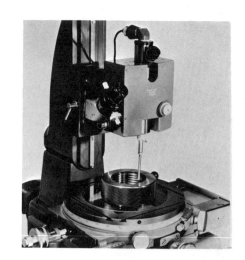

Leitz/Opto-Metric Tools, Inc.

Fig. 8-10. (Left) Operating elements of a mechanically contacting, optical-position indicator. *(1)* Light source; *(2)* graticule with hairline; *(3)* beam splitting prism; *(4)* fixed mirror; *(5)* contact shaft; *(6)* eyepiece; *(7)* graticule with bifilar index marks; *(8)* tipping mirror; and *(9)* bearing of the contact member.

Fig. 8-11. (Center) Field of view of an optically indicating mechanical contact sensor. Each graduation represents 0.002 millimeter (metric version) or 0.0001 inch (inch version).

Fig. 8-12. (Right) The mechanically contacting optical-position indicator mounted on a microscope stand in combination with a vertical displacement measuring head, which is equipped with a master scale and an optical reader.

TABLE 8-3.   REFERENCING FOR LENGTH MEASUREMENTS WITH OPTICALLY INDICATING MECHANICAL CONTACT SENSOR — 1

DESIGNATION.	DIAGRAM	DISCUSSION
Diameter of an internal surface behind shoulders		The measurement of plain linear or diametrical distances behind shoulders presents problems for most gaging systems because sensitive detection must be combined with large retraction movement. Optically indicating sensors equipped with contact disks of known diameter, when used in conjunction with the microscope's length-measuring slides, permit accurate measurements in the correctly selected locations.
Depth of a ring groove inside a bore		For the depth measurement of deep grooves on internal surfaces a very convenient and reliable method is offered by the optically indicating sensor head of the measuring microscope. Contact disks of suitable thickness must be used with the gaging height adjusted to establish actual contact with the bottom of the groove.
Measurement of the lead and pitch diameter of internal threads		The measuring microscope equipped with vertical displacement slide and with an optical indicator using a pair of ball contact tips of specific diameter and at a known distance apart, provides a unique method for selectively measuring several meaningful dimensions of internal threads, including those of thread ring gages.

TABLE 8-3.   REFERENCING FOR LENGTH MEASUREMENTS WITH OPTICALLY INDICATING MECHANICAL CONTACT SENSOR – 2

DESIGNATION	DIAGRAM	DISCUSSION
The effective pathway diameter of a barrel roller bearing outer ring		Measuring the diameter precisely in the central plane of an imaginary sphere, whose zone is constituted by the pathway of the barrel roller bearing outer ring, requires contact position adjustment in two mutually normal directions on a partially obstructed internal surface. These conditions can be satisfied with the optically indicating sensor of the measuring microscope.
Included angle of a tapered bore		Optically indicating mechanical contact, in combination with an optical height measuring device, permits making reliable diameter measurements on internal surfaces in parallel planes which are at a known distance apart. These measurements are needed to check the angle of an internal taper.
Dependable O.D. and I.D. measurement of a ring by referencing in the axial plane		The small contour curvature of large-size bearing rings reduces the resolution of optical viewing with graticule cross hairs, thereby creating adverse conditions for establishing the exactly tangential position of the referencing. When moving the cross-slide with the part normal to the measuring direction, the optically indicating mechanical contact sensor will dependably signal the location of the axial plane by the reversal of its pointer movement.

Referencing by Grazing Incidence Light Beam

When the contour dimensions of an essentially flat part are measured with the microscope, the projection of the sides that are perpendicular to the optically observed flat surface should appear as distinct boundary lines. This condition could prevail if the part being inspected were almost dimensionless in the direction of the light beams and would function essentially as a mask. However, all technical parts are made to have a thickness that is commensurate with the functional purpose.

The light beams of the measuring microscope's substage illumination are effectively collimated. Still, it cannot be entirely avoided that reflections from the side surfaces of the part should, to a small extent, interfere with the exact definition of the contour line. Under favorable conditions, the magnified contour line of the part can be sighted with an accuracy of about 0.0001 inch—rarely better. When measurements on the microscope are required to supply size information in the repeat accuracy range of 50 microinches, or even closer, a potential referencing error of 0.0001 inch cannot be tolerated.

For improving the accuracy level of microscope measurements, a purely optical referencing system was originally developed by Ernst Leitz, although a limited number of other manufacturers are now also offering comparable purpose equipment.

The principles of operation of the Perflectometer (Leitz's trade name for the system) are shown in Fig. 8-13. Figure 8-14 illustrates diagrammatically the

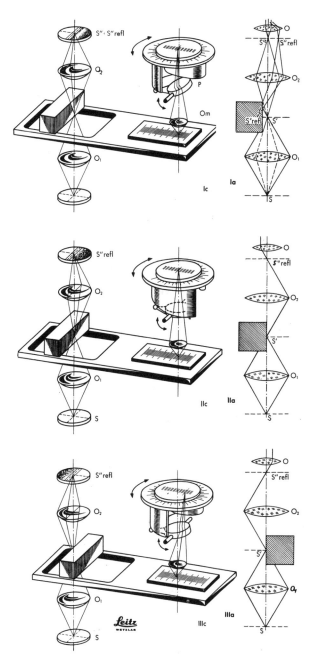

Leitz/Opto-Metric Tools, Inc.

Fig. 8-14. The operating process of an instrument (the Leitz "Perflectometer") using the grazing light-beam system for defining the boundary points of a linear distance, which is then measured by means of a master scale, read through an optical micrometer (at right). (Top) Approaching the boundary plane. (Center) Referencing at the right-hand side *boundary*. (Bottom) Referencing at the left-hand side boundary. *S* and *S"*: the projected and the reflected cross hairs; O_1 and O_2: objectives; and *O*: eyepiece.

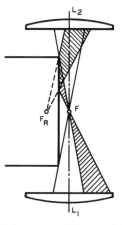

Fig. 8-13. Simplified operational diagram of the grazing incidence light-beam observation system. The cross-hair image projected from the objective L_1 is observed through the eyepiece of a microscope by its objective L_2. In the position approaching optical alignment, the reflected image F_R appears at twice the actual distance from the test piece surface as the virtual image F in the focal plane of observation.

operation of an instrument utilizing these principles. Essentially, the system consists of two rigidly connected and very precisely aligned microscope tubes, the lower of which projects the image of a cross hair. The upper microscope, with an eyepiece, has a graticule with parallel referencing lines straddling the common optical axis of the two microscope tubes. The projected hairline, when unobstructed, will lie exactly in the center of the bifilar reference mark.

When the object wall under observation is brought into the immediate vicinity of that optical plane, the reflected image of the projected hairline will also appear in the field of view of the observation microscope. Only when the object wall is moved exactly in the optical axis of the system will the directly projected image of the hairline be blocked, and the reflected image will take over the position between the bifilar reference lines.

When measuring bore diameters, for example, the reflected cross hair appears as a curve and the alignment of the sighting line at the culminating point of that contour is essential. To satisfy this requirement, the projected hairline is interrupted in its center section and has three dots. The two extreme dots serve for the alignment, and the middle one, in its bifilarly centered location, indicates the correct referencing position (see Fig. 8-15).

The level of the focal plane of the projection can be adjusted vertically to make measurements at the desired plane of the object. This adjustment permits scanning the part feature to determine the uniformity of its size.

The referencing accuracy of the Perflectometer is stated to be within ± 0.000004 inch. This value is based on the 5 second of arc average discriminating ability of the human eye and on 30 to 40 times magnification of the observation microscope.

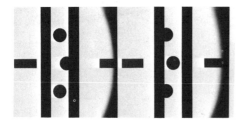

Leitz/Opto-Metric Tools, Inc.

Fig. 8-15. The three dots along the reference cross hairs of the Perflectometer as they appear in the microscope eyepiece when measuring the diameters of cylindrical parts. (Left) The two outside dots are intercepted for aligning the image. (Right) The central dot is brought into the bifilarly centered position for the final pointing.

The measuring accuracy of the Perflectometer instrument is stated to lie within a potential error range expressed by the following formula (transposed into inch units):

$$\Delta = \pm \left(20 + \frac{L}{0.2} + \frac{L}{0.5} + \frac{H}{0.1} \right)$$

where

    Δ = measuring error in microinches;
    L = measuring length in inches;
    H = height of the measuring plane over the instrument stage in inches.

The element $L/0.2$ comprises the end-referencing (sighting) error and $L/0.5$ the effect of the assumed temperature differential of about 0.5°F between the object and the master scale. The element $H/0.1$ comprises the potential tilt error of the staging.

Table 8-4 illustrates a few measuring examples that are particularly well suited for referencing by the grazing-incidence light beam.

Referencing by Double-Image Sighting

The method of referencing by double-image sighting is applicable only to configurations that can be observed in transmitted light, also known as "through" features. A lens system is used that operates with a special prism for splitting optically the observed feature into two images. The types of prisms currently available as measuring microscope accessories are the double-roof prism, which supplies centrally symmetrical images, and the single-roof prism, which causes the appearance of axially symmetrical images.

During the referencing process these images can be entirely separated, or partially superimposed, depending on the feature's element of symmetry in relation to the optical axis of the microscope tube. The prism produces these images in colors that are different but complementary, commonly orange and green. The opaque part of the observed object remains black, like the silhouette image in regular illumination by transmitted light.

Complementary colors are those that, when superimposed, cancel each other out. In this application, where the optical canceling occurs, the feature will appear in its original substage illumination, for example, white or light green. The complete superimposition of the two images results when the element of symmetry of the feature is in exact coincidence with the optical axis of the observation system. Consequently, when sections of the feature appear apart and in different colors, the staging is nonsymmetrical, and the presence of colored fringes

TABLE 8-4.   REFERENCING FOR LENGTH MEASUREMENTS BY GRAZING-INCIDENCE LIGHT BEAM

DESIGNATION	DIAGRAM	DISCUSSION
Bore diameter of a ring gage		The bore of a ring gage or the diameter of any other cylindrical bore can be measured directly with the grazing incidence light beam system of a microscope to a degree of accuracy which may be exceeded only in special comparator instruments. This method of directly obtaining size information is frequently preferred to the use of comparators which require setting by masters. (PS = Plane of Sighting.)
Wall thickness of a thin sleeve		The high degree of measuring accuracy attainable with the grazing incidence light system of a measuring microscope does not require a gaging force (which can cause distortion) and still assures the dependable location of the measuring plane. Measurements at different, closely-controlled levels, carried out in the same setup, will detect with a high sensitivity the presence of dimensional variations along the surface.
Bore diameter of a miniature ball bearing inner ring		Very small bore diameters—as found in miniature bearings—when measured individually, require highly-specialized comparator instruments. Direct measurements can, however, be made with a microscope equipped with a grazing incidence light beam measuring device. Smallest measurable diameter is 0.040 inch.
Width of narrow slots located on an internal surface		The width of very narrow slots, particularly when located inside a bore, can be measured with only limited accuracy by any other system than a measuring microscope. The grazing incidence light beam can penetrate into slots as narrow as 0.020 inch, and the separation of the feature sides is measured with the accuracy inherent in the microscope's length-measuring system.

around the image contour reveals irregularity of form.

The eventual complete disappearance of the colored images is the goal of the optical centering process (Fig. 8-16). However, this can be accomplished only when the feature is of nearly perfect symmetry and regularity, such as the precisely bored holes of a gage.

Very important applications can be found for the double-image observation in the case of interrelated features that are nominally symmetrical, yet of irregular form. The colored fringes, which are indicative of form irregularity, for example, defective circularity in the cross-sectional plane of a nominally round hole, will permit alignment by the uniform distribution of these fringes around the contour of the image or, in effect, around a visual "center of gravity." In this application of double-image viewing the maximum material condition of the feature is actually observed and used as a guide to establish the positioning of the object with reference to the center or to the axis of symmetry of the feature. Therefore it may be stated that this process of referencing complies with the concept of true position tolerancing, regardless of the size of the feature.

The sensitivity of the double-image observation is very high; displacement of the feature from the alignment position by 0.0001 inch, or even less, will produce detectable colored fringes.

A few application examples for referencing by double-image observation in length measurement

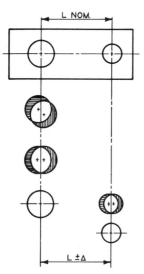

Fig. 8-16. Diagram of an optical centering process by the double-image system. The sequentially accomplished alignment and centering of the observed bore's images results in the ultimate disappearance of the colored fringes (indicated by shading).

are shown in Table 8-5. Further examples will be pointed out in conjunction with goniometric measurements.

GONIOMETRIC MEASUREMENTS WITH THE MICROSCOPE

Measurement by microscope is applicable to a wide variety of geometrical conditions that are dimensioned in angular units. Such measurements may be classified into two major categories:

1. Angular separations; and

2. Circular divisions.

The measurement of *angular separation* can be performed on conical solids and on features of solids that produce a figure bounded by two nonparallel straight lines in a specific sectional plane (for example, the angle of a thread form). In order to be measurable by a microscope, the angular feature must be either:

a. Exposed on both sides in the direction normal to the plane containing the characteristic form, whose illumination by transmitted light should create a silhouette image; or

b. Exposed on one side to present a reflecting surface when illuminated by incident light.

Less frequently, angular measurement of features consisting of optically nonexposed straight-line elements is accomplished by an indirect method. For this purpose an auxiliary member, a positively located knife edge, is brought into line contact with the surface elements to be measured, whereby the exposed extension of the knife edge will transfer the angular orientation of the hidden surface element into the line of sight of the microscope.

Feeler knives made for measuring thread angles in the significant axial plane of the threaded part are frequently supplied with engraved lines at fixed distances and parallel to the edge of the knife (see Fig. 8-17). The cross-hair graticule used for these measurements has index lines at the same distances from one of the center lines. When the knife edge is brought into line contact with one of the thread flanks, the protractor of the microscope head will indicate the angular position in which the engraved mark of the knife edge and the index line of the graticule are in coincidence.

The measurement of *circular divisions*, including polar coordinates, can be performed when the pertinent elements of the test piece are contained in a single plane or, with certain limitations, in parallel

TABLE 8-5.   REFERENCING FOR LENGTH MEASUREMENTS BY DOUBLE-IMAGE SIGHTING

DESIGNATION	DIAGRAM	DISCUSSION
Linear separation of bore centers		The center of a bore, when observed in a plane normal to the axis, is an imaginary element definable as a point equidistant from the elements of the circumference. A double-image system permits adjustment of the bore location for the referencing of its center in compliance with this definition.
Coordinate positions of holes in a plate gage		While the accurate measurement of coordinate distances is an inherent capability of a measuring microscope, the sighting of nonexistent bore centers requires the application of a double-image system. This permits establishing the "true position" of the individual bores, guided by the maximum material condition around the feature contour.
Image congruency of nominally symmetrical pierced pockets		A convenient and reliable method for checking the configuration of nominally-symmetrical "through" features is offered by double-image viewing. In this process, the position of the test-piece feature is adjusted to approach the perfect congruence of the reversed images; incongruent areas are indicative of defective symmetry.
Distance between the centers of symmetrical figures		Measuring the separation of two nominally symmetrical features— e.g., of a template gage—which is specified by the distance between the centerlines, requires the correct determination of these imaginary reference lines. The sensitive centering capability of a double-image system permits locating the "best-fitting" position of the centerline.

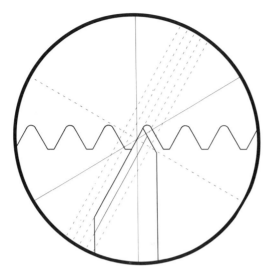

Carl Zeiss, Inc.

Fig. 8-17. Flank angle measurement on the microscope using a contact blade with engraved reference line. The alignment of the contacted part surface with the reference axis of the microscope's observation system is established by bringing into coincidence the engraved (solid) line of the blade with the dotted line adjacent to the centerline of the graticule.

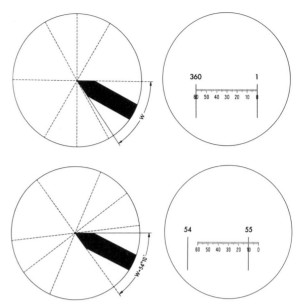

Leitz/Opto-Metric Tools, Inc.

Fig. 8-18. The operation of the optical protractor as it appears in the field of view of a microscope eyepiece. A selected centerline of the graticule is brought consecutively into coincidence with opposite sides of the silhouette image. The measured angle is the difference of the values indicated in these two setting positions.

planes. The requirements regarding the exposal of the elements to be observed are essentially the same as those for measuring angular separation.

Optical Protractors

An optical protractor for a microscope is rigidly connected to the frame of the cross-hair graticule and can be rotated 360 degrees. The protractor has a ring-shaped master scale of glass with 360 circumferential graduations in one degree increments. The individually numbered graduations can be selectively observed through a microscope at substantial magnification.

The sections of the graduated ring that appear in the field of the microscope are correlated to the angular orientation of the graticule's cross hair. The observable ring section encompasses somewhat more than one increment so that, as shown in Fig. 8-18, at least one graduation line will be visible and will appear superimposed on the readout scale of the observation graticule, which is graduated into 60 minutes. The combination of the degree graduations and the minute lines with which they coincide permits the direct reading of angular position in degrees and minutes or, by accurate estimation, in half-minutes. Some late models of measuring microscopes are equipped with angle measuring vernier scales graduated in increments of ten seconds of arc, and, for

observational convenience, the scales appear projected on a screen.

Systems that expose the readout scale with the projected degree graduation line in the field of the main observation microscope by simply flipping a switch are of great convenience and now belong to the standard equipment of several types of measuring microscopes.

The inherent accuracy of the master scale and the excellent discrimination provided by the optical readout system, permit determining the protractor angle within an error range of ±30 seconds or ±10 seconds, respectively. The ultimate accuracy of the angular measurement also depends on the alignment of the cross hair with the boundary lines of the image in which the feature being observed appears in the field of the microscope.

Application examples of the measurement of angular separations by the microscope's optical protractor are shown in Table 8-6.

Optical Rotary Table

Whereas angular separations are represented by the relative position of two nonparallel straight lines (which function as mutual datum elements), circular divisions define the relative position of two points

TABLE 8-6.   OPTICAL MEASUREMENT OF ANGULAR SEPARATION
(MUTUALLY REFERENCED ANGLES) — 1

DESIGNATION	DIAGRAM	DISCUSSION
Taper plug gage (held between centers)		Unless the tolerances specified for a taper gage are tighter than those which will be satisfied by the angle-measuring sensitivity of a microscope's optical protractor, the microscope presents a convenient method for the direct measurement of tapers.
Tapered bearing roller (held by a magnetic block)		Tapered bearing rollers are typical examples of cone-shaped parts or features whose body angle can be measured directly in angular values to a sensitivity of one minute or better with the optical protractor of a microscope. The use of knife edges is required to accurately measure large diameter parts.
Tapered bore of a roller bearing outer ring		Angles of internal features, inaccessible to direct viewing, can be measured on a microscope by using a properly located knife edge as the physical extension of the hidden side element. When assuring blade contact in the axial plane of the part, accuracy compares to measurements of external conical surfaces.

TABLE 8-6. OPTICAL MEASUREMENT OF ANGULAR SEPARATION (MUTUALLY REFERENCED ANGLES) — 2

DESIGNATION	DIAGRAM	DISCUSSION
Angular separation of two partially obstructed surface elements		Although the included angle bounded by the pathway and the flange face of a tapered roller bearing inner ring is a functionally critical and closely-toleranced dimension, because of its setback location it presents a difficult problem for most measuring systems. Microscope measurement with a knife-edge blade is a readily applicable technique.
The taper angle of the pierced pocket (of a tapered roller bearing separator)		While pierced and formed edges do not adapt themselves to establishing a correct mechanical contact for measurements, the visually observed coincidence of the feature edge with the graticule cross hairs permits angle measurements which are based on the predominant contour line.
Thread angle measurement		For thread angle measurement in the unobstructed observation plane, a microscope presents unequalled advantages. This measurement is made with the optical protractor of the microscope head either by direct-viewing of the feature or, for improved referencing accuracy in the axial plane of the screw, with the aid of knife-edge blades.

that may represent, for the purpose of positioning, specific features. For referencing, however, these points rely on a single extraneous element—the center of a circle, either actual or imaginary.

The points whose relative positions are defined by circular division may lie on a common (imaginary) circle, or may lie at different radial distances from a common reference center. In the latter case, the circular divisional value, expressed in angular units, must be supplemented by the linear value of the radial distance from the center. The resulting system of position definition is known as "polar coordinates." These fundamentals may assist in visualizing the operation of circular-division measurements on a microscope.

The test piece to be inspected must be staged on the rotary table in a well-centered position. The rotary table (see Fig. 8-19) is equipped with an angle-measuring system that is essentially the same as is used for an optical protractor.

While the circular divisions are measured by the goniometric scale of the rotary table, the radial distances are determined by the controlled translational movement of the microscope slides. For centering the rotary table in exact alignment with the optical axis of the microscope, the rotary table has an accurate reference mark that can be optically observed through the microscope and brought into coincidence with the intersection of the graticule cross hairs. The scale readings of the translational stage movements should be noted in this position, which

Leitz/Opto-Metric Tools, Inc.

Fig. 8-19. Optical rotary table mounted on the cross-slide of a toolmaker microscope. The table has an internally mounted master ring with 360 individually numbered degree graduations, which are observed through the microscope eyepiece against a graticule with 60-minute graduations.

represents the reference point for measuring the radial distances. The actual measurement of radii pertaining to individual features of the part must be carried out in alignment with one of the coordinate movements of the stage; either along the longitudinal (X) or the transverse (Y) axis. This condition is usually preestablished by rotating the table to bring the feature being inspected into the circular reference position.

Examples of measurements involving circular divisions and polar coordinates are shown in Table 8-7.

INSPECTION BY CONTOUR FORM COMPARISON

One of the original applications for early models of toolmakers' microscopes was the inspection of screw threads. A distinct parallel in applications exists between the toolmakers' microscope and the optical projector. Actually, optical contour comparison is still the most versatile method of screw thread inspection. It provides the means for inspecting and measuring all the pertinent parameters of screw threads, including linear and angular dimensions, as well as form characteristics.

The similarity between the operating principles of a toolmakers' microscope and an optical projector does not exclude several important differences that determine distinctive fields of application for both systems. Although the advantages of group-viewing as offered by the optical projector are important, the compactness of a microscope and several other significant design characteristics are frequently decisive factors in the choice of an instrument best suited to a specific category of applications.

Since microscope design permits tilting of the upright with respect to the microscope head, it enables alignment of the optical axis with the helical plane of a screw thread. At the same time, the axis of the test piece staged on the microscope remains parallel and normal, respectively, to the translational movements of the slides, by which lead and diameter are measured.

Instead of having to exchange screen templates, a microscope uses graticules that can be brought into view by simply rotating a turret containing several graticules, or by sliding in a small-size graticule cartridge. In modern microscopes having intermediate image tubes, the same graticule is used for all common magnifications (usually 10×, 20× and 30×), whereas in the optical projector, the screen template must be exchanged when different magnifications are selected (see Fig. 8-20).

**TABLE 8-7.   OPTICAL MEASUREMENT OF ANGULAR DIVISIONS
(CIRCULAR SPACINGS REFERENCED FROM A COMMON CENTER)**

DESIGNATION	DIAGRAM	DISCUSSION
Angular spacing of pocket locations in a needle roller thrust bearing separator		Holding the part on a rotary table and sighting a repetitive feature—preferably with a double-image tube—permits defining the table rotation needed to bring adjacent features into the line of sight. The cross slide is used to guide the positioning of the part on the circular table which was previously centered by an optically-sighted reference mark.
Angular spacing of roller positions in a tapered roller bearing cage		A functional check of the locational accuracy of the pockets in a tapered roller bearing cage is based on the examination of the resulting roller positions. Optical observation, combined with the controlled rotational movement of the circular table, will supply angular values in increments of arc-minutes.
Angular spacing of pierced pockets in a ribbon-type ball bearing separator		Using graticules with concentric circles superimposed on cross hairs as reference elements for the sighting of the individual round pockets in their "best-fitting," location, and then rotating the circular table sequentially to the next pocket, will supply accurate angular values for the feature positions, without incurring cumulative errors.
Polar coordinate dimensions of openings in a piercing die		Features whose positions are specified by dimensioning in polar coordinates, involving both circular divisions and linear distances, are measured correctly and with great accuracy on a microscope. This requires combining the potentials of accurate sighting, of angular measurements with the circular table, and of linear displacements with the measuring cross slides.

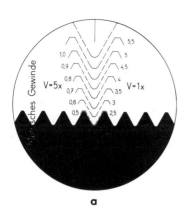 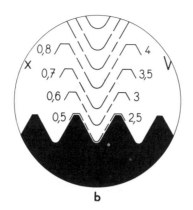 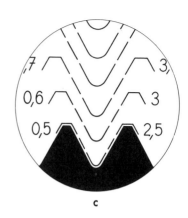

Leitz/Opto-Metric Tools, Inc.

Fig. 8-20. Contour comparison combined with referencing, for linear measurements of a screw thread, by means of alignment with the corresponding graticule tracing. The figures also illustrate the property of the intermediate image system, where the change of magnification (a, b and c equal to 10x, 20x and 30x, respectively) leaves the relation between graticule lines and image contour unaffected.

In screw-thread inspection processes that include linear and angular measurements as well, and where a high degree of repeatable accuracy is required, such as for thread gages, the capabilities of measuring microscopes offer unique advantages. By using special knife edge blades, without tilting the microscope upright, the angle and the linear dimensions of threads can be measured in the axial plane of the threaded part. The process—whose technique was described in the preceding section—will then be in a theoretically correct agreement with the screw thread standard's specifications, thus avoiding the effect of small angular errors that are introduced through observation in a direction that is tilted from the normal by the amount of the helix angle.

Optical contour form comparison is not limited to screw threads, where it serves the double purpose of visual comparison and referencing for linear and angle measurements. These capabilities are frequently required for threaded parts, threading tools, and thread gages as well. A few other categories of technical features whose inspection by optical contour comparison can prove of great advantage are gears and gear generating tools, such as hobs, cutters and so forth, splines, serrations and many other standard or special forms. The several hundred different types of graticules (some with 20 or more individual contours), which are available as standard or specially made tooling, indicate the wide application potentials of optically comparing a part's contour form to a master.

Another important form widely inspected by optical comparison is the circular arc contour. Individ-

ual experience may show, therefore, that a microscope offers the most suitable method for inspecting parts that feature nominally circular segments, including edge (corner) radii and fillets.

A limited number of examples of the microscope's use in optical form comparison are shown in Table 8-8.

SURVEY OF ENGINEERING MICROSCOPES

It was pointed out earlier that the operation of engineering microscopes comprises three basic functions, namely, magnification, indexing for reference position and staging combined with controlled and measured displacement. For surveying the representative types of these instruments, an exception will be made from this definition of the instrument category: to start with, two microscope types will be mentioned that are used in engineering production to fulfill only two of said functions, namely, the magnification and the referencing.

Centering Microscopes

Centering microscopes are used on jig borers, measuring machines and, occasionally, on other types of machine tools to align the axis of the machine spindle with a particular feature on the part surface. That alignment can serve either of the following functions:

a. Centering a point or a bore on the object surface in coincidence with the spindle axis for sub-

TABLE 8-8.   OPTICAL INSPECTION BY CONTOUR FORM COMPARISON

DESIGNATION	DIAGRAM	DISCUSSION
Radii of rounded edges and of fillets on a shaft		The form and dimensions of rounded edges and corners are generally specified by radii, implying essentially circular-arc contours. The optical comparison of a magnified image with the corresponding circular-arc line of a graticule permits checking both the general form and the size.
Corner radii of a cylindrical bearing roller		The edges of bearing rollers, while having no direct functional role, must have closely-controlled dimensions for the purposes of avoiding interference with mating members and of limiting the reduction of the effective roller surface. The dimensions can be inspected on a microscope in proper geometric relationship to adjacent contour elements.
Corner radii of a roller bearing separator pocket		The primary purpose of dimensional specifications for these features is the avoidance of harmful contact interference with the corner edges of bearing rollers. Optical comparison with graticule lines provides direct observation of the functionally meaningful maximum material condition of the fillet.
Form trueness and symmetry of an extruded bar		Features with circular-arc elements can be examined on a microscope from a number of significant aspects exemplified by this case where contour curvature, virtual center location, and form symmetry are involved.

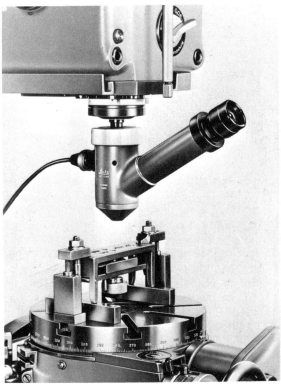

Leitz/Opto-Metric Tools, Inc.

Fig. 8-21. Centering microscope mounted into a machine tool spindle. In this application, the microscope is guiding the alignment of a bore center with the axis of rotation of the machine spindle.

Carl Zeiss, Inc.

Fig. 8-22. Machine tool microscope for observing a machine part surface and/or the tool adjustment, while held in operational position on a machine tool. The detached, telecentric light source provides substage type illumination for improved contour definition.

sequent machining operation, by substituting a cutting tool for the microscope; and

b. Establishing a fixed line of sight as an optical reference element with which different points on the object surface can consecutively be aligned, for measuring on the scales of the machine slides the distances traveled between the subsequently set alignment positions.

The centering microscope shown in Fig. 8-21 has 30× magnification, covers an object field of about $\frac{1}{4}$-inch diameter, and can be supplied with a shank in any of various standard taper sizes, for direct insertion into the bore of a machine spindle.

Machine Tool Microscopes

The observation of the magnified image of the work and tools while held on the machine tool in operational position can be of great help for inspecting size and form, and for carrying out the necessary adjustment in the tool position. While the cutting of

threads on a lathe or on a thread grinding machine are the most rewarding uses of machine tool microscopes, there are many other applications that may involve accurate form and angle measurements, or require the sensitive positioning of the cutting tool. The machine tool microscope shown in Fig. 8-22 is equipped with an optical head for goniometric measurements with one minute of arc discrimination. To satisfy various specific needs, the same basic instrument can also be supplied as a plain positioning microscope or, when used for thread measurements, equipped with a revolving graticule dial comprising a large number of different standard thread contours.

For each alternative equipment, the same basic microscope tube is used, which can be clamped on the machine tool in a position assuring about 3 inches focal distance. Illumination of the observed area is by external light source and for an exceptionally clear contour image a telecentric light fixture can be mounted beneath the work. The microscope has 30× magnification and its field of view is more than $\frac{1}{4}$ inch in diameter.

Leitz/Opto-Metric Tools, Inc.

Fig. 8-23. Plain coordinate measuring microscope for the measurement of linear dimensions on small parts. The modular design permits a wide choice of stage types and eyepieces for adapting the instrument to more versatile applications.

Plain Coordinate Measuring Microscopes

Plain coordinate measuring microscopes (see Fig. 8-23) consist of three major members:

1. A stand with a column along which the holding bracket for the microscope tube can be adjusted vertically to the proper focal position. The base of the stand supports the cross-slide;

2. The microscope tube, which is usually the plain positioning type, with cross-hair graticule and a single magnification; and

3. A cross-slide table with micrometer heads producing a measuring range of 1 by 1 inch or 2 by 2 inches.

The illumination of the observed part portion is by overhead lighting with the aid of attached lamps. The field of application of these instruments is limited to simple length or coordinate dimensions within the displacement range of the cross-slides, and the attainable measuring accuracy is on the order of 0.001 inch. Some models of coordinate mea-

suring microscopes are assembled from modules, a design concept offering great flexibility in selecting the type best suited for a particular set of applications.

Toolmaker Microscopes

Versatility is one of the characteristic attributes of toolmakers and for that reason, engineering microscopes designed to satisfy varied measuring requirements, such as resulting from the diverse tasks of toolmaking, are known as toolmaker microscopes.

Of course, there are different degrees of versatility, and this also applies to toolmaker microscopes. While there are no distinct boundaries for specific categories, it is customary to distinguish the types of toolmaker microscopes that are adaptable to an uncommonly wide range of measuring tasks, by the designation of *universal toolmaker microscopes*. Great versatility is usually inherent in the basic design of these instruments, and can be further expanded by the use of a wide choice of optional accessories.

To illustrate that distinction, Fig. 8-24 shows a plain toolmaker microscope, which is primarily intended for the measurement of threaded tools, such as taps. This instrument is equipped with all the accessories needed for that particular group of parts, however, without features and attachments that would be of no practical use within the intended application range.

The toolmaker microscope shown in Fig. 8-1 can be considered to belong to the universal category, being designed for a large variety of measuring applications, including linear, coordinate and angular dimensions, as well as the inspection of standard contour forms such as screw threads, circular arcs and so forth.

The illumination of the part portion being inspected can be provided either by a substage light source, supplying collimated and diffused light beams or by surface lighting using mono- or bichromatic lamps.

The column of the stand can be tilted by a controlled amount to within 8 degrees in either direction, for permitting the observation of the total contour of helical part features, such as screw threads, worms, hobs and so forth. Angle measurements can be carried out either with the protractor ocular head, or with the rotating stage.

Linear measurements in a single direction, or as coordinate dimensions, can be made with the micrometer stage that, in the illustrated medium-size model, has 4 inches longitudinal motion and 2

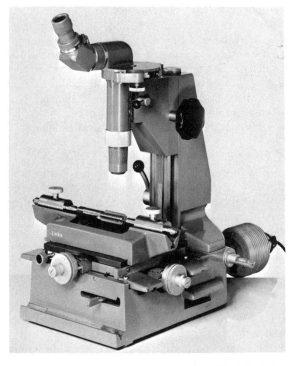

Leitz/Opto-Metric Tools, Inc.

Fig. 8-24. Toolmaker microscope equipped for the inspection of taps and similar threaded parts, in addition to the basic coordinate measurements. A workholder cradle with Vees and centers is mounted on the cross-slide. The microscope tube has a rotatable thread-contour graticule.

inches cross motion. The smallest graduation of the micrometer drum represents 0.0001 inch. The micrometer screw has a range of two inches that, for measurements in the longitudinal direction, can be extended two additional inches by inserting a 2-inch gage block between a fixed stop and the anvil of the carriage, the latter being under spring tension.

An extensive number of accessories, such as workholding devices, templates, projection or photographic attachment, substantially extend the application range of the instrument.

Measuring Microscopes

The designation measuring microscope may not be sufficiently specific without a definition of the term. Actually, the measurement of geometrically defined dimensions and forms is the primary purpose of most types of microscopes used in the mechanical phases of metalworking and related productions.

In a rather widely accepted practice, the term measuring microscope designates a category of op-

tical instruments that, by means of very accurate scales observed through sensitive optical devices with a high power of discrimination, assures a degree of measuring accuracy superior to that provided by the commonly used designs of engineering microscopes.

The micrometer screws of the generally used toolmaker microscopes fill the double function of imparting translational motion to the instrument slides and of measuring concurrently the distance of that displacement. Measuring microscopes, being equipped with master scales, do not have to rely on the pitch accuracy of the lead screws that, in this latter category of measuring instruments, serve only as the actuating members for the slide movements.

The difference in the length-measuring system may be considered the essential characteristic that distinguishes the measuring microscopes from the more widely used toolmaker version. However, there are other design features and accessories that are common to measuring microscopes, yet that would seldom be found on toolmaker microscopes, with the possible exception of the most elaborate models.

Two examples of measuring microscopes have been shown in Figs. 8-2 and 8-3. In Fig. 8-25, another type of measuring microscope is illustrated that, due to its large size (measuring range 40 by 8 inches), resembles a measuring machine. The instrument also differs from the conventional design principles of measuring microscopes by holding the workpiece in a stationary position and traversing the cross-slide, which carries the microscope head, along two axes. The distance of the slide travel along either axis is measured in relation to two built-in master scales. The scale graduation coincident with the slide position appears on the screen of an optical scale reader, which is designed to interpret the position of the projected graduation line with 0.000050-inch discrimination.

Having surveyed the capabilities of engineering microscopes, the question may arise as to where they should be used as the most appropriate type of measuring instruments. There are, of course, no general rules for selecting any specific type of multipurpose instrument for measurement tasks, many of which can also be resolved with several other types of gaging equipment. Consequently, the following outline of preferential applications for engineering microscopes must not be considered to state rigid rules—its purpose is to indicate and not to specify.

First, the limiting conditions will be reviewed. The engineering microscope is an optical instrument that should be operated by a toolmaker, a well

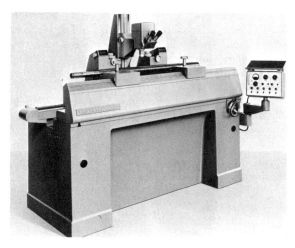

Leitz/Optio-Metric Tools, Inc.

Fig. 8-25. Universal measuring machine type microscope with 40 by 8 inches measuring range. The part stays stationary while the measuring head with the microscope tube is displaced in its guideways along master scales, whose pertinent graduations appear in large projection windows. The instrument accepts accessories for almost all measuring processes for which engineering microscopes are generally used.

trained inspector or technician, and kept in a protected location. Although engineering microscopes do not require special fixturing and masters, which involve setting and calibrating procedures, the time factor associated with microscope measurement is quite substantial in relation to the operation of comparator type gages. Therefore, the microscope is not recommended for serial measurements, with the possible exception of dimensions for which the microscope constitutes the best suited, or even the only, measuring means.

The toolmaker and measuring microscopes are unquestionably highly versatile measuring instruments, although not universal. Therefore, when the universal designation is used in conjunction with the advanced types of engineering microscopes, it is intended to convey an exceptional degree of versatility, and not a literal attribute.

However, the adaptability of engineering microscopes for technical parts in the proper size range and of suitable configuration is quite remarkable, barely being equaled by any other type of measuring instrument. Its preferred applications encompass an extended range of characteristic measurements, and whenever such tasks arise, the use of engineering microscopes deserves serious consideration.

Table 8-9 lists examples of dimensional categories for whose measurement the engineering micro-

scope can prove to be the best suited instrument. These examples show typical features and not manufactured parts. Although in the perspective of the preceding discussions some of the examples presented in this table might seem redundant, a condensed survey of its gaging potentials could serve as a guide for finding, in the judiciously selected type of engineering microscope, the proper instrument for solving many uncommon metrological problems.

Projected Image Microscopes

The projected image microscope is ideal for production inspection operations where magnification of the part or assembly is required (that is, printed circuit boards). Figure 8-26 shows the projected image microscope with its viewing screen. This new technology eliminates the fatigue problems associated with the prolonged head and body position control needed with the production use of conventional eyetube type microscopes. The projected image microscope offers a true binocular image that is brilliant and high resolution with sharp depth of field and contrast.

Magnification of up to $1000\times$ is possible within an envelope measuring 8 by 6 by 5.5 inches. The X, Y, Z resolution of this instrument is plus or minus one micron or ± 0.00004 inch. The angular resolution is plus or minus five minutes (eyepiece encoder) or plus or minus one minute (with microprocessor).

Several options and accessories are available for the projected image microscope: (1) either a Polaroid or 35-mm SLR automatic camera can be attached to the device for the purpose of permanently recording

Vision Engineering

Fig. 8-26. Projected image microscope.

TABLE 8-9.   EXAMPLES OF PREFERENTIAL APPLICATIONS FOR ENGINEERING MICROSCOPES — 1

DIMENSIONAL CATEGORY	DIAGRAMMATIC SKETCH	QUALIFYING CONDITIONS	DISCUSSION
Lengths or diameters on outside and inside surfaces		For measurements requiring incremental sensitivity not finer than about $50\,\mu$ in.	Direct measurements without requiring setting masters. Particularly well suited for diameters. Because no cumulative errors occur, accuracy in length ranges of 5 in. and more can prove superior to alternate length-measuring methods.
Very small bore diameters		Applicable to through holes; requires grazing incidence light beam sighting	Holes as small as 0.040-in. diameter are not accessible by mechanical contact members, but can be precisely sighted optically and their diameters measured with the inherent accuracy of the microscope.
Diameters behind shoulders		Using mechanical contact members with optically observed fiducial position	May constitute the highest level of practical measuring accuracy for partially obstructed bores when width of shoulder exceeds the retraction range of contact members in electronic comparators.
Separation of narrow features		Feature boundary must be exposed in the line of sight of the microscope	Narrow shoulders which do not provide secure reference surface for mechanical contact elements can still be sighted precisely by optical means and measured with the microscope.
Complete dimensional analysis		Features must be exposed and observable in a single sectional plane	Parts with diverse features including lengths, angles, radii, etc., can be measured in a single setup referenced from common datum without accumulating errors which are unavoidable in sequential measurements made in several setups.

TABLE 8-9.   EXAMPLES OF PREFERENTIAL APPLICATIONS FOR ENGINEERING MICROSCOPES – 2

DIMENSIONAL CATEGORY	DIAGRAMMATIC SKETCH	QUALIFYING CONDITIONS	DISCUSSION
Edge and corner clearance co-ordinates		Feature must be exposed to optical observation	Chamfers and fillets which are specified by their coordinate dimensions must be measured by referencing from the intersection of the extended contour line that is readily identifiable in the optical image. Chamfer angles, when required, can also be measured.
Coordinate dimensions of features		Must be observable in a common or in parallel planes	The stage displacement in mutually perpendicular directions, combined with measuring capabilities along both axes, permits the precise analysis of features dimensioned in a Cartesian coordinate system.
Separation of virtual points		Applicable to features having straight boundary lines in a sectional plane	Although most technical parts are manufactured with rounded or chamfered edges and relieved corners, the sharp intersections of boundary planes are frequently used for referencing design dimensions. In sectional planes, the imaginary intersections will appear as virtual points and their locations can be precisely sighted for referencing.
Angles of solid		Both side contours must be observable visually, either directly or by transfer	Supplying angular values directly and without auxiliary means is a valuable capability of optical angle measurement. Transferring by means of auxiliary members the slope of the concealed side into the observation field extends the applicability of the process.
Circular divisions		Object must be staged well centralized on the optical rotary table	Angular separation of feature measured with reference to the radius of a common datum circle. The basic optical measuring capabilities of the microscope are of assistance in centralizing the test piece for staging.

TABLE 8-9.   EXAMPLES OF PREFERENTIAL APPLICATIONS FOR ENGINEERING MICROSCOPES – 3

DIMENSIONAL CATEGORY	DIAGRAMMATIC SKETCH	QUALIFYING CONDITIONS	DISCUSSION
Polar coordinates		Object must be staged well centralized on the rotary table	The microscope has the combined potentials needed for measuring polar coordinates involving both angular orientation and linear distance. The center of the rotary table, in coincidence with the optical axis of sighting, is the datum for measuring radial distances.
Circular arc features		Radius range limited by the available graticule traces	Features with essentially circular arc sectional contour can be measured for radius size and inspected for the trueness of form by means of optical comparison with the best-fitting graticule tracing. Measurable features comprise arcuate sections, corner radii, fillets, etc.
Threads— complete dimensional and form analysis		Use of graticules which are available for all standard thread forms	The completeness of thread measurement by microscope is unique. Angles, all essential diameters, lead size and consistency, root and peak radii, are measured by combining visual comparison based on graticule tracings and the controlled translational movements of the instrument table.
Distance between bore centers		Applicable to through holes and other types of pierced figures	When the true position of bores of regular figures is specified, the reliability of the coordinate position measurement is predicated on establishing correctly the virtual location of the feature centers or axes. The double-image system of the microscope ideally meets these requirements.
Symmetry of internal features		Applicable to pierced figures with a nominal axis symmetry	The superimposition of inverted images displayed by the double-image system of the microscope, combined with the cancelling effect of complementary colors, provides an unequalled method of inspecting the conditions of symmetry.

an image; (2) X and Y coordinate light-emitting diode (LED) digital readouts can provide the operator with flexible measuring capabilities; (3) microprocessor outputs through an RS232 interface can send information directly to a microcomputer or printer; and (4) a variety of staging platforms are available for accurate and known linear or rotary movement of the work under the magnification lense.

Video Microscopes

Modern video technology has been married to the microscope to provide the user the advantages of both technologies. Figure 8-27 shows a microscope equipped with an optical measurement inspection system (three axes) and video output. The video display can be used for group viewing and discussion purposes. Also, the video display can be taped for formal documentation of inspection results or for future training needs.

A second type of video microscope is shown in Fig. 8-28. This style of video microscope uses an Optical Video Probe, which is the equivalent of a video camera with magnification capabilities. Magnification of up to 500× with a resolution of 0.000025 inch is possible. Here, the ability to view the work directly through conventional microscope eyetubes has been sacrificed for the versatility of the Optical Video Probe. The probe can be adapted to a coordinate measuring machine for noncontact inspection of large workpieces.

Video microscopes typically combine an X and Y coordinate workpiece staging system with an LED or LCD (liquid crystal display) digital display that is referenced to an on-screen video cross-hair overlay.

Ram Optical Instrumentation Corp.

Fig. 8-28. Direct viewing video microscope with optical video probe, digital readout and dedicated video monitor.

As the workpiece is moved below the microscope lense, the video cross hair provides the inspector with a means of visually locating exact points on the part; in turn, the digital readout permits the precise measurement of distances between features.

Remote Visual Inspection Microscopes

Remote visual inspection (RVI) is a type of nondestructive inspection that uses a family of handheld instruments called scopes. RVI scopes are necessary when the workpiece being checked is too large to place in a stationary microscope or when internal part features are to be inspected. The family of RVI scopes includes the *borescope*, the *fiberscope* and the *videoimagescope*.

Borescopes (Fig. 8-29) are shown in a variety of sizes, with one being equipped with the necessary light source. The borescope has a rigid stainless steel

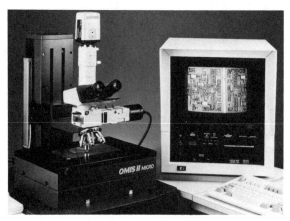

Ram Optical Instrumentation Corp.

Fig. 8-27. Video microscope.

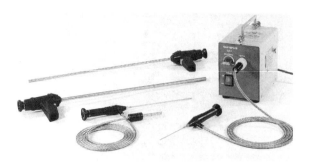

Olympus Corp.

Fig. 8-29. Borescopes with light source.

Olympus Corp.

Fig. 8-30. Borescope application.

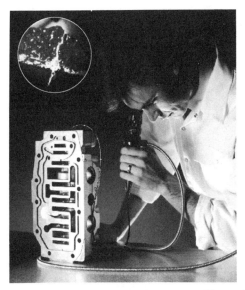

Olympus Corp.

Fig. 8-32. Fiberscope application.

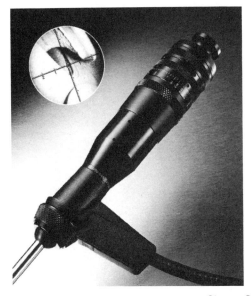

Olympus Corp.

Fig. 8-31. Direct measurement with reticle as viewed through borescope.

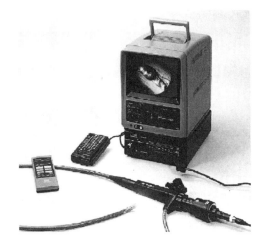

Olympus Corp.

Fig. 8-33. Videoimagescope with microcamera, monitor and VCR.

sheath that holds a series of lenses. Therefore, the borescope requires a straight shot into the area to be inspected (see Fig. 8-30). Direct measurement can be accomplished by bringing the reticle into focus (see Fig. 8-31).

Flexible fiberscopes (Fig. 8-32) use fiberoptic bundles for illumination and image transmission. The fiber-optic bundles are encased in a flexible insertion tube. A variation on the fiberscope is the video-imagescope (Fig. 8-33), which places a small video camera and lens in the distal end of the unit. Compact chip technology permits a video image to be transmitted to a video monitor where it is displayed. This RVI technology is explored completely in a book entitled *The Science of Remote Visual Inspection (RVI) Technology, Applications, Equipment,* by Peter G. Lorenz, Olympus Corporation, Lake Success, NY, 1990.

9.
Optical Projectors

The observation and measurement of objects with the aid of optical magnification is not limited to viewing through an ocular, such as used in the microscope. As mentioned in the preceding chapter, the magnified image of the object can also be projected on a glass screen, where it may be observed from a convenient distance. By the projection of the magnified image, the visual impression becomes a physical reality insofar as the dimensions and geometric forms, as they appear on the screen, can be directly compared to physical masters—graduated rulers, templates and so forth—made to the scale of the magnified specimen image.

Historically, the first optical projectors were developed shortly after World War I, for the purpose of checking the form of screw threading tools. However, the first projectors had several disadvantages, such as requiring a dark room for viewing, and the screen was too far away from the operator for the convenient comparison of the shadow image with a master chart. Subsequently, a mirror was inserted into the path of the projected shadow to reflect the image into the receiving screen in front of the operator. By enclosing the whole system in a housing and shielding the screen, the need for dark room operations was eliminated.

Although the prime users of the optical projectors were the manufacturers of threading tools and high quality threaded parts, the application of that new optical instrument did not long remain limited to a particular branch of the metalworking industry. The optical projector, as a versatile and reliable inspection tool, found rapidly expanding applications, and at present, the various models of optical projectors are indispensable standard equipment in many fields of engineering and fine mechanical production.

The main reasons for this rapid expansion of optical projector applications may be found in two concurrently evolving aspects: (a) industry's need for inspection, observation and measuring instruments capable of satisfying varied requirements for precision and consistency, and (b) the many improvements developed by the manufacturers of optical projectors and incorporated into their products.

The versatility of these instruments is usually provided by added features, comprising a wide variety of adjusting motions, built-in devices, accessories and attachments. Such additions necessarily raise the cost of the equipment, and economic considerations could hamper their applications. Fortunately, the makers of optical projectors or, as some manufacturers designate them, optical comparators, offer these instruments in many varieties of basic design with a wide range of optional equipment.

Optical projectors, in their applications as inspection instruments, present many favorable properties, several of which are listed below:

a. The projected image can be observed by several people simultaneously; thus this method of inspection lends itself to analyses by group.

b. Several dimensions and form characteristics of a specimen can be observed, compared and evaluated in a single setting.

c. The number of dimensions to be inspected on the part, whether individually, or in their interrelations with other dimensions of the same part, can be increased without needing additional instruments or tooling, as long as these dimensions are contained in a common observational plane.

d. Standard comparator charts, particularly for such repetitive forms as circular arcs with different radii, angles, thread forms, gear contours, and so on, can be used on optical projector screens. Such standard charts are made of plate glass, precisely fitting the chart ring of the screen, and permit the rapid, precise comparison of the projected image with the basic design form.

e. Direct measurements of various length dimensions can be made on the image as it appears on the screen, by using either graduated rulers or, for more accurate measurements, glass scales with etched graduations. Similarly, angles can be checked, by using regular drafting protractors.

f. Mechanical gages cause contact pressures, or may be forced over the work, resulting in distortions. Measurements on optical comparators, where no physical contacts are needed, are free from such potential errors.

g. Mechanical gages, particularly fixed gages, are subjected to wear, requiring regular checking and, at intervals, replacement. Wear does not enter into measurements by optical comparison.

h. Correlated surface elements, whose relative positions can be laid out in a two-dimensional system of coordinates, may easily be checked on optical projectors equipped with measuring tables, without the need for expensive special fixturing. Similarly, direct measurements of angles can be carried out by means of the built-in screen protractors, with which many models of optical projectors are equipped.

i. The use of workholding and locating fixtures in conjunction with the optical projection provides means to measure individual dimensions or features in a positively determined relation to reference elements on the part. Fixtures designed for translational indexing can be used for transferring the work between subsequent measurements.

j. The tracing of optically obstructed surface elements and the application of optical cross-sectioning provide means for the accurate measurement of dimensions, whose inspection by any other method than optical projection is extremely difficult or impractical. Another technique for inspecting on a projector the optically hidden surface sections is by means of replicas, using one of several conveniently applicable replica compounds.

k. The observation of surface characteristics by light reflection, using either normal or oblique illumination, substantially widens the scope of inspection procedures that can be carried out by means of optical projectors.

The enumeration of several useful characteristics of the process carried out by modern optical projectors is not intended to provide a complete listing. However, it does convey a general idea of the capabilities of optical projection as a means of inspection and dimensional measurement. At the same time, it indicates the fields of application where optical projectors represent the type of instrument best suited for particular tasks. More detailed information regarding the capabilities of optical projectors should result from the following description of typical equipment, their functions, accessories and characteristic applications.

To visualize the great variety of inspection processes for which the appropriate models of optical projectors are adaptable, a list of potential applications is presented in Table 9-1. This condensed listing contains titles only, without complementary details. It should serve as a preview of the following discussions whose sequence, however, will differ, as it is guided by aspects related to equipment and not by processes, on which Table 9-1 is based.

THE OPERATING PRINCIPLES OF OPTICAL PROJECTORS

To produce an undistorted and magnified shadow or reflected image of an object on a screen, where it can be conveniently observed, is the primary purpose of an optical projector. To accomplish this objective, the following basic elements are needed:

1. *The light source*, usually of great intensity, to produce a clearly defined projection even at a high rate of magnification;

2. *The collimating lens*, whose role is to refract the light into a beam with parallel rays of almost uniform intensity on the entire area of object illumination. Sometimes this element is also called the *light condenser*;

3. *The projection lens system* that magnifies and transmits the object contour or image resulting from the collimated parallel light rays; and

4. *The viewing screen*, on which the projected contour or image of the object appears and is displayed for inspection.

The schematic arrangement of these principal elements is shown in Fig. 9-1, where the path of travel of the light is indicated by arrows. Light rays originating from the light source hit the object, whose physical body creates a shadow bounded by the actual contour of the object when viewed in the direc-

TABLE 9-1. POTENTIAL APPLICATIONS OF OPTICAL PROJECTORS

CATEGORY OF INSPECTION	TYPES OF OPERATIONS
Optical system by which screen image is produced	(a) Projection—shadow silhouette (b) Reflection—mirror image of surface (c) Combination of both—projection and reflection
Application for inspection by observation	(a) Surface properties—texture, finish, surface conditions (flaws, scratches, nicks, cavities, etc.) (b) General contour—straightness, consistency of curvature, blending, etc. (c) Contact patterns with mating parts
Inspection by comparison to master charts	(a) Screen Charts, interchangeable with regular screen, for the inspection of standard forms, e.g., angles, radii, screw threads, gear forms, etc. (b) Overlay Charts, prepared for specific applications either as simple contour charts or charts with tolerance zones
Inspection by direct measurements on the screen image	(a) Linear measurements (distances)—using graduated rulers or glass scales (b) Angle measurements—using drafting or toolmaker protractors (c) Radii—using transparent templates (d) Making contour line tracings on translucent paper placed on the screen—for subsequent analysis
Inspection with measuring devices built into the optical projector	(a) Coordinate table movements (along X and Y axes) by reading the displacement distance on micrometer heads—for linear dimensions (b) Screen protractors—graduated screen rings combined with vernier—for angle measurements
Inspection with the aid of fixturing and special attachments	(a) Helix angle adjustment of table to project thread forms, etc., normal to contour plane (b) Transferring dimensions by means of work holding fixtures and charts with reference points (c) Optical sectioning with special illumination

tion of the light rays. This shadow is then magnified by the lens system and projected onto the viewing screen. In the particular system shown in Fig. 9-1, an auxiliary element, a relay lens is used to transfer the shadow on the projecting lenses.

Mirrors placed ahead and behind the projection lenses have the role of diverting the direction of the light rays with the purpose of reducing the overall size of the equipment. The total length of the light ray path, or "throw," is fixed for any given lens system and magnification. By folding up the light path by means of mirrors, that fixed throw can be squeezed into a smaller space, thus contributing to the more compact design of modern optical projectors. The mirror system provides the additional advantage of bringing back the object image into the proximity of the operator, thus making the observation of the object more convenient and precise.

Considering the principal elements more closely, various particular requirements may be pointed out. The extent to which these are satisfied affects the operational characteristics of the optical projector.

The *light source* is usually a powerful lamp of up to 1000 watts or more, placed before a curved reflecting mirror. For specific applications, when exceedingly sharp and well defined shadows are needed for accurate inspection of the image, the regular tungsten filament lamps may be replaced, in some models, with high pressure mercury arc lamps. The light intensity obtained from mercury arc lamps surpasses by several times the performance of even the best tungsten filament lamps. Modern mercury arc lamps produce a steady light without flickering, but require a rather expensive separate power unit.

The lamp houses of the light sources are designed so as to avoid harmful heat transfer to the optical system, the work-staging and the operating elements of the projector. For this purpose the lamp house is usually externally mounted to isolate convected heat and is equipped with a powerful blower fan. It also has special heat absorbing glass filters to keep back the heat rays that might affect the dimensional stability of the object.

The details mentioned with respect to the shadow projection light source are essentially valid for the reflection light units as well. The purpose of these units is to illuminate the front side of the object, which faces the lens system, and thus to generate reflected light, which is also then magnified and projected on the screen as the object image.

Figures 9-2, 9-3 and 9-4 illustrate the optical systems of different makes and models of optical projectors equipped with reflection units. A different arrangement of essentially similar elements is found on optical projectors using a horizontal staging plane, which requires a vertical optical axis. Usually projection and reflection can be operated alternately, or simultaneously, depending on the particular type of inspection task to which the illumination of the object is applied. More details regarding the use and potentialities of surface exploration by light reflec-

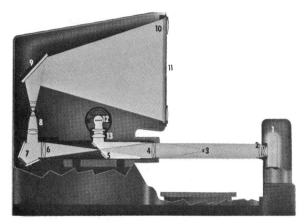

Ex-Cell-O Corporation

Fig. 9-2. Arrangement of the optical system in a bench type projector, equipped for illumination of projected and reflected images.
Direct Projection: (1) light source; (2) collimator; (3) object inspected; (4) first relay lens; (5) mirror with telecentric stop; (6) second relay lens; (7) penta-mirror; (8) magnification lens; (9) main mirror; (10) field (fresnel) lens; (11) viewing screen.
Surface Projection: (12) internal light source; (13) collimator; (5) mirror with telecentric stop; (4) first relay lens; and (3) object inspected.

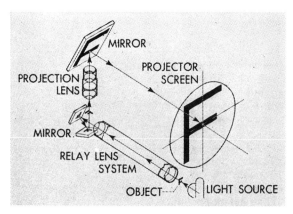

Ex-Cell-O Corporation

Fig. 9-1. Schematic view of the optical system in a typical projector, designed to produce a magnified shadow of the object on the screen.

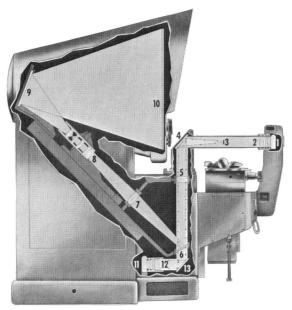

Ex-Cell-O Corporation

Fig. 9-3. Schematic view of the optical system inside a floor type projector, providing for both shadow projection and reflected-surface projection.
Elements of Shadow Projection: (1) light source; *(2)* collimator; *(3)* object; *(4)* front mirror; *(5)* relay lens; *(6)* reversing mirror; *(7)* second relay lens; *(8)* projection lens; *(9)* projection mirror; *(10)* viewing screen.
Elements of Reflected-Surface Projection (complementing the elements listed above): *(11)* reflection light-source; *(12)* light collimator; and *(13)* reflection illumination-mirror.

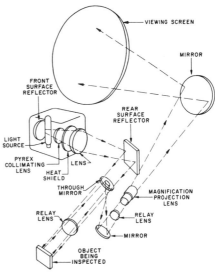

Jones & Lamson/Textron

Fig. 9-4. Schematic view of a reflected light projecting unit. Arrows indicate the path of the light and of the reflected surface, through the optical system, which finally projects the magnified image of the illuminated object face on the screen. This drawing does not show the basic shadow projection system, some elements of which are common with the reflected light unit.

tion will be discussed in conjunction with optical projector applications.

Modern optical projectors, which provide a choice of different magnifications, are also equipped with light switches to regulate the light intensity, thereby producing the best level of illumination for any particular magnification.

The collimator lens is located in the lens housing and is made of heat resistant glass, because of the radiated heat of the close-by lamp. In some models, provision is made for the collimator to accept polarizing filters for photo-elastic stress analysis purposes, or special light filters where the application calls for colored light.

The magnifying lens system is designed to provide an erect, unreversed image of the part on the screen. The plainer types of optical projectors have interchangeable lens systems to accomplish different magnifications. The more elaborate models, which are designed to serve as measuring machines, are equipped with several lens systems for different

magnifications. These can be brought selectively into the operational position by a power drive, simply by actuating a conveniently placed switch.

The lenses used for the better makes of optical projectors have high resolution to present a clear definition of the object and are coated for extra light transmission. When several lens systems are contained in a common turret, the individual lenses are solidly mounted into appropriate positions, to obviate the need for repeated focusing for different magnifications. The change of magnification is accomplished by the rotation of the turret for bringing the required lens into the operational position.

The *screen* is made of ground glass, with finely grained texture, to provide a bright, glare-free image. A good quality screen, when combined with a well designed optical system, should have no "hot spots" and must present an image easy to measure with great accuracy without causing fatigue to the operator. The screen of a good optical projector must permit observation of the image without distortion, even from different angles, as when viewed by a group. The image also should appear uniformly sharp over the entire screen area.

The basic operating principles are essentially common to most types of optical projectors, whether bench models (Fig. 9-5), or pedestal models (Fig.

Jones & Lamson/Textron

Fig. 9-5. (Left) Bench type optical projector mounted on a plain base and set up for comparing the shadow image of the object to the contour lines of an overlay chart that is attached to the projector screen.

Fig. 9-6. (Right) Floor type optical projector equipped for linear measurements along two coordinate axes, and for angular measurements by means of a screen protractor. The box at the left-hand side is the externally mounted light source.

9-6). While the preceding illustrations of systems and instruments show arrangements with a horizontal illumination path resulting in a vertical focal plane, it is possible, and often advantageous, to rotate the arrangement by 90 degrees to obtain a horizontal focal plane. This variation can be accomplished either by adapting the instrument so as to produce a horizontally located focal plane (Fig. 9-7) or by designing the optical projector originally for horizontal part staging (Fig. 9-8). However, the latter design requires an externally attached light source for surface illumination to produce a reflected image. For such applications a semitransparent mirror set at 45 degrees is mounted on the objective to deflect the illumination light on the object surface without obstructing the entrance of the reflected light into the projection system.

THE PRINCIPAL ELEMENTS OF OPTICAL PROJECTORS

Although optical projectors are manufactured in a great variety of sizes, with diverse equipment, serving a wide range of different purposes, there are several design characteristics that are basically common to most models of this category of measuring instruments. Some of the principal design elements of the optical projectors will be discussed from this aspect

of common characteristics, without concentrating on any particular make or model.

General Arrangement

The optical system of the projector works indifferently of the position in which the instrument is set up, whether vertical, horizontal or at any intermediate angle. Consequently the choice of the most favorable position depends on other than optical considerations, such as the following:

a. *Space requirements.* Since the length of the optical path, the "throw" of the system, is of fixed dimension, the governing consideration is the least space commensurate with adequate work staging at a level where convenient manipulation is assured. For the majority of optical projectors the horizontal optical path is preferred, whereas the application of the vertically arranged optical axis (Figs. 9-7 and 9-8) is generally limited to small- and medium-size instruments.

Starrett Precision Optical

Fig. 9-7. Floor-standing vertical axis optical measuring projector.

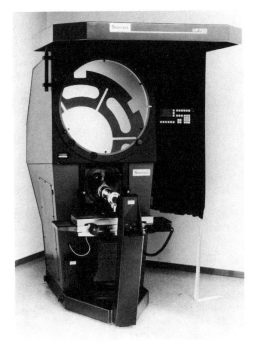

Starrett Precision Optical

Fig. 9-8. Floor-standing horizontal axis optical measuring projector.

b. *Convenience of screen observation.* When the screen image appears in an essentially vertical position (Figs. 9-5, 9-6 and 9-8), group observation is easier than when group members must lean over a horizontal screen; for this reason, most types of optical projectors have basically vertical screens. On the other hand, the essentially horizontal screen, usually arranged in a sloping position, presents some advantages when handmade contour sketches of the screen image have to be prepared frequently.

c. *Work staging.* In the case of the horizontal projection path, the focal plane of the optical system is in the vertical position. For many work-staging tasks this is a convenient position, particularly with heavy objects. However, the horizontal projection path, requiring vertical staging, has its drawbacks when thin objects with little rigidity must be held. The vertical staging of the object may call for fixturing, whereas in the case of horizontal work positioning, such thin objects can simply rest on the top of a glass staging table. A characteristic example of optical projectors designed for a horizontal staging plane, with a vertical projection path combined with vertical screen position, is shown in Fig. 9-8. Some models of optical projectors with basically horizontally arranged

optical path can be equipped with optional attachments for object staging in the horizontal plane, thereby adapting an essentially heavy instrument to very small or nonrigid workpieces (see Fig. 9-14).

Instrument Table for Work Staging and Positioning

In the plainest types of optical projectors that may advantageously be used as inspection tools for long production runs, the work-staging table consists essentially of a base plate on which workholding fixtures can be mounted. In this case, the table top has a T-slot or a rail, to locate the fixture and to guide it while focusing.

The majority of optical projectors built for universal applications are equipped with cross-slide tables, and have controlled traverse motions by means of micrometer screws. Usually, movements along three axes are provided for the stage; two of these, designated as X and Y axes, are for measuring, and the third, the Z axis, is for focusing. Some manufacturers offer measuring means along the focusing axis as well, the accuracy of measurements in that direction, however, being dependent on the keen perception of the optimum focusing position. The actuating elements, or conjointly functioning members, provide means to control the measuring motions based on distinct position indications. For the focusing motion most instruments are equipped with a screw having a wheel or knob, to permit the sensitive adjustment of the inspected surface into the focal plane, as directed by the sharpness of the projected object image.

Depending on the general arrangement of the projector, whether the focal plane is in the horizontal or in the vertical position, the means of translational movement will be different. On optical projectors with a horizontal focal plane, where the object is lying flat, supported on a glass plate, the movement along the X and Y axes can be accomplished by regular cross-slides (see Figs. 9-7 and 9-8). A different translational mechanism is needed for optical projectors whose focal plane is in the vertical. Since most of the large-size optical projectors are of the latter design, power drives for the vertical table movement are frequently applied. Such power elevation units may have quite elaborate controls, considering that, in addition to the rapid motion, fine adjustment, known as jogging, must also be provided. Some models have electronic controls for regulating the speed of the vertical table adjustment; such controls provide a great convenience of operation together with excellent sensitivity for fine positioning.

Jones & Lamson/Textron

Fig. 9-9. Vertical measurements on the floor type optical projector (see Fig. 9-6) can be carried out by raising or lowering the staging table. Large-size machines have power elevating equipment and the graduated ring attached to the table elevating screw permits the direct reading of the position changes in increments of 0.0001 inch.

Figure 9-9 shows the graduated measuring ring of a vertical table elevating device; the ring is keyed on the elevating spindle.

Although some of the extremely heavy machines have power movements for the traverse adjustment as well, that movement is usually accomplished by hand through the lead screw of the slide. To avoid stick-slip of the slide motion and to assure ease and sensitivity of the adjustment, the staging tables of the heavier types of modern projectors rest on ball or roller slides.

For small-size cross-slides, the screws of the micrometers also serve as the actuating members of the table movement. The large-size optical projectors, with heavy tables and long displacement range along a single axis, have independent table traverse by screw and handwheel, supplemented with a micrometer head for measurements over a reduced range. In this case the table slide is equipped with a spring that makes the slide bear against the face of the micrometer spindle. The anvil of the micrometer is adjustable along the edge of the table and can also be entirely disengaged when the table traverse screw is used for rapid adjustment (see Fig. 9-10).

Although most micrometer screws have a direct measuring range of only one inch, accurate measurements over greater distances can readily be accomplished by extending the measuring range of the micrometer. For this purpose gage rods of one inch or multiples of one-inch length can be inserted between the face of the micrometer spindle and the anvil, which is attached to the table slide (Fig. 9-10).

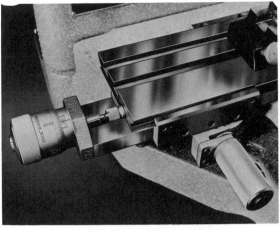

Jones & Lamson/Textron

Fig. 9-10. Horizontal measurements are accomplished by controlled table traverse. The large barrel micrometer reads in 0.0001-inch increments and has a one-inch measuring range. This range can be further extended by inserting end measuring rods between the spindle of the micrometer and the adjustable anvil block of the table.

There are now available measuring tables for particular models of optical projectors, which are free from the inherent inaccuracies, and eliminate the potential reading errors of micrometer screws. Such systems also avoid the inconvenience of using gage rods when the measuring distance exceeds the travel range of the micrometer. One model of these tables has a measuring range of 4 by 8 inches, operates by master scales mounted inside the table and provides digital readout in 0.0001-inch increments over the entire measuring range.

In addition to the right angle coordinate movements, optical projectors designed for universal applications are usually provided with a swivel movement of the staging table as well. This swivel adjustment, on the order of 12 to 15 degrees in either direction, is needed for the optical inspection of objects for which helical planes represent the optically accessible section of the contour. Typical examples of such parts are cutting tools with helical grooves, for example, threading taps, hobs, helical groove milling cutters and so forth. Inspection normal to a helix angle is not limited to cutting tools; there are also other types of parts whose unobstructed contour observation requires the rotation of the stage around an axis normal to the optical path. The swivel movement of the stage can either originate from the table column, or an additional table top may be provided for that purpose (see Fig. 9-11). In each case the ro-

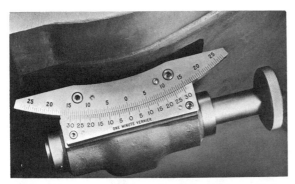

Jones & Lamson/Textron

Fig. 9-11. Table-angle adjustment is needed when the observation plane differs from the work-locating plane, such as in the case of screw threads. Graduated segment permits controlling the swing of the table to fractions of a degree, to precisely duplicate the helix angle of the object.

tational adjustment of the table, which must conform to the helix angle of the part, is guided by a graduated segment with the least reading in fractions of a degree. When this segment is supplemented by a vernier, the angle adjustment can be read in increments of one minute of arc.

The Optical System—Magnifications

When choosing the most favorable magnification for the viewing of the object, two opposing aspects must be brought into proper balance:

a. The higher the magnification, the better definition may be obtained of the intricate details of the object.

b. The lower the magnification, the larger will be the field diameter, that is, that area of the object that can be projected onto the screen.

This relationship may be expressed by the following simple formula:

$$\text{Field diameter} = \frac{\text{screen diameter}}{\text{lens magnification}}.$$

Although for specific applications there will be only one optimum magnification, optical projectors in general are intended for the inspection of diverse objects requiring different magnifications. For this reason, most types of optical projectors are built with an optical system that permits interchanging the lenses, in some models even without affecting the focusing of the object.

The most frequently used magnifications in optical projectors are $5\times$, $10\times$, $20\times$, $25\times$, $30\times$,

$31.25\times$, $50\times$, $62.5\times$, $100\times$ and $125\times$. This listing refers to magnifications available in different makes of optical projectors, but never are all of these used concurrently in a single model. Closely similar magnifications are those chosen, or offered as options, by different manufacturers. As an example, one maker or user may prefer the $30\times$, whereas another the $31.25\times$ magnification. The former rate is based on the decimal system, whereas the latter has the property of being a complementary factor to real fractions: 31.25 is the $\frac{1}{32}$ part of 1000. Consequently, when using the $31.25\times$ magnification, each $\frac{1}{32}$ inch on the image is equal to a 0.001-inch distance on the actual object. This relationship permits the use of a standard scale with $\frac{1}{32}$-inch graduations to measure the screen image details in 0.001-inch increments, without the assistance of a measuring type coordinate table. At $62.5\times$ magnification, 0.001 inch on the object appears as $\frac{1}{16}$-inch distance on the screen image.

For the purpose of selecting and actually using the most favorable magnification available on the projector, with the least effort and time requirement, modern optical projectors have built-in lens changers. These devices contain several lenses in a slide or on a turret and the desired lens can be brought into the operative position either by a slide movement, or by rotating the turret. Figure 9-12 shows

Jones & Lamson/Textron

Fig. 9-12. Turret type lens selector with lenses for six different magnifications. The indexing motion that brings the selected magnification lens into operating position is carried out by a built-in electric motor controlled by a multiple-position switch in the front of the projector.

such a lens turret that can be indexed by remote control and power drive into the desired position, simply by operating a multiple-position switch in the front of the instrument.

Concurrently with the mechanical lens changers, modern optical projectors are also equipped with light intensity selector switches. This permits the use of the most favorable illumination for each magnification, without causing glare or reducing unnecessarily the life of the bulbs due to protracted use with excessive intensity.

The Projector Screen

The screen onto which the image of the object, whether only a shadow or combined with the reflected face, is projected, is usually made of ground glass. It has fine grain texture on its surface to prevent glare and to permit the distinct viewing from different angles. Depending on the design of the optical system, the position of the screen may be exactly horizontal (inconvenient for viewing) or vertical (frequently used), but many manufacturers of instruments prefer designs with slightly tilted—usually about 10 degrees—screen position in relation to either of the preceding basic varieties.

Most makes and models of optical projectors have round screens because this form utilizes with the best effect the image produced by the optical system. In exceptional cases only, particularly on certain makes of large-size projectors, rectangular screens are used. The most frequently, although not exclusively, used screen sizes, which are generally characteristic for the size category of the optical projector as well, are 8, 10, 14, 20, and 30 inches in diameter.

The glass screens have precisely ground edges that fit into the screen ring of the projector. The use of precisely ground screen edges makes it possible to: (a) locate the screen in the correct optical plane of the system; (b) reliably maintain the basic coordinate position for measurements; and (c) assure the interchangeability of the regular glass screens with different types of chart screens.

The screen ring can be fixed or rotatable. The latter design is found in most optical projectors, which are designed for universal applications. For the rotatable types, the screen ring is precisely guided to retain its plane during rotation. The screen ring has angular graduations around its entire circumference, usually in increments of one degree, or one half of a degree. The numbering of the circumferential graduations is often from 1 to 90 degrees in each of the

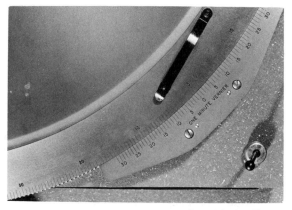

Jones & Lamson/Textron

Fig. 9-13. The screen protractor consists of a rotatable, screen-holder ring, graduated in ½-degree increments around its entire circumference. When used in conjunction with a 2 by 30 division vernier segment, readings to one minute of arc can be made.

four rectangular sectors, because the precisely perpendicular cross-hair lines of the basic glass screen eliminate the need for rotating the screen ring by more than 90 degrees for measuring any angle. An additional screen line at 30 degrees is useful particularly in the measurement of standard screw thread forms. By complementing the ring graduations with a fixed segment vernier, replacing a plain index line, angle measurements with an indicating sensitivity of one minute of arc may be accomplished.

The described equipment, currently designated as a screen protractor (see Fig. 9-13), permits rapid angular measurements on the screen image. However, it must be remembered that besides the readout increments, the ultimate accuracy of angle measurements also depends on the precision with which the screen hairline is brought into coincidence with the image contour lines, which represent the arms of the angle being measured.

The protractor ring of the screen can be rotated directly by hand, using the attached knobs. For the purpose of fine adjustment, the screen mounts of some models have peripheral worm gears to be driven by a worm with serrated handling knob; the worm can be swung out of engagement for the hand rotation of the mount. Clips on the screen ring serve to conveniently attach overlay charts over the basic glass screen.

Staging and Measuring Capacity

For the selection of the proper size of optical projector, the *screen diameter* may be regarded as the most characteristic single dimension. The envelope dimensions of the optical system, and thereby the minimum outside dimensions of the entire instrument, are a function of the screen diameter.

From an application point of view, it is that characteristic dimension, generally the diameter of the round screen that determines the area of the object that can be contained in the screen image at any specific magnification. Although it is quite feasible to move the object and thereby to bring its various sections into the field of view of the system, for many applications it is desirable to have the entire object, or a whole detail contained in the screen image. This is particularly the case for comparison inspection by means of contour charts.

The *size of the largest object* that can be mounted on the projector stage may also need to be investigated. This is the case when the object has at different locations, details that must be explored by optical projection, or even measured in controlled dimensional relation to each other. Factors that need to be checked in such cases are the following:

a. The *working distance of the optical system*—the distance of the focal plane from the relay lens. This is the lens behind the object stage, which picks up the image of the part and transmits it to the magnifying lens system. It is also called focal clearance.

b. The *throat clearance*—the free distance between the light source condenser and the relay lens. For all practical purposes, this is the sum of the condenser clearance, which is the distance between condensing lens and focal plane, and of the working distance.

c. The *vertical clearance*—the maximum distance between the table top and the lens centerline.

d. The *range of coordinate movements* is still another significant aspect for evaluating the capacity of optical projectors. It is also called the measuring range, and it is not necessarily identical with the total table travel that, in some models, may exceed the distances of the measuring movements.

The purpose of the coordinate movements is to bring, by means of controlled displacement, distant details of the object into the focal field, and also to establish coincidence between pertinent image elements and screen index lines or chart outlines.

The following are typical capacity data for different sizes of standard models of a well-known make of optical projectors:

Screen Diameter	10 (in.)	14 (in.)	30 (in.)
Lens centerline above table:			
Maximum	$3\frac{1}{4}$	$5\frac{1}{4}$	$5\frac{1}{4}$
Minimum	$1\frac{1}{4}$	$1\frac{1}{4}$	$\frac{1}{4}$
Measuring range:			
Lateral	4	8	8
Vertical	2	4	5
Table dimensions:			
Length	16	30	30
Width	5	6	6

COMPLEMENTARY EQUIPMENT OF OPTICAL PROJECTORS

Because of the varied applications of inspection by means of optical projectors, the equipment requirements may differ over a rather wide range. The size and measuring capacity of the projector, although a major consideration, represent one aspect only. Depending on the selected purpose, the optical projector might be required, in one case, to perform the whole year round only a single task, for example, that of comparing object contours to screen charts. In other cases, universal type measuring machine applications may be planned, or again, special gaging with single purpose fixturing is the operation the optical projector will have to perform.

To satisfy the diversity of intended applications without the burden of unnecessary and costly accessories, optical projectors are usually available as base models with a wide selection of optional equipment. The following survey, which attempts to organize the varied optional items into categories, may serve as a guide when making the selection of optical projector equipment.

Optional Projector Equipment

These are usually built-in devices that may become integral elements of the base machine.

1. *Measuring movements.*

a. The coordinate movement of the staging table along X and Y axes. Depending on the basic arrangement of the projector, these axes are contained either in the vertical or in the horizontal plane. Micrometer heads or indicators are commonly used to measure the amount of displacement in the direction of the coordinate axes.

b. Focusing movement with cross-slide actuated by a screw. When this device is not part of the machine, the focusing must be performed by moving the staged object along the axis of the optical system; a groove in the table plate may be provided to guide a fixture base when carrying out this free-hand focusing operation.

c. Swivel adjustment of the staging table. This adjustment is primarily needed when parts with contours normal to the helix angle are being inspected. Most universal types of measuring projectors have this adjustment as a built-in feature.

2. *Projector lenses.* The intended application of the optical projector must be the deciding factor in the choice of one of the following alternative designs:

a. Single magnification by a fixed lens. This is selected when the projector is to be used continuously on a single specific application that always requires the same magnification.

b. Variable magnification by interchangeable lenses. This is the proper equipment when diverse inspection tasks requiring different magnifications will have to be performed, yet the frequency of changes does not call for mechanized lens changer.

c. Variable magnifications by lens selector (slide or turret). Time saving and very convenient equipment for optical projectors that are intended for universal applications where changes in the desired magnification may follow in rapid succession.

3. *Illumination.* The following frequently selected alternatives may be listed:

a. Frontal illumination by filament lamp. This is the basic equipment of optical projectors to produce the shadow image of the object on the viewing screen.

b. Reflection illumination—directed against the rear side of the object that faces the condenser (relay) lens, and produces the reflected image of that illuminated face. This is an optional equipment for the moderately priced types of instruments, yet standard in the universal types.

c. Oblique surface illumination—optional equipment that may produce meaningful images when the regular reflection illumination, which is directed normal to the object face, would fail. Ex-

amples of application: nonreflecting object surfaces or where the surface texture should be studied under the optimum angle of incidence of the light rays.

d. Mercury arc light source—to produce extra high intensity of illumination for magnification in the upper range (62.5 times and over), or where critical details of the object must be observed on an image that must have exceptionally distinct definition.

e. Vertical illumination for normally horizontal light path projectors (Fig. 9-14). This is usually an optional equipment for staging small objects by supporting them on a horizontal glass plate. It is used on heavy projectors where the regular arrangement requires the object to be staged in an erect position.

4. *Special equipment.* Leading manufacturers of optical projectors are in a position to supply various special equipment designed for specific purposes. Such accessories are helpful in making feasible the application of inspection by optical projection where regular equipment would fail. Typical examples of this kind of special equipment are:

a. Optical sectioning—by diverting concentrated light obliquely, from one or two directions, on the object and masking part of the light beam with a knife edge blade. The effect of this arrangement is as if the object were cross-sectioned along the boundary of the blade shadow and the image

Jones & Lamson/Textron

Fig. 9-14. (Left) Optical projectors with a normally horizontal light path can also be used with vertical illumination, when mounting a workholder with built-in mirror. This special arrangement permits staging small objects on heavy projectors in a horizontal position, lying flat on the glass top of the fixture.

Fig. 9-15. (Right) Tracing attachment for exploring the contour of optically hidden surface elements. The device is equipped with a contact feeler and an indicator stylus, both mounted on a common coordinate slide. The movement of the projected stylus-tip is compared to the contour line of an overlay chart.

resulting on the screen shows the contour line of the object in the sectioned plane. This method of inspection is particularly useful for exploring objects whose cross-sections vary, such as turbine blades.

b. Tracing of hidden surfaces and observing the momentary position of the indicating tip on the screen where it can be compared to the reference line of a screen chart representing the magnified design contour of the explored surface section. (Fig. 9-15). The device consists of a slide moving freely along two coordinate axes (*X* and *Y*) and carrying two identical arms. One of these bears the feeler tip, or stylus, which is guided by the physical contour of the optically hidden object surface. The other arm, with the indicating tip located in the optical path of the light beam and in the focal plane of the optical system, moves in conformity with the guided feeler tip.

Object Staging Fixtures

In some applications it may suffice to place the object on the top of the staging table in a position where its contour to be explored is in the optical plane of the projector. Simply placing the object on a horizontal glass plate is one of the current applications of this kind of staging.

In most cases, however, some kind of workholding device is desirable, or even mandatory. With the understanding that frequently there is more than a single reason for the application of fixturing, a few typical circumstances that require object holding fixtures are listed below:

a. To hold the object securely when gravity alone does not prove sufficient;

b. To orient the object so as to bring its area to be explored into the optical plane of the projector;

c. To reduce setup time for repetitive inspection or series of identical parts;

d. To locate the object from a physical reference surface that is outside the field of view of the projector; and

e. When the workholding fixture is required also to accomplish the moving of the object in a controlled manner over a specific distance to check the relative positions of interrelated surface elements.

The diverse workholding fixtures used on optical projectors may be listed with regard to their functions under the following categories:

Jones & Lamson/Textron

Fig. 9-16. (Left) Fixture base, fitting the groove of the projector table, where the base can be locked into position by means of thumb-screws. The cross groove on the base top serves to guide fixtures during focusing.

Fig. 9-17. (Right) Vise mounted on a fixture base, the latter fitting the projector table. Different types of vises can be used in the same base.

1. *Fixture bases* (see Fig. 9-16) that may serve two purposes:

a. Locating the holding fixture in controlled relation to the surface and guide grooves of the staging table; and

b. Raising the level of the holding fixture, when needed, to bring small objects into the optical plane of the projector lens system.

2. *Vises* (see Fig. 9-17) with various capacities and jaw forms.

3. *Clamping fixtures* (see Fig. 9-18) other than vises, such as simple and double Vee-block clamps, centers, special part locating and holding devices.

4. *Rotating tables* (see Fig. 9-19), also with angle graduations, to orient the object to be inspected in different planes and angle positions.

Jones & Lamson/Textron

Fig. 9-18. (Left) Vee-block with clamp to hold cylindrical objects at a fixed level. Position of the Vee-block can be adjusted on the base to bring the object into the focal plane of the optical system.

Fig. 9-19. (Right) Circular table with angle graduations, which permits rotating the object over a controlled angle between subsequent inspection steps. The sharpness of the screen image will indicate whether the angle relationship of the object details coincides with the applied indexing position.

5. *Special fixtures* with built-in locating and reference surfaces and, when needed, with provisions for displacement over controlled distances and angles.

Accessories for Screen Image Analysis

Examples of accessories for screen image analysis:

1. *Chart rails* (Fig. 9-20) used to locate and to retain overlay charts on the screen in a properly aligned and secure position. The rail may be single or double, the latter being preferable for large frame sizes and for chart materials of moderate stiffness.

2. *Glass scales and templates* permit rapid measurements directly on the screen image. These scales are available with both fractional and decimal divisions to permit checking distances in any currently used projector magnification. Glass templates are used to check radii of circular arc curvatures on the projected image. Radius templates are made to comprise a range of different radii, but separate templates are needed for each magnification, unless they are in an integral multiple relation to each other.

3. *Photographic attachments* when placed in the screen ring replacing the glass screen, operate in a

manner comparable to a contact printer, to make exact duplicates of the screen image. The attachment is essentially a film holder, and by making use of the optical projector's optics and illumination, printable film pictures of the screen image are produced for permanent records.

4. *Screen charts* (see Table 9-2) can be made to be interchangeable with the original glass screen of the projector. To be interchangeable, the screen chart must be made of glass ground exactly to the diameter of the screen ring bore. However, screen charts are more frequently of the *overlay chart type*, which can be attached over the basic glass screen in either of two ways:

a. *By guide rails* (Fig. 9-20). This device has the advantage of locating the chart in a controlled coordinate and angular position.

b. *By means of clips.* In this case, reference lines on the chart must be aligned with corresponding targets on the workholding fixture or the object image, by means of optical observation.

For screen chart material, either of the following alternatives can be selected, whichever is considered best suited for the specific application:

(1) *Glass screen charts* have the advantage of very clear chart lines and are less sensitive to routine cleansing with any standard agent. However, they are relatively expensive and require great care in handling and storage.

(2) *Plastic charts* made of vinyl type materials can either be rigid for use in guide rails or flexible for attaching by means of clips. Plastic charts are light, less sensitive in handling, easily stored and little affected by temperature and humidity variations. Although not entirely equivalent in line definition to glass charts, charts made of plastic have the advantage of being less expensive. Plastic chart materials can also be used for making special charts by current drafting room processes, permitting the application of drawing ink lines as fine as 0.005 inch.

(3) *Translucent paper charts* are made for rapid inspections in short-run production and are prepared in the drafting room to provide reference outlines for image comparison. Translucent paper is also used for copying the outlines of the screen image to obtain a permanent record for subsequent study or filing.

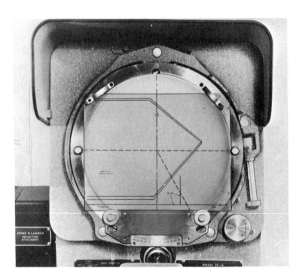

Jones & Lamson/Textron

Fig. 9-20. Chart rails to locate and to hold overlay charts have cams for alignment by fine adjustment. Reversing the rails provides alternate height position when using overlay charts of different frame dimensions.

TABLE 9-2.   TYPES AND APPLICATIONS OF OPTICAL PROJECTOR SCREEN CHART

CATEGORY	GENERAL CHARACTERISTICS	TYPICAL APPLICATIONS
Reference line charts	Usually cross hairs through center of the screen, also with an additional 30/60 degree radial line.	Used to align that contour section of the screen image which is chosen for reference base. Then object is relocated by moving staging table along one or two axes until the interrelated contour line of the image coincides with the same screen hairline. Distance of displacement is read on micrometer heads. For angle measurements the screen ring is rotated and dimension read on the screen protractor.
Basic form charts	Radii in the form of complete circles, step radius charts or circular arcs for long radii. Protractor charts. Grid charts.	Radius charts aid in checking conformity of curvatures with circular arcs of known radius. Other basic form charts permit angle and co-ordinate measurements, by reading directly from the screen.
Combined reference and form charts	Radii and cross hairlines combined on the same chart.	These contain both, basic forms for comparison and hairlines for referencing in connection with coordinate and protractor measurements. Figure 21 shows a typical application example.
Standard form charts	Contain all critical parameters of the object which can be observed in a single plane.	Dimensions and forms of standardized parts with several specified parameters are inspected with these charts, which are available, e.g., for most of the standard thread forms, in different magnifications. Either single-line charts (Fig. 22), or double-line charts providing tolerance zones (Fig. 23) are used.
Combination form charts	Combine characteristic forms of several sizes of parts pertaining to the same type of product.	The application of optical projector inspection can be limited to parameters for which no other convenient method of inspection is known. In that case the critical form sections—e.g., thread angles, roots, etc.—of a series of a specific type of product may be combined on a single chart (Fig. 24).
Specific object inspection charts	Made by drafting room process or from layout by specialized supplier to show all pertinent dimensions of any object fit for optical projector inspection.	Used for repetitive inspection to reduce the time required by coordinate measurements. Permits observation at a single glance by comparing image outline to chart lines. Can be prepared either as a single line chart or as a tolerance zone chart.

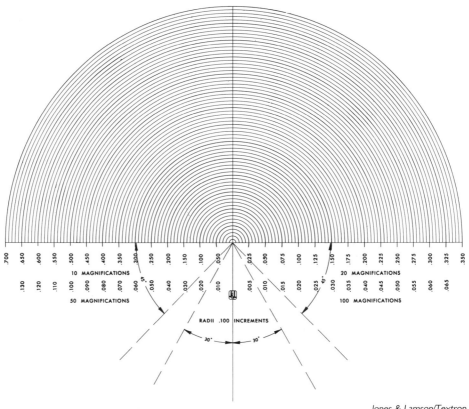

Jones & Lamson/Textron

Fig. 9-21. Combined reference and form chart for checking in a single setup: (a) form (radius), by comparison; (b) angle, with the screen protractor; (c) length dimensions, by using the coordinate movement of the staging table and aligning image contour with chart hairlines.

APPLICATIONS OF OPTICAL PROJECTORS

The inspection operations and dimensional measurements that can be carried out with the optical projector, in many respects, are similar to those discussed in detail in the preceding chapter dealing with engineering microscopes. However, there are still substantial differences between these two families of optical measuring instruments, with regard to capabilities and suitability for particular application tasks and conditions. These differences will be discussed first from a general viewpoint, disregarding exceptions, which may be numerous.

Microscopes are instruments intended primarily for toolroom and gage room applications and require a certain degree of skill and experience for their efficient operation. Thus, when used to inspect critical dimensions or for measurements requiring special procedures, they should be assigned to a competent toolmaker or a specialized metrologist.

Optical projectors are basically production-oriented instruments and, in fact, are often used on the shop floor by machine tool operators or trained inspectors. Optical projectors are generally sturdy instruments, less sensitive than microscopes, easier to operate and even unskilled personnel can quickly be trained to carry out simple inspection processes on these instruments.

Optical projectors are not adaptable to various types of special accessories designed for microscopes; consequently, certain categories of measurements listed in the preceding chapter do not apply to optical projectors. While the power of resolution for linear measurements in the highest grades of measuring microscopes subdivides the ten-thousandth of an inch, resulting in increments of 50 or 20 microinches, the least scale graduation of even

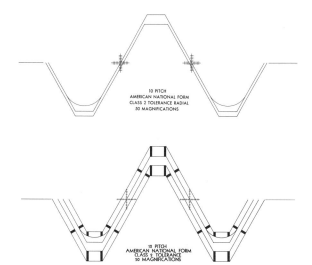

Jones & Lamson/Textron

Fig. 9-22. (Top) Standard-form chart of a screw thread with single contour line.

Fig. 9-23. (Bottom) Standard-form chart of a screw thread with double boundary lines creating a tolerance zone to aid inspection.

the most advanced types of optical projectors is in 0.0001-inch units.

On the other hand, optical projectors provide application advantages in many other respects, in comparison to the capabilities of engineering microscopes. Examples of such characteristics and operating conditions are listed below, also with the purpose of pointing out the preferential areas of optical projector usage.

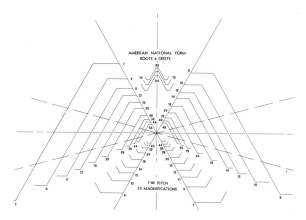

Jones & Lamson/Textron

Fig. 9-24. Combination-form chart containing all root and crest forms of American National Screw Threads, from 7 to 80 pitch.

1. *Volume and weight of the test piece.* Optical projectors are available in sizes that can accept for staging and inspection heavy parts of considerable outside dimensions.

2. *Field view on the object.* Even for medium-size optical projectors this exceeds the comparable capacity of microscopes in general. A large object field permits the synoptic viewing of extended areas on the work surface, reduces the need for object displacement or indexing and offers wider applications for inspection by contour comparison without actual measurements.

3. The *open screen,* commonly at eye level, permits group viewing and the observation of the image in unrestricted position under more natural conditions than the viewing through a microscope eyepiece.

4. *Machine tool applications* for the continued observation of the work progress, guided by screen charts, without impeding the movement of machine tool members or the handling of the controls by the operator.

5. *Individual screen charts* for purely visual inspection of toleranced part features can be prepared and mounted on the screen according to the requirements of the scheduled inspection operation. On such charts, the tolerance ranges appear as graphically laid out zones that must contain the pertinent contour of the part being inspected.

6. *Reproduction by photography* requires only the exchange of the screen against a plate frame, for preparing a silhouette photo exactly of the size in which the image appeared on the screen. The silhouette image of the part can also be reproduced superimposed on the pertinent screen chart, resulting in photos of value for subsequent analysis or for reference purposes.

7. *Duplication by contour drawing.* By mounting vellum paper on the glass screen, the contour of the inspected part can be traced with a pencil and the so-prepared tracing may serve record purposes, or may be overlaid on the screen image of the subsequently inspected mating part for observing the resulting contact conditions.

8. *Exploring internal surface contours by stylus device.* A stylus attached to a tracer arm follows the optically hidden contour of the part's surface and the image of the parallel moving duplicate stylus, which lies in the optical plane of the instrument, is pro-

TABLE 9-3.   EXAMPLES OF PREFERENTIAL APPLICATIONS FOR OPTICAL PROJECTORS – 1

CONDITION INSPECTED	DESIGNATION OF THE OBJECT	SCHEMATIC DIAGRAM	DISCUSSION
Linear and angular dimensions	Extruded shape		The measurement of several linear and angular dimensions can be accomplished in a single setup, thus assuring the proper interrelation between the consecutively inspected dimensions.
Coordinate locations toleranced by true positions	Drill plate		A layout chart on the screen with tolerance zones which were established based on the coordinate positions of the feature centers, permits the inspection of the part in accordance with the principles of true position tolerancing.
Multiple dimensions of a large component	Gear housing		Multiple dimensions on large parts, to the limits of the instrument's measuring range, can be inspected by observing consecutively the adjoining sections of the object mounted on the stage of the projector. Features contained in different, yet parallel planes, can also be measured.
Sizes combined with geometric interrelations	Shaft with several bearing seats		By establishing the starting stage of the measuring process an appropriate datum, e.g., the axis of the part's image, geometric interrelations, such as parallelism, squareness, coaxiality etc. can be reliably inspected, together with the measurement of the individual dimensions.

TABLE 9-3.   EXAMPLES OF PREFERENTIAL APPLICATIONS FOR OPTICAL PROJECTORS — 2

CONDITION INSPECTED	DESIGNATION OF THE OBJECT	SCHEMATIC DIAGRAM	DISCUSSION
Complete screw thread analysis	Ground thread tap		Threaded parts, from small taps to machine components of substantial size and weight, can be inspected for all major dimensions of external screw threads, such as diameters, flank angle, lead, root and crest form, etc. Thread form screen charts greatly assist such inspections.
Stamping with uncommon yet tightly toleranced shape	Gear segment with lever arm		An enlarged contour drawing prepared as a screen chart, either with or without tolerance zones, permits the rapid checking of the part's conformance with the design, regarding both form and size.
Functionally interrelated surfaces	A housing bore inspected for parallelism with the base and squareness with the face		The direction of a bore is functionally determined by the position it provides for a snugly fitting cylindrical arbor. The extending section of the latter is optically exposed and can be compared to the contour line of the part's face as well as to the general position of the part established by its base.
Complete analysis by visual observation	Printed circuit		The proper location and connections of element, leads and terminals of modern printed circuits, particularly in miniature sizes, must be inspected at high magnification. On optical projectors special screen charts assure the speed and the completeness of such inspection.

TABLE 9-3.   EXAMPLES OF PREFERENTIAL APPLICATIONS FOR OPTICAL PROJECTORS – 3

CONDITION INSPECTED	DESIGNATION OF THE OBJECT	SCHEMATIC DIAGRAM	DISCUSSION
Mating members in functional contact	Tapered bearing roller and ring pathway		Functionally meaningful mating conditions, such as the location of the contact areas on the flanges, can be measured and also photographed for future reference.
Internal surface dimensions by replica technique	Outer ring of a roller bearing		Replica compounds are available which are rapidly curing and convenient to use, yet provide a faithful duplication of the surface contour with excellent detail definition. Optically inaccessible areas of a part can be inspected by mounting the replica on an optical projector.
Internal surface contour by tracer attachment	Ball joint socket with seal groove		The contours of optically hidden surfaces can be inspected in comparison to a screen chart line by using a tracer attachment on an optical projector. The tracer has two styli which move conjointly; one rides on the part surface and the other lies in the focal plane of the projector.
Play due to clearance in an assembly	Fit conditions in a cylindrical roller bearing	Clearance gap	Clearances in assemblies which are accessible to optical viewing can be inspected by applying a force consecutively in opposite directions, or alternating free drop and total support, while the assembly is being observed on the projector screen.

jected on the screen; there its position can be compared to the tracing line of a screen chart displaying the basic contour of the traced element.

9. *Outputting of electronic linear measurement information.* Through the use of a simple two-axes digital electronic display (Figs. 9-7 and 9-8), information can be gathered for output to computerized video and other statistical process control systems.

Examples of typical applications, utilizing the characteristic properties of optical projectors, are shown in Table 9-3, with specimen sketches illustrating diagrammatically the dimensions being inspected.

Projected and Reflected Object Image

An essential function in the operation of optical projectors is the illumination of the specimen for the purpose of producing an adequate image on the screen. That image should have characteristics best suited for the inspection of the meaningful form, size or surface texture of the part. As indicated earlier, such an image can be a contour projection, a surface reflection or a combination of both. Although an inspector experienced in the operation of optical projectors determines either in advance or by experimentation the most suitable type of object illumination, a few general considerations are presented in the following to complement the general discussion of the uses of optical projectors.

The most common application of the optical projector is based on projecting the *shadow image* of the object on the viewing screen. The shadow, or silhouette, represents the contour of the object in that plane that is brought to coincide with the focal plane of the optical system. For round objects, this should be identical with the diametral plane, where the projection of the silhouette faithfully represents the cross-sectional contour of the part. For flat pieces of essentially uniform thickness the projected image is, for most practical purposes, also a true replica of the object contour. This claim is particularly valid for relatively thin objects; however, in the case of flat parts with greater thickness the projected shadow could suffer in sharpness of the contour. Regular bodies of revolution, or substantially round objects whose surface contour is either continuous or repetitive, for example, the teeth of thread cutting taps, gears, milling cutters, and so forth, are primarily suited for the inspection of the geometric characteristics by means of a projected shadow (see Fig. 9-25).

Considering the wide applications of the shadow projection, and because of the relatively simple op-

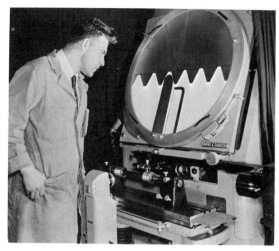

Jones & Lamson/Textron

Fig. 9-25. Inspection by projection of a shadow image. Checking the lead error of a screw thread with 0.0001-inch sensitivity.

tical system and moderate light intensity, by which such projections can be produced, the plain and less expensive types of optical projectors are designed solely for presenting shadow images.

The *reflected image* of the object is actually an unreversed and magnified mirror picture of that face of the object that has been located in the focal plane of the optical system. It may also be compared to a magnified photograph, although different in a few essential respects:

(a) the image appearing on the screen is the replica of that portion of the object surface that was brought into the focal plane. Although the better optics of modern projectors provide a certain depth of focus, it is far less than the depth common with regular photographic cameras.

(b) The reflected screen image is produced with the aid of parallel light beams without the perspective effect created in photography. The use of parallel light beams serves the important purpose of producing a reflected image of undistorted proportions and in precisely controlled magnification.

Reflected images substantially expand the field of optical projector applications by permitting the inspection of object characteristics that are not detectable on a shadow image (Fig. 9-26).

Observation by reflection usually combines both methods of illumination: from the back of the object, to produce the contour silhouette; and from the front, to reflect the light from the face of the object.

TABLE 9-4. OPERATIONAL CHARACTERISTICS AND ADAPTABILITY OF OPTICAL PROJECTORS – 1

CATEGORY	CONDITION	DISCUSSION AND EVALUATION
Measuring accuracy	Pointing precision	Pointing to establish reference positions for the purpose of dimensional measurements, is usually accomplished by bringing into coincidence one of the screen hairlines with the selected element of the projected part image. At 30X magnification, as an example, and with a well defined image outline, pointing with a precision of 0.0002 inch can be expected. By increasing the magnification the precision of the pointing can be improved.
	Dimension definition	The transfer of the location and orientation of the dimension to be measured on the physical part in a manner specified on the drawing, is often aided by the cross hairs of the basic screen plate, when these latter are in precise alignment with the translational movements of the staging slides. In certain cases the contour lines of special screen templates are used for reliably defining the location of the dimension to be measured.
	Resolution of distance indications	The micrometers for the table traverse and the lead screw dials for the vertical displacement, which are used for most types of optical projectors, are graduated in 0.0001-inch increments. Considering, however, the backlash inherent in screw type measuring devices, also the use of complementary end bars and the deflection of the staging members, a linear measuring accuracy in the order of 0.0002 inch may be expected. Directly sensing displacement measuring devices, also such devices with numerical readout, and accurate to 0.0001 inch, became available more recently, for certain models of optical projectors. The graduation of most types of screen protractors is in full or half degrees, and are complemented with vernier segments graduated in minutes of arc.
	Repeat accuracy of measurements	A factor which affects all contact type measurements, i.e. the gaging force and the deflections which it causes, is not present in optical measurements. For this reason the attainable repeat accuracy is governed only by the combined capabilities of the two above-discussed factors, namely the precision of pointing and the accuracy of the distance measurement.

TABLE 9-4. OPERATIONAL CHARACTERISTICS AND ADAPTABILITY OF OPTICAL PROJECTORS – 2

CATEGORY	CONDITION	DISCUSSION AND EVALUATION
Inspection capabilities	Inspection of toleranced dimensions	Routine inspection in production requires minimum operation time and a virtually error-free process which does not require special skills. For part dimensions which are accessible to optical viewing, the optical projector equipped with special work locating fixtures and screen charts showing tolerance zones, will satisfy these requirements, permitting the inspection by attributes of several concurrently observable dimensions.
	True position inspection	The location of several features in relation to specific datum elements is frequently toleranced by the "true position" concept. Screen charts with the enlarged contours of the features, surrounded by tolerance zones, and laid out at nominal locations from the datum which is also represented, often provide the best means for true position inspection.
	Comparison of form trueness	Forms with representative contours which can either be projected or viewed by reflection, are particularly well adaptable to inspection by optical projectors. Standard forms, such as those of screw threads are viewed against enlarged screen chart contour lines, arranged in groups of several sizes for the same family of forms, or prepared individually for special forms.
	Verification of mating forms	Parts whose form congruence or contour relationship are important for assuring the functionally required mating conditions, can be inspected concurrently by direct observation, without the need of correlating subsequently made dimensional measurements.
	Appraisal of surface conditions	The reflected image of the observed surface area viewed at an adequate magnification can reveal significant surface conditions which are not apparent to the unaided eye and may not be represented by tracing type surface texture measuring instruments. Such conditions are machining patterns, grinding checks, faults, pits, blemishes, etc. The use of oblique illumination striking the surface at an appropriate angle can even further increase the discriminating capacity of such inspection.
Applicational aspects	Measuring range	The usually large screen permits observation of a field which alone could contain the whole object or its area of interest. Coordinate tables and work locating fixtures with controlled transfer steps, can extend the practical measuring range to a multiple of the field of view.

TABLE 9-4. OPERATIONAL CHARACTERISTICS AND ADAPTABILITY OF OPTICAL PROJECTORS — 3

CATEGORY	CONDITION	DISCUSSION AND EVALUATION
Applicational aspects (cont.)	Result interpretation	Direct visual observation permits grasping several characteristics of the object at a single glance. Observation by group adds to the ease of interpretation.
	Permanent records	Photographic pictures of the image projected on the screen furnish permanent records, adaptable for analysis. However, that process is rather time consuming.
	Versatility	Very wide range of different inspection and measuring operations, to suit specific requirements, are readily available. Special attachments further extend that range to comprise inspection problems which can best be solved by optical projection.
Operating conditions	Space requirements	The physical size of the projector is primarily a function of the optical system (the "throw" of the light beam) and of the screen size. When equipped with a coordinate staging table, the size of this member must be added to the over-all length of the optical system. Powerful light sources and their heat shielding further increase the volume of optical projectors.
	Need for utility supplies	Most models of optical projectors can be plugged into the regular electric light circuits. Some larger models, particularly when equipped with extra powerful light sources, may require higher capacity lines. No other kind of utility supply is needed to operate the equipment.
	Environmental requirements	Optical projectors are designed to operate under average inspection room, or even shop conditions. These generally sturdy pieces of equipment are comparable to precision machine tools, and are less sensitive to vibrations, voltage surges, lack of delicate handling, than most other types of instruments with comparable measuring accuracy.
	Transportability	Optical projectors are basically stationary units which are not designed to be transported to the work. However, being self-contained without the need for foundation or costly installation, relocating within the plant is a matter of routine equipment moving.

TABLE 9-4. OPERATIONAL CHARACTERISTICS AND ADAPTABILITY OF OPTICAL PROJECTORS — 4

CATEGORY	CONDITION	DISCUSSION AND EVALUATION
Human factors	Operators' skill	No special skill is required to operate the equipment for routine tests by comparing screen images to master chart outlines. To operate the projector as a measuring machine requires average inspector experience only.
	Potential gaging errors	Comparison inspection, particularly with tolerance zone charts, provides so obvious information that the possibility of error is extremely faint. Coordinate table measurements are in the same category as inspection with micrometer. For critical measurements, the possibility of observation by group also reduces the incidence of error.
	Operators' strain	Unless improperly low or high power illumination is used, or the selected magnification is inadequate, eye strain is moderate. The convenient mode of observation and, on bench type models, operating in a seated position, greatly reduces operators' fatigue.
Economics	Initial cost	Universal optical projectors in larger sizes are rather expensive types of equipment. They are comparable to precision machine tools, without drive, but with complex optical systems. Initial cost can be reduced substantially by limiting equipment to actual application requirements.
	Operating cost	The optical system is not subjected to wear. The lack of substantial forces acting on any functional member during regular operation, reduces the potential wear of moving parts to a negligible level.
	Tooling cost	No disposable tools, etc. are needed to operate optical projectors. Master chart for special applications are made by drafting room procedures, are relatively inexpensive and continuously re-usable. Special fixturing for work holding is moderate because of the availability of versatile universal tooling.

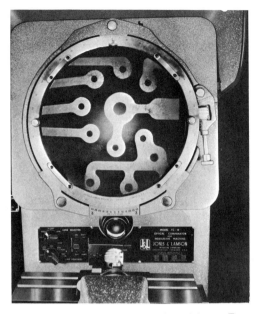

Jones & Lamson/Textron

Fig. 9-26. Inspection by reflected image—checking critical dimensions of a printed circuit at high magnification.

A few typical applications of reflected light viewing on optical projectors are:

a. Surfaces where the dimensions and coordinate positions of holes, grooves, notches and so forth are to be inspected and measured, for example, die plates and flat gages;

b. Surfaces with embossed or depressed patterns, for example, character dies or stamped letters produced on work surfaces;

c. Surfaces where a machined or otherwise applied complex pattern must be compared to a master at substantial magnification, for example, printed circuits (see Fig. 9-26);

d. Cross-sections of deep drawn or extruded components, where wall thicknesses, corner radii, various length dimensions, and so forth can thus be observed and compared in a single setup;

e. Bottoms of cavities, blind holes, countersunk bores and so forth can be observed by aligning the hole axis with the direction of the light beam and focusing on the plane to be observed; and

f. Surface properties, texture, finish, presence of flaws, and so forth can be observed either under

light directed normal to the surface, or at an acute angle of incidence; the latter type of illumination—usually called oblique light—often provides a better definition, particularly in the case of nonreflecting surfaces.

APPRAISAL OF OPTICAL PROJECTOR CHARACTERISTICS

The foregoing discussion of optical projectors in general, and of their adaptability to varied measuring tasks, may convey the impression of broad advantages over most other systems of measuring instruments.

Without intending to subtract from the valuable characteristics of optical projectors as inspection tools, a realistic appraisal of the optical projector's position in the field of inspection instruments will be attempted.

Such an appraisal should prove helpful in determining the appropriate areas of application for optical projectors as alternates for other types of measuring instruments that may also be technically suitable for carrying out a given measuring task.

Table 9-4 surveys several aspects of measuring instrument evaluation, indicating the general characteristics of optical projectors as inspection tools. The information supplied should aid a comparison with other types of measuring instruments in aspects that are pertinent to a specific set of inspection conditions.

No category of measuring instruments, and the least such diversified one as comprised under the designation of optical projectors, can be defined in terms equally valid to all members in a group. For that reason, the evaluations regarding the potential of the category are based on advanced types of instruments. Consequently, the instrument capabilities indicated in this table must not be considered as universally applicable; plainer types of optical projectors will provide a lesser degree of application potential, and the continuous progress in design leads to models exceeding those on which this evaluation is based.

It follows that this tabulated survey is not intended to serve as a rigid guide, but may point out diverse aspects that should be considered in selecting for specific inspection tasks a certain category of measuring instruments, such as optical projectors.

Angle
Measurements

UNITS OF ANGLE MEASUREMENT

While for the measurement of length the adoption of an ultimate standard, either material or natural, is needed, angular dimensions can be established by the subdivision of the circle.

The angular magnitudes that are derived from the subdivision of the circle have for a basic unit the right angle, resulting from the intersection of two mutually perpendicular straight lines. The point of intersection is then considered to coincide with the center of the imaginary circle being subdivided.

There are two known systems for the subdivision of the right angle into angular units: the sexagesimal and the centesimal.

The sexagesimal is the generally employed system in this country for the technical measurement of angles. This system divides the right angle into 90 units, known as degrees (°). Each degree is further divided into 60 minutes (') and each minute into 60 seconds (").

The centesimal system, which is rarely used, divides the right angle into 100 units, called grades (ᵍ), each grade containing 100 minutes (ᶜ), and each minute 100 seconds (ᶜᶜ).

In many theoretical applications the angle is measured as part of a circle, that is, it is subtended by an arc of defined length and the magnitude of that angle is expressed as the ratio

$$\frac{\text{the length of the arc subtending the angle}}{\text{the radius of the circle}}.$$

Angles so defined are expressed in circular measure, having as a unit the *radian*. Since the circumference of a circle is π times its diameter, 2π radians equal 360 degrees.

A system based on the radian is also used in this country for military purposes. This is known as the *millieme* or *mil* system, having as a unit an angular dimension equal to approximately a thousandth part of a radian. In practical applications, the mil is defined as a 6400th part of a circle, instead of dividing the circle into 6283.18 . . . parts, which would exactly correspond to one-thousandth part of a radian.

In technical measurements it is often very convenient to determine or to express an angular magnitude in terms of its trigonometric functions, the sine, the cosine and the tangent being the most frequently used. For that reason the operation of several systems of angular measurement is based on the application of trigonometric functions.

GEOMETRIC CONDITIONS MEASURED IN ANGULAR UNITS

Two intersecting straight lines form an angle; these lines may be elements of two actually or virtually intersecting planes (Fig. 10-1). When these elements are also contained in a third plane that is normal to the line of intersection of the two non-parallel planes, then the angle separating the two element lines will also represent the size of the dihedral angle created by the intersecting planes.

In metalworking and allied industries many parts and features are bounded by, or have for tangents, intersecting planes whose projection on a third plane will result in two intersecting straight lines; the mutual position of these planes is defined in angular units.

To express the separation of intersecting lines in angular units, the point of their intersection is con-

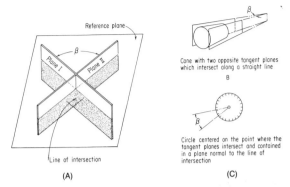

Fig. 10-1. Examples of geometric conditions that are defined as angular separations.

sidered the center of an imaginary circle that is contained in the common plane of the intersecting straight lines. The size of the central angle that the two intersecting lines intercept will be the measure of the *angular separation* that defines the relative position of said lines or planes.

The geometric conditions of compound angles are also determined by measuring the angular separation of the bounding planes. However, in the case of compound angles more than a single reference plane needs to be established, using one for each individual angle measurement, and subsequently determining the angular separation of the reference planes. The individual measuring steps applied to inspect the compound angle on a technical part being essentially identical with the measurement of plain dihedral angles, the case of the compound angles is mentioned here only for the sake of completeness.

Angular units are also used to measure spacing conditions that are related to a circle. *Angular spacing* is frequently applied in engineering to specify conditions that result from the division of a circle.

A further application where angular units are used to define geometric conditions is the *measurement of digressions* from a straight line, a plane or a nominal interrelation, such as the mutual parallelism or perpendicularity of straight lines or planes.

Because the characteristics of specific geometric conditions that are to be measured in angular units govern the applicable methods and equipment, a survey of these conditions should precede the discussion of processes and instruments of angle measurements. Due to the very large array of geometric forms that are adaptable to measurements in angular units, no attempt is made toward completeness. The main purpose of the survey presented in Table 10-1 is to visualize the great variety of geometric condi-

tions that are determined by angular measurements. The examples shown in Table 10-1 also explain the need for many different types of instruments that are used to measure the meaningful magnitudes of diverse geometric conditions in the common angular units of degrees, minutes and seconds.

SURVEY OF ANGLE MEASURING SYSTEMS AND TECHNIQUES

The wide variety of geometric conditions that are defined in angular units calls for different types of angle measuring methods and equipment. The diversity of geometric conditions, however, is only one factor, although an important one, that governs the selection of the suitable system of measurement. There are, as well, many additional factors, for example, the size and general shape of the part, the location and accessibility of the angular feature to be measured, the expected range of angle variations, the required sensitivity and accuracy of the measurement and various others.

The many conditions determining the required or desirable capabilities of the applied measuring methods and equipment, prompted the development of many different systems and techniques of angular measurements. These techniques are based on the use of diverse measuring equipment, which ranges from general-purpose measuring tools, perhaps supplemented with auxiliary devices, to special angle measuring instruments designed for very accurate operations.

To review, at least, the essential varieties of the different techniques of instruments of angular measurement in a systematic manner, certain characteristics common to individual groups have been selected for this discussion. Based on such common characteristics, the major angle measuring techniques are assigned to specific groups, representing a particular system of measurement. This, to some extent, arbitrary grouping, is being used both in Table 10-2, which presents a concise review of the systems and processes of angle measurements, and in the subsequent, more detailed discussion of the major angle measuring methods and instruments.

ANGLE MEASUREMENTS BY COMPARISON TO ANGULAR REFERENCE BODIES

One of the most widely used general groups of angle measurements is based on comparing the angular conditions of a part or of a feature to an angle

TABLE 10-1. EXAMPLES OF GEOMETRIC CONDITIONS MEASURED IN ANGULAR UNITS — 1

CONFIGURATION OR CONDITION	GEOMETRIC DEFINITION	DIAGRAMMATIC SKETCH	TYPICAL TECHNICAL PART
Angular Separations			
Bounded by two non-parallel faces which are normal to a third common plane	Body forming a dihedral angle		Wedge
Bounded by two or more planes which are mutually inclined in several directions	Body forming a compound angle		Turning tool
Body of revolution bounded by non-parallel yet straight line side elements	Tangent planes to diametrically opposite elements of the surface form a dihedral angle.		Component parts of a tapered roller bearing
Feature characterized by the angular separation of its meaningful boundary elements	Observed in a reference plane the significant surface elements are bounding a plane angle.		Screw thread

TABLE 10-1.   EXAMPLES OF GEOMETRIC CONDITIONS MEASURED IN ANGULAR UNITS – 2

CONFIGURATION OR CONDITION	GEOMETRIC DEFINITION	DIAGRAMMATIC SKETCH	TYPICAL TECHNICAL PART
Angular Spacing			
The features of a part located around a common reference circle	Circumferential spacings determined by the central angles between the radii pertaining to individual features.		Ball bearing separator
Features along a common reference arc of circular form	The spacings are determined by the central angles between representative radii.		Teeth of a gear sector
Rotatable body with non-circular periphery	Each contour element is defined by an angular and a linear dimension, both related to a common reference circle.		Cam
Disk shaped part with features located on concentric circles with specific angular spacings	The features are represented by points whose positions are defined by a spacing angle and by a radial distance from the center of the reference circle (polar coordinates).		Drill jig (fixture)

TABLE 10-1. EXAMPLES OF GEOMETRIC CONDITIONS MEASURED IN ANGULAR UNITS – 3

CONFIGURATION OR CONDITION	GEOMETRIC DEFINITION	DIAGRAMMATIC SKETCH	TYPICAL TECHNICAL PART
Digressions from a Basic Direction or Position			
Deviations from the basic horizontal or vertical position	Basic directions determined by gravity whose force also establishes the fiducial position of the instrument's indications.		The positions of machine tool members as established in the erection
Deviations from straightness or alignment	A perfectly straight line must coincide with the significant element—actual or virtual—of the inspected members.		The alignment of the axes of bearing bores
Flatness	All surface elements must be contained in a common plane.		Surface plate or optical flat
Geometric inter-relations of directional or positional parameters	The significant elements of the inspected surfaces must have positions in compliance with the specified geometric relationships.		The squareness of the guideways of machine tool slides —perpendicularity

TABLE 10-2. SYSTEMS AND TECHNIQUES OF ANGULAR MEASUREMENTS — 1

SYSTEM CATEGORY	METHOD	DIAGRAMMATIC SKETCH	EXAMPLE OF EQUIPMENT	DISCUSSION
Comparing directly or by characteristics to an angular reference body	Direct contact comparison—mating with a reverse reference body		Taper gage for machine tool tapers	Comparison is guided by feel or assisted by applying bluing dye.
	Indirect comparison—transferal by means of a comparator gage		Taper setting gage (master) used in conjunction with an air gage plug	Comparator is set to indicate zero for a part which is congruent with the master.
	Compensatory incline with slope angle equal to the nominal taper		Sine block on a surface plate—used in conjunction with an indicator height gage	The indicator scans the top element of the part to measure digressions from the parallel to the datum plane.
	Comparing the ratio of significant dimensions (attribute) to those of the basic taper		Auxiliary elements (pins and gage blocks) with length measuring instrument	The conceptual attribute: the ratio of two diameters contained in specific parallel planes must be transferred into surface elements which are accessible for length measurements.

TABLE 10-2.   SYSTEMS AND TECHNIQUES OF ANGULAR MEASUREMENTS — 2

SYSTEM CATEGORY	METHOD	DIAGRAMMATIC SKETCH	EXAMPLE OF EQUIPMENT	DISCUSSION
Direct measurement of the angle which is bounded by the part's opposite surface elements	Direct contact measurement		Hand protractor, either with directly or optically observed angular scale	Operates by establishing mechanical line contact simultaneously with two opposite elements of the object surface which, for bodies of revolution, must be contained in the axial plane of the part.
	Optical comparison		Optical projector with protractor ring around the screen, or measuring microscope with goniometer eyepiece	Measures the angular separation between the contour lines of the object's image which must be produced by staging the part with its axis at right angles to the optical axis of the instrument.
Producing rotational displacement controlled either mechanically or optically	Incremental rotational positioning		Dividing head or indexing rotary table	Indexes to discrete positions with the aid of a circularly divided mechanical element.
	Continuous rotational positioning		Worm driven rotary table with either mechanical or optical measuring system	Sensitive rotation produces angular spacings, either controlled at the input stage or guided by an optically observed ring scale.

TABLE 10-2.   SYSTEMS AND TECHNIQUES OF ANGULAR MEASUREMENTS — 3

SYSTEM CATEGORY	METHOD	DIAGRAMMATIC SKETCH	EXAMPLE OF EQUIPMENT	DISCUSSION
Optical measurement of the angle of directional digressions	Measures by reflection the small deviations of a plane from a selected reference position.		Autocollimator—either purely optical or combined with electronic sensing	Measures with great sensitivity the small deviations from the perpendicular position of a reflective target.
	Uses a line of sight to measure the angular value of displacement in the target's position.		Theodolites—with different degrees of sensitivity	Measures in several planes the angular positions of a line of sight connecting the instrument to a target on the object, referencing the angular displacements mutually or from a pre-established reference plane.
Angular measurement of deviations from a gravitation-related position	Sensitively senses and, to a limited extent, measures deviations from the horizontal (normal) to the gravitational vector.		Spirit level with directly observed bubble or with coincidence vial	Measures the small deviations from the horizontal of a flat or cylindrical supporting surface.
	Measures the slope of a supporting incline in angular values from the horizontal zero to right angle.		Clinometer—a rotatable angle measuring instrument combined with a spirit level for establishing the reference plane	Measures the incline angle of the supporting surface over an extended range, commonly from the horizontal to the vertical.

of known size. The inspection then consists of checking the compliance, or of determining the magnitude of discrepancy between the master and the object. For the latter function (the measurement of the amount of disagreement between the reference angle and the part) additional measuring instruments are usually needed.

The reference angle applicable to this type of measurement can be represented by a number of different elements that, for the purpose of this analysis, may be termed physical, virtual or imaginary bodies. It is well to remember that most types of angular measurements based on the use of reference bodies rely on similar sensing techniques: deviations from a basic condition, which is represented by the reference body, are observed on an indicating instrument as linear variations whose values must then be converted into angular units.

To assign positive meanings to terms used in the discussion of angular reference bodies, definitions of a few important terms are listed in Table 10-3. These definitions, although related to a cone, are also applicable to other types of symmetrically tapered figures and bodies.

Because of the diversity of angular reference bodies that may be used for technical measurements, a grouping by a certain distinctive characteristic of the employed reference masters is adapted in the course of the following discussion; this grouping is reviewed in Table 10-4. However, the selected characteristic represents only one of the many aspects that distinguish the angular reference masters used in industrial measurements. Therefore, each group may comprise several types of gages, examples of which will be discussed in greater detail.

Fixed Type Angular Reference Bodies

For checking the angular conditions of parts that were manufactured to standardized dimensions, or to any other positively specified size, the dimensional inspection with the aid of fixed type reference bodies provides distinctive advantages. These are the high degree of built-in accuracy that can be accomplished and also consistently maintained (barring substantial wear), and freedom from setting errors, which are potentially present wherever the human element is involved in establishing a reference dimension by assembly or adjustment. In cases where length or diameter must also be represented by the reference master, in addition to a single angular dimension, very often certain types of fixed reference bodies, used directly or for gage setting, constitute the only practicable choice.

Fixed Taper Gages

Machine tool tapers are the type of technical part most frequently inspected with fixed taper gages. Generally, standard dimensions are used; however, there are several systems of standards that are accepted in practice. As an example, tool shanks are made with:

a. Brown & Sharpe tapers (approximately $\frac{1}{2}$ inch per foot taper except for one size that has a slightly large taper);

b. Morse tapers (approximately $\frac{5}{8}$ inch per foot, yet not consistently the same for all current sizes);

c. Metric tapers standardized by the International Standards Organization, the ISO (1:20 taper).

Within each system there are different sizes of standardized diameters and lengths, identified by consecutive numbering. In addition to tool shanks, the machine tool industry is also using standardized tapers for connecting other kinds of detachable machine tool elements. Examples for these applications are the spindle tapers for mounting chucks, face plates, arbors, cutter heads and so forth.

In practical applications the machine-tool taper gages operate as direct checking tools, and are designed to represent the mating element of the part being inspected (plug gages for tapered bores, and ring or sleeve gages for tapered shanks; see Fig. 10-2). In the most frequent usage, common bluing is applied and the mating parts are lightly rubbed against each other, for displaying the condition of compliance or deviation. With skill and care, deviations from perfect conformance as small as one or a few ten-thousandths part of an inch can be detected. However, this method does not provide numerical information regarding the magnitude of angular deviations between the master and the part.

Morse Twist Drill & Machine Co.

Fig. 10-2. Taper plug and ring gage for inspecting the angle of machine tool tapers. The scribed ring serves for checking the larger diameter of internal tapers.

TABLE 10-3. TAPER TERMINOLOGY

Cone	Frustum of cone

TERM	SYMBOLS OF THE DRAWINGS		DEFINITION
	CONE	FRUSTUM	
Taper	$D : l$	$(D - d) : h$	Taper designates the relationship of the diameter D to the axial length l of a cone. The taper is usually expressed by reducing D to 1 (one).
	Reduced to unity: $1 : X = 2 \tan \dfrac{\gamma}{2}$		
Taper angle	γ		The term taper angle (γ) refers to the angle intercepted by the cone's two side elements which are contained in a common axial plane.
Slope	$\dfrac{D}{2} = \mathrm{l}l$	$\dfrac{D - d}{2} : h$	The slope of a cone expresses the relationship between half of its diameter and its axial length. The term is also applied to a non-symmetrical angular figure or body, designating the rate of separation of the side element from the base line.
	Reduced to unity: $1 : 2X = \tan \dfrac{\gamma}{2}$		
Slope angle	$\dfrac{\gamma}{2}$		The slope angle ($\gamma/2$) is intercepted by a side element and the axis of a cone. When applied to a non-symmetrical angular figure or body, it expresses the angle of separation between the side element and the base line.

TABLE 10-4. EXAMPLES OF ANGLE MEASUREMENTS BY COMPARISON TO AN ANGULAR REFERENCE BODY

CATEGORY	TYPICAL TOOL AND SETUP	DISCUSSION
Measurements with fixed reference bodies	Taper plug and ring	Reference bodies are made in widely different degrees of accuracy depending on the intended application. Examples in the order of increasing accuracy: limit gages for inspection by manual insertion, setting gages for pneumatic comparators, grand masters used as reference standards.
Multifaced block with calibrated interfacial separations to reference angular spacings	Polygon block	Intended primarily for inspecting the accuracy of circular divisions by the optical observation of the polygon's reflecting faces. Polygons are made to the highest technically practical degree of angular accuracy.
Measurements with combinative reference bodies	Angle gage blocks	In metrological laboratory work where angles of many different sizes must be measured by comparative methods, angle gage blocks can be assembled by combination to provide masters for angular separations in increments as small as one second of arc.
Locating the part by means of an adjustable compensating angle	Sine block	Functionally the sine blocks and plates create a controlled slope to a datum plane. The angle of that slope will be equal to the basic taper angle of the part, but in an axially reversed direction resulting in the nominal canceling of the part's angular condition in relation to the datum plane.
Creating a virtual body—defined by discrete points only—for the purpose of a referencing gage	Assembly of gage rollers and balls	Although correct in its principles, the method of creating a virtual cone by combining the line contacts of two gage rollers and the point contacts of two gage balls, is only exceptionally used because of practical limitations. Other examples of virtual bodies used for angle referencing are shown in Table 10-6.

Another important use for fixed angle gages is as setting masters for comparative type measurements. For the inspection of angular bodies and features, on both external and internal surfaces, the comparative measurement can provide unexcelled advantages. For many types of tapered parts, comparative gaging permits great accuracy to be obtained even with very short operational time. It also permits combining the checking of the angle with the inspection of linear dimensions.

The setting of comparator gages for angle measurements is accomplished by using for masters, fixed angular reference bodies that are made to represent the part being inspected in its basic design dimensions. The instrument, which usually measures two diameters along the axis of the part, indicates the effect of taper variations between the master and the part in linear and not angular values. While it is feasible to use two gages and correlate their indications, the differential type comparator gage is preferred for taper measurements. The differential gages, which are generally of the pneumatic and less frequently of the electronic type in this kind of application, indicate only the amount of that variation from the master that affects the ratio of the two simultaneously gaged diameters, without regard to their absolute sizes. By knowing the axial separation of the two measured diameters, the gage indications can be easily translated into angular values. It is customary to spell out on the product drawing the allowed angular tolerance in linear values also, with reference to a specific gaging length, that is, the axial separation between the measured diameters on which the linear tolerances are predicated.

Angle Gage Blocks

A physical body that constitutes a reference angle of specific angular dimension can also be produced by a combinative method using appropriate individual members. The process of assembly is similar, both in principle and in technique, to preparing a gage block stack by wringing. The individual members of an assembly, which are to represent a specific angle, are the angle gage blocks.

Whereas in assembling conventional gage blocks the purpose is to create a positive linear distance represented by the stack's two end faces, angle gage blocks are assembled to produce a specific angular separation between the open end faces, without regard to the linear distance over which the assembly extends.

Angle gage blocks are usually supplied in sets (see Fig. 10-3) that contain individual blocks se-

The L. S. Starrett Co.

Fig. 10-3. Angle gage block set containing 16 individual blocks for setting any angle, from zero to 90 degrees, in increments of one second of arc.

lected to permit the development by assembly of any angle from 0 degrees to 90 degrees. The number of individual blocks contained in a set depends on the smallest increment in which any angle within said limits must be made available. Examples of such sets, composed with the objective of particular incremental potentials, are shown in the following tabulation:

Smallest Angle Increment that can be Produced	Number of Individual Blocks in the Set	Detailed Listing of the Blocks in the Set
1 degree	6	6 blocks of 1, 3, 5, 15, 30 and 45°
1 minute	11	6 blocks of 1, 3, 5, 15, 30 and 45° 5 blocks of 1, 3, 5, 20 and 30 min
1 second	16	6 blocks of 1, 3, 5, 15, 30 and 45° 5 blocks of 1, 3, 5, 20 and 30 min 5 blocks of 1, 3, 5, 20 and 30 sec

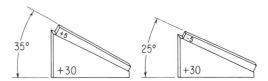

The L. S. Starrett Co.

Fig. 10-4. Diagram showing the additive and the subtractive assemblies of the same pair of angle gage blocks.

The relatively small number of individual blocks needed to produce such a large variety of different angles, for example, 16 blocks for a total number of 324,000 angles, is partly assured by the judicious selection of the nominal sizes for the individual blocks. Another factor that contributes to limiting the number of the required individual members is the capability of angle gage blocks to be used in either of two positions: (a) adding to the combined angular value of the preceding blocks in the stack, or (b) when turned over end-to-end, subtracting from the total angular value (see Fig. 10-4).

Angle gage blocks are available in different accuracy grades, which are chosen in accordance with the intended application. Since Federal Specification GGG-G-15a, which was developed for length measuring gage blocks, does not apply to the angle gage blocks, the accuracy grades are determined by the choice of the individual manufacturers. A prominent U.S. producer of angle gage blocks uses two accuracy classifications: "Tool Room" grade, with a 1-second maximum error, and "Laboratory Master" grade with a $\frac{1}{4}$-second maximum error.

When several angle gage blocks are wrung together, small errors can occur in the resulting angle due to compounding if the blocks are not properly aligned. In case of perfect alignment, the theoretically extended operating surfaces will intersect in a single line, which may be considered a virtual vertex line common to all elements in the stack. The extent of such compounding error can be negligible when, in building up the stack, the longitudinal side faces of the individual blocks are contained in an essentially common plane. To make that visual guide effective, the side faces of the angle gage blocks are made perpendicular to the working faces, to $\frac{1}{2}$ minute or better.

The applications of angle gage blocks are multiple, although by far not comparable to that of the regular blocks that are used as length standards. A few examples of angle gage block applications will be named:

a. Checking the circular dividing accuracy of rotary tables and dividing heads; this application is comparable to that of the polygon block shown in Table 10-4;

b. Setting a revolving workholding table or a magnetic chuck into the required tilt position (see Fig. 10-5);

c. Inspecting and refining the setting accuracy of tilt tables with graduated segments; and

d. Providing a reference angle, or a compensating angle, for inspecting angular features on workpieces, whose configuration makes them adaptable for inspection with the aid of angle gage blocks (see Fig. 10-6).

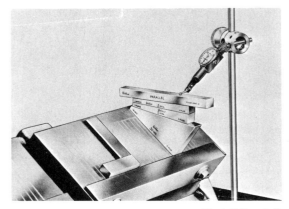

The L. S. Starrett Co.

Fig. 10-5. Application of angle gage blocks in setting a tiltable magnetic chuck.

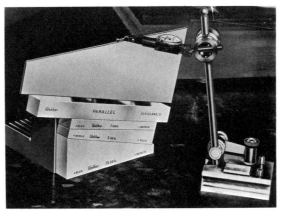

The L. S. Starrett Co.

Fig. 10-6. Angle gage block stack supporting an angular part for the inspection of its actual angle.

Adjustable Angular Reference Bodies

The most important members of adjustable angular reference elements are the sine bars and the sine blocks. Tangent bars, because of their limited use, are pointed out only for the sake of completeness.

In order to make an angle adjustable, one of its bounding elements must be movable. To qualify an angle as a reference body, said movement of the boundary element must be precisely controlled; furthermore, any selected mutual position of the boundary elements must result in an angular separation that is positive and of accurately known magnitude.

These conditions are satisfied by a device known as a sine bar, when it is used in conjunction with a flat plate and gage blocks. The operation of the sine bar, as well as of its workholding fixture variety, the sine block, is based on known trigonometric relationships between the sides and the angles of a right-angled triangle, where dimensions of two sides determine the size of the third side and of the two acute angles.

A convenient method for physically establishing the controlled conditions of a right-angled triangle is the use of an essentially horizontal and precisely flat plate on which gage blocks are stacked in a direction normal to the plane of the plate, thus establishing a right angle. When the positive height of that gage block stack is supporting on one end, a beam of known length, whose other end rests on the flat plate, a right-angled triangle of controlled dimensions is created. To assure that the supporting points of that beam are actually at the intended positive distance apart, the mutually parallel axes of two rolls (buttons) of equal diameter are substituted for the impracticable concept of dimensionless points. The angular conditions thus created will not be affected by the final diameter of the rolls, as long as these are perfectly round, of equal diameter and mutually parallel, as well as with the working surface of the sine block.

In actual application of the sine bar it was found practical to use a beam of fixed length a (see Fig. 10-7) and to vary the height of the gage block stack b according to the size of the desired angle B, based on the simple trigonometric relationship:

$$\sin B = \frac{b}{a},$$

hence the designation *sine bar*.

Sine bars and blocks (see Fig. 10-8) are commonly manufactured in 5-inch or 10-inch lengths, repre-

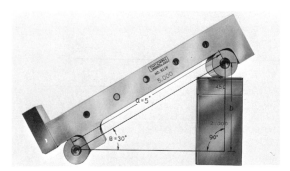

The Taft-Peirce Mfg. Co.

Fig. 10-7. Contour lines of a triangle superimposed on the photograph of a sine bar set for $B = 30$ degree angle, to illustrate the applied basic trigonometrical relationships.

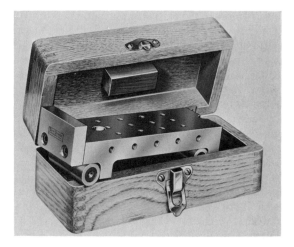

The Taft-Peirce Mfg. Co.

Fig. 10-8. Sine block equipped with end plate to assist work-positioning. The holes on the top and side faces are tapped for attaching auxiliary locating and clamping elements.

sented by the distance between the axes of the two supporting rolls. The fixed distance in integrals of the inch facilitates the setting of the bar or the block to the desired angle, based on values found in regular sine tables. In practice it is found more convenient, because it presents readily applicable values, to use special tables in which the effective length of the bar is substituted for the unity.

Since gage blocks, which constitute one group of elements in the setup, incorporate a very high degree of linear and angular (squareness) accuracy, the reliability of angle measurements by means of sine bars depends essentially on the accuracy of the sine bar itself.

In view of the important role of sine bars and related devices in the field of dimensional measurements, a Commercial Standard (No. 141-47), which was issued by the U.S. Department of Commerce, in cooperation with the National Bureau of Standards, defines the required accuracy of several functionally essential elements of sine bars. The Standard distinguishes two qualities or grades, designated as commercial class and laboratory class, respectively. The tolerances of these two grades are generally established in a ratio of 2:1. The actual tolerance limits, which are specified by the standard and adopted by reputable manufacturers of inspection equipment, are listed in Table 10-5.

Sine blocks, which are extensively used for angle measurements on technical parts, operate by physically establishing the basic size of the angle to be checked in relation to a supporting surface plate. In the measuring process, the sloping platform of the sine block surface is used to support the tapered part whose vertex—actual or virtual—must point toward the elevated end of the sine plate, in a position where the part's axis is contained in a plane precisely perpendicular to the hinge line of the sine block (see Fig. 10-9).

The essentially simple elements of which a sine block setup is composed permit a high degree of accuracy to be accomplished by relatively inexpensive

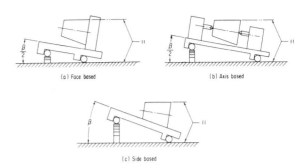

Fig. 10-9. Cone angle measurement by supporting the part on a sine block. The setting of the sine block angle in relation to the part's taper angle β varies, depending on the selected system of work locating.

means. The slope of the top surface of a laboratory grade sine block in conjunction with gage blocks of Grade A or better, would have a possible error of 5 to 10 seconds of arc, or even less, with respect to the selected basic angle size. This accuracy level refers to angles in the 0- to 30-degree range; above that size the setting accuracy of the sine block setups gradually decreases, as a consequence of trigonometric relations. Sine blocks should not, in principle, be used of the direct setting of angles in excess of about 45

TABLE 10-5.   TOLERANCES FOR SINE BARS, BLOCKS AND PLATES

(Commercial Standard 141-47)

SIZE	Bar	Buttons or cylinders		
	Working surface to be flat, square with sides and parallel (if double) within—	Cylinders to be alike, round and straight, within—	Cylinders to be parallel with each other and with working surface of bar within—	Cylinders to be at nominal center distance (±)—
Commercial Class				
Inch	*Inch*	*Inch*	*Inch*	*Inch*
5	0.000 10	0.000 05	0.000 10	0.000 2
10	.000 15	.000 05	.000 15	.000 3
20	.000 20	.000 06	.000 20	.000 4
Laboratory Class				
Inch	*Inch*	*Inch*	*Inch*	*Inch*
5	0.000 050	0.000 03	0.000 050	0.000 10
10	.000 075	.000 03	.000 075	.000 15
20	.000 100	.000 04	.000 100	.000 20

degrees. For measuring acute angles above that range, a square mounted on the sine block can serve as a secondary reference plane, and the sine block proper will have to be set to the complementary angle of the basic angle.

It should be remembered that the sine block is not a measuring instrument, but only a link, although an important one, in the angle measuring process. The actual measurement consists in comparing the plane of the part's top element to the plane of the supporting surface plate. For carrying out this measurement, height gages, mechanical or electronic, are used. Height variations between the different points along the top element of the part are indicative of angular errors. For determining the magnitude of these errors, the ratio of the measured height difference e to the length of the explored section m expresses the angular value δ of the deviation from the basic angle by the simple relation

$$\tan \delta = \frac{e}{m} \quad (e \text{ and } m \text{ in inches}).$$

The meaning of the measured angular deviation will depend on the system by which the object is located on the sine block (see Fig. 10-9); it may reflect the variations of the slope angle, in the case of face-based or axis-based methods of workholding, or it may express the variations of the taper angle when the part's location is side-based.

The accuracy with which angular variations can be measured with the aid of a sine block is, probably, less than the previously stated setting accuracies of the block itself. The end result of the measurements is affected by the mounting of the specimen on the sine block (misalignment causes cosine errors); the flatness of the basic surface plate; the sensitivity and accuracy of the height gage; and the precision with which the gaging distance (the m factor) is measured.

Sine blocks or, as frequently termed in their wider varieties, sine plates are also used to support centers. Sine centers (Fig. 10-10) may hold the part either directly, or mounted on an arbor having center holes on both ends.

By superimposing two sine plates, compound angles may be created (Fig. 10-11). Such compound angle sine plates are also frequently used—in addition to measurements—in manufacturing processes, particularly in toolmaking as workholding devices.

Measurement of Cone-Shaped Technical Parts by Comparison to a Virtual Body or by Attributes

The cone is the basic form of a tapered body of revolution. The taper angle of a regular cone can be determined by the interrelated position of a limited number of points that are contained in a common axial plane of that cone or cone frustum.

It follows that for gaging the angle of a regular cone, either on its outside or inside surface, a specific number of properly located physical points can represent the practical equivalent of a reference body. To distinguish the functionally satisfactory substitute arrangement from a solid reference master, the term "virtual body" will be used to designate a set of discrete points that is used to determine the taper angle of a cone.

There are several practical reasons for using, under specific conditions, virtual bodies instead of solids as reference bodies in angular gaging. Such reasons are:

1. Virtual bodies can be built up from standard elements, whereas solid gages must be manufactured separately for each particular angle and size group of limited range.

2. Virtual bodies may permit the exploration of dimensions, particularly hidden ones, which would not be accessible for measurements by fixed gages.

3. Virtual bodies can be used for measuring meaningful linear dimensions, which reflect angular or diametrical size variations of a cone-shaped body.

Elements that are used to create virtual bodies for gaging conical parts are balls, rollers and gage blocks. The true geometric form and dependable size of these elements is essential; for this reason the use of precision balls and gage quality pins or rollers is recommended.

Another generally used method of cone angle measurement is based on the trigonometric relationship that exists between the taper angle and the dif-

The Taft-Peirce Mfg. Co.

Fig. 10-10. (Left) Sine centers, consisting of a long (10- or 20-inch) sine plate with guideways for two adjustable center heads.

Fig. 10-11. (Right) Double-level sine plate with two mutually perpendicular hinge axes for producing a compound angle surface.

ference of two diameters located a specific distance apart. Although, for practical reasons, to assure controlled contact positions, gage pins are commonly inserted between the part's surface and the contact tips of a length measuring instrument, the essential function here is the direct measurement of diameters, as distinct from a virtual master that is built up by using physical elements. The relationship of the specific diameters being measured is considered a distinctive attribute of the taper angle; for this reason the designation "attribute gaging" has been selected for that particular method of cone angle measurement.

Cone-shaped technical parts that can be measured by means of a virtual body or by attributes can have different forms and other characteristics. The applicable method will vary depending on the meaningful dimensions that must be checked or measured. Comparable results may, in certain cases, be accomplished by different methods, varying the types and the combinations of gage elements. While a complete survey of these methods would be impractical, a few examples of frequently used techniques, which are shown in Table 10-6, will indicate the potentials, as well as the limitations, of this system of angle measurement.

ANGLE MEASUREMENTS WITH DIRECTLY INDICATING INSTRUMENTS

The size of an angular separation can be measured directly by means of an instrument that is equipped with a scale graduated in angular units. Because the angular measure is derived from the subdivision of the circle, a scale graduated in equally spaced angular units must have a circular form. Angle measuring instruments that are equipped with circular scales graduated in angular units and having sensing members to embrace the meaningful contour elements of an angle are designated by the term protractors. The scale may extend over a complete circle, comprising 360 degrees, or over a part of a circle, for example, a segment of 90 or 45 degrees.

For analyzing the operational principles of the process, it must be remembered that in measuring linear dimensions the elements of the part's surface whose separation is to be measured must be selected and then related to the scale graduations of the measuring instrument. This alignment of the boundary elements with the scale graduation may be effected by direct contact, as in the case of graduated rules

and vernier calipers, or by optical alignment, as on optical projectors and toolmaker microscopes.

For the direct measurement of angular separations, essentially similar principles are applied, yet the referencing techniques must differ. While two single points define a plain linear distance, two straight line elements are needed to intercept an angle; consequently, the aligning members of the angle measuring instrument must possess functional elements that represent a straight line. The instruments used for direct angular measurements may be considered to belong to either of two basic categories: (a) measuring with mechanical contact, and (b) measuring with optical alignment.

Direct Angle Measurement by Mechanical Contact

The tools used for direct angle measurement by mechanical contact are commonly known as mechanical protractors. Some types of mechanical protractors are simple instruments, having limited accuracy only. However, the instruments generally used by toolmakers are the universal bevel protrac-

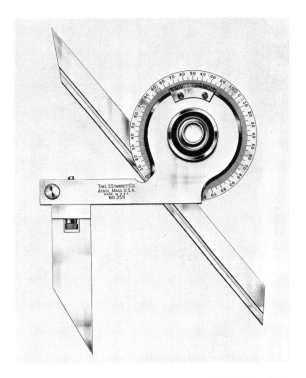

The L. S. Starrett Co.

Fig. 10-12. Universal bevel protractor with thumb nut for fine adjustment and locking. The vernier segment permits angular reading in 5-minute ($\frac{1}{12}$th-degree) increments; the auxiliary blade assists in measuring acute angles.

TABLE 10-6.   METHODS OF TAPER ANGLE MEASUREMENT BY COMPARISON TO A VIRTUAL CONE

DESIGNATION OF THE MEASURED DIMENSION	DIAGRAM	APPLICABLE FORMULA	AUXILIARY ELEMENTS
The angle of a taper plug by the ratio of two diameters which are at a known distance apart		$\tan \dfrac{\theta}{2} = \dfrac{D_1 - D_2}{2(H_1 - H_2)}$	Gage pins of equal diameter and two pairs of gage blocks (stacks of H_1 and H_2 lengths)
The angle of an internal taper by the ratio of two specific diameters		$\tan \dfrac{\theta}{2} = \dfrac{D_1 - D_2}{2H}$	Precision balls of equal size with two sets of gage blocks, of H length each
The angle of a slender taper bore by the axial separation of two balls of known diameters		$\sin \dfrac{\theta}{2} = \dfrac{r_1 - r_2}{(H_1 - r_1) - (H_2 - r_2)}$	Two different size precision balls of appropriate diameters
The angle of an internal taper by the axial separation of two balls, each leaning on a roller; all elements of known diameters		$\sin \dfrac{\theta}{2} = \dfrac{D_1 - D_2}{2(H_1 - H_2)}$	Two equal balls and two gage quality rollers with different diameters
The angle of a Vee-groove by the axial distance between the top elements of two gage rollers		$\sin \dfrac{\theta}{2} = \dfrac{d_1 - d_2}{2H \cdot (d_1 - d_2)}$	Two gage rollers of different diameters
Slope angle of a groove with one inclined and one vertical side, with balls or rollers leaning on gage blocks		$\tan \theta = \dfrac{H_1 - H_2}{L_1 - L_2}$	Two balls or rollers of equal diameter and gage blocks of appropriate lengths

The L. S. Starrett Co.

Fig. 10-13. (Left) Bevel protractor equipped with a sliding auxiliary blade in use for the direct measurement of a part having an acute angle.

Fig. 10-14. (Right) Bevel protractor resting on a surface plate, which also supports the workpiece. The angle being measured is related to that common reference plane.

tors, one form of which is shown in Fig. 10-12. The designation universal refers to the capacity of the instrument to be adaptable to a great variety of work configurations and angular interrelations. The measurements can be made either by embracing the two bounding elements of the angle (see Fig. 10-13) or by extraneous referencing, for example, both the part and the instrument resting on a surface plate (see Fig. 10-14).

The universal bevel protractors are generally equipped with a full circular scale graduated into 360 degrees, and complemented by a vernier scale applied to a segment. The segment has 12 graduation lines with special spacing in either direction (see Fig. 10-15) to permit the subdivision of each one-degree interval on the main scale into 12 equal parts, corresponding to $\frac{1}{12}$th part of a degree, that is, 5 minutes.

The inherent limitations of long-range verniers, such as the uncertainty in determining the correct

The L. S. Starrett Co.

Fig. 10-15. Vernier scale segment of a bevel protractor with twelve graduations, in both directions from the center, representing $\frac{1}{12}$ degree = 5-minute steps. The vernier scale line, which coincides with one of the main scale graduations, indicates the significant value. In the illustrated example, 20 minutes must be added to the 50-degree position of the main scale, which the 0 reference mark of the segment has just passed.

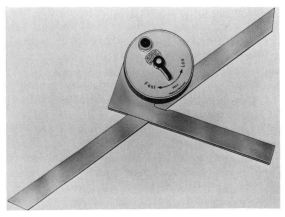

Carl Zeiss, Inc.

Fig. 10-16. Optical hand protractor. This protractor has an enclosed circular master scale, graduated into four 90-degree segments, with degree intervals subdivided into 12 increments of 5 minutes each. The lens system provides a large field of view with distinct, direct reading to the least graduation.

coincidence of the significant graduation lines and the potential error of interpretation, can be avoided by the use of the optical bevel protractor (see Fig. 10-16). This instrument is similar in outside appearance, as well as in the technique of its application, to the mechanical bevel protractor. The optical bevel protractor comprises a graduated circular glass scale with individually numbered degree marks and a fixed subdivision scale with 12 graduations, representing 5-minute intervals. The circular glass scale is enclosed for protection and also for masking the nonpertinent graduations, with only that scale segment being exposed that carries the degree mark corresponding to the angular position of the contact arms. The exposed graduation of the main scale, in optical coincidence with the subdivision scale, is observed through a magnifying window, in which the individual graduation lines of the subdivision scale appear to be about 0.035 inch apart.

Direct Angle Measurement by Optical Alignment

The accuracy of direct angle measurement by optical alignment exceeds that attainable with mechanical contact instruments by a ratio of five-to-one, or even higher. The improved accuracy is due to the combined effect of several factors, such as the optical alignment by observing a substantially magnified image of the part; the larger scale whose graduation provides a better resolution; and the stationary character of these instruments, assuring a much higher stability in the measuring process than that attainable by a hand-held measuring tool.

This group of instruments, which comprises the toolmaker microscope and the optical projector, was not designed originally for the purpose of angle measurements, but most models are equipped to measure angular separations. With the aid of special accessories—optical rotary tables—that are available for certain types of microscopes, circular divisions can also be measured.

The alignment of the two bounding elements of the angle is accomplished by establishing coincidence successively between the significant elements of the angular feature's image and a reference hairline either in the optical plane of the microscope, or on the screen of the optical projector. Considering that the visual observation of the alignment is carried out on an image 10, 20, 30 or even more times the original part size, while holding and adjusting the object on a stationary stage, it will be obvious that the consistently attainable accuracy becomes much higher than when using a hand protractor. When measuring the taper angle of conic parts or features, the optical viewing presents an image with contour lines closely corresponding to those of the theoretical axial section of the part.

Due to the stationary character of microscopes and optical projectors, the protractor scales are larger and provide a better resolution than found in hand protractors. For this reason, the verniers or optical subdividing scales of the protractors on stationary instruments are usually graduated in single minutes.

Additional details on the protractor devices of engineering microscopes and of optical projectors are presented in Chapters 8 and 9, respectively, dealing with these categories of optical measuring instruments.

MEASUREMENT OF ANGULAR SPACINGS BY CIRCULAR DIVISION

Many engineering parts are designed with features arranged in locations that result from the division of a common reference circle. The spacing of these features is determined by associating them with the pertinent radii of the reference circle and specifying the central angles that these radii intercept.

To translate such design concepts into manufactured parts, the work must be rotated in a precisely controlled manner to bring the individual features into the plane of action of the metalworking tool in accordance with the required angular spacings. Similarly, for checking the correctness of their angular position, these features must be presented to the sensing member of a measuring instrument in pre-

Brown & Sharpe Mfg. Co.

Fig. 10-17. Milling machine type dividing head connected with driving mechanism, for generating spirals. The dividing head can be used for direct indexing by means of plate on spindle nose; plain indexing through a 40:1 ratio worm-drive and index plates; differential indexing by change gears.

cise correspondence with the basic values of the specified angular spacings.

For rotating the part by the required angular increments, whether for production or for inspection, a workholding device is needed whose rotational movement can be positively controlled in angular units.

Such devices are known as dividing heads (Fig. 10-17) and rotary tables (Fig. 10-18). Both are de-

Walter/Dapra Corporation

Fig. 10-18. Rotary table to be used either for continuous rotational movement with readings on a graduated drum or for incremental setting by means of the index plate.

signed for essentially the same function, that of imparting controlled rotational displacements. The difference is primarily in the position of the rotational axis: horizontal for the dividing heads and vertical for the rotary tables. This difference, however, is not exclusively distinctive, because there are certain designs of dividing heads and rotary tables that may be used with their axes in either the horizontal or the vertical position. Actually, there are models of dividing heads and rotary tables (Fig. 10-19) that rest on a tiltable base, providing controlled angular setting around an axis that is normal to that of the rotational movement.

Considerations related to the work to be measured will frequently guide the choice between dividing head and rotary table. Examples of these aspects are the weight and the bulk of the part that may require full support on a rotary table, the shape of a long part causing it to be more manageable in a horizontal position between centers and, generally, the accessibility of the features to be machined or inspected, making either a vertical or a horizontal position for the part's axis more suitable.

The *applications* for dividing heads and rotary tables are multiple, in view of the many different kinds of engineering parts whose design specifications are related to circular divisions. A few examples of such parts, listed at random, will assist in visualizing the great and important territory where precise angular spacings are essential: gears, splines and other transmission elements; rotating cutting tools; cams and other mechanical timing elements; parts with an-

International Machine & Tool Corp.

Fig. 10-19. Tiltable rotary table with 360-degree rotational movement and 90-degree tilt movement. The angular value of the setting can be read for the rotational positioning in seconds of arc, combining the peripheral graduations of the table with incremental values on the drum; for the tilt in $\frac{1}{4}$ minute on the sleeve of the axle.

gular cross-sections; components with equally spaced circumferential pockets, such as antifriction bearing separators, and many more, including a large array of parts used in less common applications.

When discussing the use of circular dividing devices for the purpose of angle measurements, it is well to review the geometric conditions that are specified in angular units and are inspected with the aid of said instruments. Circular divisions are used to specify two categories of geometric conditions:

1. The angular separation of features whose location is directly associated with radii of a common reference circle, such as the teeth of a gear, the pockets of a ball bearing separator and so forth; and

2. The included angles between mutually inclined boundary planes, such as the sides of a polygon or of a multisided bar.

In the case of angular separations, the amount by which the circular dividing device must be rotated is identical with the nominal size of that separation. However, for measuring the mutual incline angle of the bounding faces of an angular part that is mounted in a centered position on a circular dividing instrument, the required rotation must equal the supplementary angle of the incline angle. As an example, assuming the inspection of a polygon whose faces are mutually inclined by 108 degrees, the corresponding angle of rotation for the circular dividing device will be $180 - 108 = 72$ degrees.

The *inspection of the angular spacing* of individual features on parts, which are rotationally positioned by means of a dividing head or a rotary table, may be carried out in different ways. The primary goal of the inspection is generally to check the coincidence of the feature with the nominal condition, which is established by the part's controlled rotation. In the case of digression, the amount of noncoincidence must be measured in units that can be correlated with the tolerance specifications of the product drawing.

Most frequently, the inspection is carried out with indicators, mechanical or electronic, referencing from one of the features at a point contained in a circular path that is common to all other features to be inspected.

The indicator's sensing direction should be tangential at the point of contact to the circular path of inspection, in order to obtain linear indications that can be converted readily into angular values. For assuring these conditions in practical applications it will be necessary to hold the indicator on an arm that can be retracted to clear the part during the indexing to the next feature and then to bring back the indicator precisely into the once established gaging position.

For certain types of parts the sensing direction of the indicator will be in line with the radius of the reference circle because the essential condition to be checked is the amount of radial variations as a function of the angular position of the part. The periphery of such parts is the functionally meaningful surface and the dimensioning is expressed in polar coordinates. Cams are the most important engineering elements belonging to this group of inspection objects. Figure 10-20 illustrates a cam checker instrument, having for principal members a dividing head and a device for measuring the length of the part's radius in different, closely controlled, angular positions. The inspection of axial cams is also based on correlating rotational positions with linear dimensions of the operating surface; in this case, the sensing direction of the indicator is parallel to the axis of rotation.

Another method of inspection, characterized by a very high degree of sensitivity and accuracy, is the observation of angular coincidence with the aid of an autocollimator. This process is applicable primarily to angles whose boundary planes are reflecting flat surfaces and are accessible to direct optical observation. In the case of a cone it is possible to use two elements of the bounding surface, which are contained in a common axial plane, for locating special reflectors. These reflectors function as auxiliary members and represent tangent planes to the selected line elements (generatrices) of the cone surface. An arrangement for cone angle measurements operating on these principles is shown in Fig. 10-21. Predicated on proper fixturing with regard to design and workmanship, the accuracy of this angle measuring method is of an order comparable to the angular spacing accuracy of the rotary table used in the process.

When measuring with a toolmaker's microscope or an optical projector, either of which is equipped with a rotary table, the coincidence between the individual features and the nominal angular spacing can be directly observed optically, using the cross hair or a template figure for referencing. This is a convenient and dependable method, particularly for small parts, such as watch movement elements, whose size would not permit the use of mechanically contacting sensing members of indicator gages. The wider applicability of optical observation in the inspection of circular divisions is limited by the ac-

(Left) Carl Zeiss, Inc.
(Right) Leitz/Opto-Metric Tools, Inc.

Fig. 10-20. (Left) Camshaft checker combining, on a common stand, with two parallel guideway systems, an optical dividing head and a linear distance measuring instrument that points toward the centerline of the dividing head and its tailstock.

Fig. 10-21. (Right) Measuring the taper angle of a cone by the combined use of a rotary table (for angular setting) and an autocollimator, for establishing an optical reference plane. Significant side elements (generatrices) of the cone surface are represented by tangent planes produced with the aid of contacting reflector blocks. The exposed sides of the reflectors can be observed with the autocollimator, which indicates with great sensitivity the amount of any deviation of the actual angle from its nominal value.

ceptable part size, the need for special accessories to complement the basic toolmaker microscope or optical projector and, finally, the accessibility of the feature to direct optical observation in transmitted or reflected light.

Systems of Instruments for Controlled Rotational Movement

Several different systems can be used, individually or in combination, to control the rotation of dividing heads and rotary tables, for accomplishing angular displacements whose size is determined in basic units of angular measurement. The operation of these systems can differ in several respects, which will guide the choice for a particular application. Examples of such operational characteristics are: the cycle time for a change in angular position; the accuracy of the positioning; the incremental limitations of the rotational movement; the sensitivity of the angular spacing measurement; the adaptability to mechanization; and various others.

The basic systems of controlling the rotational movement in circular dividing devices are surveyed in Table 10-7, together with each system's pertinent performance characteristics.

The applications of the different systems of angular displacement control in instrument design are reviewed in Table 10-8, which describes the general types of circular dividing instruments.

TABLE 10-7. SYSTEMS OF CIRCULAR DIVIDING INSTRUMENTS – 1

CATEGORY	DESCRIPTION	DIAGRAM	DISCUSSION
Incremental indexing	Direct indexing with the aid of a ring with peripheral notches, or of a plate with a single circle of holes, mounted on the main spindle and engaged by a fixed position tongue or pin.		Each index ring or plate provides a limited and fixed number of positions around a circle representing a complete rotation. For different divisions the index ring must be exchanged. Commonly used index rings or plates have 24 or 36 equally spaced positions.
	Indirect indexing by rotating an input shaft which carries a worm in engagement with a gear keyed on the main spindle.		Uses interchangeable index plates, each having several dividing circles with different numbers of holes. Indexes by engaging the locating pin of the input shaft crank into the selected hole of the index plate. With the use of three index plates over 100 different indexing divisions can be accomplished.
	Differential indexing is similar to the preceding type, except for a gear train whose variable ratios supplement the selections provided by the worm based indirect indexing.		The capabilities of the preceding type are expanded by the interposition of a gear train with change gears to modify the effect of the input by a differential action. Some models accomplish any number of indexing positions to 400, and even beyond this level. The gear train may also be used to produce sustained rotation, e.g., for helical milling.
	Direct indexing by face-geared mating dividing rings which produce an error-compensating multiple engagement by the simultaneous meshing of all the teeth.		The possible indexing positions are determined by the number of gear teeth (usually 360, exceptionally 720 or 1440). Auxiliary devices, such as a sine bar, may be used to produce intermediate positions. The simultaneous engagement assures a high degree of accuracy and sustained reliability.

TABLE 10-7. SYSTEMS OF CIRCULAR DIVIDING INSTRUMENTS – 2

CATEGORY	DESCRIPTION	DIAGRAM	DISCUSSION
Continuous rotational positioning	Worm drive with micrometer drum. Continuous positioning guided by a scaled drum on the shaft of the worm. Similar, in the principles of design, to the indirect indexing type, except for the graduated drum which replaces the index plates, and provides an almost infinite resolution of the positioning.		Depending on the model, different worm ratios are used from 40:1 to 180:1. The higher ratio improves the setting sensitivity and scale resolution, but requires more cranking for the same displacement. The scale graduation values vary, depending on the design, from about 6 minutes (1/10th degree) to one-half minute, and when equipped with vernier the readout resolution may be extended to a practical maximum of 2 seconds of arc.
	Optical dividing head or rotary table. Continuous positioning guided by a graduated master ring which is observed through an optical system (microscope or projection screen). That system also comprises a micrometer device for subdividing the optically magnified spacings of the main scale.		Indicates the actual position of the instrument's rotatable member, independently of the approach direction or the condition of the actuating elements. The double pickup optical system, when applied, compensates for minor concentricity errors in the mounting of the master scale. The accuracy of the instrument is not affected by wear in the mechanical members. Maximum discrimination in the order of one second of arc.
	Angle (rotational) encoders with optical or inductive gratings provide digital readouts for the angular spacings of rotational displacements. Considered to be practically a continuously positioning device in view of its high resolving capacity.		Depending on the selected system of resolver the discriminating capacity can vary from one degree to a fraction of a second of arc, with accuracy in the one second order. The encoder is not a measuring instrument, but incorporated into an appropriate mechanical device can provide unique measuring and recording capabilities for circular divisions.

TABLE 10-8. SYSTEMS OF ANGULAR DISPLACEMENT CONTROL IN CIRCULAR DIVIDING DEVICES

SYSTEM CATEGORY	PRINCIPLES OF DESIGN	PERFORMANCE CHARACTERISTICS	APPLICATION NOTES
Indexing Rotational displacement by discrete increments	Directly keyed dividing plate with circumferential notches or holes	The peripheral notches engaging a single pawl or pin to provide the most direct method of circular division.	The fastest method of indexing to a limited number of positions, usually not exceeding 30 per circle.
	Dividing disc on an auxiliary input shaft with worm for translating the motion on main spindle	The worm reduces the input shaft rotation by a fixed ratio, e.g., 1:40, thus providing a more sensitive positioning than by direct division.	This is the basic design for milling machines, providing a great variety of divisions, as well as continuous rotation through gear trains at rates adjusted to table traverse.
	Complete circle of identical radial teeth on the faces of matching index plates	The averaging effect of simultaneously engaged teeth assures unique indexing accuracy which is self-improving by random indexing.	The most accurate system for producing circular divisions in increments not smaller than one-quarter degree.
Continuous rotational positioning The input member indicates the angle of the imparted rotation	Operates by precision worm with low pitch angle resulting in a high reduction ratio (e.g., 1:180); the graduated drum is keyed directly on the worm shaft.	Depending on the design and on the accuracy of execution of the worm, rotary tables of this system can have discriminating capacities for rotational setting in increments of, e.g., 12, 5 or 2 seconds of arc, read on the drum in combination with a vernier.	To assure smooth rotational movement a reasonable amount of play is needed between worm and gear; therefore the direction of rotation must be identical in establishing the starting and the terminating reference positions of the rotational setting.
Continuous rotational positioning Graduated master ring keyed on main spindle indicates the circular spacing position	Rotational displacement by worm assuring sensitive setting, but the circular position is determined by the circumferential graduations of the master ring, which are optically observed through a micrometer device for indicating the incremental positions between the master scale graduations.	The master ring, usually of optical grade glass, carries numbered graduations in one degree or one-third degree intervals, observable through microscope or projected on a screen. The observation system incorporates a micrometer or a secondary scale providing discrimination in seconds of arc (e.g., 5, 2 or 1 seconds).	Eccentricity between the axis of rotation and the center of the mounted master ring affects the measuring accuracy, however this can be corrected by double microscopic pick-up with relayed image. The resolution obtained with the optical observation system determines the sensitivity. The system is free from effects of backlash or mechanical wear.

Description of Characteristic Dividing Instruments

In Table 10-8, the operating principles of diverse dividing instruments were only examined, without discussing specific models. How these principles are applied in the design of the actual instruments can best be demonstrated by describing a few models that are typical and widely used representatives of various system categories.

Indexing Tables with Meshing Radial Serrations

The essential elements in indexing tables with meshing radial serrations, a system of circular dividing devices—known also as face-geared indexing tables—are two mating rings with an equal number of precisely machined radial serrations on their opposing faces. When in meshing position these serrations provide a positive locking engagement of the rings, which are solidly attached to two concentrically guided members of the dividing device. By lifting the top member the serrations become disengaged and the raised member can be rotated by any circular distance whose angular value represents an integral number of serration spacings. After lowering the top member the serrations again become engaged and establish a locking position whose positiveness and accuracy are unaffected by either the direction or the distance of rotation. The mechanics of serration engagement and disengagement is shown in Fig. 10-22.

The system of meshing serrations, due to its basic mechanical simplicity, lends itself to the manufacturing of very precise circular dividing devices, whose operational accuracy exceeds the degree accomplished in the manufacturing phase. This increased accuracy results from the following important characteristics of this system:

a. Dividing errors that occurred in the production of the serrated rings tend to compensate due to the simultaneous engagement of all the serrations around the periphery of the rings; and

b. The repeated process of disengagement-indexing-engagement during the operation of the device, particularly when random divisions are produced, creates a lapping action that gradually corrects the residual dividing errors and improves the meshing properties of the opposing serrations. The resulting continuous upgrading of the indexing accuracy is a remarkable property of this system, particularly when compared to other mechanical dividing systems, whose initial accuracy tends to suffer degradation in usage in consequence of wear.

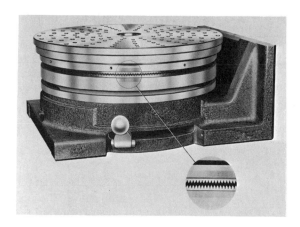

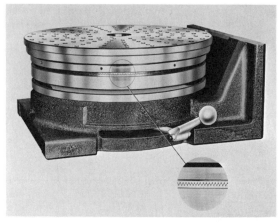

A A Gage Div., USI

Fig. 10-22. Indexing table with meshing radial serrations; the inset drawings show diagrammatically the disengaged and the engaged positions of the indexing serrations.
(Top) Open position.
(Bottom) Closed position.

As the result of the high degree of manufacturing precision, combined with the inherent beneficial properties of the system, the indexing accuracy of rotary tables with meshing radial serrations is frequently on the order of $\pm\frac{1}{4}$ second of arc, on special models even as high as ± 0.1 second of arc. This high degree of accuracy is valid for any circular division that represents an integer multiple of the serrational spacing.

Most models of rotary tables of this system have 360 serrations and consequently can be indexed by increments of one degree or by any multiple of this spacing, as guided by a graduated ring and an index mark, respectively, on the periphery of the mutually rotating members. For particular applications, indexing rotary tables with 720 or 1440 serrations, representing spacings of from $\frac{1}{2}$ to $\frac{1}{4}$ degree, respectively, are also available.

The disengagement of the top member can be effected either by hand, aided by cam action, or by power, air or hydraulic systems being generally used. The indexing system is susceptible to partial or complete automation, as well as to remote control.

Although this system of indexing is most generally used for devices with vertical axes, with certain limitations it may also be applied to dividing head type—horizontal axis—instruments. In the latter case, however, an extra sturdy center bearing is required, which is assisted by support rollers on which the cylindrical periphery of the movable instrument part rides (Fig. 10-23).

One of the applicational limitations of this system, common to all types of fixed indexing instruments, is the finite number of the available rotational positions. To compensate for that inherent limitation, certain models of radial serration type indexing tables are supplied with a special sine base. This additional member serves to subdivide the least indexing spacing into smaller increments—minutes and seconds of arc—with the use of interchangeable gage blocks. Figure 10-23 illustrates such a special model of indexing table, using a very sensitive mechanical indicator for fiducial gage in the final stage of the setting.

Rotary Tables with Continuous Rotational Positioning by Precision Worm

Although similar in design concept to the generally used milling machine dividing heads, both using the worm as a reduction element, the circular dividing devices assigned to this category are different in several respects, such as the purpose of application, various design aspects and the accomplished degree of accuracy.

The conventional dividing heads are actually indexing devices that can be rotated by fixed increments equal to specific fractions of the circle. Although many different fractions are usually available, their number is finite. The worm drive of the common dividing heads serves as a reduction mechanism to step down by a fixed ratio (most frequently 40:1) the movement introduced by the rotation of the indexing arm. This mechanism permits the separation of the index plate holes by central angles that are many times larger than the effective angular displacement they establish.

The rotary table shown in Fig. 10-24 represents an exceptionally accurate model of this group of circular dividing devices, the vernier permitting the resolution of the angular setting to 2 seconds of arc. The high level of workmanship, including the corrective lapping of the fine pitch worm, gives this type of rotary table an operational accuracy in which the error of angular setting around the entire 360 degrees rotation does not exceed ±2 seconds of arc. This accuracy can be accomplished both in the horizontal and in the vertical position of the table and, assuming proper care inclusive of adequate lubrication, it can be maintained over a considerable period of usage.

A A Gage Div., USI

Fig. 10-23. Indexing table of radial serration type, specially designed for applications requiring arrangement with the horizontal axis. In its disengaged state, the table is supported with the aid of two external rollers. This model is also equipped with a sine device for setting intermediate positions between the adjacent integral degree steps.

Moore Special Tool Co.

Fig. 10-24. Rotary table with worm drive for continuous rotational displacement. The periphery of the table carries the graduation in degrees, and the incremental positions can be read in minutes and seconds from the drum graduations, in combination with a vernier.

The worm type rotary tables with graduated drum attached to the input shaft have an inherent property that must be remembered in order to avoid serious measuring errors. The smooth operation of the worm requires a certain amount of clearance that will produce backlash when the direction of rotation is reversed. Therefore, an operating direction (it is inconsequential whether clockwise or counterclockwise) should be selected for any measurement, then starting with the establishing of the initial position, the selected rotational direction must be retained during the entire measuring process.

Optical Dividing Heads and Rotary Tables

Optical dividing heads and rotary tables use graduated master rings or disks, commonly of glass, concentrically mounted and rigidly attached to the rotating spindle of the circular dividing instrument. The specific graduations of the master scale, which indicate the actual circular position of the rotatable member, can be observed through a microscope, representing an integral part of the instrument; it is equipped with an adjustable graticule or scale carrying the reference mark for the scale reading. The adjustment of the graticule is used to establish coincidence between the reference mark and the observable scale graduation. The amount of graticule adjustment is displayed on an auxiliary scale, the vernier, which indicates the fractional value to be added to the observed main scale graduation. These latter marks are usually spaced at $\frac{1}{3}$ or $\frac{1}{6}$ degree intervals, with full degrees consecutively numbered.

The process can also be reversed by first setting the vernier scale to the required fractional value and then rotating the dividing device until the desired graduation mark is brought into coincidence with the adjusted reference mark. One of the major advantages of optical dividing devices using graduated master rings is the unrestricted reversal of the setting process or change in the direction of rotation, without the occurrence of backlash or hysteresis, which could interfere with the inherent accuracy of the instrument elements.

There is one potential weak point in the use of master rings that are manufactured separately and then mounted on the rotating part of the dividing device: even a very small amount of eccentricity between the axis of rotation and the center of the mounted scale ring will affect the accuracy of the instrument's operation. Eccentricity in the mounting causes a sinusoidal division error that can be quite substantial even for a relatively small shift in the position of one of the centers. As an example, an eccentricity of 0.0003 inch on a 6-inch diameter circular scale will cause, at its most pronounced position, an error in excess of 20 seconds of arc between two diametrically opposed reading locations.

This possible interference with the potentially high accuracy of optical dividing devices can be effectively avoided by the use of a dual optical observation system. Such a system, also called double pickup, is shown on the diagrammatic drawing of Fig. 10-25. The principles of the error-compensating effect that is provided by the double observation system are explained in the diagram of Fig. 10-26.

Circular master scales used in more advanced designs of optical dividing devices are accurate to a maximum error of plus or minus one second of arc. Instruments equipped with double pickup for ob-

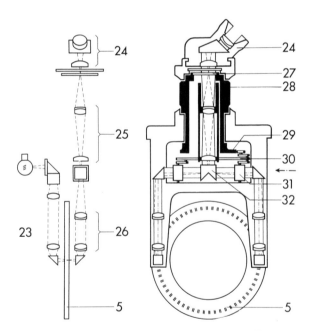

Leitz/Opto-Metric Tools, Inc.

Fig. 10-25. Diagram showing the operating principles of the double observation system used in the Leitz optical dividing head. The master ring has bifilar graduations at 20-minute intervals, which are viewed simultaneously at two diametrically opposite positions. The images of the graduations can be brought into coincidence by deflecting the light paths, and an optical micrometer indicates the angular value of the introduced displacement. *(5)* Master ring; *(23)* light source; *(24)* rotating eyepiece; *(25)* intermediate image system; *(26)* objective; *(27)* fine adjustment mechanism, operated with *(28)* knurled ring, which actuates *(29)* cam, through *(30)* transmission rods, causing the *(31)* optical flat to tilt (optical micrometer); *(32)* double pickup prism.

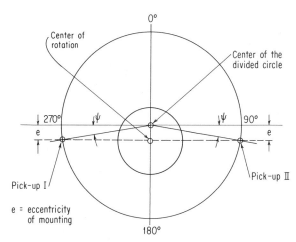

90 degrees rotation from the 0°–180° plane produces the following reading:
$$\frac{(90° + \psi) + (270 - 180 - \psi)}{2} = 90°$$

Fig. 10-26. Diagram showing the error-compensating principles of the double pickup observation system in circular dividing instruments.

O M T/Bentley Industrial Corp.

Fig. 10-27. Optical rotary table with projection equipment on whose screen the angular setting of the table in significant degrees, minutes and seconds of arc, in relation to a starting reference position, can be easily observed in a single, distinct display.

servation and with microscopes having appropriate resolving power permit obtaining an operating accuracy expressed by the following formula:

$$\Delta = \pm \left[2 + \frac{8}{R} \times \sin \frac{\alpha}{2} \left(1 + \frac{A}{8} \right) \right]$$

where

Δ = working uncertainty, in seconds of arc;

R = radius of the divided circle on the part, in inches;

A = axial distance of the measured surface from the face of the dividing head, in inches;

α = angle of spacing, in degrees.

To reduce the operator's eyestrain when continuously using an optical dividing head or rotary table, in certain models the microscope eyepiece is replaced by an illuminated screen (see Fig. 10-27). In these instruments, too, the essential measuring element is the circular glass scale, solidly mounted on the spindle of the rotating member. The graduation lines of the scale, together with the corresponding degree number, are projected onto a screen on which the two graticule images of the minute and the second scales also appear. These graticules are parts of an optical micrometer whose scales are adjustable by means of knurled knobs located close to the screen.

Optical rotary tables are also made with power drive and digital readout; however, the use of these sophisticated types of instruments is warranted for special applications only.

Calibration of Circular Dividing Instruments

All measuring instruments, those designed to produce circular divisions being no exception, can possess a finite accuracy only. The errors that arise from instrument limitations are usually not of a magnitude that could significantly affect the expected accuracy of specific measuring processes. Nevertheless, for assuring the reliability of measurements, it is necessary to determine the operating accuracy of circular dividing instruments by means of a dependable calibration process.

For calibrating rotary tables the most generally used, and at the same time very accurate method, is based on the use of *polygons* and *autocollimators*. Polygons are multifaced blocks, which are usually made of steel or glass, either as a single piece, or consisting of rigidly assembled elements. Polygons, which represent master gages of a very high degree of accuracy, are made with different numbers of sides, those most commonly used having 3, 5, 6, 9, 12, 18, 36 or 72 sides (see Fig. 10-28). The sides are equally spaced around a common circle, producing angular spacings between the adjacent sides, which are equivalent to 360 degrees divided by the number of sides. Polygons having any other number of equally spaced faces (not exceeding 72) are commercially available and also, exceptionally, polygons with unequal or mixed angular separation between the sides.

The faces of the polygons are made flat to a very high degree of accuracy, with errors usually not ex-

The L. S. Starrett Co.

Fig. 10-28. Optical polygon. The sample shown is a solid type made of chromium carbide, and has 12 equally spaced sides. Such polygons are also available with different numbers of sides in a range from 3 to 72.

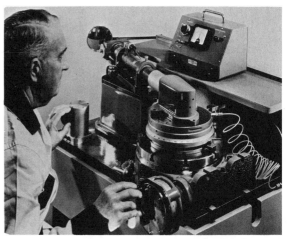

Moore Special Tool Co.

Fig. 10-29. Calibration of a rotary table with the aid of a precision index (face-geared indexing table) and autocollimator.

ceeding ¼ light band per inch, and have a reflective finish to permit a well defined optical observation. The accuracy of the angular spacings is, for the majority of types, better than 2 seconds of arc. Polygons used for critical measurements are supplied with certificates of calibration, spelling out the actual error in the spacing between each pair of adjacent faces in increments of 0.2 second of arc.

The autocollimator is an optical measuring instrument, similar to a telescope, yet emitting a light beam whose reflection from the target surface, such as the face of a polygon, can be precisely observed. The light beam leaving the autocollimator passes through a reticle, which also receives the reflected image. That reflected image will coincide with the original reticle marks only when the reflecting surface is truly perpendicular to the emitted light beam; the amount of noncoincidence displayed in angular units can be read through the microscope eyepiece or on the microscope dial of the instrument. (Additional details on autocollimators will be discussed in the next chapter.)

For carrying out the calibration process, the polygon is mounted centrally on the face of the rotary table, and the autocollimator is set up to face one of the polygon sides in a precisely perpendicular direction. By noting the rotary table's graduation value at the starting position, the table is now rotated by the amount of the nominal polygon side spacing, and the reflection of the next polygon side is observed by the autocollimator. Noncoincidence, if present, can be detected and measured very accurately, directly in angular values. The process is then repeated for the subsequent sides of the polygon.

The development of very accurate *face-geared indexing tables* supplied a tool that can in many applications replace the polygon as a calibration standard for angular spacings (see Fig. 10-29). Actually, the face-geared indexing table, with a single reflector

mounted on its top face, has several valuable advantages over the polygon as a calibration tool. These are: the availability of a much greater number of indexing positions—360 or even 1440, compared to a practical maximum of 72—and an actual indexing accuracy that is better than that usually accomplished in rigid polygons. On the other hand, polygons, except those having a large number of faces, are less expensive than a calibration grade indexing table. Furthermore, polygons can be used where space limitations do not permit the setting up of indexing tables with a mounted reflector.

Yet another method of rotary table calibration, rather infrequently used, but possessing the advantage of not requiring polygons, relies on the use of angle gage blocks. However, this method has several limitations, partly connected with potential errors resulting from the repeated exchange of angle gage blocks; neither is it applicable to full circular inspection.

Advanced Automated Master Angle Calibration

Recently, the state-of-the-art in master angle calibration was extended with the development of the advanced automated master angle calibration system, AAMACS (see Fig. 10-30). In a project sponsored by the Strategic Defense Initiative, the AA-MACS was developed and placed at the Precision Engineering Division, National Institute of Standards and Technology in Gaithersburg, Maryland. This piece of equipment makes angle measurements

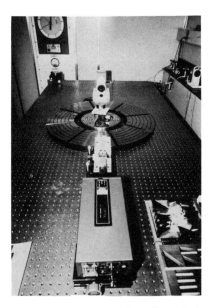

A.G. Davis Gage & Engineering Co.

Fig. 10-30. The Advanced Automated Master Angle Calibration System (AAMACS) in service at the Precision Engineering Division of the National Institute of Standards and Technology, Gaithersburg, Maryland, USA.

that are ten times more accurate than that of previous technologies.

Through the use of a proprietary triple differential index system (for example, three indexing tables stacked concentrically), 379,080,000 discrete mechanical positions are possible. This system of discrete positions yields a measuring accuracy of 0.25 second of arc, a resolution of 0.0034 second of arc, and a repeatability of 0.005 second of arc. The AAMACS is now used to calibrate autocollimators, optical polygons, angle blocks and indexing tables, which in turn are used in the production of machine tools, coordinate measuring machines and telescopes.

The AAMACS was designed to specifically address four main problems that have affected the accuracy of angle measurement in the past. These problems are: (1) refraction and air turbulence affecting the light path of photoelectric autocollimator systems; (2) seismic and acoustical disturbances causing movement between mechanical components of optical systems; (3) thermal drift caused by the proximity of an operator during a long period of measurement; and (4) autocollimator errors caused by calibration problems typical with this system of angle measurement.

THE MEASUREMENT OF INCLINES

Spirit Levels

For inspecting the horizontal position of surfaces, as well as for evaluating the direction and magnitude of minor deviations from that nominal condition, spirit levels are the most commonly used instruments. The spirit level has for an essential element a closed glass tube of accurate form, the vial, which is produced by bending or, for the more accurate instruments, by grinding and polishing. Controlled curvatures for precision vials are also produced by grinding the bore of a glass tube to the shape of a barrel with a calculated radius. The tube is filled almost entirely with a liquid, leaving only a small space for the formation of an air or gas bubble. Since the liquid, due to its greater specific weight, tends to fill the lower lying sections of that closed space, the position of the bubble will indicate the topmost section of the curved vial, as long as the incline causing the nonsymmetrical position of the bubble stays within the sensing range of the instrument. Although water can be used for filling the vial, low viscosity fluids, such as ether, alcohol or benzol, are preferred for accurate instruments.

The vials of spirit levels are sensitive to ambient conditions; higher temperatures cause the liquid to expand, restricting the volume of the bubble, whereas at low temperature the bubble will grow in size. Actually, there are certain specifications for the length of the graduated vial section, requiring it to be greater than the extent of the bubble at −10°C (14°F) temperature.

Temperature variations also cause the glass vial to expand or to contract; for this reason the better grades of spirit levels use an elastic mount for the vials to avoid harmful strains. To reduce the effect of heat transfer in handling, the precision type spirit levels are made of a relatively stable casting and are equipped with thermally insulated handles.

Figure 10-31 illustrates a type of sensitive spirit level equipped with a cross vial; this model is supplied in lengths of 6, 8, 10 and 12 inches. For precise measurements, the level should be placed on the surface to be inspected with a minimum of side tilt, causing a bubble displacement of less than one division on the cross vial, which is less sensitive than the main vial. Depending on the selected grade of sensitivity, the illustrated type of spirit levels are available with the following indication values for each graduation of the main or longitudinal vial:

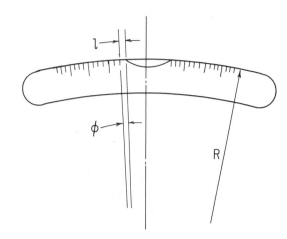

Mahr Gage Co.

Fig. 10-31. (Left) Sensitive spirit levels with cross vials. These precision levels have thermally insulated handles.

Fig. 10-32. (Right) Frame levels with cross vials. The sides have the form of Vee-grooves for locating on cylindrical surfaces. The large-size frame levels have handles made of thermally insulating material.

Fig. 10-33. Diagram showing the basic geometry of the graduated vial of a spirit level. The bubble is in the center (zero incline) position. The curvature is grossly exaggerated to illustrate the principle. Each *l* graduation spacing is equivalent to ϕ-degree incline variation when the vial curvature has a uniform *R* radius.

0.001   in./ft—approx. 17 arc seconds
0.0005   in./ft—approx. 8 arc seconds
0.00025 in/ft—approx.   4 arc seconds

The bottom surface of mechanics' levels usually has an inverted Vee-groove of 120 degrees included angle, running parallel with the precisely machined sides, to permit the use of the level on shafting or similar cylindrical surfaces.

For measuring the position of nominally vertical flat or cylindrical surfaces, the frame level (Fig. 10-32) is often of great usefulness. Precision instruments of this type are also equipped with thermal insulation and cross-vials. The above-mentioned sensitivity grades also apply to the illustrated type of instruments.

The sensitivity of vials used in spirit levels is commonly expressed in either of the following two ways:

a. Each graduation line representing a specific slope is defined by a tangent relationship, for example, 0.0005 inch per foot.

b. An angular value is assigned to the vial length covered by the distance of two adjacent graduation lines; the unit of second of arc is used in precision instruments.

For calculating the radius to which the curvature of the vial must be ground in order to comply with the assigned graduation value, each graduation distance is considered to be an arc subtending an angle

of tilt ϕ (see Fig. 10-33). When expressing the tilt ϕ in radians (one second equals 0.000,004,85 radian) the relationship is

$$l = R\phi$$

where

l = length of each graduation distance, in inches;
R = mean radius of the vial curvature, in inches.

As an example, examining a vial with 0.100-inch graduation line distance, to be used in a level where each division represents $\frac{1}{10}$ minute (6 seconds of arc) tilt angle, the radius of the vial curvature is

$$R = \frac{l}{\phi} = \frac{0.1}{0.000,029,1} = 3436 \text{ inches or 286.4 feet.}$$

The actual operating accuracy of spirit levels, even when used under rigorously controlled laboratory conditions and at 68°F reference temperature, will usually be limited by an uncertainty on the order of one graduation value. In addition to the technical difficulties of grinding the curvature of vials exactly to the calculated radius, the uncertainty of bubble position observation in association with specific graduation lines must also be considered.

The coincidence type spirit levels, an example of this system being shown in Fig. 10-34, possess unique design characteristics that greatly increase the precision of observation as well as the calibration

Metra-Tech Corp.

Fig. 10-34. (Top) Coincidence type spirit level with compensating adjustment by means of a special micrometer, graduated to indicate slope per inch, to 10 microinches least increment (about two seconds of arc). This value corresponds to the discriminating capacity of the coincidence bubble. The auxiliary vials serve the approximating adjustment in the main measuring plane and the checking of side tilt (roll).

Fig. 10-35. (Bottom) The observable image of the coincidence bubble in different positions; the illustration in the center shows the well-adjusted reference position.

accuracy of these instruments. This is accomplished by disassociating the deviation indication from the bubble position, using that position only to establish a definite reference position. By means of a special optical system, using two prisms, the two ends of the bubble are made to appear in a single magnified image showing two semicircles (see Fig. 10-35). The closing of the two contour images into a continuous circular line indicates the attainment of the reference position. The vial is supported in a compound lever system that can be adjusted by means of a micrometer screw to compensate for deviations of the inspected surface from the true horizontal plane. The amount of adjustment is indicated by the special micrometer, which has a double thread with differential pitch. Each graduation line of the micrometer represents 2 seconds of arc, and the level system permits adjustments over the range of approximately $\pm\frac{1}{8}$ inch per foot ($\pm 0°35'$) from the basic horizontal.

Electronic Levels

The essential element of the electronic level is a pendulum suspended inside a frame or housing. That frame is mounted on a pedestal with feet at a specific distance apart, commonly extending over a length of 5 inches. The base of the instrument constitutes a plane that represents, in principle, the true horizontal when the pendulum is at its central position.

Electronic levels produced by different manufacturers vary in details of design, yet are common with regard to the use of induction coils as sensing elements. The principles of operation of a domestic make of electronic level are shown diagrammatically in Fig. 10-36.

The sensing frame is connected by a cable with a remote amplifier that is similar to those used for electronic length comparator gages (see Chapter 7). The amplifier can be operated by line current, and some designs are also available with a mercury cell pack for power source. When a reference plane is used that does not coincide exactly with the geological horizontal, a zeroing adjustment can be made within the range of a few minutes of arc, either mechanically on the suspension of the pendulum, or electrically, on the amplifier, depending on the particular design of the instrument.

The use of electronics in sensing and displaying in angular values the mechanical deflections of the pendulum assures several favorable properties for this type of angle measuring instruments, such as the following:

a. Remote observation—usually a few feet from the sensing element's location, although there is no reason why longer distances, up to a hundred feet, cannot not be used, if needed,

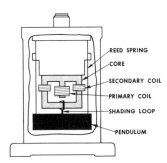

REED SPRING
CORE
SECONDARY COIL
PRIMARY COIL
SHADING LOOP
PENDULUM

Federal Products Corp.

Fig. 10-36. Diagram illustrating the principles of operation of one type of electronic level, using a shading loop attached to the pendulum to translate into electrical signals the gravitation-induced movements of the latter.

b. Rapid response, with high repetitive accuracy,

c. High amplification for sensitive measurements

or, alternately, extended range, for example, for setting, selected by simple electrical switching, without affecting the zero position of the once set reference plane. These conditions of amplification and range are shown in the following tabulation, with reference to the model explained in Fig. 10-36:

Setting Position	Full Scale Range		Value per Graduation	
	Angular	Slope	Angular	Slope
A	400 sec	0.002 in./in.	20 sec	0.000100 in./in.
B	100 sec	0.0005 in./in.	5 sec	0.000025 in./in.
C	20 sec	0.0001 in./in.	1 sec	0.000005 in./in.

Clinometers

Clinometers are instruments for measuring, in the vertical plane, the incline of surfaces in relation to the basic horizontal plane, over an extended range. The functional elements of these instruments are the sensitive vial mounted on a rotatable disk, which carries a graduated ring with its horizontal axis supported in the housing of the instrument. When the clinometer is placed on an exactly horizontal surface, the bubble of the vial is in its center position as long as the scale of the rotatable disk is at zero graduation. Placing the instrument on an incline, the bubble can again be brought into its central position by rotating the disk; the amount of rotational displacement can be read on the scale, indicating the deviation from the horizontal plane of the surface being measured.

Clinometers are manufactured in several designs with different degrees of sensitivity and measuring accuracy. Clinometers with 10-minute graduations are available for shop uses, and the range of currently manufactured models extends to include high precision instruments. These high precision instruments possess optical readout in increments of one second of arc and are equipped with a coincidence-

Reading: Main Scale   61° 25'
Micrometer   3' 55.2"
61° 28' 55.2'

Hilger & Watts Inc./Engis Equipment Co.

Fig. 10-37. (Left) Precision clinometer with direct reading to 0.2 second of arc. It permits precise zero referencing with the aid of a coincidence-indicating bubble.
(Right) Field of view.

displaying bubble having a sensitivity of the same order (see Fig. 10-37).

Clinometers may be used for either of the following two categories of measurement, which also determine the technique to be used, namely:

a. The measurement of an inclined surface in relation to the basic horizontal. By placing the instrument on the surface to be measured and rotating its graduated disk to produce zero indication on the fiducial bubble, the scale value of the disk position will be equal to the angle of incline; and

b. The measurement of the relative position of two mutually inclined surfaces. By placing the clinometer on each of the surfaces in turn, and taking the readings with reference to the basic horizontal, the difference of the two readings will indicate the angular value of the relative incline.

In carrying out either of these measurements proper care must be applied to keep the axis of the rotatable disk parallel to the hinge line of the incline; this positioning is assisted by the cross vial of the instrument.

II.
The Measurement of Straightness, Flatness and Perpendicularity

The measurements discussed in this chapter primarily concern the digressions from a nominal direction or position of characteristic elements of the inspected surfaces. For expressing the amount of directional digressions, angular units are often used as a means to define the measured geometric condition by a single dimension.

It should be remembered, however, that such use of angular units differs from their role in actual angle measurements. These angle measurements were discussed in the preceding chapter, having for a major subject the mutually referenced angular separations, as well as angular spacings resulting from the subdivision of the circle.

The scope of measurements related to directional or positional digressions is extremely wide and varied. For the convenience of discussion, the methods and instruments of measurement to be discussed in this chapter will be associated with three basic conditions, namely:

a. Straightness, related to a straight line;

b. Flatness, related to a perfect plane; and

c. Perpendicularity, related to the edges or faces of a perfect cube.

This delineation of conditions, however, should not be regarded as a rigid categorization, but rather an expediency to facilitate the reviewing of that complex area of dimensional metrology. Actually, the concept of straightness, flatness and perpendicularity are closely interrelated, and many of the instruments discussed are used interchangeably for the measurement of more than a single category of conditions.

Straightness, flatness and perpendicularity are basic conditions of many geometric concepts applied in engineering design and manufacture. Consequently, the measurements associated with any of these conditions are varied and differ in many respects. Such measurements can be directed toward checking the object's compliance with the specified geometric requirements, or one of these basic conditions is established to serve as a means for the measurement of associated conditions. Typical examples of the latter kind of measuring application are the establishment of a straight line to check alignment, or of flatness, to inspect perpendicularity.

In view of that diversity of metrological usages, a detailed categorization would be impractical. However, to indicate the scope of the metrological processes that are directed at or related to straightness, flatness and perpendicularity, the individual sections of this chapter are accompanied by tabulated surveys of the frequently used methods of measurement. While making no pretense of completeness, these surveys should assist the reader in visualizing the broad field of measuring techniques that are associated with these basic geometric conditions.

Optical Tooling

The term optical tooling is frequently used for a specific category of engineering measuring processes in which optical observation is substituted for mechanical contacting and referencing members. Optical tooling is particularly advantageous for large objects, where substantial distances are involved between the interrelated surfaces of a part or a structure, or when a remote location of the measuring instrument is required. Optical tooling is widely used in the manufacture of machine tools, engines and general machinery, in the aircraft and shipbuilding industries, also for the production and erection of aerospace equipment. However, the use of optical tooling is not restricted to large scale metrology; the measurement of critically dimensioned parts of medium and small size also offers a growing number of preferential applications.

These processes rely on certain physical properties of light, such as its straight path of propagation, the parallelism of light beams that are collimated with the aid of appropriate lenses, its reliable characteristics of reflection and refraction, and various others. Advances in optics, both as a science and as a branch of manufacture, originating many new types of instruments based on optics alone or in combination with electronics, largely contributed to the rapid expansion of optical tooling in many areas of dimensional measurements. More recently, the many applications of lasers, as a source of dependably coherent and high intensity light, promise further advances both in the accuracy and in the scope of optical tooling.

In the process of optical tooling, definitely located and directed virtual lines (the line of sight) and planes (by a sweep motion of the line of sight or by the wavefronts of collimated light beams) are established and then observed either directly or in their reflection. Selected elements on the object surface are observed in their relation to these optical lines or planes and, in the case of noncoincidence, the amount of digression can be sensitively measured with the aid of the optical observation instrument.

The viewed elements can be specific points or areas of the object's actual surface, although more frequently, intermediate members are used for presenting the feature or surface element being investigated, in a manner susceptible to dependable observation with the applied optical instrument.

The important instruments of optical tooling are alignment telescopes, jig transits, optical levels, theodolites and autocollimators. These instruments are used generally in conjunction with different auxiliary devices, such as targets, collimators, reflectors and prisms. Several of these instruments and various accessories are discussed in detail in the following sections of this chapter, and some indications regarding their typical applications are given.

A concise survey of the more common applications for optical tooling is presented in Table 11-1, illustrating process principles based on the use of various instrument systems. Although diverse uses of autocollimators also fall under the general category of optical tooling, these will be reviewed together with other autocollimator applications in Table 11-7.

STRAIGHTNESS AND ALIGNMENT MEASUREMENTS

The straight line represents the path of all linear dimensions. Considering the premise that the shortest distance between two points is along a straight line, that path is not necessarily present in a physical sense on the part being measured for size, but it must be incorporated in the length measuring instrument. Straightness, which is a fundamental concept of linear measurements, is also a functionally important condition of many engineering products. As an introduction to the discussion of straightness measurements, a survey of a few basic methods is made in Table 11-2.

In commonly used length measuring tools the straightness of the measuring path is physically represented by the straight form of a graduated rule or scale, by the bar of a vernier caliper or by the guideways of a length measuring instrument. In some other types of linear measuring tools the straight line is functionally assured, such as by the axis of a micrometer spindle or of an indicator spindle. Deviations from the straight line of said elements affect the reliability of the length measurement carried out with the aid of these tools.

The obvious method for establishing or checking the straightness of a surface element is by means of direct contact comparison with a tool of known and adequate straightness. The sensitivity of such direct comparison is, of course, limited. The *straight edge* represents the most commonly used tool for that type of limited sensitivity straightness measurement. It consists of a rectangular steel bar of relatively narrow cross-section, with a precisely machined edge, which is straight within standardized or guaranteed limits.

The straightness accuracy will depend on the intended use of the straight edge. The following formulas, having foreign standards (DIN) for source, but expressed in inch units, will be indicative of the degree of straightness that can be expected from this type of tool:

Tool Accuracy Grade	Maximum Deviation from Mean Plane of Any Point along Measuring Surface
Reference quality straight edge	$\pm\left(0.000,040 + \dfrac{L}{200,000}\right)$ inches
Toolmaker quality straight edge	$\pm\left(0.000,080 + \dfrac{L}{100,000}\right)$ inches

TABLE 11-1.   EXAMPLES OF ENGINEERING MEASUREMENTS BY OPTICAL TOOLING — 1

METHOD OR PURPOSE	EQUIPMENT	DIAGRAM	DISCUSSION
Alignment testing	Alignment telescope with collimator and intermediate targets		Measuring the alignment of nominally coaxial bores by establishing an optical reference line between the two extreme bores, one holding the telescope and the other the collimator. A second target is placed sequentially into the intermediate bores; the amount and direction of lateral displacement, when present, is measured with the eyepiece micrometer of the telescope.
Straightness of the translational movement of a machine element	Alignment telescope with collimator and target		The straightness of the displacement path of a machine tool (lathe) carriage with reference to the line of sight representing the common axis of the main spindle and of the tailstock is inspected by observing the degree of coincidence of the target on the carriage at different positions along the guideways.
Optical transfer of coordinate distances established in a remote measuring plane	Cathetometer—a telescope mounted for vertical displacement along a graduated column which is attached to a horizontal slide		A vertical bar with graduations and vernier for 0.001-inch reading guides a bracket into which a telescope is mounted. The bar can be displaced horizontally along the guideways of a slide holding the bar, the distances of this movement being also measurable. The alignment of a selected part feature with the coordinate position of the telescope is established remotely with the aid of a line of sight.

TABLE 11-1.   EXAMPLES OF ENGINEERING MEASUREMENTS BY OPTICAL TOOLING – 2

METHOD OR PURPOSE	EQUIPMENT	DIAGRAM	DISCUSSION
Establishing optically a horizontal reference plane	Optical (sighting) level (Telescope rotatable around a vertical axis)		For setting up large machine members in positions related to the horizontal, an optical reference plane can be established by the sweeping motion of a telescope which is rotating around a precisely vertical axis; scales or targets resting on the part's surface are observed and digressions from the datum plane are accurately measured.
Establishing optically a vertical reference plane	Jig transit (Telescope rotatable around a horizontal axis)		The erection of heavy machinery and of similar structures can be dependably guided by a vertical datum plane established optically by means of a jig transit. The orientation of that plane normal to another datum direction is assured by an axle mirror mounted on the jig transit. An accurate plumb line can also be established.
Establishing directionally controlled reference lines at any tilt angle and in any orientation	Engineering theodolite (Telescope rotatable around two axes—vertical and horizontal—both rotations measurable with high accuracy)		When several vertical reference planes must be established at different angular spacings, and the elevation angles of diverse part features in those planes must be measured in relation to a common datum plane, the engineering theodolite should be used. Its angular accuracy is unique and it can also be equipped for measuring small linear digressions.

TABLE 11-2.   EXAMPLES OF METHODS FOR INSPECTING STRAIGHTNESS — 1

OPERATING PRINCIPLES	TYPICAL INSTRUMENTS	DIAGRAM	DISCUSSION
Straightness of a surface element determined by its parallelism to a straight edge of known accuracy	Straight edge and gage blocks		Unequal height of the supporting elements create a controlled deviation from true parallelism. Gage blocks are used to check whether the digressions from the parallel at intermediate points correspond to the calculated values.
The straightness of a surface element inspected by direct contact with a tool of calibrated straightness	Knife edge rule		A knife edge rule brought to bear against the surface element to be inspected will indicate by the presence and width of light gaps the lack of contact caused by deficient straightness.
Straightness of a shaft determined by rotation on fixed supports and measuring the runout of its surface	Bench centers or Vee-blocks and indicator stand		Using the part's axis (between centers) or its surface (supported on Vee-blocks) for a datum, straightness defects will cause runout when the part is being rotated; the indications must be evaluated in relation to the selected datum element.
The straightness of a cylinder inspected by an indicator supported on a surface plate	Surface plate and an indicator held in a movable stand		Using a surface plate for a reference plane the entire length of a surface element of the part can be inspected in relation to the selected supporting points.

TABLE 11-2. EXAMPLES OF METHODS FOR INSPECTING STRAIGHTNESS — 2

OPERATING PRINCIPLES	TYPICAL INSTRUMENTS	DIAGRAM	DISCUSSION
Straightness of a shaft checked by comparing it to the horizontal	Block level with Vee-base		By bringing two reference sections of a long shaft into a horizontal position, the straightness over its entire length can be checked by correlating the slope, when present, of the intermediate sections with the basic horizontal, using the indications of a Vee-base precision level.
Straightness of a surface element determined by measuring the locations of discrete points	Measuring machine with sensitive indicator in the displaceable measuring head		The tracking accuracy of measuring machine slides permits determining, with a high degree of resolution, the locations of individual points along the selected surface and obtaining numerical data on the out-of-straight condition.
Straightness measured by comparison to the translational path of a precise slide	Precision slide and sensitive indicator		Precision slides made for instruments possess a straightness of translation whose degree of accuracy might be much higher than that expected from the part. When two distant points of a surface element of the part are aligned with the slide track, an indicator will display variations from the straight.
Straightness measured by observing a supported target in relation to a line of sight (Optical tooling)	Alignment telescope and a target on an appropriate base		A base in contact and guided by the surface to be inspected will transmit to the target it carries the variations from the straight of the supporting surface. Displacements of the target are measured with the ocular microscope of the alignment telescope.

where *L* represents the length of the tool in inches. The use of these formulas is illustrated by the following example.

A reference quality straight edge 20 inches in length should have a straightness accuracy of

$$\pm \left(0.000,040 + \frac{20}{200,000} \right) = \pm 0.000,140 \text{ inch.}$$

The actual use of straight edges for checking the straightness or flatness of technical surfaces usually relies on such toolmaker techniques as bluing for seeing the ratio of contact, or feeler gages for the appraisal of gap widths. Another method consists in supporting the straight edge on the surface to be inspected with the aid of two gage blocks of slightly different lengths (for example, 0.150 and 0.151 inch), which are placed at a known distance apart. The wedge-shaped gap thus created can then be explored with the use of gage blocks of intermediate sizes, which should fill the gap at specific distances from the support points. In order to reduce the effect of sag on the effective straightness of the tool, it is recommended that the straight edge be supported at two points, each at 22% of the total length from the ends.

Other types of tools for the direct contact checking of straightness are the *knife edge* (Fig. 11-1) and the *triangular straight edge* (Fig. 11-2). Both types of tools have narrow measuring edges and are ground to a small radius to avoid chipping of the edges. These tools are used for checking the straightness of a surface, or the coincidence of tangent lines to two parts of nominally equal lengths, by observing the light gap that may appear between the part surface and the knife edge at certain sections of the contact length. With proper skill and adequate lighting, preferably a diffused light, gaps of 0.0001 inch width can be distinctly detected. With decreasing gap width, the color of the filtering light changes, appearing red from 0.00007 to 0.00005 inch, and then changing to bluish when the gap is reduced to about 0.00003 inch. The diffraction of the white light used for illuminating the background causes these color changes, which can serve as a guide for estimating the width of the observed gap.

The following formula expresses the straightness accuracy, which can be expected from knife edges and triangular straight edges made by reputable manufacturers:

Maximum deviation from a mean plane:

$$\pm \left(0.000,020 + \frac{L}{500,000} \right) \text{ inch.}$$

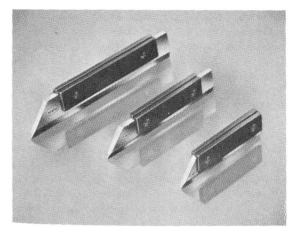

Mahr Gage Co.
C E J Gage Co.

Fig. 11-1. (Upper) Toolmaker knife edges.

Fig. 11-2. (Lower) Triangular straight edges.

As an example: a knife edge 4 inches in length should have a straightness accuracy of

$$\pm \left(0.000020 + \frac{4}{500,000} \right),$$

that is, ±0.000028 inch.

Apart from the widely utilized roles discussed above, the straight line has many other important functions in the field of dimensional measurements. The subsequent descriptions will be concerned with these less common functions of the straight line, as well as with the instrument sand methods of its measurement.

The straight line represents the axis of symmetrical bodies or features, such as cylinders, cylindrical bores, square or rectangular bars, cones and many others. Assuming a regular form, the location and the orientation of the axis will reflect the position of that body or feature in relation to specific datum elements.

A very frequent case of axis position determination is known as the alignment of two or more axes, such as those of bores supporting a common shaft. Alignment expresses a condition where the examined axes lie on a common straight line. Deviations from that theoretical condition can be of two basic types (see Fig. 11-3), either of which may occur singularly or in combination. These deviations from the basic condition are distinguished as: (a) lateral displacement (shift or positional error); and (b) angular displacement (directional error).

A straight line along a surface or at a specific distance from that surface may also serve as a datum for measuring the form, the location and the orientation of the surface elements that are being related to that straight reference line. For such applications, as well as for alignment measurements, a widely applied method consists in establishing a virtual line, known as the line of sight, by means of an appropriate optical instrument, such as an alignment telescope.

It is possible to aim the telescope at specific points on the surface being investigated; however, in the majority of cases such direct observation is not practical or even feasible. Rather, auxiliary members are used to mechanically contact the selected surface element of the object and to translate its position into the field of view of the sighting instrument. Such auxiliary members are targets, collimators, scales and their complementary accessories.

Telescopes can be mounted remotely from the object, by using a frame with guides for slides for which translational movement over controlled distances is provided. In such arrangements, the optically observed surface elements of the object are related to specific slide positions along the guideways. These are the operating principles of optical tooling bars and cathetometers.

By supporting the telescope in a manner permitting precise rotational movement of the instrument around an axis, which intersects the optical axis at right angles, a sweep motion can be produced establishing a virtual plane as the scanning area of the line of sight. Such optical planes may be used as datums for dimensional measurements by optical tooling. The optical datum plane may lie in the vertical (established by a jig transit) or in the horizontal (resulting form the sweep motion of an optical level).

Finally, it is also possible to measure by means of optical observation the angular separations between lines of sight directed at discrete points, or between planes normal to the plane of sighting, by considering the vertices of these angles coincident with the station of observation. The instrument used for such directional measurements is the engineering theodolite.

Several of these basic instruments of optical tooling, as well as various important accessories, will be discussed, referring to representative models to indicate typical characteristics. The instruments discussed in this section have in common the use of the line of sight, either in a fixed position or conjointly with its controlled translational or rotational displacement.

The Alignment Telescope

The alignment telescope was developed for industrial applications from the basic sighting telescope. Its primary purpose is to establish a straight line of sight in a particular location to serve as a basic reference line or datum for dimensional measurements.

The optical elements of the alignment telescope are contained in a tubular housing, known as the barrel, which is ground precisely cylindrical to a standard diameter, nominally 2.25 inches, and its center line is coincident with the instrument's optical axis within very close limits (Fig. 11-4). The optical system comprises an adjustable objective lens that can be focused from nearly zero to infinity. The obtainable magnification varies with the focusing distance, being about $4\times$ in front of the lens and about 35 to $50\times$, depending on the particular model and eyepiece, when set to infinity. The eyepiece is usually in line with the direction of sighting, although some models can be equipped with angle eyepieces as well. Also contained in the optical system is a cross-hair graticule, which serves as the reference element in aiming at the target.

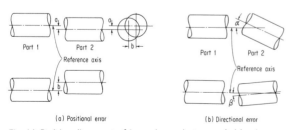

(a) Positional error   (b) Directional error

Fig. 11-3. Nonalignment of axes shown in top and side views.

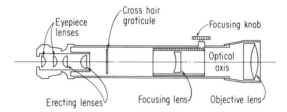

Fig. 11-4. Diagram of the principal optical elements in an alignment telescope.

The graticule can be traversed over small distances in two, mutually normal directions for measuring the displacement of the target from the line of sight. The traverse is either an actual translational movement, or a virtual shift of the cross-hair image, depending on whether the instrument is equipped with a mechanical or an optical micrometer. Commonly used values for that adjustment are a range of ±0.050 inch or ±0.100 inch, in increments of 0.001 inch, although micrometers with a reading to 0.0002-inch least graduation are also available

Industrial telescopes are also made without graticule adjustment for specific applications, such as to function as a basic reference line of sight. These instruments find further applications to optically translate a defined slide position that has been established by a cathetometer or by an optical metrological bar system. These optical instruments are known as sighting, or line-of-sight telescopes.

The line of sight being a mechanically intangible, although physically defined element, it must be positively associated with the mechanical surfaces of an engineering structure to possess practical usefulness. This requirement is assured by the mechanical presentation of the optical axis in the form of its radial extension resulting in a cylinder of known diameter, namely, the barrel or the tube of the telescope.

The excellent cylindrical form of the telescope barrel and its precise alignment with respect to both concentricity and parallelism in relation to the optical axis of the instrument, permit the latter to be guided by the barrel in establishing a line of sight parallel to a machine surface. An example of this type of use is holding the telescope barrel in the bore of a fixture whose outside surface conforms to the Vee-ways of a lathe.

For many optical tooling applications it is more advantageous to establish a line of sight between two selected distant points, one contained in the optical axis of the telescope and the other represented by an appropriately located target. For establishing that line of sight, the telescope must be aimed without losing the selected reference point from its position in the optical axis. This is accomplished by rotating the telescope around that point with the aid of a spherical mount, usually of 3½-inch diameter, which holds the barrel in a concentric position. As shown in Fig. 11-5, additional supporting members are a base with a vertically adjustable hollow column in its center, and the extension bracket, with fine adjustment screws, which provides a support for the barrel between the spherical mount and the eye-

Rank-Taylor-Hobson/Engis Equipment Co.

Fig. 11-5. Alignment telescope held in a mounting sphere, supported by a horizontal base equipped with adjusting bracket.

piece. The vertical column, known also as the candle, has a ball seat to accept the spherical mount, which then may be secured with the aid of a spring clip (see Fig. 11-5).

It may be required to establish the line of sight precisely in a horizontal plane, and for that purpose to adjust vertically either the target or the telescope. A very precise level, frequently of the coincidence type with sensitivity on the order of 2 seconds of arc, is used as a guide for such adjustment. The level is mounted on the telescope by means of a special base whose legs straddle the barrel (see Fig. 11-6).

An optical square (pentaprism) can be mounted on the objective lens end of the alignment telescope to establish an observation line that is perpendicular to the basic line of sight. Certain types of optical

Rank-Taylor-Hobson/Engis Equipment Co.

Fig. 11-6. Coincidence level with striding base mounted on the barrel of an alignment telescope.

squares are accurate to one second of arc and permit observation along either the basic line of sight or at 90 degrees to it. Depending on the type of optical square, the deviation of the line of sight can originate in the geometric center of the spherical mount, or it can be offset from this point by a specific distance, for example, 4 inches. The latter type permits a full optical sweep of 360 degrees.

The eyepiece micrometers. The alignment telescope represents one of the several categories of optical instruments whose operation requires the precise measurement of the linear displacements of a magnified reference mark or of a reflected image with respect to a basic position, such as the optical axis of the instrument. Essentially similar systems of optical length measuring devices are used in most types of optical instruments designed for dimensional measurements.

These devices operate either by the mechanical translation of the reference graticule actuated directly by the micrometer screw or by the use of the optical micrometer. The operation of the mechanically translated graticule is shown diagrammatically in Fig. 11-34, in conjunction with an autocollimator.

The operation of the optical micrometer is based on the known refractive properties of optical glass. A plane and parallel-sided glass block placed in the path of a light beam will permit the light to pass straight through the block as long as the face of the block is in a plane normal to the direction of the light. However, when the light strikes the block at other than a right angle, it will be refracted and will follow, during its passage through the block, a path different from its original. When leaving the block, the light will again be refracted by an amount equal to that refracted when it entered, but in the opposite direction. Consequently, the light beam will resume its original direction, although it will have been shifted laterally by an amount that is a function of the block material's refractive index and the length of the passage through the block (see Fig. 11-7).

For practical purposes the linear distance of the shift can be considered proportional to the angle of light incidence, as long as that angle remains small, that is, in the order of a few degrees.

In the optical micrometer, the plane-parallel block is rotatable around an axis that is perpendicular to the direction of light. A micrometer device is used to produce a controlled and measurable tilt or tipping of the block, causing the path of the light beam passing through the block to be shifted by a known amount in either direction from the basic and in a plane normal to the rotational axis of the block.

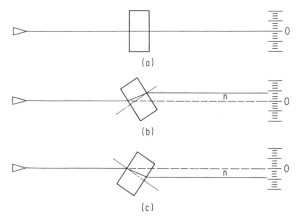

Fig. 11-7. Operating principles of the optical micrometer. (a) Light rays striking perpendicularly to the sides of a parallel-sided block continue in a straight line. (b and c) The tilt of the block causes displacement n of the image, due to the refraction of the light passing through the block.

One form of optical micrometer was shown in Fig. 8-6, illustrating diagrammatically the manner in which the optical principles explained above are utilized in instrument design.

A high degree of accuracy can be achieved with the optical micrometer, due to the fixed pivot axis and the inherently constant refractive properties of optical glass of known composition.

Jig Transits

The jig transit is an optical instrument for establishing a vertical sighting plane. Its main application in optical metrology is to determine the location of specific surface elements of a large object in relation to an optically scanned vertical reference plane. Optical scales are most commonly used for auxiliary target members to correlate selected elements of the object surface with the optically established reference plane. Another method consists in mounting the jig transit on a horizontal bar system, which permits the translational displacement of the instrument over specific distances and in a direction that is precisely perpendicular to the established vertical reference plane.

The jig transit (see Fig. 11-8) consists of a telescope solidly mounted on an axle that is supported in a rigid frame in such a manner as to permit the free rotation of the axle. The frame can be leveled to assure the horizontal position of the pivot axis around which the rotation of the telescope axle is effected.

The vertical reference plane established by the rotation of the jig transit's telescope can be located in

Keuffel & Esser Co.

Fig. 11-9. Optical level with tilt adjustment to establish a plane-in-space level to one second of arc.

Keuffel & Esser Co.

Fig. 11-8. Jig transit to establish optically vertical reference planes and plumb lines. It is equipped with a mirror mounted on the axle to determine squareness to a line of sight.

an orientation precisely normal to a horizontal line of sight, when guided by autoreflection mirrors, which are mounted exactly square on either end of the axle.

Certain types of jig transits have a hollow base that permits the use of the telescope for sighting down through the instrument and, in that position, functions as a guide in setting up the jig transit over a specific station point.

An optical micrometer, when mounted on the telescope, will measure the displacement of the observed target point from the plane of sight. The common range of such micrometers is ±0.100 inch, with indications in increments of 0.001 inch.

When required by the work for which the jig transit is used, various accessories can be added to the basic equipment, thereby extending the instrument's adaptability. Examples of such accessories are different types of sensitive levels, an autoreflection target and an autocollimator conversion unit.

Optical Levels

The operational purpose of the optical, or sighting, level is similar to that of the jig transit, however, with the essential difference that this instrument establishes the optical reference plane in the horizontal. The particular model of instrument shown in Fig. 11-9 is designed to establish, by its rotation around a vertical axis, a plane in space that is level to within one second of arc. The instrument is equipped with a two-stage level system, consisting

of a circular level with 10 minutes sensitivity for rough leveling, and a coincidence type level for fine leveling within one second of arc.

Optical levels are used for the precise leveling of large-size machines, foundations, engineering structures and also for dimensional measurements in relation to a single, or several parallel, horizontal reference planes. For the latter type of measurement, the optical level must be displaced vertically over controlled distances with the aid of a vertical metrological bar system equipped with an integral scale. Such bar systems have vertically adjustable, counterbalanced brackets serving as the support for the optical level.

The telescope used for the illustrated type of instrument has a focusing range from 4 inches to infinity, and permits the consecutive observation of each particular scale graduation, which is in alignment with the optical axis of the instrument, and also of the target on a distant object element.

For measuring the distance of noncoincidence between the optical reference plane and a slightly displaced target point, the telescope of optical levels can be equipped with an optical micrometer, similar to that used for the jig transit.

The Engineering Theodolite

In large scale metrology, specific tasks may require a more versatile method of measurement than provided by the alignment telescope whose operation is restricted to a fixed line of sight or, when mounted on a jig transit or optical level, to the scanning of a single plane. By using a single point of observation, complemented with the potential referencing from a precisely horizontal plane, a wide range of measurements can be carried out with an optical instrument that is rotatable both in the horizontal (azimuth) and in the vertical planes (from the zenith to the nadir), around two mutually perpendicular and intersecting axes.

Such an instrument is the engineering or surveyor theodolite (see Fig. 11-10). The theodolite combines two optical dividing instruments in a stand that holds, as an integral member, an alignment telescope for viewing specific targets. In Fig. 11-11, a cross-section of the optical system of the illustrated theodolite shows the correlation of the optical axis of the telescope with the rotational axes of the theodolite stand.

In view of the critical character of the applications for which engineering theodolites are intended, the operating elements of these instruments are usually designed to assure a very high degree of measuring accuracy. For angular position measurements around both rotational axes, very sensitive optical dividing devices are provided that, in several types of theodolites, possess a readout resolution to one second of arc least increment. In such high precision models, the effect of slight eccentricity errors in the mounting of the dividing circles is eliminated by the use of a dual viewing system, similar in prin-

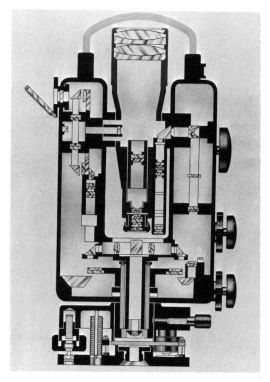

Keuffel & Esser Co.

Fig. 11-11. Cross-sectional diagram of the theodolite shown in Figure 11-10.

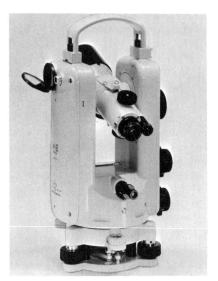

Keuffel & Esser Co.

Fig. 11-10. Engineering (terrestrial) theodolite with 30× magnification telescope and directional setting, with angular reading to one second of arc on both the vertical and the horizontal circles, viewed simultaneously through the microscope adjacent to the telescope.

ciple to that described in the preceding chapter in conjunction with the optical dividing heads.

The telescope of the instrument shown in Fig. 11-10 has a 30× magnification and 1.75 inches effective aperture. That telescope is easy to focus from about 6 feet (shortest distance) to infinity, possesses excellent resolving power and a field of view specified as 1½ degrees. Adjacent to the telescope is the eyepiece of the microscope through which the setting angles both in the vertical and in the horizontal planes can be read simultaneously.

Theodolites are commonly mounted on the leveling plate of a sturdy tripod so that the wide adjustment range of that versatile type of optical viewing and measuring instrument can be fully utilized. The independent erection assures measurements from an extraneous datum point, without relying on mounting and referencing surfaces available on the object being measured.

Auxiliary Equipment for Optical Tooling

To establish appropriate links between the optically produced and observed lines of sight and the pertinent surface elements of the object being measured, various pieces of auxiliary equipment are used in optical tooling. A few of these that are considered as instrument accessories, for example, the supporting members of alignment telescopes, have already been mentioned. Other equipment, which is used in conjunction with several of the described optical sighting instruments, will be discussed briefly in the following. While the range of auxiliary equipment for optical tooling is quite extensive, the more important functions are covered by targets, including collimators, by scales and by supporting devices such as stands and metrological bar systems.

Targets. The primary function of targets used in optical tooling is to provide an optically observable point at a specific distance from the object surface. Because of the distances over which telescopic observation must frequently be made, the aiming at the center of the target is aided by reference lines applied to the face of the target. These are arranged in a pattern that is helpful in establishing the optical coincidence of the target center with the cross hair of the telescope graticule (see Fig. 11-12).

When mounted in a holder, the aiming center of the target is at a precisely maintained specific distance from the base that will be supported by the selected portion of the object surface. Target holders are commonly of rigid design, although holders with ball sockets are also available to accept targets mounted into spherical adapters that permit the target to be rotated around its center (Fig. 11-13).

The targets are made of glass, plastic, aluminum or other appropriate material, depending upon

Rank-Taylor-Hobson/Engis Equipment Co.

Fig. 11-13. Circular pattern target enclosed in a spherical mount and equipped with illuminator.

whether transparent, translucent or reflective properties are required. For uses where the target must be observed at several positions along a line of sight, open targets can be used; these have nylon-monofilament cross-hairs mounted in steel rings fitting standard target bases. An illuminating unit can also be mounted behind the target to improve the visibility, particularly in poorly lighted areas or for observation over long distances (see Fig. 11-13).

For alignment measurements involving large diameter bores, special target holding and centering devices, known also as spiders, are used (see Fig. 11-14).

Collimators. Collimators are target instruments commonly used in conjunction with telescopes for alignment measurements. The principles of operation are shown diagrammatically in Fig. 11-15. Collimators provide reference elements for both lateral and angular coincidence with a fixed line of sight.

The collimator housing is a precisely ground steel tube, similar in appearance and of the same diameter as the tube of the telescope for which it provides the target (see Fig. 11-16). In its front part, the collimator carries a target with radial lines for defining the center. Behind the front target is a collimating lens with an illuminated target mounted in its focal plane. The image of that second target is projected through the lens as parallel beams. The centers of the two targets represent the optical axis of the collimator that, to a

Keuffel & Esser Co.

Fig. 11-12. Glass alignment targets.
(Left) Unmounted.
(Right) In a hardened and ground steel ring of standard mounting diameter.

Rank-Taylor-Hobson/Engis Equipment Co.

Fig. 11-14. Spider fixture for holding a target within a large bore, in a precisely centralized position.

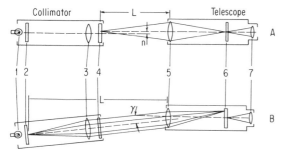

Fig. 11-15. Operating principles of a collimator viewed through an alignment telescope.
(A) Measurement of lateral displacement (locational shift).
(B) Measurement of angular displacement (directional nonalignment).
L = Target distance; n = lateral displacement; γ = angular displacement.
Elements: (1) light source of the colllimator; *(2)* directional graticule; *(3)* collimator objective; *(4)* target graticule; *(5)* telescope objective; *(6)* micrometer graticule; and *(7)* telescope eyepiece.

high degree of accuracy, is produced parallel with the outside surface of the housing.

In alignment testing the telescope is first focused on the front target to determine the lateral coincidence of the features being inspected. Then the focus of the telescope is set to infinity for observing, in the same direction of sighting, the image of the second target that is projected through the collimating lens by parallel rays of light.

Rank-Taylor-Hobson/Engis Equipment Co.

Fig. 11-16. Instruments of optical alignment measurement.
(Upper) Collimator with cylindrical barrel and light source for target illumination.
(Lower) Alignment telescope, also with cylindrical barrel.

Collimators are used for alignment testing over long distances, particularly inside bores. With the telescope they permit the precise measurement of both types of alignment errors, lateral and angular, as long as the extent of these errors does not exceed the observation range of the alignment telescope for which the collimator serves as target.

Scales for optical metrology. There are two major categories of application for scales designed to be observed through a telescope:

1. Target scales, used for the measurement of the distance between a line of sight and the object surface against which the scale is set; and

2. Travel distance measuring scales, for determining the separation between the consecutively observed points along an optical tooling bar.

Both scales are made with markings 0.050 or 0.100 inch apart, numbered in full and decimal inches, with graduations accurately located within ±0.001 inch. However, the application purposes call for different scale patterns.

The target scale is usually observed remotely from different distances, varying from a few feet to 100 feet or more, and for that reason it has four rows of bifilar lines with different separations ranging from 0.004 to 0.060 inch. For each operating distance that type of marking can be selected that has the least separation (for accuracy of aiming) and that is still distinctly observable through the telescope (for error-free discrimination). The pattern of this type of scale is shown in Fig. 11-17.

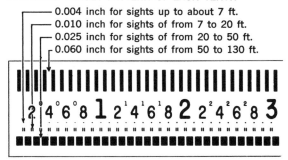

STANDARD TARGET PATTERN OF EACH TENTH

- 0.004 inch for sights up to about 7 ft.
- 0.010 inch for sights of from 7 to 20 ft.
- 0.025 inch for sights of from 20 to 50 ft.
- 0.060 inch for sights of from 50 to 130 ft.

Keuffel & Esser Co.

Fig. 11-17. (Top) Optical metrological scale held in a leveling base.
(Bottom) A close-up of the graduation pattern.

The travel distance measuring scale is generally attached to the metrological bar tracks and observed from a close distance. Therefore, this type of scale has a single row of bifilar graduation lines only. However, two sets of numbers are applied to facilitate the reading from either side of the bar system.

Scales for optical metrology applications are made from different materials. A widely used type is manufactured of hardened tool steel with matte white surface to assure maximum contrast for the block line marks and figures.

Metrological bar systems. To provide means for extraneous referencing by setting up the sighting instruments detached from the object, metrological bar systems are frequently used in optical metrology. These can be assembled from standard components to suit specific requirements.

Bar systems, which are usually set up around the object, provide a self-contained framework for holding and moving optical tooling instruments along fixed paths and over controlled distances either in a horizontal or vertical direction. Figure 11-18 is a diagrammatic presentation of typical measuring capabilities, represented by scanning planes, which can be accomplished by optical tooling operated in conjunction with metrological bar systems.

The instruments commonly mounted on metrological bar systems are the alignment telescopes, jig transits and optical levels, equipped with various ac-

cessories and used in combination with such auxiliary devices as plain or illuminated targets, reflectors, scales and so forth.

The principal components of metrological bar systems are vertically adjustable stands that support tracks or rails, providing guideways for carriages on which the optical instruments can be mounted. For measuring the distances of the translational movements, optical scales are installed into the tracks. These can be complemented with index bars when length measurements over distances exceeding the length of the graduated scales (generally 10 inches) are required.

The use of metrological bar systems permits remote measurement of distances between selected object elements by optical sighting, without mechanical contact. For these measurements very precisely controlled datum lines and planes can be maintained, without the need for instrument mounting surfaces on the object being measured.

While the elements of the bar system provide a reasonable degree of tracking accuracy, the measurements by optical tooling are controlled by inherent targeting, aligning, leveling and squaring capabilities of the sighting instruments. For that reason the final positioning adjustment of the instruments following their mechanical displacement is usually carried out by relying on optical means.

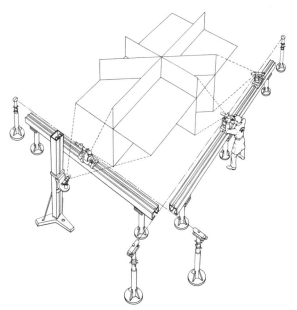

Keuffel & Esser Co.

Fig. 11-18. The application of a metrological bar system. Schematic illustration based on a typical setup.

For distance measurements to fine increments, a combination of the bar system and instrument capabilities is used. The optical scales of the bar system will indicate the distance traveled in steps of 0.050 or 0.100 inch. Once the instrument has been moved to the scale graduation nearest to the target, the remaining distance is bridged by the adjustment of the sighting instrument's eyepiece micrometer, which is commonly graduated in 0.001-inch increments. While, exceptionally, eyepiece micrometers with 0.0001-inch graduations can also be supplied, their practical usefulness is limited considering that the sighting discrimination of the telescope is generally on the order of 0.001 inch.

THE MEASUREMENT OF FLATNESS

Flatness represents the geometric concept of the plane, when applied to a solid surface where it becomes measurable by mechanical means, either directly or in combination with optical instruments.

Flatness is a functionally meaningful condition in many technical parts for reasons related to the intended service of that part. Controlled flatness may be required to provide full contact with a mating part, such as in seals; as a precondition for the parallelism of nominally flat surfaces; as a reliable boundary plane for linear dimensions; to provide locating planes for the dependable mounting or assembly of manufactured parts; and many others.

Considering the functional role of flatness, the measurement of that condition on the operating surfaces of manufactured parts is often an important operation of the dimensional inspection process. The choice of the best suited method of flatness measurement is governed by various factors, such as the size and the shape of the part; the area to be inspected; its accessibility and interrelations with other surfaces; and the desired degree of measuring accuracy.

Table 11-3 presents examples of different methods of flatness measurement on technical surfaces; these encompass a wide range of instrument categories and degrees of sensitivity. Although the listing comprises methods from the shop level to the most accurate laboratory procedures and instruments, its purpose is to indicate the variety of possible choices, rather than to attempt completeness.

In dimensional metrology, various types of auxiliary equipment are used to provide a dependable flat surface that can serve as a datum plane for the measurement of linear dimensions or geometric conditions. Two basic types of equipment that constitute mechanical representations of a plane surface

are of major importance for dimensional measurements, namely:

    a. Toolmaker and the optical flats; and

    b. Surface plates.

These are now described as well as some aspects of their use and inspection.

Toolmaker Flats

Toolmaker flats are steel disks, hardened, ground and lapped on both faces to a high degree of flatness, comparable to the measuring surface of a gage block (see Fig. 11-19). The primary purpose of toolmaker flats is to serve as a mounting surface for gage block assemblies and similar high precision elements of dimensional measurements. The flatness and the finish of an inspection quality toolmaker flat should permit the direct wringing of gage blocks to either surface, thereby obviating the need for mechanical clamping. Frequently used sizes of toolmaker flats are the 2-inch and 4-inch diameters, in thicknesses of $\frac{1}{2}$ inch and $\frac{3}{4}$ inch, respectively.

Optical Flats

Optical flats used for metrological applications are usually made of clear fused quartz, although commercial qualities made either of quartz or pyrex (borosilicate glass) are also available. One of the important uses for optical flats is the measurement of flatness by laying the transparent flat on the surface to be inspected in such a manner as to create interference bands observable under monochromatic light; the resulting band pattern permits the object's flatness conditions to be evaluated (see Fig. 11-20).

Dimensional metrology offers many other applications for optical flats when referencing from a highly reliable flat surface is required, particularly in length measurements. Another use for optical flats in

The Van Keuren Co.

Fig. 11-19. Toolmaker flat, essentially a steel disc, hardened, ground and precision lapped to a flatness within a few millionths of an inch.

Fig. 11-20. Optical flats made of clear, fused quartz, stacked in pairs and observed under monochromatic light. The straightness of the visible light-interference bands—also known as fringes—indicates the excellent flatness of the mutually contacting surfaces.

TABLE 11-3.   EXAMPLES OF METHODS FOR THE MEASUREMENT OF FLATNESS — 1

PRINCIPLES OF THE PROCESS	ESSENTIAL INSTRUMENTS	DIAGRAM	DISCUSSION
Direct contact with a reference plate of known flatness	Cast iron surface plate		A shop method long used by the toolmaker; still being applied for producing hand-scraped flat surfaces and also in the manufacture of surface plates. In this method, three plates are alternately brought into contact, using blue dye to indicate the location of high spots.
Translating the plane of a surface plate of known flatness by means of an indicator	An indicator stand and a surface plate		Measures the combined effect of errors in flatness and parallelism. For flatness measurements the part must be set up with top surface essentially parallel to the reference plane.
Measuring the digressions of individual points on the surface from a tangent plane	Surface plate with air gage probe in recessed mount		For a part resting on the surface plate the latter will constitute a tangent plane, in relation to which the location of individual areas of the part's surface are measured while passing over the extending contact point of a recessed air gage probe.
Correlating the surface being inspected to the horizontal	Precision block level or electronic level		The general flatness condition of a large surface which is in an exactly, or close to horizontal position, may be checked with a sensitive level instrument, placed on the surface in different positions and orientations. Inconsistency of the slope angle is indicative of poor flatness.

TABLE 11-3.   EXAMPLES OF METHODS FOR THE MEASUREMENT OF FLATNESS – 2

PRINCIPLES OF THE PROCESS	ESSENTIAL INSTRUMENTS	DIAGRAM	DISCUSSION
Measuring optically the variations in the angular position of a mirror base displaced on the surface	Autocollimator and reflector with base having appropriate pad separation		While a perfectly flat surface will support a member in unchanged angular position at any location, as long as the orientation remains the same, digressions from the flatness can be measured by an autocollimator viewing a reflector which is placed consecutively in different positions along several tracks on the surface.
Producing light interference fringes originating from the surface being inspected	Optical flat with monochromatic light or plano-interferometer		An air wedge of very small angle between an optical glass flat and a reflective object surface produces a fringe pattern with a spacing equal to one-half wavelength of the monochromatic light used. Deviations from perfect flatness of the object surface cause digressions in the parallelism of the fringes, proportional to the flatness error.

metrology is to guide a moving instrument member in its translational displacement, which must be strictly confined to a specific plane; such conditions are required for the profile measurements discussed in Chapter 13.

Optical flats are made in a wide range of sizes, commonly from 1 inch to 12 inches in diameter, or larger, and in thicknesses adapted to the diameters, varying from $\frac{1}{4}$ inch to 2 inches. Several grades or qualities are produced, having different degrees of flatness, expressed in microinch maximum deviations from the perfect plane. The industrywide adopted grades are the reference grade with 1-microinch, the master grade with 2-microinch, and the working grade with 4-microinch flatness. There are also commercial grade optical flats whose accuracy is commonly one half of that of the working grade, that is, 8-microinch flatness.

Depending on the intended use, optical flats are made available with single or double measuring sur-

faces. Double-surface optical flats are often required to possess, in addition to the high degree of flatness on both faces, a comparable parallelism accuracy as well, to provide plane-parallel measuring elements.

Federal Specification GG-O-635 (issued May 1960, amended May 1961) covers the optical flats used in laboratories and inspection departments.

Surface Plates

The concept of flatness, in the minds of those engaged in metalworking production, is readily associated with surface plates. While surface plates are only infrequently used for measuring flatness directly, they represent the most generally applied means for establishing a solid datum plane for the purpose of a very wide range of dimensional measurements.

Table 11-4 surveys, in a rather condensed manner, a few of the more common applications of surface plates in dimensional measurements. Several of

**TABLE 11-4.   EXAMPLES OF SURFACE PLATE APPLICATIONS FOR
DIMENSIONAL MEASUREMENTS — 1**

DESIGNATION OF APPLICATION	CONJOINTLY USED EQUIPMENT	DIAGRAM	DISCUSSION
Referencing plane for length measurements	Indicator stand and gage blocks		Comparative length measurements can be carried out without a comparator gage, by using the surface plate to locate all three elements of the process: the part, a gage block stack or master and the indicator stand. Frequently applied method for parts whose size or form makes staging on a comparator gage table difficult.
Referencing for parallelism measurements	Indicator stand and supporting blocks		Supporting a part on one of its sides on a surface plate—directly or with intermediate parallel blocks—offers the part's opposite side for measuring its distance from the surface plate. Uniform distance indicates parallelism of the opposite part side.
Establishing a datum plane for dimensional layout	Height gage with scriber inserted into the adjustable head		Layout for guiding the subsequent machining operations is usually dimensioned from datum lines which can often be correlated with the plane of the surface plate, while setting up the part. Auxiliary members, such as parallel blocks, angle irons and tiltable plates extend the adaptability of the process.
Referencing for the measurement of linear distances	Height gage, either a directly measuring type, or one that is used for transfer from an adjustable step gage		The separation of features or surface elements of a part is frequently measured by resting the part correctly located on a surface plate, the supported side being considered the datum plane and the pertinent distances are measured with the height gage. Repeat after turning the part by 90 degrees to complete the measurement of the coordinate positions.

**TABLE 11-4.   EXAMPLES OF SURFACE PLATE APPLICATIONS FOR
DIMENSIONAL MEASUREMENTS – 2**

DESIGNATION OF APPLICATION	CONJOINTLY USED EQUIPMENT	DIAGRAM	DISCUSSION
Referencing for squareness measurements	Squareness gage with indicator	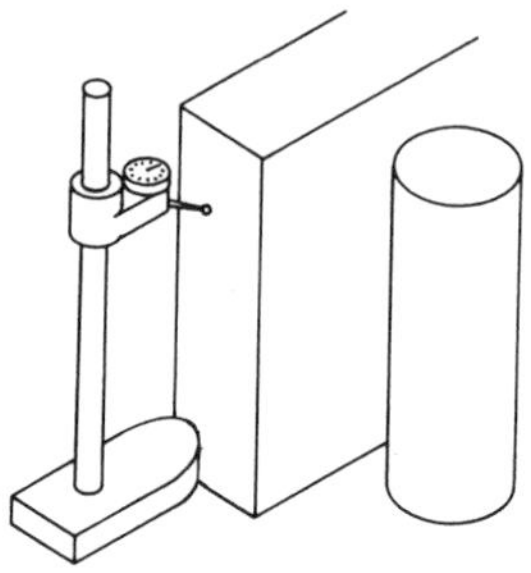	A special gage stand, whose base front functions as a referencing contact in line with the indicator point, is adjusted to zero indication with a setting master which represents a plane normal to the surface plate. The gage will indicate the out-of-squareness of two simultaneously contacted points which were selected to represent the side of the part.
Referencing for angle measurements	Sine plate with gage blocks and indicator height gage stand	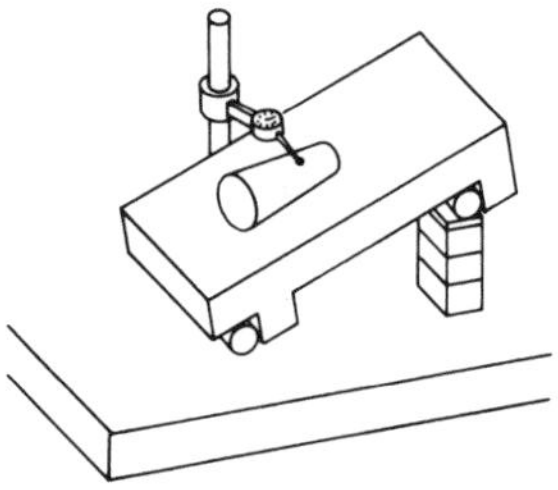	Surface plates are commonly used as datum planes for measuring angles with the aid of sine plates, where a tapered part with angle nominally equal to the sine plate setting should offer a top element parallel to the surface plate. Deviations are measured by an indicator whose stand is supported on the surface plate.
Datum plane for autocollimator measurements	Rotary table and autocollimator and reflector where needed	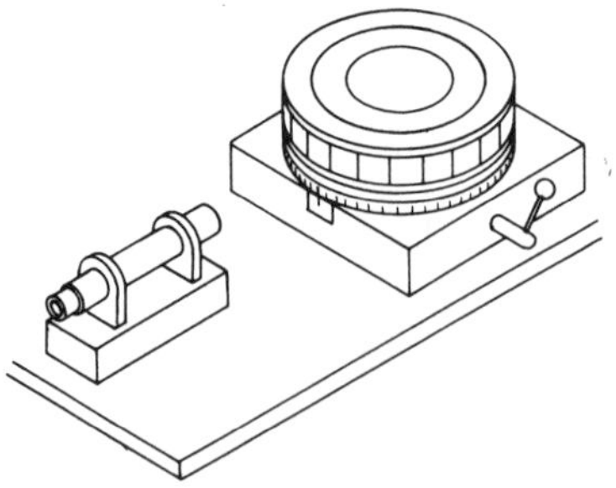	When measuring angular spacings correlated to circular divisions produced by a rotary table and observed with an autocollimator, the two detached gaging elements must be functionally connected by a common reference plane, such as a surface plate. The diagram shows the checking of a rotary table with the aid of an optical polygon.
Structural base for instruments	Functionally interrelated instrument elements	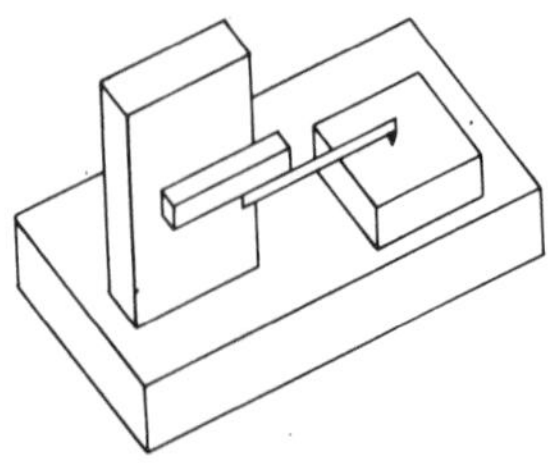	An application where the surface plate is not directly participating in the measuring process. Surface plates can replace machine bases supporting solidly attached instrument elements in a manner assuring excellent stability of critical positional inter-relations.

these measuring methods are discussed in other chapters of this book, in conjunction with the type of instrument used for implementing the process. However, all these methods are characterized by having, for a connecting link between the object and the measuring instrument, a common reference plane provided by the surface plate.

Surface plates were originally manufactured of cast iron with cross-ribbed bases to reduce weight and to increase the resistance against warpage. Although cast iron surface plates, particularly in the smaller sizes, are still in use and are being manufactured, stone surface plates made of different types of hard and homogeneous granite have become standard equipment for conducting dimensional measurements based on a solid reference plane.

Granite surface plates possess several valuable properties that justify their preferential use in many applications. Such properties are: great hardness, superior wear resistance, dimensional stability, no rusting or deterioration under common environmental conditions and freedom from burrs when indented.

Federal Specification GGG-P-463b on Granite Surface Plates. Although this specification was intended primarily to assure compliance with specific requirements of the surface plates supplied to federal agencies, several details of this publication are of interest to the metrologist because they discuss functionally meaningful quality conditions, particularly accuracy requirements. Until the time when the U.S. standards, now in development, will be available, the Federal Specification provides the only generally accepted guidelines for evaluating surface plates used for dimensional measurements.

The following highlights from this publication have been selected with particular regard to the functional adequacy of surface plates.

1. *Material.* Granite from different sources is considered and two classes are established based on the rock type, physical properties (for example, compressive strength) and mineral components.

2. *Hardness.* For the measurement of hardness a Scleroscope test is prescribed and different hardness numbers are specified for grades A and B. For grade A, which is intended for high quality inspection work, an average hardness of 90 is required. The minimum hardness varies according to the class, that being related to the homogeneity of the material; Class 1 should have a minimum hardness of 90, which is equal to the average of Grade A; for Class 2 a minimum hardness as low as 64 is permitted.

(*Note:* In the Specification, the term *class* refers to material properties, such as wear resistance; the term *grade* designates the level of accuracy.)

3. *Wear resistance.* This is still another significant property of the granite material. The recommended test consists in tracing a not less than 1-inch long straight line path on the plate surface with a Rockwell C penetrator diamond under 3 pounds load by causing it to perform reciprocating movements over a three-hour period to complete 1238 cycle strokes. The resulting wear groove is then measured; for Class 1 its depth should not exceed 0.0071 inch at any spot and 0.0035 inch on the average.

4. *Sizes.* These are not defined in the Federal Specification as standard requirements. The listing of the work surface dimensions in the Specification, which is reproduced in Table 11-5, serves to establish categories for minimum requirements of work surface accuracy. The minimum thickness, a dimension affecting resistance to sagging of the work surface, is defined differently for the various grades.

5. *Clamping edges.* These are required by the Specification for all four sides of the surface plates, unless otherwise specified in the contract. The proportions of the edge dimensions are also spelled out as a means of assuring adequate weight supporting properties around the borders of the surface plates.

Since the publication of the Specification, a growing trend can be observed toward reducing the use of ledges for holding fixtures and instruments by means of clamps. The flatness of the surface plates near the perimeter is generally not equal to the average flatness of the rest of the plate surface. That flatness defect, "slough-off" in shop language, is recognized in the standard practices for surface plate flatness inspection. When using clamps bearing against the ledges of the surface plate, the clamping force is exerted in that peripheral area of questionable flatness accuracy. For that reason preference is frequently given to threaded, stainless steel sockets, cemented in recessed positions into blind holes drilled from the plate surface and to the use of special hold-down clamps secured by bolts fitting the socket thread (see Fig. 11-21).

6. *Accuracy of the work surface.* The following is the wording of the Specification: "No point on the work surface shall vary from a mean plane thereof by more than the amount specified" (see Table 11-5).

This requirement, although logical in its concept, raises the question of whether digressions from the

TABLE 11-5.   CURRENTLY USED SIZES OF GRANITE SURFACE PLATES WITH VALUES OF FLATNESS ACCURACY REQUIRED BY FEDERAL SPECIFICATION GGG-P-463b

WORK SURFACE DIMENSIONS in Inches		WORK SURFACE TOLERANCES in Microinches			
Width	Length	Grade AAA	Grade AA	Grade A	Grade B
12	12	25	50	100	200
12	18	25	50	100	200
18	18	25	50	100	200
18	24	50	100	200	400
24	36	50	100	200	400
24	48	75	150	300	600
36	36	75	150	300	600
36	48	100	200	400	800
36	60	125	250	500	1000
36	72	150	300	600	1200
48	48	100	200	400	800
48	60	150	300	600	1200
48	72	175	350	700	1400
48	96	250	500	1000	2000
48	120	350	700	1400	2800
60	120	375	750	1500	3000
72	96	300	600	1200	2400
72	144	550	1100	2200	4400

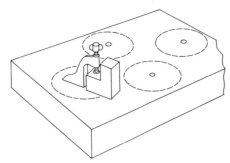

Fig. 11-21. Hold-down clamp, bolted to a threaded insert in the surface plate.

theoretical mean are to be considered equally significant, when appearing as protruding ridges or as depressed valleys. Obviously, the former will affect the serviceability of the plate, while narrow valleys, which are easily bridged by the contacting elements of fixtures and instruments, may have no functional effect on the general usefulness of the granite surface plate. For that reason, in general industrial practice that wording is not followed literally, and it is interpreted to require that at no point should the *effective supporting level* of the plate surface deviate from the prescribed limits. The currently used methods of surface plate inspection also reflect this interpretation of the flatness requirements.

The Inspection of Surface Plates for Flatness

Flatness is the primary functional characteristic of surface plates and will directly affect the accuracy of measurements based on that condition. For that reason it is necessary to determine the degree of flatness of surface plates, both in the original inspection process and at periodic recalibrations.

The high accuracy requirements of surface plates that are used currently in dimensional measuring processes exclude the application of the earlier toolmaker methods of flatness inspection, relying on straight edges or on the mutual checking of several, usually three, surface plates. Two systems of surface plate flatness inspection will be mentioned in the following, with particular emphasis on the most widely applied autocollimator method.

Indicator type surface plate testers. These instruments operate, in principle, by selecting a portion of the plate surface as the reference plane and using it to support one member of the instrument. Another member of the same instrument either carries or serves as the anvil of a sensitive mechanical indicator with 50- or 20-microinch least graduations. The probe of that indicator is brought into contact, either directly or through an intermediate element, with the surface portion adjoining the one selected for a reference plane. Deviations between that latter and the scanned area will cause indications by the instrument that may be used as a quick check of the uniform flatness of the plate surface.

The supported member may be a granite straight edge, offering a gap above the selected element of the plate surface for measurement by the indicator; a steel block resting on carbide pads and carrying an extending arm on which an indicator is mounted; or a hinged type of surface tester.

An example of a hinged surface plate tester is shown in Fig. 11-22. This instrument consists of a granite block that is slotted to a depth permitting small relative deflections between the two sections. One section is heavier and functions as the referencing member, with a rigid extension holding a sensitive indicator. This, in turn, is in sensing contact with the top of the hinged section of the block. That hinged member follows the rise and fall of the surface plate area on which it rests. When the slopes of the areas supporting the two members differ in direction or extent, the hinged member deflects,

Rahn Granite Surface Plate Co.

Fig. 11-22. Indicator type, surface plate flatness tester.

causing the indicator's pointer to signal a corresponding linear value.

These instruments are not recommended for the accurate measurement of flatness conditions over the entire plate surface. However, they can be quite useful for rapid checks whose results could indicate the need for applying a more comprehensive calibration process, such as those discussed in the next section.

Inspection of surface plates for flatness by autocollimation. The flatness of a surface plate is a condition that should be inspected in accordance with the purpose that the plate is intended to serve, namely: as a supporting member in a measuring process that provides a common and near-to-perfect plane, Consequently, the effective surface, and not the actual surface of the plate in all its minute details, is the meaningful characteristic. The effective surface can be visualized as a tangent plane to the physical boundary of the plate surface.

The bases of measuring equipment, as well as of workpieces and holding fixtures, are made to extend over areas large enough for supporting the weight of the member without undue pressure resulting from excessive load concentration. Such a base surface, even when reduced to three or four pads, will not penetrate into the narrow valleys or minute craters of the plate surface, but ride on its most protruding elements, that is, it will take the position of a tangent plane.

These functional considerations are taken into account in the most widely used and generally accepted method of surface plate inspection: flatness measurement by reflectors and an autocollimator. A detailed description of these tools is given in the subsequent section; the following is a condensed explanation of the process principles and techniques.

In this method the angular deflections of adjoining sections are measured along specific tracks on the plate surface, and the results are evaluated in relation to a selected reference plane. Figure 11-23 shows the eight lines that are commonly used as the tracks of the angular deflection measurements. These are the four perimeters, two diagonals and two center lines. The instrument employed for the measurement of angular deflections is the autocollimator, which is used to observe a reflector placed in consecutive positions along each of the measuring tracks. As indicated in Fig. 11-23, the autocollimator is not aiming at the displaceable reflector directly, but through an intermediate mirror placed close to the corner of the plate, hence the frequently used designation of corner mirror.

This arrangement of two reflectors has the advan-

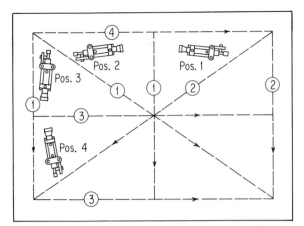

Fig. 11-23. The pattern of tracks explored in testing the flatness of surface plates. Four positions of the autocollimator instrument, directed toward specific corners on the plate surface, are shown. A deflecting mirror placed consecutively in those corners permits the observation of the reflector as it is moved along two tracks originating from each of the corners, as indicated by directional arrows.

tage that the autocollimator may be placed in that particular location on the plate surface where the instrument can be most conveniently operated, but without interfering with the exploration of the entire track length by the sequential repositioning of the displaceable reflector. Figure 11-24 shows an autocollimator setup with a corner mirror and a rectangular base reflector, arranged for making deflection measurements along a perimeter of the surface plate.

The rectangular base of the displaceable reflector is supported on three pads, two of which are mounted parallel to the guided edge, in line with the reflector's direction of displacement. The third pad, halfway between the two former ones and at the opposite edge of the base, will support the reflector at right angles to the direction of its translation, thereby constraining roll, that is, deflections in a plane parallel with the surface of the mirror. This support ac-

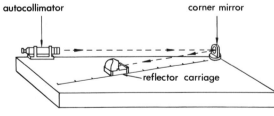

Hilger & Watts/Engis Equipment Co.

Fig. 11-24. The arrangement of the autocollimator and the corner mirror with respect to the reflector, which is being displaced along a measuring track on the surface of the plate. The dotted line indicates the line of sight.

tion does not interfere with the free positioning of the two pads along the line of displacement, where changes in the relative levels of the pads will cause mirror deflections (pitch movement) proportional to the departures from flatness of the inspected surface element.

The center distance between the two aligned pads is usually an integer inch value, for example, 4, 5 or 6 inches, selected to permit the subdivision of the measuring track length into a specific number of equal sections. For guiding the reflector during the measuring process a straight edge is clamped to the plate surface along the track being inspected and the reflector is placed step by step into adjoining locations, the trailing pad taking the position of the leading pad in the preceding setting.

Various methods, closely related in principles, were developed and are being used for recording, computing and evaluating the series of angular readings of the autocollimator. The readings obtained along the individual tracks are correlated to a reference plane in the computing phase of the evaluation. That reference plane comprises three significant points on the plate surface, such as three corners of the plate, or the center and two corners at the ends of one of the longer perimeters. Finally, the correlated angular values are converted, by the use of a sine function taking into account the separation of the pad centers, into linear units representing elevation differences. In the plot, displayed in a "Union Jack" pattern, the elevation differences with respect to the reference plane are usually expressed in units of 1/100,000th part of an inch, without decimal points. Such a chart can then be developed into a contour map and also permits visualizing the topography of the plate surface, in a manner illustrated diagrammatically in Fig. 11-25.

The described type of map is also useful for determining that portion of the plate surface that has the best flatness conditions for use in measurements requiring the highest degree of equipment accuracy. The technique for implementing the discussed method of surface plate flatness measurement, including the required calculations and the recommended designs of tabulation forms, are described in detail in technical publications of several manufacturers of autocollimators.

Mechanized surface plate flatness inspection by autocollimation. This method, a recent development, is based on the application of automatic autocollimators, supplemented with a chart recorder. The scanning of the plate surface is carried out along the same eight significant tracks used in the manual

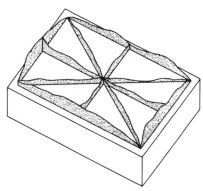

Fig. 11-25. A relief map, limited to eight cross-sections, sketched to visualize the flatness conditions of a surface plate using the height variation values obtained by converting angular deviations into linear units. The variations of the surface from a reference plane are highly exaggerated to illustrate the significant dimensions.

method. However, instead of transferring the reflector by hand to consecutive measuring positions in the selected path, the reflector is drawn along that track by a cable and winch device, actuated by a synchronous motor.

Deflections of the reflector while traversing the measuring path will cause signals to be generated by an automatic autocollimator that is equipped with a servo-driven nulling system. These signals are transmitted to a recorder, whose paper speed is in a constant ratio to the translational movement of the reflector. The resulting tracing line will be representative of the slope conditions along the scanned track, the linear scale being calibrated to indicate angular deflections in seconds of arc.

This method requires less time for the actual measurements than the manual transfer of the reflector and offers the advantage of complete inspection by means of continuous scanning, in contrast to the step-by-step inspection obtained by the conventional system. The cost of the equipment, however, limits the application to plants where a considerable number of surface plates must be inspected. The evaluation of the continuous chart tracing differs from that based on the tabulation of discrete measurements.

THE MEASUREMENT OF PERPENDICULARITY

The right angle is considered to be the creation of the human mind, because in nature its presence is not deliberate. In the products of human endeavor, however, whether in architecture, in trades or in-dustry, in civil engineering and even in agricultural activities, the right angle, resulting from the mutual perpendicularity of lines and surfaces, has been present since ancient times. The reasons for that varied and extensive application of the right angle are many. They may be related to technical convenience and practicality, but also to less distinct concepts; they could reflect the human mind's search for the expression of orderliness or mathematically definable regularity.

While the originating circumstances could be a matter of conjecture, the presence of the right angle and its basic role in engineering design are very real. Consequently, perpendicularity represents a condition that must be measured.

There are quite a number of terms in current usage to designate the positional relationship of two or more physically defined elements at right angles to each other. These terms are *perpendicularity, normalcy, squareness* or, expressed in angular units, *90 degrees*. Concepts related to natural forces are also used to specify the mutual position of perpendicularity, namely, *plumb* in relation to the horizontal. Actually, gravitational force is used to create in measuring instruments a very reliable representation of the horizontal; that is, the surface planes of liquids under the effect of gravity. (*Note:* Plumb and level can be considered as perpendicular for most practical purposes, disregarding the spherical shape of the globe.)

Even in engineering products that do not have mutually perpendicular surfaces, the right angle is present in the dimensioning of the part; in the majority of engineering drawings the linear dimensions are interrelated in a manner comparable to a system of rectangular coordinates.

While it is simple to conceive dimensioning in a perfect system of rectangular coordinates, its implementation in manufacturing and measurements is never perfect; it can only be accomplished with a lesser or higher degree of perpendicularity. The measurements needed for assuring perpendicularity to a degree of accuracy that is a commensurate with the requirements of the product or of the process constitutes the subject of this section.

To indicate the variety of methods that are used in engineering production for inspecting the perpendicularity of interrelated surfaces on manufactured parts or equipment, a few processes were selected as examples; these are presented in Table 11-6.

Perpendicularity, as a positively defined geometric relationship, also offers an important link in measuring processes applied to the inspection of other

TABLE 11-6.   EXAMPLES OF METHODS FOR THE MEASUREMENT OF PERPENDICULARITY — 1

PRINCIPLES OF THE PROCESS	ESSENTIAL INSTRUMENTS	DIAGRAM	DISCUSSION
Direct contact with reference part of known squareness	Toolmaker square		Commonly used shop method for squareness checking with the aid of a toolmaker square, both the part and the tool resting on a surface plate. Gap between the edge of the square and sections along the contacted object surface is either visually observed or measured with a feeler gage.
Checking digression from a master cylinder by direct comparison	Cylindrical square, with one face off angle and with graduations on the surface		A gage cylinder with deliberately produced out-of-squareness brought into contact with the surface to be measured for perpendicularity. Curved graduation lines on the cylinder surface indicate the tilt angle of the part surface, by rotating the cylinder to an orientation producing the least light gap.
Indicating the deviations from a straight line path which is normal to the reference plane	Height gage stand with guideways for traveling indicator head		A special height gage stand with guideways to provide straight and vertical guidance to a balanced head which travels with very little friction and carries an indicator in contact with the object surface. A surface plate should support both the workpiece and the instrument.
Checking perpendicularity by measuring runout	Indicators and supporting members for the part		The lack of perpendicularity of two surfaces which are related to a common axis can be measured by rotating the part around that axis and indicating the runout of selected circular paths. The distance from the axis correlated with the runout supplies data for the angular value of the squareness error.

TABLE 11-6.   EXAMPLES OF METHODS FOR THE MEASUREMENT OF PERPENDICULARITY — 2

PRINCIPLES OF THE PROCESS	ESSENTIAL INSTRUMENTS	DIAGRAM	DISCUSSION
Squareness checked by comparison to cross-slide movements	Measuring machine or jig borer with excellent perpendicularity of the cross slide tracks		The mutual squareness of two or more surface elements may be measured by comparison to the tracks of precision cross slides, using a solid member such as the machine spindle for holding the sensing element. This latter is either an indicator or a spindle microscope with cross hair graticule.
Checking perpendicularity with collimated light beams which are turned at right angles	Autocollimator, optical square (pentaprism) and reflector		Aligning an autocollimator with the datum plane and turning the emitted light beams at a right angle with the aid of an optical square, permits the measurement of perpendicularity in the position of an interrelated plane, by observing the reflection from a mirror moved along that second plane.

basic geometric conditions, such as straightness, alignment, flatness and parallelism. With the aid of the appropriate mechanical members, the direction of a specific section of a basically straight surface element, or an area of an essentially flat surface, can be translated into a plane at right angles to the original surface. That secondary plane may be presented in a position that permits the very accurate measurement of its orientation in relation to a reference direction. This selective translation of specific surface elements is accomplished by means of special reflectors mounted on supporting bases. Such reflectors may be visualized to function as a bell crank, one arm of which is a mechanical and the other an optical element, offering the amplification potentials of a lever with very great ratio of arm lengths. This utilization of perpendicularity for the measurement of related geometric conditions is discussed more thoroughly in the following section on the systems and applications of autocollimators.

Mechanical Means of Squareness Inspection

Steel squares. An obvious method for measuring the perpendicularity of interrelated, generally adjoining surfaces is by direct comparison with tools of known squareness. Such tools are the different types of steel squares (see Fig. 11-26) that are manufactured from alloy steel, hardened, then stress relieved for stability. Steel squares are made either of one piece, or with a thin blade solidly mounted into

Mahr Gage Co.

Fig. 11-26. Precision steel squares with mounted thin blades; also in single-piece design with knife edge blade.

a wider beam. The single piece squares usually have beveled edges (knife edge), while the edges of the mounted blades are flat.

Depending on the intended use, the steel squares are selected on the basis of different accuracy grades. The accuracy is measured by the amount of maximum deviation of the blade edge from the perpendicular to the plane of the supporting beam, and it is expressed as a linear dimension in the function of the blade length.

The following are examples of grades (not standardized in this country) expressing the permissible maximum deviation from squareness by a formula:

Grade	Formula for Maximum Squareness Error	Expressed as Angular Error for 8-inch Blade Length
I	$\pm\left(0.00008 + \dfrac{L}{100,000}\right)$	0° 0′ 33″
II	$\pm\left(0.0002 + \dfrac{L}{50,000}\right)$	0° 1′ 14″
III	$\pm\left(0.0004 + \dfrac{L}{20,000}\right)$	0° 2′ 45″
IV	$\pm\left(0.0008 + \dfrac{L}{10,000}\right)$	0° 5′ 30″

L = length along the blade from the reference plane. All linear values in inches.

Steel squares are seldomly used for direct measurements in the sense of bringing both arms into contact with two nominally perpendicular surfaces of a part. A more common method of use consists in resting both the beam of the steel square and one of the interrelated surfaces of the part, on a surface plate or toolmaker flat. In that position of the part and of the tool, an element of the vertical part surface to be inspected is brought into contact with the blade of the square. In the case of incomplete contact due to squareness errors in the part, either the resulting light gap can be observed visually and its width estimated, or a feeler gage may be used for more positive measurement. A skillful toolmaker or inspector may resort to other techniques of comparison, perhaps better adapted to the particular configuration of the part.

It must be realized, however, that the attainable accuracy by any of these processes is limited, in addition to the errors in the steel squares, by several other factors, particularly the uncertainties in the measurement of the contact gaps and the great reliance on the skill of the operator.

Granite right-angle blocks are referencing equipment for squareness measurements with the aid of an indicating instrument. In such measurements, a surface plate is used to locate all three participating elements, namely, the part, the indicator stand and the right-angle block. Figure 11-27 illustrates a granite right-angle block of stepped design, which is preferred because it offers additional reference surfaces and also reduces the weight of the block without limiting its effective height. The illustration shows a block with air connection to introduce pressurized air from the shop line, through a hole in the block leading to radial grooves on the bottom surface. An air cushion can be thus created that assures the sensitive displacement on the surface plate of even a heavy block, without friction or wear that could affect the original accuracy of the equipment. Such blocks are made in heights of from 6 to 18 inches with a squareness accuracy of ±2 seconds of arc.

Cylindrical squares. A rather widely used method of perpendicularity measurement on surface plates consists in determining whether two selected points on the vertical surface of the part are contained in a common plane at right angles to the plate surface. The two points on the part are contacted simultaneously by two surface elements of a special gage stand, namely, the spherical front of the base and the probe of an indicator held in a vertically adjustable bracket, as shown in Fig. 11-28. The gaging process involves the comparison with a reference plane of the mutual positions of the contacted points on the part surface, that reference plane being established in the setting phase of the gaging with the aid of a cylindrical square—also illustrated in Fig. 11-28.

The Herman Stone Co.

Fig. 11-27. Right-angle block of granite in stepped design, with connection for pressurized air to produce an air cushion while moving the square on a surface plate.

The Taft-Peirce Mfg. Co.
Brown & Sharpe Mfg. Co.

Fig. 11-28. (Left) Cylindrical steel square used as the setting master for the squareness testing indicator stand.

Fig. 11-29. (Right) Direct reading cylindrical square for determining perpendicularity errors in 0.0002-inch increments over a length of about 6 inches.

Cylindrical squares for indicator setting are made of alloy steel, in different diameters and heights, 5 inches by 12 inches being a preferred size for precise measurements. The squareness accuracy is on the order of 2 seconds of arc. A recessed design of the cylinder base has the purpose of providing self-cleaning properties while moved along the top of a surface plate.

For the direct measurement of perpendicularity on a surface plate with the aid of a cylindrical square, a special type of tool is available. This is a direct reading cylindrical square (see Fig. 11-29) that is ground square on one end; the other end is ground to a fixed angle in relation to the sides. When the cylinder is standing on its slanting face it may be rotated to an orientation where an element of its side is in closest contact with the part surface that is not exactly perpendicular to the surface plate. The side of the cylinder has dotted curves that are numbered in increments of 0.0002 inch, indicating the amount of out-of-squareness of the bounded area. The number of the topmost dotted curve in contact with the part surface indicates the amount of squareness error.

THE SYSTEMS AND APPLICATIONS OF AUTOCOLLIMATORS

Autocollimators are sensitive and inherently very accurate optical instruments for the measurement of small angular deviations of a light-reflecting flat surface. The autocollimator resembles the telescope, although its operation is not restricted to the observation of an extraneous target point. The autocollimator has its own target that is projected by collimated light beams on a remotely placed surface and the reflected target image is observed in the ocular of the instrument.

Figure 11-30 illustrates the optical principle of autocollimation. The illuminated hairline target is projected through a collimating objective in the form of a parallel rays onto a flat surface that acts like a mirror. The reflected image, passing through the objective of the autocollimator, appears on a graticule and is observed through the magnifying ocular of the instrument. The position of the reflected image, in relation to the graticule's reference mark, indicates whether the reflector surface is exactly normal to the projected light beams, or tilted relative to that reference direction.

When the reflector surface is tilted, the direction of the returning light beams are deflected from the projected ones by twice the angle of tilt α; the resulting deflection is the sum of the incident and of the reflected angles. This angular deflection causes the reflected image, which appears in the focal plane of the objective lens, to be shifted by the linear distance y, which is a function of the reflector's tilt angle ($y = 2f \tan \alpha$, where f is the focal length of the objective, in inches).

The basic operating elements of the autocollimator are the following (see also Fig. 11-32).

1. A point source of light in the focus of a lens illuminating a target cross hair.

2. A collimating objective lens having the target wires in its focus and emitting, by parallel beams,

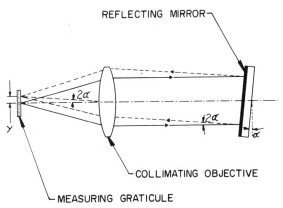

Fig. 11-30. The optical principles of autocollimation: collimating objective; reflecting mirror; and measuring graticule.

the image of the illuminated target onto an extraneous reflector (object surface).

3. A reference line graticule, in the focal plane of the objective, on which the reflected image of the target cross hair appears in a location governed by the angular position of the reflector with respect to the direction of the collimated light beams.

4. A micrometer microscope eyepiece for the magnified observation of the graticule and for measuring the distance of the image shift caused by the tilt of the reflector from the reference position.

These operating elements of the autocollimator can be arranged in different ways, resulting in several distinctive instrument designs. Actually, it is possible to attach a few additional optical elements to an alignment telescope and with these accessories convert the telescope so it will function as an autocollimator.

In their outer appearance, several models of autocollimators resemble the alignment telescope in that most of the elements of their optical system are arranged in line and are mounted in a cylindrical housing, the barrel. This system is called a *barrel type autocollimator*.

Other types of autocollimators are of designs that assure greater compactness by creating an internally reflecting, or folded, light path. In subsequent discussions, these are referred to as *folded beam type autocollimators*.

For measuring the distance of the image displacement on the graticule, either of two essentially different design systems is used: (a) an eyepiece micrometer serves to adjust the position either of the image, or of the graticule to establish coincidence of the cross-hair image with the reference mark; or (b) the graticule has, instead of a reference mark, a complete scale by means of which the position of the target image can be measured. Both systems have graduations in angular units to indicate, directly, the amount of tilt that produced the observed linear shift of the target image. Additional details of these measuring systems are discussed later in conjunction with the instrument types for which they are actually used.

The autocollimator detects any angular deflection from the observed plane's basic position, which is perpendicular to the direction of the emitted light beam. The angular deflections of the reflecting plane can occur around a horizontal pivot axis (tilt), a vertical pivot axis (rotation), or a combination of both (compound angular deflection). Any of these deflections will cause a shift of the target image on the graticule, and these displacements are usually expressed with respect to graticule axes, namely, tilt to the vertical and rotation to the horizontal axis. (*Note:* These directional designations are based on the generally used autocollimator setup where the collimated beams are traveling in the horizontal, while the basic position of the reflecting plane is in the vertical.)

In many measuring applications, the angular deflections of the object occur in one direction only, or the concurrent deflections in other directions are not considered to be significant. In such cases the displacement of the target image is evaluated along a single measuring axis only. The operation of many autocollimators is limited to this type of measurement, and such instruments are known as single-axis autocollimators. To permit the measurement of either tilt or rotational displacement with a single-axis autocollimator, the entire barrel of the cylindrical type autocollimator can be rotated in its holder around the instrument's optical axis, by 90 degrees, usually established by mechanical stops. Certain models of autocollimators are equipped with rotatable eyepiece assemblies, which provide a more convenient way of changing the measuring plane.

For applications where both the tilt and the rotation must be measured concurrently, two-axis autocollimators are used. These instruments can measure both the vertical and the horizontal shift of the target image in the same setting. For this purpose double axis autocollimators are equipped either with two independent micrometers or, when graticule scales are used, with a single micrometer screw for adjusting the graticule in a symmetrically diagonal direction.

Due to the collimating action of the objective lens the emitted light rays are parallel, and the reflected light also consists of parallel rays. It follows that all elements of the reflected image are carried by reflected rays that strike the objective lens at the same angle; consequently, the distance of the reflecting surface from the autocollimator has no effect on the measured value of the angular deflection. For this reason, while capable of detecting angular deflection with a very high degree of sensitivity, autocollimators are almost insensitive to distance variations. This insensitivity to distances that change proves to be of great advantage for many measuring tasks, as evidenced by the selected application examples.

The distance of the reflecting surface from the autocollimator does affect the measuring process, however, in at least two technical aspects:

1. When the distance is extended, and/or when the tilt angle of the object becomes greater, more of the returning rays will pass the objective without striking it. This condition may reach a degree where the reflected image becomes barely perceptible, or actually disappears from the graticule of the instrument. Consequently, distance and object deflection are conjugately limiting factors with regard to the measuring range of the autocollimator; and

2. The brightness of the reflected image will decrease at a rate equal to the square of the distance. This condition has detrimental effects on the discrimination needed for the precise alignment of the target image with the reference mark of the graticule.

For these reasons the practical limit of the object's distance from the objective of a regular type autocollimator with visual observation is about 100 feet, the actual limit being dependent on such factors as the effective aperture of the objective; the power of the light source; and the size and the reflectivity of the object surface. It is a good practice to keep the distance between the instrument and the object as short as practical. However, with a laser adapted as the autocollimator light source, the usable distance between the instrument and the object is substantially extended.

The sensitivity of the autocollimator in detecting angular deflections and the level of accuracy it provides for measuring tilt angles are widely utilized in metrology for solving diverse measuring problems. The growing need for every accurate measurements is stimulating many imaginative metrologists to develop new useful applications for autocollimators. For that reason it would be impractical to compile an even approximately complete survey of the autocollimator-based measuring processes. The purpose of the application examples presented in Table 11-7 is, therefore, to indicate only the variety of metrological tasks that are successfully handled with the aid of autocollimator instruments.

Barrel Type Autocollimators

As a general category, the barrel type comprises most of the currently used types of autocollimators. These instruments have as a common characteristic the resemblance to an alignment telescope, the majority of the optical elements being arranged in line, mounted into a tubular housing that also carries the eyepiece with the micrometer screw for the translational movement of the reference graticule.

In the earlier designs of the barrel type autocollimators, the cross wires of the target also appeared

in the field of view of the eyepiece, but had to be ignored when evaluating the position of the reflected target image. The inconvenience caused in the graticule reading by the presence of the actual target wires has been eliminated in autocollimators of more advanced design, an example of which is shown in Fig. 11-31. These instruments have, as a design element, a beam-splitter prism interposed in the path of the target image. The target wires are located outside the line of sight and the prism deflects the image of the illuminated target wires in the direction of the objective, yet does not cause the wires to appear in the field of view of the eyepiece. The reflected image of the target, however, can travel unrestrictedly through the beam splitter and be observed in relation to the reference lines of the micrometer graticule.

Figure 11-32 shows the design principles of the optical system in an autocollimator with suppressed target-wire view, and Fig. 11-33 illustrates the field of view and the measuring micrometer of the same instrument. The described type of autocollimator can be used of measuring small, angular deviations from a reference plane, which can be either horizontal or vertical, simply by rotating the micrometer eyepiece through 90 degrees.

The same basic types of autocollimators are also available for the concurrent measurement of angular deviations both in the horizontal and vertical planes. For that purpose the eyepiece unit has two micrometers for setting the readout graticules in two directions, at right angles to each other. Figure 11-34 is the close-up view of the two-directional micrometer

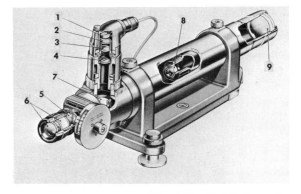

Hilger & Watts/Engis Equipment Co:

Fig. 11-31. Partially sectioned view of a barrel type autocollimator showing the principal operating elements.
(1) Filament lamp; *(2)* diffuser; *(3)* condenser; *(4)* target graticule; *(5)* setting graticule; *(6)* eyepiece lenses; *(7)* beam splitter; *(8)* rear component (objective); and *(9)* front component (objective).

TABLE 11-7. EXAMPLES OF AUTOCOLLIMATOR APPLICATIONS IN DIMENSIONAL METROLOGY — 1

GEOMETRIC CONDITION	APPLICATION EXAMPLE	DIAGRAM	DISCUSSION
Straightness of a surface element	The straightness of a round shaft		A Vee-base carrying a reflecter can be moved to different positions along the shaft while being observed through an autocollimator. Sag or bend of the shaft will be readily detected and measured as angular deflection of a section equal to the length of the Vee-base, in its relation to the selected reference section which was used for zeroing in the autocollimator.
Straightness of a translational movement	The effective straightness of a machine tool guideway		A reflector mounted on the carriage of a lathe and viewed through an autocollimator will permit inspection of the effective straightness of the guideways, by measuring the deflections of the reflector while the carriage is being moved to different positions along the lathe bed.
Flatness of a surface	The flatness of a surface plate assessed by correlating the straightness of interrelated tracks		By measuring the straightness of several tracks established in a specific pattern on a nominally flat surface and referencing from a common datum plane, the angular deflections of a reflector of known base length are established, permitting the corresponding differences in elevation to be calculated.

TABLE 11-7. EXAMPLES OF AUTOCOLLIMATOR APPLICATIONS IN DIMENSIONAL METROLOGY — 2

GEOMETRIC CONDITION	APPLICATION EXAMPLE	DIAGRAM	DISCUSSION
Squareness of interrelated surfaces of a technical part	Measuring the squareness of the sides of a machine element		The squareness of two sides of a part may be inspected by setting up the part with one of the sides resting on a surface plate and attaching a plane parallel mirror to the other side for sighting with an autocollimator whose optical axis is parallel to the surface plate. Small squareness errors are measured precisely with the micrometer of the autocollimator.
Perpendicularity of a translational movement	The tracking of a slide in guideways of an upright in relation to the guideways of the table		The translational movement of a slide along vertical guideways can be checked for squareness to the machine tool bed by first aligning the autocollimator with the horizontal surface and then observing the reflector attached to the slide through a pentaprism resting on the bed guideways.
Squareness of a rotational axis to a reference direction	Consistent perpendicularity of a machine tool spindle to the direction of a slide movement		A front mirror mounted on a precisely square test plate which is inserted into the spindle bore and observed by means of an autocollimator either directly or through a pentaprism, will detect alignment errors when the spindle is brought to different rotational positions.

TABLE 11-7. EXAMPLES OF AUTOCOLLIMATOR APPLICATIONS IN DIMENSIONAL METROLOGY — 3

GEOMETRIC CONDITION	APPLICATION EXAMPLE	DIAGRAM	DISCUSSION
Parallelism of external surfaces	Gage blocks wrung to a master cube to produce a setting gage for internal measurements		Gage blocks of calibrated parallelism are used to translate opposite surfaces into a common direction of sighting, where they are viewed consecutively with an autocollimator, by temporarily shielding the non-observed surface. Parallelism is needed to assure the accuracy of length transfer for which this assembly is used.
Parallelism of internal surfaces	The parallelism of two mutually opposite surfaces such as the contact faces of a large micrometer		After establishing a reference plane by means of a plain reflector and an autocollimator, the two surfaces which are facing each other are consecutively checked for squareness with the reference direction, by rotating the pentaprism. The results of the two squareness measurements are correlated to determine the degree of parallelism.
Inspection of circular spacings	Checking the setting accuracy of a dividing head with the aid of an optical polygon observed with an autocollimator		An optical polygon mounted on the face plate of a dividing head or on the top surface of a rotary table, is observed by means of an autocollimator to determine the correlation or error between the actual rotational movement and the indicated displacement of the device, in steps corresponding to the central angles of the polygon.

TABLE 11-7. EXAMPLES OF AUTOCOLLIMATOR APPLICATIONS IN DIMENSIONAL METROLOGY — 4

GEOMETRIC CONDITION	APPLICATION EXAMPLE	DIAGRAM	DISCUSSION
Inspection of angular separation	Checking the included angle of a cone by optically comparing the spacing of the sides with the corresponding settings of a rotary table		The angular separation between opposite faces or side elements of a tapered body can be measured by rotating the body when mounted on a circular dividing device through an angle supplementary to the body's nominal angle. The presence and amount of discrepancy between the nominal and actual supplementary angles is measured by the autocollimator.
Angular variations due to deflection	Sag of a machine tool bed due to weight and/or force		The sagging of a machine tool bed due to cutting force or weight of the loaded member can be determined by measuring the resulting angular deflection as displayed by a mirror and observed with an autocollimator, in relation to a reference condition which was established prior to the action of said forces.
Length comparison	Height difference between a master and a specimen measured by the tilt of a mirror which both objects support concurrently	$\Delta = \text{SIN } \psi \times \ell$	A mirror base having for legs two rollers, parallel and of equal size, at known distance apart, will permit the comparison of an object's length to that of a gage block set when each of these members is supporting one of the legs. Height difference causes angularly measurable mirror tilt, determined by an autocollimator whose sighting direction must stay parallel to the datum plane.

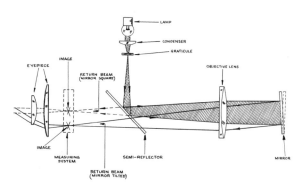

Hilger & Watts/Engis Equipment Co.

Fig. 11-32. Diagram illustrating the general optical principle of the autocollimator shown in Figure 11-31.

Hilger & Watts/Engis Equipment Co.

Fig. 11-34. Close-up view of the double micrometer eyepiece of a two-directional autocollimator.

ROTATE MICROMETER EYEPIECE UNIT THROUGH 90° FOR PERPENDICULAR DISPLACEMENT

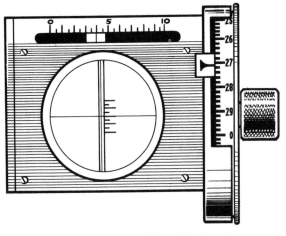

Hilger & Watts/Engis Equipment Co.

Fig. 11-33. Field of view of the measuring micrometer of a single-axis autocollimator. The significant target line (in this case the vertical) is straddled by the bifilar setting lines. The turncounter (top) reads 4 minutes, and the micrometer shows 27.1 seconds of arc.

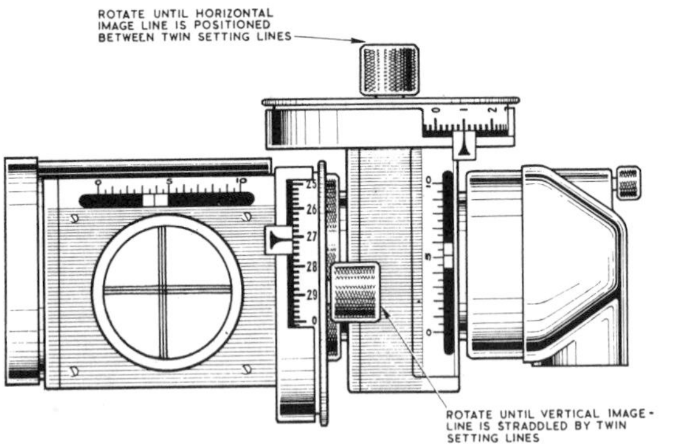

ROTATE UNTIL HORIZONTAL IMAGE LINE IS POSITIONED BETWEEN TWIN SETTING LINES

ROTATE UNTIL VERTICAL IMAGE-LINE IS STRADDLED BY TWIN SETTING LINES

Example: Horizontal displacement	Vertical displacement
Micrometer = 27·1 sec.	Micrometer = 1 sec.
Turn counter = 3 min. 30 sec.	Turn counter = 5 min.
Reading = 3 min. 57·1 sec.	Reading = 5 min. 1 sec.

Hilger & Watts/Engis Equipment Co.

Fig. 11-35. Field of view of the two-directional autocollimator illustrating the method of reading the micrometers.

eyepiece unit equipping the barrel type autocollimator. Figure 11-35 shows the diagram of the readout process with two micrometers and of the field of view where both bifilar reference lines are straddling the cross-hair image. Each graduation line of the micrometer drum represents 0.2 second of arc; the equal width of the alternating lines and spaces permits subdividing the reading into 0.1 second of arc increments.

Folded Beam Type Autocollimators

A representative model of this type of autocollimator is shown in Fig. 11-36, and Fig. 11-37 illustrates diagrammatically the major optical elements of this instrument presented in a cross-sectional view. The folded beam design provides an effective focal length of 20 inches with a short and compact instrument, which is sturdy enough for shop use and can be carried easily. Due to the folded beam design,

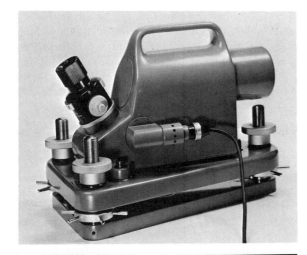

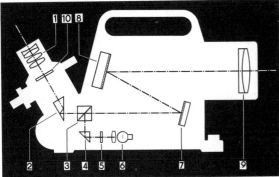

Leitz/Opto-Metric Tools, Inc.

Fig. 11-36. (Upper) Folded beam type autocollimator mounted on a vertically adjustable, double-tier leveling base.

Fig. 11-37. (Lower) The optical system of a folded beam type autocollimator: *(1)* measuring eyepiece; *(2)* deflecting prism; *(3)* beam-splitting cube; *(4)* deflecting prism; *(5)* hair-line target; *(6)* light source; *(7)* first beam-deviating mirror; *(8)* second beam-deviating mirror; *(9)* objective; and *(10)* graticule.

the eyepiece of the instrument can be arranged at a slant angle that is most convenient for viewing in the majority of autocollimator applications.

This type of autocollimator has an exceptionally wide measuring range: 30 minutes for readout in 0.5 second of arc increment, or 16 minutes when 0.1 second of arc least graduation of the graticule is required. The central graticule of the eyepiece is fixed, with the exception of a limited range zero setting adjustment at the start of the measuring process. This fixed graticule carries a scale for the entire range of the minute graduations, which are represented by bifilar lines or blocks providing narrow gaps for the precisely straddled positioning of the reflected cross-hair image. That image can be translationally dis-

placed by the screw of the optical micrometer to bring it into a central position in the nearest graduation gap. The magnitude of that positional adjustment is indicated by a complementary graticule of annual form, which is rotatable around the central graticule. The circumferential graduations of the annular graticule are in 0.5 or 0.1 second of arc increments and can be read concurrently with the scale of the central graticule.

This type of autocollimator is made as a single-axis or a double-axis model, both using essentially identical systems of oculars that, however, differ in the following respects:

a. In the single-axis version, the ocular can be swivelled over 90 degrees, to measure either the tilt or the rotational displacement of the object.

b. For the double-axis instrument a fixed ocular is used that is equipped with a coordinate graticule, carrying graduations along both the vertical and the horizontal axes (see Fig. 11-38). The cross-hair image that appears on the graticule can be brought consecutively into straddled graduation position by means of the optical micrometer. By this process both the vertical and the horizontal variations of the cross-hair image, with respect to an initial setting, can be determined by a mechanical action, without affecting the image position.

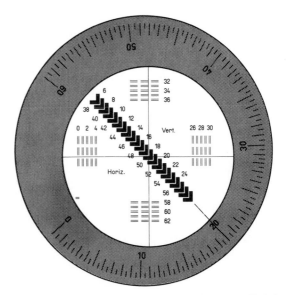

Leitz/Opto-Metric Tools, Inc.

Fig. 11-38. The eyepiece graticule of a double-axis autocollimator equipped with optical micrometer for positioning the cross-hair image.

The illustrated type of autocollimator can be mounted on a leveling base, unless it is placed directly on a surface that represents the datum plane of the measurement, such as a surface plate, which supports both the object and the autocollimator. The vertical adjustment of the autocollimator instrument can be accomplished with the aid of double-tier base (see Fig. 11-36), which can also be clamped on a tripod for independent mounting of the autocollimator.

Comparison Autocollimator

This instrument, shown in Fig. 11-39, incorporates an uncommon principle of autocollimator design. Its purpose is the simultaneous observation of two close-to-parallel reflecting surfaces for measuring the angle of deviation from mutually referenced parallel planes. Whereas in regular autocollimator measurements the instrument establishes the reference direction and indicates the deviation of the observed reflecting plane from the true normal to that direction, in the comparison autocollimator the instrument references from an extraneous plane that can be represented by a stationary mirror.

Therefore, it is not necessary to secure the position of the autocollimator, for example, by clamping during the measurements. The instrument can be moved, inadvertently or intentionally, without affecting the accuracy of the measurements, as long as the remotely mounted reference mirror is maintained in an unaltered position.

These operational conditions present distinct advantages, particularly for processes requiring several measurements in relation to a common reference plane. Preferential applications may be found in the inspection of surface plates, or in the measurement of angular separations in setups that combine a rotary table with a stationary reference mirror. The optical tipping plate by means of which the coinci-

dence of the reflected images can be established is actuated by a micrometer graduated in 1/10th second of arc and numbered for each full second of arc. The measuring range of this particular instrument extends over 120 seconds, with an accuracy of 0.5 second for the total range, and 0.1 second of arc repeat accuracy for resetting to any specific position.

Photoelectric Autocollimators

As discussed previously, the measurement of small angular deviations with the aid of an autocollimator requires the visual observation of the reflected cross-hair image through the eyepiece of the instrument. In measuring processes involving the continuous use of autocollimators, such as for example the calibration of polygons, the checking of rotary tables and so forth, protracted viewing through the eyepiece can become quite tiring.

The photoelectric autocollimator (Fig. 11-40) makes the viewing through the eyepiece unnecessary; the pointer on the meter face of an amplifier unit indicates by its zero position that the reflected cross-hair image is exactly centered in the field of view. The zero indication is accomplished by means of a photoelectric detector built into the eyepiece unit. That device comprises a vibrating slit that, when exactly centered over the image line of the cross hair, transmits intermittent light signals to the photocell. A frequency discriminator receives the amplified output of the photocell and produces a null indication on the meter when the signal frequency is identical with the frequency of the reference voltage. The centered position of the cross-hair image is accomplished in the measuring process by making adjustments, when needed, with the micrometer screw, whose drum graduations then show in angular values the deviation of the reflecting plane from the setup position.

Davidson Optronics Inc.

Fig. 11-39. Comparison autocollinator for measuring the angular difference between two close-to-parallel reflecting surfaces.

Hilger & Watts/Engis Equipment Co.

Fig. 11-40. Photoelectric autocollimator with micrometer adjustment on the instrument, and zero-position indication on the meter of a remotely located amplifier.

Besides avoiding operator eyestrain, with the accompanying conditions of increased incidence of error and reduced performance, the photoelectric zero indication provides further advantages in certain types of autocollimator usage. As an example, when measuring the angular separation of the reflecting faces of an object mounted on a laboratory type rotary table with one second of arc graduations, the inspector can rotate the table—slowly approaching or overtraveling the nominal angle—until the occurrence of the null indication is observed on the remote meter. The total value of the rotation displacement will then be read on the table scales without the need of adjustments on the micrometer of the autocollimator. The repeat accuracy of the instrument's photoelectric setting is claimed to be 0.05 second of arc.

For extremely critical applications, photoelectric autocollimators with monochromatic light source are also available. The *monochromatic photoelectric autocollimator* of Hilger & Watts has a high intensity lamp with monochromatic filter to produce a light of 7100 Å wavelength, which corresponds to the peak spectral response of the photocell. While in the photoelectric autocollimator the vibrating slit detector is attached to the micrometer screw, the monochromatic model has a special optical micrometer that displaces the optical path of the reflected image, thus avoiding a mechanical linkage between the photoelectric sensing assembly and the micrometer.

The achievable repetition of setting with this type of instrument is claimed to be 0.015 second of arc (at 2σ standard deviation), with 0.25 second accuracy over the entire measuring range. However, this range extends over 100 seconds of arc only, as compared to the 10 minutes of arc range of the regular photoelectric autocollimator.

Automatic Position Sensing Autocollimator

While the photoelectric autocollimator indicates the position of exact perpendicularity of the observed plane with respect to the optical axis of the instrument, any deviation from that basic condition must be compensated by the adjustment of the angle measuring member, either on the autocollimator or on the work-staging end of the setup. Therefore, the use of the photoelectric autocollimator, similar to autocollimators operated by visual observation, is limited to the measurement of discrete angular spacings in such applications where adequate time is available for carrying out the necessary adjustment and for reading their values.

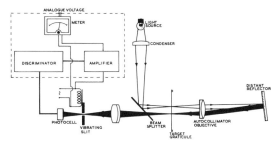

Hilger & Watts/Engis Equipment Co.

Fig. 11-41. Diagram of the major elements of an automatic position-sensing autocollimator, which directly indicates, on a meter, the value of the observed angular displacements.

However, for applications requiring the uninterrupted measurement of small and slowly changing angular displacements, a continuously indicating instrument with automatic position sensing is needed.

Figure 11-41 shows diagrammatically the major elements of the automatic position sensing autocollimator. The instrument is also equipped with optics for visual setting. This is needed for adjusting the position of the reflected cross-hair image into the sensing range of the automatic operation; a lamp on the amplifier unit indicates when the preadjustment was accomplished and when the automatic detecting system was brought into operation.

Some of the basic components of the automatic position sensing autocollimator are similar to those of the previously described photoelectric type, namely, the slit, vibrated by an electromagnetic coil that transmits light impulses to a photocell with the essential difference, however, that the slit position can be biased. When the oscillations of the slit are not symmetrical to the reflected cross-hair image, that condition will be detected by the discriminator that produces a DC output with amplitude and polarity proportional to the sensed misalignment. The amplified output of the discriminator is superimposed on the coil of the vibrating device and is also used to indicate on the meter, in seconds of arc units, the angular value of the sensed displacement. The output being an analog voltage, it can also be channeled into a digital voltmeter or into a chart recorder.

Automatic Autocollimator with Continuous Servo-Setting

The arrangement of the functional elements of the illustrated instrument are also shown diagrammatically in Fig. 11-42. A collimated light beam, which originates from a single source, is projected onto a mirror representing the object surface being inves-

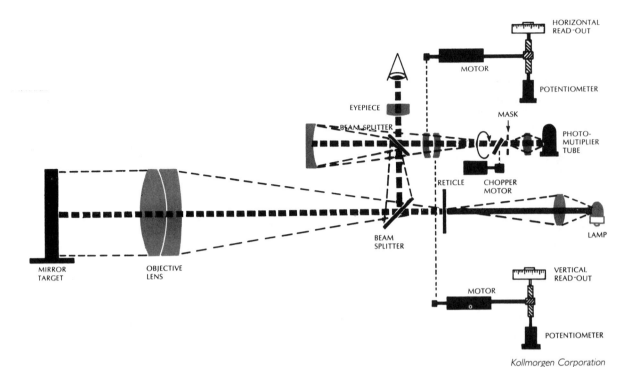

Kollmorgen Corporation

Fig. 11-42. Optical schematic diagram of the dual-axis automatic autocollimator showing the paths of the emitted and the reflected light beams.

tigated. The reflected light is refocused through servo-driven lenses into an image plane where its centering position can be observed visually, as well as with the aid of a photomultiplier tube. Any off-center position actuates the servo-motors to adjust the micrometer lenses and to indicate the angular value of the momentarily required centering adjustment, that is, the angular displacement of the target.

For the continuous recording of the sensed and compensated angular deviations, the servo-motors also drive individual potentiometers whose output can be channeled into appropriate chart recorders.

While Fig. 11-43 shows a dual-axis type of automatic autocollimator, similar purpose instruments are also made for single-axis measurements only, and are available with different degrees of sensitivity. These vary, according to the particular model of instrument, from about 1 second to the remarkably high degree of 0.01 second of arc sensitivity. The measuring range, of course, is in an inverse relation to the degree of sensitivity and it varies, in the case of the instrument models referred to, from about ± 10 minutes to ± 1 minute of arc. With regard to the particular model shown in Fig. 11-43, the respective values are: sensitivity better than 0.1 second of arc,

Kollmorgen Corporation

Fig. 11-43. Dual-axis automatic autocollimator with continuous servo-setting.
(Left) Electronics unit.
(Right) Optical sensing unit.

and the angular measuring range ± 2.5 minutes of arc.

The major applications for these instruments are found in the inspection of guidance systems. However, there are various less sophisticated uses for automatic autocollimators in engineering metrology in general. A very interesting application was discussed earlier in this chapter in conjunction with the automatically recorded inspection of surface plates; similar principles of measurement can be applied for inspecting the straightness of machine tool guideways.

The Optical Square (Pentaprism)

The reflectors used in conjunction with the operation of autocollimators are of various designs as far as their size, the form of their housing and the location of the supporting pads are concerned. However, with regard to the optical principles involved, all designs comprise a plain mirror, although of excellent flatness and reflectivity. Actually, in many instances the flat reflective surfaces of precision parts or of gage blocks are being substituted for mirrors in autocollimator measurements. For this reason only one particular type of reflector is discussed in detail, one that combines reflection with deflection (bending of the light by a constant angle).

Many optical measurements require the turning of either the line of sight, or of the projected and reflected image, at exactly right angles. A few of such applications are indicated in the examples shown in Table 11-7. This precisely controlled deflection of the light's direction is accomplished with the aid of a particular type of five-sided prism, hence the designation *pentaprism*. When mounted in an appropriate housing, having supporting pads normal to the planes of two sides of the prism, it is generally known as an *optical square*, with reference to its functional use (Fig. 11-44). Pentaprisms are also used in mountings for attachment in front of the objective lens of optical viewing instruments, whereas the optical square is used in a detached manner, usually supported on the object surface or on an auxiliary stage.

The construction of the pentaprism is based on an optical and a geometric concept. namely:

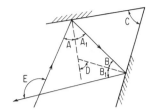

A and B = angles of incidence
A_1 and B_1 = angles of reflection
D = A + B = exterior angle of triangle
C = D
E = 2C = angle of light turn

Fig. 11-45. The principles of light reflection from two opposite mirrors that include an acute angle.

1. The optical law of reflection, according to which the angle of incidence and the angle of reflection are equal and symmetrical with respect to the perpendicular to a flat reflecting surface;

2. The geometrical theorem that the exterior angle of a triangle is equal to the sum of the two opposite interior angles (Fig. 11-45).

The function of an optical square in reflecting the light at right angles could be accomplished with the aid of two mirrors that are set at exactly 45 degrees to each other. However, it is much more practical and reliable to produce these mirrors as a single piece, in the form of a prism with two opposite sides including a 45-degree angle, and making these sides, by silvering, reflective toward the inside of the prism. The resulting optical device is a constant deviation prism, which produces an invariant angle of reflection. For most practical purposes that angle remains the same even when the prism is slightly rotated in either direction around its vertex, or an axis parallel to the vertex (Fig. 11-46).

The functional purpose of two of the other sides of the five-sided prism is secondary only. These sides are mutually square and symmetrical with respect to the reflective sides, in order to avoid the interfering effect of unbalanced refraction on the light passing through the prism. Because of the symmetrical design, the refraction of the light beam entering and leaving the prism are of equal value but opposite

Davidson Optronics Inc.

Fig. 11-44. Optical square containing a pentaprism in a housing with precisely aligned locating pads. Its purpose is to deflect by exactly 90 degrees the collimated light beams entering one of the open faces within ±5 degrees from the normal.

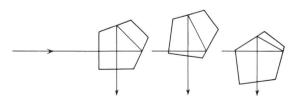

Fig. 11-46. A pentaprism will turn the collimated light beams at right angles, even when rotated around its vertex within a few degrees in either direction.

in direction, with a resultant canceling effect. For the same reason, it is advisable to set up the pentaprism with the receiving face approximately perpendicular to the incident light. The fifth side of the pentaprism has no functional role at all; it results from keeping the length of the reflecting sides shorter than the vertex distance, in order not to extend needlessly the overall dimensions of the prism.

There is one more point that must be considered in the use of pentaprisms. Although the squareness of the reflected light beam is only insignificantly affected by refraction when, due to rotation of the prism, the incoming light beam is not exactly perpendicular to the side facing the light, a tilt of the vertex edge from the normal to the direction of the light will alter the turning angle, compounding the basic 90 degrees with the tilt angle. To avoid errors from such causes the housings of optical squares available for industrial applications are provided with locating pads and guides for the correct positioning of the pentaprism.

Earlier in this section it was stated that pentaprisms are made of a single piece. Actually, most pentaprisms have a correction wedge solidly attached to one of the faces for the purpose of correcting minor errors that are almost unavoidable in the grinding and lapping of the reflecting faces to an exactly 45 degree angle. As this glass wedge has the property of bending the direction of the light by a slight amount, by rotating the wedge, the minor error in the mutual angle of the reflecting sides can be balanced to a degree that satisfies the expected accuracy of the pentaprism. Such adjustments are commonly on the order of one or two seconds of arc and, once adjusted, the wedge is solidly cemented to the face of the pentaprism. This correction process is carried out by the manufacturer of optical squares, and is in no way the responsibility of the user.

12.
The Systems and Applications of Measuring Machines

The configuration and the size of technical parts are determined by more than a single dimension, except for the perfect sphere. Most of these dimensions are interdependent with regard to location and position. The geometric interrelations, which are either inferred or actually specified and toleranced, permit the establishment of functionally meaningful dimensional conditions for a manufactured part with a minimum of data.

On engineering drawings, the required interrelations of the geometric dimensions are usually expressed by means of datum elements, such as points, lines, or planes on the surface of the part; the specified dimensions must then be measured in relation to the pertinent datum. In common design practice, the datum can be physically present on the part's surface—a face, an edge, a generatrix—or be represented by a virtual element. Examples of the latter are the axis of a bore or of a body of revolution, the center of a circular arc element, the line of intersection resulting from the imaginary extension of side surfaces and so forth.

In the inspection phases of continuous manufacturing, such geometric interrelations are: (1) assumed to be inherent and consistently assured by the manufacturing process; or (2) are checked in separate inspection operations; or (3) the means of their inspection may be incorporated into the staging for dimensional measurement.

A classical method of inspection in the cases last mentioned consists in establishing auxiliary datum surfaces from which the measured dimensions can be referenced. Surface plates are most commonly used for such purposes. However, the limitations of adaptability and accuracy, as well as the tediousness and the potential errors of surface plate work in the inspection of complex parts with closely toleranced dimensions, are well known and need not be discussed here.

The measuring machine, by combining the functions of locationally controlled work staging and dimensional measurement into a single unit, can provide a unique means for measuring linear dimensions, and in some cases, angular dimensions also. These measurements are carried out by locating and orienting the part in compliance with theoretical design requirements.

Such characteristics as inherent accuracy, adaptability to diverse part configurations and versatility with regard to the applicable processes, can often result in the superiority of measuring machines in comparison to other types of measuring instruments.

In the past few years, the applications for measuring machines have been expanding at a growing rate in many branches of industrial production. It is obvious that advances in instrument design as well as the widening range of available instrument types and models have had a major role in that expanded usage; conversely, it is also likely that technical need has been an important factor in this development.

The following are a few highlights indicative of recent trends:

1. Combining mechanical instruments with electronic sensing, indicating and recording devices, for fiducial as well as for actual measuring functions;

2. Advances in optical displacement measuring devices, including better discriminating optical micrometers and improved pointing capabilities;

3. Numerical display of displacement distances based on optical fringe-producing or inductive grat-

ings that are capable of originating discrete impulses for each incremental part of the translational movement by the instrument's traveling members; and

4. The development of highly stable structural materials, as well as advances in manufacturing and inspection techniques, for the mechanical elements of measuring machines.

The individual instruments to be discussed were selected to illustrate various categories of measuring machines. Although of recognized excellence, each of the specifically mentioned models is generally not unique, but represents one of several comparable makes and types.

The purpose of the following review of measuring machines is to analyze the application potentials of this category of instruments for diverse tasks in dimensional metrology.

DEFINITION AND GENERAL EVALUATION

Measuring machines are stationary instruments that employ their own standards for obtaining dimensional measurements, and are designed for work at a high degree of accuracy. Reliance on standards that are integral elements of the machine implies the assurance of precisely controlled displacement paths for the movable sensing members.

Measuring machines that operate along more than a single axis must also possess translational and—when an integral part of the system—rotational movements whose paths are in mutual geometric relationship controlled to a degree commensurate with the intended measuring accuracy of the instrument.

For surveying a general group of instruments, comprising a large array of models that differ in many respects, it is convenient to select a major characteristic for establishing categories. In the case of measuring machines, one aspect that stands out over many other properties is the intended scope of operation, expressed by the number of axes, along or around which measurements can be carried out.

Based on the major characteristic of measuring axes, supplemented by a few other distinctive aspects, a survey of measuring machines classified into major categories is presented in Table 12-1. Several well-known types of measuring machines that, by this classification, belong to the same comprehensive category differ in the principles of their own design as well as in their capacities; still, the basic purpose of application of these machines may be considered to be similar.

It is of interest to note, from the information compiled in this table, that whereas the range of measurable geometric conditions varies—from plain length to many linear and angular dimensions—the attainable sensitivity, reflected by the least increment of the measured dimensions, is at an essentially comparable level. That power of resolution, assessed on the basis of the most sensitive standard models of instruments of each major category, is indicative of the current capabilities of dimensional measuring processes that can be performed on industrial measuring machines.

SINGLE-AXIS MEASURING MACHINES

Plain linear distances on external and internal surfaces, are the most frequently occurring dimensions in engineering measurements. On technical parts, the linear dimension to be measured is generally defined as the separation of two mutually parallel surfaces or surface elements, such as the length, width and thickness of a rectangular block or the outside and inside diameter of a ring.

Disregarding the comparative type of measurement, which necessitates the setting of the instrument with the aid of extraneous masters, it may be stated that, for the purpose of plain linear measurement, the best currently available measuring instrument—short of measuring machines—is the classical screw micrometer. Although micrometers have provided invaluable service in the past, and are finding many useful applications in current practice, the inherent accuracy limit of the instrument equipped with a vernier is 1/10,000 inch.

It follows that for measurements where the required repeat accuracy is 1/10,000 inch or better, and/or where the locating of the measuring plane is more critical than the degree that can consistently be assured by a hand tool, a higher grade instrument than the screw micrometer must be used. It is at this point that the domain of application for the single-axis measuring machine begins.

Single-axis measuring machines are, of course, not simply upgraded substitutes for screw micrometers, but incorporate many distinctive characteristics and also provide—depending on the particular type—additional capabilities. These properties permit the measurement of specific linear dimensions with a consistently high degree of accuracy and in locations that precisely agree with the geometric concepts of the design.

Three general types of single-axis measuring machines are considered, representing distinctive sys-

TABLE 12-1.   MEASURING MACHINES AND GENERAL CHARACTERISTICS

CATEGORY AND RANGE	DIAGRAM OF TYPICAL ARRANGEMENT	DISCRIMINATION	GENERAL PERFORMANCE CHARACTERISTICS
Single-axis measuring machines. One measuring axis, with stage adjustment in a plane normal to the main axis		10 microinches	For the measurement of plain length dimensions, generally on external surfaces; a few types of machines are equipped for internal measurements too. Can also be used for the over ball or wire measurement of threads, tapers, etc.
Coordinate measuring machines. Measures along two mutually perpendicular axes parallel to the staging plane. Special models equipped for third axis measurements		100 microinches	For measuring distances in relation to a common system of rectangular coordinates (generally two, exceptionally three-dimensional), by aligning the movable measuring head with reference points on the part with the aid of mechanically contacting probe. Indicates distances numerically, permits data acquisition besides visual display. Characterized by rapid action and simple operation.
Jig borer type measuring machines. Measures along three mutually perpendicular axes, as well as rotational displacement		20 microinches and 5 seconds of arc	Using electronic gage probe for contact member it measures linear distances in three dimensions also on surfaces which are inaccessible to most other systems of measuring machines. Can be equipped for angular and rotational specimen staging. Linear and circular surface elements can be traced and recorded. Characterized by great versatility and adaptation.
Optical measuring machines. Measures generally along two axes in a plane parallel to the stage; also 'angles and circular divisions		20 microinches and 10 seconds of arc	Uses optical targeting for referencing from points, line elements or entire feature contours on the observable object surface. Measures angular separations, circular divisions and polar coordinates. Exceptionally, a third axis capacity may be added when complemented with mechanically contacting probe. Referencing by microscope, without mechanical contact and by observing an area of the surface.

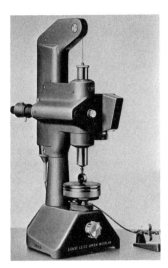

Leitz/Opto-Metric Tools, Inc.

Fig.12-1. Vertical length measuring machine with graticule micrometer.

Bausch & Lomb, Inc.

Fig. 12-2. Vertical length measuring machine with optical vernier.

tems and different ranges of measuring capabilities. These types differ primarily in two basic respects, namely, the staging of the specimen and the system of length measurement.

The first group comprises instruments that utilize a master scale attached to the axially movable vertical measuring spindle. The vertical arrangement permits locating the specimen on a staging table that also serves as the datum plane for the length measurement. Typical of this group is the Leitz optical length measuring machine (see Fig. 12-1), which will be described later, pointing out certain design characteristics of similar instruments currently manufactured in England (Watts) and, more recently, in this country also (Bausch & Lomb; see Fig. 12-2). Incidentally, instruments of this group closely resemble the measuring machine designed in 1890 by Ernst Abbe, the originator of a measuring principle, still frequently quoted and of unchallenged correctness, which calls for the aligned arrangement of the specimen and of the master scale.

The second group comprises the horizontally arranged mechanical length measuring machines. These machines, having a threaded measuring spindle for master element, resemble in the basic principles of their operation, a stationary micrometer, although in a highly refined variety, with substantially upgraded discriminating properties. Important features include the integral staging table and the variable referencing position of the tailstock, which incorporates the measuring anvil. A typical

representative of this group is the standard measuring machine of Pratt & Whitney (Fig. 12-3), which will also be discussed in greater detail later.

The third group combines the characteristics of purely optical distance measurement with the advantages of a horizontal staging and independent datum member. This system of measuring machines originates from Carl Zeiss and strictly adheres in its design to the above-mentioned length measuring principles. The Zeiss plants in Germany are currently manufacturing improved versions of this type of measuring machine; the model selected for detailed description is the Horizontal Metroscope (see Fig. 12-4).

From the user's point of view, one of the essential aspects in selecting measuring equipment, is the scope of its applicability, that is, the array of basic measuring processes that can be carried out with any specific category of instrument. It is from this particular point of observation that the various types of single-axis measuring machines were assigned to specific categories. In Table 12-2, the functionally meaningful application potentials of the different categories of single-axis measuring machines are pointed out with the aid of typical measurement examples. The different types of single-axis measuring machines are presented in the order of increasing capabilities.

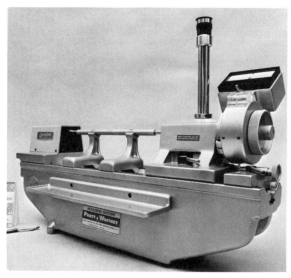

Pratt & Whitney Cutting Tool and Gage Div., Colt Industries

Fig. 12-3. Horizontal length measuring machine with super-micrometer, electronic fiducial indicator for the tailstock anvil and microscope tube for the observation of reference marks at interger-inch intervals.

Carl Zeiss, Inc.

Fig. 12-4. Horizontal length measuring machine with purely optical readout using a 4-inch-length master scale. Applicable for both external and internal measurements, also in combination with diverse accessories.

Optical Length Measuring Machines in Vertical Arrangement

In the following description, which is based on the Leitz optical length measuring machine, references are also made to distinctive design characteristics of comparable instruments of different makes.

The precisely guided and counterbalanced measuring spindle is a hollow cylinder and carries in its center plane a glass master scale of 4-inch length,

with numbered graduations at 0.05-inch intervals. The individual graduation that represents the momentary position of the measuring spindle, can be visually observed by being projected, at about 50 times magnification, onto a screen. On some other models of similar instruments, particularly those of earlier designs, the observation of the scale is made with the aid of a microscope tube. The screen carries intermediate graduations in 0.005-inch increments, individually numbered and represented by bifilar marks.

To operate the instrument, the measuring spindle must first be raised to clear the stage; after the part has been located, the spindle is released to advance toward the part. When the tip of the measuring spindle contacts the work surface and the approach movement comes to a standstill, the projected scale line can be at any position along the screen, depending on the actual size of the part. For bringing the visually observed graduation line of the master scale into a central position between the nearest bifilar screen marks, either the screen can be traversed in a direction parallel to the main scale's graduations (Leitz), or the optical path of the scale line can be deflected by tilting an interposed parallel-plate deflector block (Watts). Either of these adjustments is carried out manually by means of a thumb screw, which is mechanically coupled with an optical micrometer indicating the amount of centering adjustment. The graduation lines of that micrometer, or vernier, represent increments of 0.0001-inch (B & L) or of $12\frac{1}{2}$ microinches (Leitz). For determining the measured size, the screen scale and the vernier must be read conjointly.

The gaging force of this type of instrument is essentially constant and maintained at a level of 7 to 10 ounces, depending on the particular model. This is accomplished either by a counterweighted spindle whose downward advance rate is regulated by a hydraulic dashpot (Leitz), or by the controlled action of the electric motor driving the spindle (B & L).

Length measuring machines using optically observed master scales have several potential advantages over many models of similar-purpose mechanical instruments. The more important of these characteristics are the following:

a. The entire effective measuring range of the optical system is covered by a single master scale along which measurements can be made at any section. Most models of mechanical length measuring machines operating with micrometer screws are limited to shorter distances and require the repositioning of

TABLE 12-2.   CHARACTERISTIC CAPABILITIES OF SINGLE-AXIS MEASURING MACHINES — 1

The application examples are listed in the order of increasing capabilities. Each individual process described also applies to all machines following in the order of listing.

INSTRUMENT TYPE	DIAGRAM OF THE PROCESS	DESCRIPTION OF THE MEASURING CAPABILITY
Vertical length measuring machine (optical)		Plain external length dimensions. The part is located on one of the boundary planes of the distance being measured.
Mechanical length measuring machine in horizontal arrangement		External length measurements in a plane parallel to the staging surface on the part; measurements at different levels.
		Pitch diameter of an external thread measured over gage wires, using special staging fixture.
		The taper angle of a cone determined by diameter measurements over pins which are supported in two parallel planes at known distance apart.

TABLE 12.2.   CHARACTERISTIC CAPABILITIES OF SINGLE-AXIS MEASURING MACHINES – 2

INSTRUMENT TYPE	DIAGRAM OF THE PROCESS	DESCRIPTION OF THE MEASURING CAPABILITY
Optical length measuring machine in horizontal arrangement		Internal diameter of a plain ring gage measured in its axial plane by means of a floating stage.
		Pitch diameter of an internal thread measured over balls. The part is supported on a self-aligning stage and special setting gages are used to establish the basic size of M which is in a definite ratio to the pitch diameter.
		Taper thread pitch diameter measured over two wires and inside a ball-tipped probe in planes at a specific distance from a datum surface (the face of the part).

TABLE 12-2.   CHARACTERISTIC CAPABILITIES OF SINGLE-AXIS MEASURING MACHINES – 3

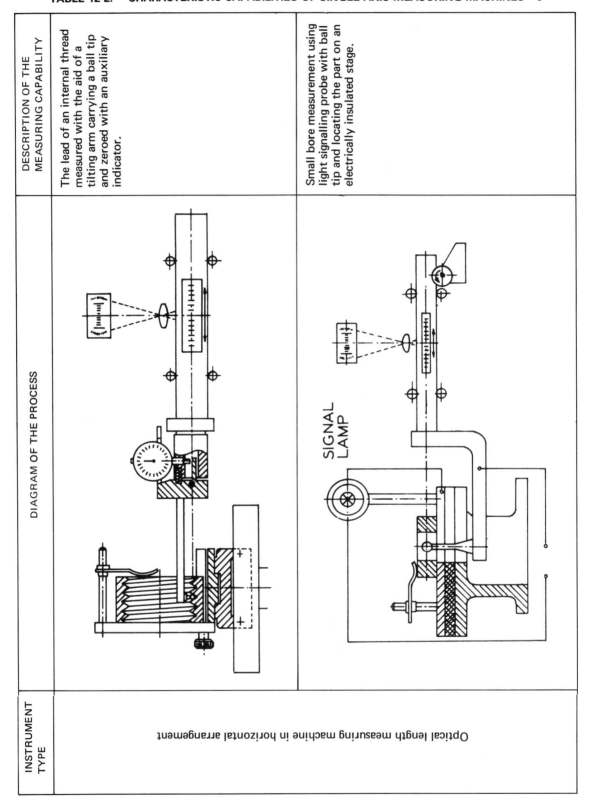

DESCRIPTION OF THE MEASURING CAPABILITY	DIAGRAM OF THE PROCESS	INSTRUMENT TYPE
The lead of an internal thread measured with the aid of a tilting arm carrying a ball tip and zeroed with an auxiliary indicator.		
Small bore measurement using light signalling probe with ball tip and locating the part on an electrically insulated stage.		Optical length measuring machine in horizontal arrangement

the anvil member for measuring lengths in excess of the micrometer's range.

b. Movable mechanical devices, such as screws, must operate with a specific minimum clearance between the mating members, consequently, they are not free from backlash, the hysteresis effects of which must be avoided by conscientiously observed process rules.

c. Micrometer screws, like all mechanical devices, are also subject to wear, a phenomenon nonexistent in optical devices.

For evaluating the application potentials of the described type of optical length measuring machines, it is well to also consider the limitations of the system such as: (1) the achievable accuracy of the master scale graduations that, for commercially available instruments, is claimed to be ±5 or 10 microinches, depending on the particular make; (2) the power of resolution of the optical observation that, at about 50× magnification, may be on the order of ±10 microinches; and (3) the mechanical and optical limitations of the micrometer used for determining the last meaningful digit of the mechanical indications.

Mechanical Length Measuring Machines in Horizontal Arrangement

To illustrate mechanical length measuring machines in horizontal arrangement, the standard measuring machine manufactured by Pratt & Whitney has been selected. This machine operates by combining the functions of two masters, the measuring bar and the super-micrometer. The measuring bar is attached to the bed of the machine and serves as a step gage, with graduations in one-inch intervals. The graduations are represented by hair lines engraved on stainless steel plugs along the master bar, and are observed through a microscope that is solidly mounted into the measuring head. The measuring head houses the super-micrometer, distinguished by its large graduated drum with individual marks for 1/10,000th parts of an inch, and supplemented by a vernier indicating 10 microinch increments.

A setting screw mounted on a displaceable block permits the sensitive adjustment of the measuring head's position along the bed to a point where the hair line of a plug in the master bar coincides with the center mark in the field of view of the microscope. The tailstock is adjusted with respect to that reference position either by direct contact between

the opposing measuring tips or by the interposition of a gage block of integer inch size.

To keep the contact force constant—a major requirement for mechanically sensing measuring instruments—a milliammeter is used for the purpose of a gage, indicating by the zero position of its pointer that the selected gaging force has been applied. That force can be adjusted to a required operating level of from 2 to 48 ounces.

As long as the position of the tailstock and of the anvil that it carries remains stable, the measuring range of the instrument can be altered by moving the measuring head to an appropriate base position, as indicated by the coincidence of the reference marks on the master bar plug and in the microscope reticle. Experience indicates that the potential inaccuracies of having the master bar and the line of measurement in parallel arrangement are kept under sensible control by the accuracy of the instrument's mechanical elements.

Optical Length Measuring Machines in Horizontal Arrangement

Horizontal length measuring machines using optically observed master scales provide means for an extended scope of measuring operations, as indicated in Table 12-2. The instrument's adaptability to a large variety of measuring tasks is the result of combining two important features, viz., the horizontal arrangement to assure the separation of the staging and of the referencing functions, and the use of optical scale reading whose accuracy is unaffected by the direction of travel of the gage head. The latter property is particularly important for internal measurements.

The horizontal measuring machine illustrated in Fig. 12-4 has 4 inches of effective travel, but can accommodate parts up to 24 inches long by traversing the tailstock along the bed to an appropriate position. For measuring external length dimensions not exceeding 4 inches, the machine can be set up by applying a self-zeroing process, that is, by direct contact between the gage tips of the head and of the tailstocks. However, for outside lengths over 4 inches, and also for all internal measurements, appropriate setting masters must be used to establish reference positions. The setting masters generally used are gage blocks or gage rods for outside measurements, and gage rings for internal dimensions.

For internal measurement, special contact arms are used and the direction of the contact force must be reversed. However, as long as the zeroing was carried out by using the same force action that is being applied for the actual measurement, the scale indications will retain their original accuracy.

The reversibility of the transfer movement without effect on the measuring accuracy is finding unique application in the internal measurement of very small holes, which is carried out with almost zero gaging force. The essential elements used in this process are shown in the last illustration of Table 12-2.

The probe, being attached to the measuring spindle, and having at its tip a ball of known size, is brought sequentially into contact with the opposite side of the bore being measured. The distance traveled between the two contact positions is read on the scale of the machine; this value must then be supplemented by adding the known diameter of the ball. To establish contact without applying appreciable force that could deflect the probe, a low voltage electrical circuit is established, feeding a signal lamp whose flickering indicates that contact has just been reached between the probe and the object surface. A special insulated staging platform is provided for operations involving referencing by electrical contact.

The 4-inch-long master scale of the machine is graduated and numbered at 0.05-inch intervals. The particular scale graduation that corresponds to the position of the gage head is projected at 46 times magnification on a screen in the readout window. The screen also incorporates the optical micrometer, which subdivides the main scale graduations into 50 microinch increments displayed with a degree of resolution permitting further subdivision by estimation.

Various adjusting, elevating and translational movements of the machine table, and the availability of additional staging fixtures to accommodate parts of different configurations, assure the great versatility of this particular type of single-axis length measuring machine.

The Zeiss–Jena Length Measuring Machine

To benefit from the inherent accuracy of graduated master scales in the measurement of linear distances that exceed the 4-inch practical length of densely graduated scales, special devices are needed for combining the reference functions of two master scales. In such systems, the longer scale carries graduations at 4-inch intervals represented by bifilar lines. By superimposing optically these bifilar marks on the observed section of the 4-inch scale, an optical coupling is created that results in a single virtual master scale of substantial length, yet provided with fine graduations on any particular 4-inch sec-

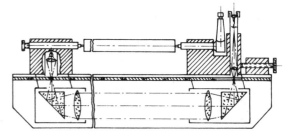

Fig. 12-5. The optical compounding system of the Zeiss–Jena length measuring machine. The projected image of the long-range scale graduations serves as index mark for the high-resolution readout micrometer.

tion, which can be selected by the appropriate positioning of the instrument's tailstock.

The schematic drawing in Figure 12-5 illustrates how this optical coupling is accomplished in a functionally effective accordance with the Abbe principle, which requires the in-line arrangement of the object and the master. For setting the instrument to measure an approximately known length dimension, the tailstock must be moved along the table to the particular 4-inch mark, which corresponds to the distance being measured. The protective enclosure of the long master scale has small openings at the positions of the 4-inch graduations lines, whose level coincides with the focal plane of a projection system inside the tailstock. When the optical axis of the projection system is aligned to a specific graduation line, the line's image appears in the field of view of the microscope through which the short master scale, graduated at 0.05-inch intervals, is observed. For subdividing the 0.05-inch graduation intervals, the instrument's observation microscope is equipped with an optical micrometer, similar in design to that discussed in the preceding description of the Horizontal Metroscope.

COORDINATE MEASURING MACHINES

The development of displacement measuring devices that provide digital readout for the distance traveled by a movable machine member brought into being a new breed of measuring machines. In these instruments, the traveling member, or gage head with the mounted sensing probe, is guided along two straight line paths that are contained in a common plane and are mutually perpendicular, representing the X and Y axes of a rectangular system of coordinates. Optionally, a third (Z) axis, in a direction normal to the plane of the X and Y axes, can also be provided in several models of these machines.

The coordinate measuring machines considered in this section are equipped with staging tables of substantial size to accommodate a wide range of parts. As an example, the table size of the machine shown in Fig. 12-6 is about 42 by 36 inches. The guideways of the traveling head are, depending on the particular model, either of a cantilever or of a bridge type design and permit a nearly frictionless displacement of the head by applying a very small force.

For the function of referencing, that is, locating on the part the boundary points of the distances being measured, contact probes of a tapered plug form are the most commonly used elements. These probes are supplied in an extensive range of sizes and taper angles, to accommodate holes of different diameters. Jig plates with bores located in a network of rectangular coordinates represent typical examples of parts whose meaningful dimensions can very effectively be measured by using tapered probes for locating the individual positions. It is interesting to note that the tapered probe, by seating itself in the best fitting center with respect to the maximum material condition of the bore, automatically locates the individual reference points in agreement with the concept of true position dimensioning.

This inspection operation can also be reversed, and the coordinate measuring machines may be used for laying out the location of holes to be drilled or bored in positions specified by means of a system of rectangular coordinates. For this application, scribes or automatic punches can be substituted for the contact probes.

The essential design characteristic of these instruments is the application of a noncontacting type displacement measuring device that senses, in discrete increments of an inch, the separation between the individual points of referencing. The length of the sensed displacement is displayed in digital form representing integer and decimal fractional inches.

There are several comparable systems of instrumentation for accomplishing the incremental sensing of displacement; the two systems having found the widest acceptance will be discussed briefly.

Displacement Measuring Devices Using Diffraction Gratings

Diffraction gratings are transparent plates, usually made of glass, having a large number of equally spaced parallel lines. Such gratings have long been used for spectroscopic work and are commercially available. However, actual production is limited to a few companies in possession of very special manufacturing equipment.

When two equal gratings are placed on each other, with their lined sides facing, but off parallel by a very small angle, the periodic crossings of the lines on the superimposed gratings causes the appearance of relatively wide fringes—alternately light and dark—in a direction normal to the original grating lines. The width and spacing of these alternating shadings, known as "moiré fringes," are functions of the grating spaces and of the angle of slant.

If one of the two gratings is moved in a direction normal to that of the grating lines, then the fringes will appear to move in the direction of the grating lines. The ratio of the displacement of the fringes to the lateral displacement of the grating lines is the same as the ratio of the fringe spacing to the grating line spacing. The optical phenomenon produces a relatively simple, yet very reliable, tool for presenting an amplified image of a translational movement (see Fig. 12-7).

By using a photocell, the array of fringes passing a line of observation during a specific period or action can be counted with the aid of appropriate electronic instruments. When the grating spacing and the slant angle are selected to produce fringe distances directly associated with the standard unit of length measurement, the resulting count expresses numerically the distance of displacement.

The basic system of length measurement can be refined by more sophisticated instrumentation. A

Giddings & Lewis Measurement Systems

Fig. 12-6. Coordinate measuring machine of bridge type design with three axis digital readout.

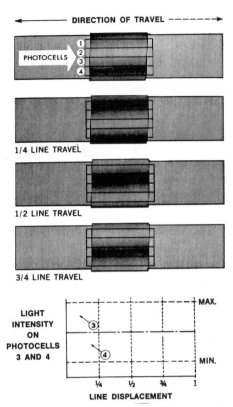

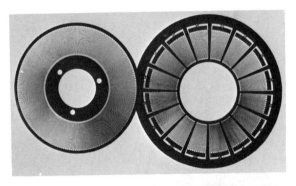

Giddings & Lewis Measurement Systems

Fig. 12-7. Principles of translation displacement measurement by moiré fringes. The illustrations show diagrammatically the master grating with the superimposed index grating in different relative positions, causing a shift in the fringe pattern locations, which are sensed by the photocells.

few important developments, their purposes and principals of operation are reviewed briefly below:

1. The basic system is insensitive to the direction of displacement; actually, the photocell makes counts when the relative motion between the two gratings is just a reciprocating, and not a continuously advancing one. By operating two photocells instead of one, and placing them at a 90 degree phase separation, the pulses received by these elements will differ according to the direction of the fringe displacement.

2. A plain count of the fringes permits discrimination of the displacement distance in increments equal to that of the grating's spacing. However, by using phase sensitive devices, which are capable of sensing various phase portions of the fringes while passing through the area of observation a subdivision of the grating spacing, such as to its $\frac{1}{4}$th or $\frac{1}{10}$th

part, becomes feasible. The resulting incremental readout makes it possible to use moiré fringe count devices for measurements in 1/10,000 part of an inch by utilizing gratings having 1000 lines per inch.

3. The principles of fringe pattern-producing gratings, combined with photosensitive electronic counting devices, are also used for measurements of rotational motions to indicate the amount of displacement in angular values.

The Inductosyn

Another group of incremental displacement measuring devices, known by the name of "Inductosyn," utilizes the inductive coupling between conductors separated by a narrow air gap. The system is adaptable for measuring and controlling the distance of linear displacements or the angle of rotational movements.

The Inductosyn produces output signals that result from averaging the total number of poles in a linear system, thereby, compensating for small errors that might be present in the spacings of the individual conductors (see Figs. 12-8). Because the signals are purely electrical, they can be displayed in digital form, recorded on a chart, printed out in numerals or punched on tape. One of the distinctive advantages of the inductive system is the rapid response,

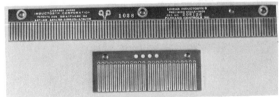

Farrand Controls, Inc.

Fig. 12-8. Inductosyn gratings.
(Top) Angular measurements.
(Bottom) Linear measurements.

permitting traverse speeds of the measuring head up to 6000 inches per minute.

Inductosyns, like other electronically counting and rapidly responding incremental displacement measuring systems, can be used in measuring machines for either of two basic methods of inspection: (1) traversing or rotating the movable machine member to a point where the boundary line of the separation to be measured is in coincidence with the reference position of the instrument's sensing element, and then reading out the distance of displacement (applicable for measuring unknown distances); or (2) causing the movable member of the machine to carry out a translational or rotational displacement that corresponds to the nominal value of the separation to be inspected, and then checking the presence of coincidence or the amount of deviation with the aid of an appropriate sensing device that incorporates the reference position (preferred method for certain types of inspection operations). Examples of sensing devices are the electronic indicator for length and autocollimator for angle.

Independently of the selected pulse generating system, which is the basis of the incremental distance measuring process, coordinate measuring machines possess several distinctive characteristics, a few of which are mentioned below.

The convenience and rapidity of operation of the various systems of increment counting measuring machines is greatly enhanced by the "floating zero" system. This designation refers to a switching device by means of which the zero reading of the numerical displays can be started at any position along the effective traversing or rotating range of the movable machine head.

Distinctive advantages of the described systems of coordinate measuring machines are the speed of operation and the potential freedom from errors, even when less experienced operators are assigned to inspection work involving a large number of interrelated dimensions. On the other hand, to reap these advantages, a specific method of locating the individual positions on the part is generally applied, namely, locating by means of tapered probes. There are certain limitations to this locating process because the probe can deflect; it is moving with a finite clearance in its guide bushing and, being attached to a heavy, although easily moving machine member, the sense of feel is affected.

The effective use of this type of instrument is predicated on parts whose critical dimensions are specified by a system of rectangular coordinates referenced from a plane that is parallel to that contain-

ing the X and Y axes of the measuring head travel. Finally, the significant gaging points on the part, even when at different levels, must be accessible from a single plane and represented by physical configurations, such as holes, precise shoulders and so forth, which are susceptible to position identification by the single probe selected for a specific measuring process.

In view of the inherent advantages of this relatively new breed of measuring machines, it appears appropriate to consider the range of potential uses in a more detailed and specific manner.

Coordinate Measuring Machine Types

Coordinate measuring machines (CMMs) are typically placed in one of three basic categories based upon their mountings. The mounting categories are: (1) floor-mounted, (2) bench-top, and (3) portable. The floor-mounted variety (Fig. 12-9) is a large capacity stationary machine of the highest accuracy. Floor-mounted CMMs are further subdivided into two styles of construction, bridge and cantilever. The bridge style of construction shown in Fig. 12-9 is characterized by a bridge structure that supports the vertically mounted arm-with-probe for maximum stability. The cantilever construction style of CMM shown in Fig. 12-10 provides the user with easier access to the worktable by supporting the arm-with-probe via a cantilever design. The bench-top variety (Fig. 12-11), also referred to as a *personal CMM*, is designed to provide low cost CMM capability for companies whose parts will fit into this ma-

Giddings & Lewis Measurement Systems
Fig. 12-9. Floor-mounted coordinate measuring machine of bridge style construction.

TABLE 12-3.   EXAMPLES OF COORDINATE MEASURING-MACHINE APPLICATIONS – 1

DIMENSIONAL CATEGORY	DESIGNATION	DIAGRAM OF THE PRINCIPLES OF MEASUREMENT	DISCUSSION
Linear distances	The linear separation of external or internal straight sides in planes normal to the direction of measurement		External and internal distances with boundaries which are directly accessible to the probe can be measured with cylindrical contact tips of known diameter, in planes parallel to the machine table.
	The linear separation of external or internal sides in recessed positions, either parallel or slanted to the gage head axis		The location of points which are not directly accessible because recessed behind flanges or shoulders, may be measured by using disk type probe tips of sufficient diameter to reach the significant surface area. Spherical contour disks are selected for slanting surfaces.
Two-dimensional coordinate positions	The coordinate positions of round holes in a flat surface		The relative locations of round holes contained in a common plane are eminently adaptable for inspection by coordinate type measuring machines. The reference point of the measurements is usually the center of one of the holes, with a second hole selected for the orientation of the measuring axis.
	Coordinate positions of diverse features, including round holes, radii and fillets measured by optical targeting		The coordinate positions of external and internal features bounded by straight lines, or circles, also such not directly defined due to radii or fillets, can be determined by optical targeting using for a guide the eyepiece graticule of a spindle microscope or the screen chart of an optical projector.

TABLE 12-3.   EXAMPLES OF COORDINATE MEASURING-MACHINE APPLICATIONS – 2

DIMENSIONAL CATEGORY	DESIGNATION	DIAGRAM OF THE PRINCIPLES OF MEASUREMENT	DISCUSSION
Slopes and Tapers	Measurement of the slope angle (slant in relation to the datum direction)	$$\frac{d}{L} = \tan \alpha$$	The slope (angle α) of a selected section in relation to one of the measuring axes to which the part has been aligned, can be determined by measuring the positions of two points along the surface section and then correlating their difference in distance to the separation of the points along the measuring axis.
	Measurement of the taper angle (angular separation of two opposite sides)	$$\frac{(D_1 - D_2)/2}{L} = \tan \frac{\beta}{2}$$	The measurement of the taper angle requires the staging of the part with its axis in the horizontal plane and also parallel to one of the measuring axes. The correlation of the difference in diameters with the axial separation of the gaging planes provides the taper angle (β). Machines equipped for Z axis measurement simplify the staging requirements.
Height relationships	Vertical distances —the height difference of part surfaces which are parallel to the horizontal datum plane		The measurements are predicated on essentially flat surfaces which are staged parallel to the horizontal measuring plane and are accessible to a probe moving along the Z axis. The simultaneous indication of the coordinate positions along the X and Y axes permits precise location of the gaging points.
	The consistency of the height difference between interrelated surfaces indicates the degree of parallelism.		The parallelism of surfaces which are located on opposite sides of a part can be measured by height comparison by staging the inferior surface of the part parallel to the horizontal reference plane and making Z axis measurements on the top surface.

TABLE 12-4.   EXAMPLES OF COORDINATE MEASURING-MACHINE APPLICATIONS – 3

DIMENSIONAL CATEGORY	DESIGNATION	DIAGRAM OF THE PRINCIPLES OF MEASUREMENT	DISCUSSION
Curved contours	Curved (irregular) contours can be approximated using closely spaced points determined by coordinates		The resulting coordinate values represent elements of a line parallel to the actual part contour, the individual points being at r (= radius of the probe) distance from the physical surface in a direction normal to the tangent at the point of contact. Measuring point spacing must assure proper resolution.
Three-dimensional coordinate positions	The locations of discrete points of a surface measured in a system of three-dimensional coordinates by contact with spherical probe tip		Three-dimensional inspection of an irregular surface (e.g., of a forging) by measuring the locations of specific points at predetermined spacings in several parallel gaging planes. The part must be staged to assure the horizontal position for the datum plane of the surface being measured.
Fixture staged inspection (Examples of uncommon uses for coordinate measuring machines)	Assuring by special support fixture the horizontal position of the measured surface		Support and orientation for parts of irregular shape can be assured by special holding fixtures, which may also serve as transfer members for datum elements not directly accessible to the contact probe.
	Establishing controlled interrelation between the staging and the measuring surfaces		For presenting in the horizontal plane the part surface which contains the dimensioned elements, the slanted supporting surface must rest on a holding fixture which provides the needed positional compensation.
	Presenting sequentially the angularly interrelated surfaces of the part by means of a circular dividing device		By checking the actual alignment with a datum direction of adjoining surfaces of an object which is consecutively rotated by the nominal spacing angles, the mutual orientation of the sides can be measured. Rotary tables with numerical read out of angular positions are also available.

Giddings & Lewis Measurement Systems

Fig. 12-10. Floor-mounted coordinate measuring machine of cantilever style construction.

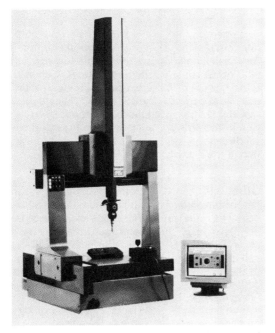

Brown & Sharpe Manufacturing Co.

Fig. 12-11. Personal (bench-top) bridge style coordinate measuring machine.

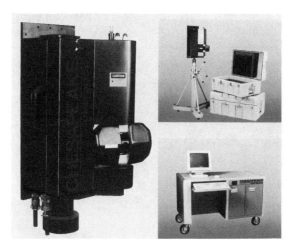

Chesapeake Laser Systems Inc.

Fig. 12-12. Portable laser (noncontact) coordinate measuring machine.

chine's smaller measurement envelope. The portable (noncontact) variety of CMM shown in Fig. 12-12 is used to take measurements on objects too large to fit into a mechanical CMM. This portable system uses laser interferometry to measure X, Y and Z coordinates with submicron resolution. The head of the CMM emits a single helium–neon laser beam, which is targeted at a retro-reflector mounted on the object being measured. The target reflects light back to the laser head for encoding and conversion to ASCII format. This information can then be compared to the original computer model of the part. A section found later in this chapter and entitled "The Purchase Decision and CMM Standards" will speak to factors involved in selecting the type of CMM best suited for a given situation.

Probes

There are quite a variety of targeting or sensing members currently being used with coordinate measuring machines. Examples of several types of probes and sensing members are shown in Table 12-4. However, unless sensing members are used that permit positive alignment of the measuring head axis with the pertinent element of the part surface (that is, direct targeting), correction factors must be applied in evaluating the indicated dimension. For example, in linear measurements the radius of the contact tip must be considered, adding its value to the indicated distance for internal measurements, and subtracting it for external sections. All modern CMMs are equipped with special correction factors that automatically compensate for the diameter of

TABLE 12-4.   EXAMPLES OF TARGETING (CONTACT) MEMBERS FOR COORDINATE MEASURING MACHINES — 1

DESIGNATION	SKETCH WITH SIGNIFICANT DIMENSIONS	ALIGNMENT CHARACTERISTICS	APPLICATION NOTES
Tapered probe tip	D_1 D_2	Essentially self-aligning in a hole of round form and free from burrs at the edge contacted by the probe.	Best suited probe shape for measuring the distances between hole centers. The datum for the coordinate relationship should be represented by one of the holes, using a second hole for establishing alignment with the measuring axes. Complies with the concept of true position dimensioning.
Cylindrical probe tip (with parallel sides)	$D/2$	Offset—equal to half the cylinder diameter—must be calculated when laying out the measuring process or evaluating the indicated values.	Preferable shape for contacting outside surfaces as well as inside surfaces with radii larger than that of the probe when the surface is square to the plane of measurement. Will ride parallel to the envelope surface of the sides being scanned and is insensitive to interruptions e.g., grooves normal to the probe axis.
Ball tip or probe tip with spherical end	$D/2$	Alignment in the horizontal plane, for the ball only, with offset equal to half its diameter. Both permit repeatable location in the vertical direction (along the Z axis).	The ball is the proper shape for measurements along three axes and preferable for vertically slanting surfaces, such as an erect conical body or tapered hole. The ball shape is sensitive to local variations of a physical surface from its general envelope.
Disk, plain cylindrical or with spherical contact surface	$D/2$ $D/2$	Offset equal to half of the diameter. The rounded perimeter (e.g., spherical zone) will cause shifting of the effective contact distance relative to the axis when used on slanting surface.	Permits sensitive contact to be made on areas inaccessible to other probe shapes because not open toward the machine head, such as surfaces behind shoulders, with receding slant or in the bottom of grooves. The rounded (spherical or toroidal) contact zone is required for surfaces which are not parallel to the probe axis.

**TABLE 12-4.   EXAMPLES OF TARGETING (CONTACT) MEMBERS FOR
COORDINATE MEASURING MACHINES — 2**

DESIGNATION	SKETCH WITH SIGNIFICANT DIMENSIONS	ALIGNMENT CHARACTERISTICS	APPLICATION NOTES
Off-set cylinder, or Cut-off cylinder		Basically aligned in a single orientation; deviation from the correct contact direction causes error which increases gradually for the offset cylinder, and rapidly for the cut-off design.	Its use requires the shaft to be precisely guided in the bushing of the head, to assure that the probe axis will remain unchanged in any orientational position which could result from the probe's rotation.
Centering indicator (preferably electronic with rotatable mount)		Has no inherent centering ability but will rotate around an aligned axis and indicate the radial value of concentricity errors. (A probing process also known as tramming.)	Used on coordinate measuring machines in special cases only, such as centering the head relative to large holes. By setting and locking the machine head to a position corresponding to the basic hole center, the scanning of the hole surface by rotating the indicator will show as runout error the direction and amount of axial non-coincidence.
Centering microscope		Alignment of the selected surface element is assisted by the pattern of the target graticule and the magnified image of the surface area.	Targeting the centerline of the measuring head on a particular point of a surface feature bounded either by straight or circular lines is achieved without mechanical contact when brought into focus and the significant area can be encompassed by the microscope's field of view. Useful for centering on intersections undefined due to radii.
Optical projector with lens system and screen mounted on the measuring head		Alignment by establishing coincidence of the enlarged feature image with the reference marks on the screen chart.	Similar in function to the centering microscope; while bulkier in size, it permits remote observation of the targeted area. Of advantage when measuring the relative locations of special features whose enlarged contours traced on screen charts assist rapid and precise targeting.

Renishaw Inc.

Fig. 12-13. Electronic touch probe for use on coordinate measuring machines.

the touch-probe ball as inputted to the electronic controller. The electrical signals resulting from the measuring machine's sensing head, whether originating from photoelectric or inductive systems, are typically channeled into a computer control for electronic digital display and hard copy printout via an attached printer for recording the distances of the individual measuring points from selected datums.

The most popular and versatile probe on the market today is the electronic touch probe (see Fig. 12-13). As the stylus ball of the probe comes into contact with the workpiece, its deflection is recorded electromagnetically with high resolution. This deflection is perpendicular to the workpiece; therefore, the measuring force is proportional to the deflection of the probe and can be determined. Each measurement is electronically recorded and stored to be printed in hard copy. The permanent numerical information produced in printed form can be of distinct value in diverse inspection operations. Benefits that can be derived are (1) the availability of authentic measurement records; (2) the reduction of operation time and incidence of error by eliminating the manual recording; and (3) in particular applications, the data obtained can be directly compared to computer-aided design (CAD) and or computer numerical control (CNC) information previously prepared.

Accessories

A variety of new accessories have made the CMMs of today more useful in the quality control laboratory as well as in the factory environment. The modern coordinate measuring machine is computer-controlled and is often equipped with advanced statistical process control applications software, graphics supporting software and a touch screen operator interface. Better record keeping, increased accuracy, fewer hand calculations, increased up-time and reduced training requirements have all been allowed through the integration of computer software with the modern coordinate measuring machine.

Two additional accessories have allowed the CMM to take its place on the factory floor alongside other production gages. The first of these accessories is an environmental enclosure as shown in Fig. 12-14. This positive air-flow enclosure protects and keeps the CMM and its work surface free from airborne contaminants found in the factory environment. Computer-controlled automatic temperature compensation is the second important advance, allowing for the shop floor use of the CMM. Dimensions and tolerances on drawings apply at 20°C. Since it is not cost-effective to maintain the factory at 20°C (Centigrade), a means of compensation for the differences in the temperatures of the workpiece and the CMM is needed. Computer software is now available for CMMs that will convert measurement information into the 20°C standard. The temperature of the CMM and the part being measured are moni-

Brown & Shape Manufacturing Co.

Fig. 12-14. Environmental enclosure for personal CMM permitting high precision shop floor gaging.

tored through the use of sensors. This temperature and measurement information are automatically converted to the standard gage temperature allowing for known coefficients of thermal expansion previously inputted by the inspector. However, this software cannot adapt to rapid temperature changes and is therefore most effective at temperatures between 15°C and 35°C. The use of both the environmental enclosure and temperature compensating software will provide optimal accuracy in shop-floor type CMM workpiece inspection.

The Purchase Decision and CMM Standards

The versatility, accuracy and documentation capability of the modern CMM makes it the inspection tool of choice to an ever-increasing number of precision parts manufacturers. Complex manufactured parts and tools (that is, molds and dies and so forth) are often accepted, rejected or repaired solely on the strength of CMM data. In fact, more and more manufacturers are demanding that their suppliers provide formal dimensional documentation of quality via CMM as a condition of doing business.

Presently, there are two standards sanctioned by the American National Standards Institute (ANSI) that relate directly to coordinate measuring machines: (1) B89.1.12M-1985, which covers "Methods for Performance Evaluation of Coordinate Measuring Machines" and (2) ANSI/CAM-I101-1990, which relates to "Dimensional Measuring Interface Specification" or (DMIS) and is the culmination of a multiyear effort to develop an industrywide standard interface between computer-aided design systems and coordinate measuring machines. Both of these standards should be consulted as the research leading up to a CMM purchase decision begins.

Several factors must be considered when the purchase of a CMM is imminent: (1) the required capabilities of the machine, including the size of the work envelope, the accuracy of the unit and the kind of information that can be generated by the system's software; (2) the expertise of the vendor; (3) service after the sale, which must include availability of replacement parts and a service technician; (4) the simplicity of the machine, which relates to operator training, type of controls and the number of different people expected to use the CMM; and (5) price, which in 1993 dollars ranges from a low of $15,000 to upward of $1,000,000. With this kind of price range, special attention must be paid to the factors listed above, as well as potential future uses and innovations that may render a given CMM inadequate or obsolete.

JIG BORER TYPE MEASURING MACHINES

The accuracy of jig borer type machining operations has greatly advanced in the past decade mainly as the result of improvements in integral measuring devices, supported by developments in the materials and in the manufacturing processes of the mechanical machine elements. Jig borers are designed to position the object in the plane of a system of rectangular coordinates, and possess complementary controls in directions normal to that plane, as well as for rotation around different axes, with all these movements being carried out in relation to common datum elements.

The positioning flexibility of jig borers, combined with the use of very accurate measuring devices for setting the object and for controlling the tool advance, brought about a situation where the conventional measuring instruments that were available to inspectors proved to be inadequate, dimensionally for checking complex parts machined on a modern jig borer. These conditions led to the practice of inspecting the part while still on the jig borer. However, such procedures often caused costly tie-ups of the jig borers required for machining operations and were still poor expedients considering the temporarily mounted sensing devices that replaced the cutting tools. Another shortcoming of this practice, not only in principle but also affecting the results, is attributable to the use of the same measuring devices in the inspection operation that controlled the original production process. Furthermore, no recourse could be made to such methods in the receiving inspection of parts that were machined by an outside supplier, or on complex and critical components that were manufactured by means of special fixturing without the use of jig borers.

In view of the described conditions, it would have seemed an obvious solution to the problem to create measuring machines that by incorporating the jig borer's characteristic flexibility of adaptation, would possess the essential properties of a measuring instrument as well. In this respect, important additions would include excellent tracking and orientational accuracy of the traverse and rotational movements, very sensitive and accurate displacement measuring devices as well as sensitive locating instruments, preferably with indicating capabilities.

In retrospect, it may appear surprising that, although the practice of using jig borers as a means for checking the accuracy of complex parts dates back to the late 1930s, it was only in recent years that jig borer type measuring machines were created, primarily as an American contribution. Jig-borer type measuring machines are now manufactured in this country by several machine tool manufacturers including the Moore Special Tool Company of Bridgeport, Connecticut, who pioneered that type of instrument. The available range covers sizes from 9 by 14 inches to 24 by 48 inches. These dimensions express the effective table movements along the X and Y axes, respectively.

The practice of using jig borers for making measurements has probably not been generally discontinued. However, the advantages of single-purpose measuring machines for operations involving no chip removal, but requiring the highest attainable level of accuracy, are becoming more widely recognized.

Upgraded workmanship in producing machine components and the refinement of the displacement measuring devices of the basic jig borer, together with the addition of complementary equipment, constituted the first phases of development (see Fig. 12-15). However, some of the larger-size measuring machines of this category represent original designs and were developed for the sole function of dimensional measurements (see Fig. 12-16). In contrast to jig borers, measuring machines are not required to resist the forces applied to, and the dynamic effects resulting from, heavy chip removal. However, the concept of inspection implies measuring and positioning accuracies that represent a higher grade than those available for the production phases. In some of the measuring machines of more recent origin, these requirements have been satisfied to a remarkable degree by means of unusual manufacturing methods and as the result of distinctive design characteristics.

The discussion of jig-borer type measuring machines would be incomplete without mentioning the "Superoptic" measuring machines of SIP (Société Genevoise d'Instruments de Physique), which utilize optical master scales for measurements along the three axes (see Fig. 12-17).

The reason for discussing these machines in con-

Moore Special Tool Co.

Fig. 12-15. Jig borer type measuring machine in compact design (Moore No. 1½). The translational movement of the tables is actuated and measured by a special lead screw. Equipped with a highly precise rotating spindle, it can also be provided with power drive for both the translational and rotational movements.

Moore Special Tool Co.

Fig. 12-16. The Moore No. 5, universal measuring machine, equipped for operation by numerical control and for the digital display of the setting position coordinates.

American SIP Corporation

Fig. 12-17. "Superoptic" (SIP) type measuring machine with a rigid, double-column frame that was originally developed for jig borers capable of heavy chip removal in machining operations. The machine has a power drive, and measurements are made by optically observed master scales along three axes.

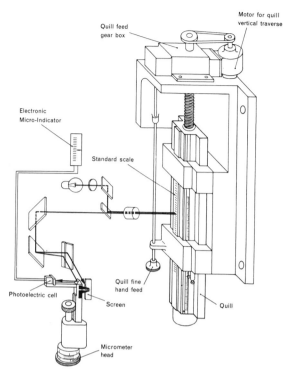

American SIP Corporation

Fig. 12-18. Diagram of the positioning and measuring system for the vertical quill movement in the Superoptic measuring machine.

junction with the jig borer type, instead of in the category of the optical measuring machines, is that the basic design of the Superoptic is essentially identical with that of the widely used Hydroptic jig borers of SIP. An additional feature for the measuring machine version is the electronic microindicator. That device complements the displacement measuring system and assists the operator in the precise, to 10-microinch, centering of the projected scale lines between the reticle indexes of the optical scale reading screen (see Fig. 12-18).

The Superoptic measuring machines are unique with regard to their large dimensions, resulting in the following measuring ranges along the three axes, which is accomplished by retaining the full support of the table over its entire length of travel:

Direction of Travel	Moving Machine Member	Working Range (in inches)	
		Superoptic-6	Superoptic-7
Longitudinal	Staging table	40	55
Transversal	Headstock saddle	28	40
Vertical	Quill in headstock	18	18

The targeting function requiring the alignment of the quill axis with the boundary elements of specific dimensions on the object surface is usually carried out with the aid of an electronic indicator held in a rotatable mount. The indicator device is retained in a high precision spindle that can be rotated in the quill either manually or by a motor drive. For centering in a bore the power rotation is used, and for picking up specific points on the part surface the spindle can be locked in one of eight fixed orientations. A high degree of skill is needed for accomplishing the essential targeting operation with a precision commensurate with the very high measuring accuracy that is assured for the positioning of the machine elements along each of the three measuring axes.

When considering the means of continuous displacement measurement used in jig-borer type machines, two systems appear to dominate the field, viz., the optically observed master scale type and the precision lead screw type. Both systems possess characteristic advantages, but they do not assure the definite superiority of either system. Several factors, such as the degree of workmanship, the selection of materials, the application conditions and various others, can have a decisive effect on the ultimate accuracy and suitability of the selected system.

Optical devices are essentially free of water, strain and hysteresis. The actual position of the displacement machine element is displayed independently of the conditions in the traversing mechanism, and of the direction of the translational movement.

Lead screws producing very smooth and sensitive displacement are necessary elements of all length measuring devices, including the optical systems. Modern technology permits the manufacture of threads that are practically free of lead errors, and advances in metallurgy have provided highly stable and wear-resistant materials for lead screws and nuts. The power of resolution of a fine pitch lead screw equipped with a large graduated drum, which is observed in conjunction with a vernier, can exceed the level currently achievable with optical devices. Further, lead screw operated displacement systems lend themselves to programmed positioning by using numerically controlled instrumentation developed for other systems of machine tools.

The parallel advances in two technically competing systems proved to be mutually beneficial as demonstrated by the progress achieved to this date, and additional improvements may be anticipated. Although no single "best system" resulted, the choice available for the selection of a length measuring device most suitable for specific applications is of undisputed value to dimensional metrology.

To visualize the characteristic measuring potentials of jig borer type measuring machines, three inspection operations are discussed and the respective machine setups are shown in Figs. 12-19 to 12-21. The equipment used is a No. 3 Moore measuring machine with motor drive for the translational movements of the tables and for the rotation of the spindle. The electronic indicator, which functions as the sensing element, has its output signals channeled into either the polar chart or the strip chart recorder, selected in accordance with the particular tracing operation. The movements of the machine elements and of the recorders are synchronized to produce

Fig. 12-20. Measuring machine setup for the measurement of the curvature of a bearing adapter and for recording the deviations of the traced element from its basic form.

Fig. 12-19. Measuring machine setup for the angle measurement of a tapered, roller bearing outer race, resting on a sine plate. A strip chart recorder is connected to the electronic gage head through the amplifier.

Fig. 12-21. Measuring machine setup for the measurement of the coaxiality of bearing seats bored in an automotive differential housing. A rotary recorder is used for preparing polar charts, which indicate the relative conditions of the traced bore surfaces.

continuous line tracings in 1:1 ratio for the polar chart recorder or at a selected amplification ratio for the strip chart recordings. These systems serve to complement the indications of the meter's pointer in processes where chart recordings are useful for more informative and dependable measurement evaluation.

The selected examples demonstrate the role of multiple-axis measuring machines in inspection operations where, in addition to the size conditions of defined dimensions, their geometry and relative locations must also be determined. Tables 12-5 to 12-7 describe briefly the purpose and the techniques of these inspection operations, using diagrammatic drawings to illustrate the individual dimensions and other geometric conditions that must be measured in the process.

The versatility of jig borer type measuring machines is further extended by the use of rotary tables or dividing heads. Circular spacings and polar coordinates can be measured, both axially, on the face, or radially, on the perimeter of a round object. By combining the rotary table with a sine plate, such as shown in Fig. 12-19, circularly spaced features on a

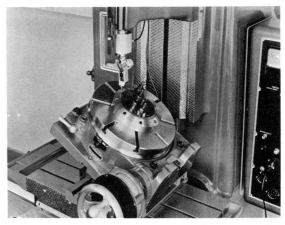

Moore Special Tool Co.

Fig. 12-22. Rotary table supported on a sine-plate, which is mounted on the table of a jig borer type measuring machine, for checking circular divisions and axis directions of features on a conical surface. Targeting is accomplished with an electronic indicator.

conical surface can be inspected for location and axis angle (Fig. 12-22).

TABLE 12.5.   EXAMPLE OF MULTIPLE-AXIS MEASURING-MACHINE APPLICATION
(Machine equipped with strip chart recorder)

PART DESIGNATION: Double row tapered roller bearing cup

DIMENSION TO BE MEASURED: The slope angle of the pathway

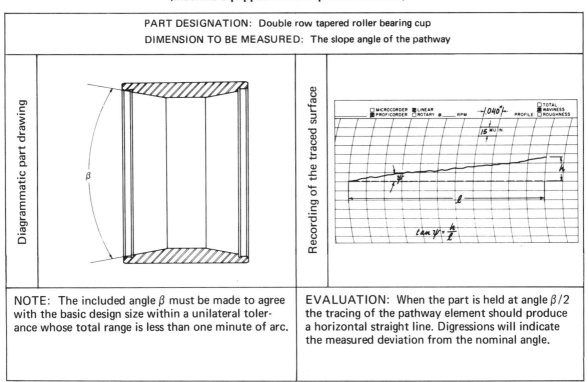

NOTE: The included angle β must be made to agree with the basic design size within a unilateral tolerance whose total range is less than one minute of arc.

EVALUATION: When the part is held at angle $\beta/2$ the tracing of the pathway element should produce a horizontal straight line. Digressions will indicate the measured deviation from the nominal angle.

TABLE 12.6.   EXAMPLE OF MULTIPLE-AXIS MEASURING-MACHINE APPLICATION
(Machine equipped with strip chart recorder)

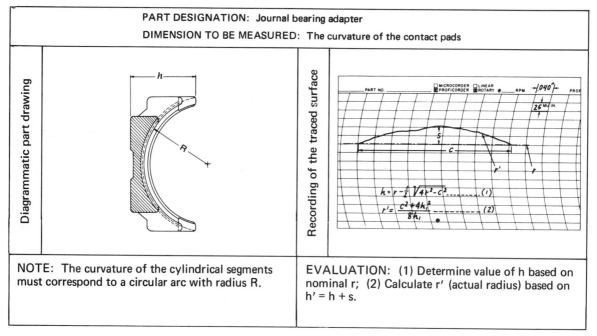

PART DESIGNATION: Journal bearing adapter

DIMENSION TO BE MEASURED: The curvature of the contact pads

NOTE: The curvature of the cylindrical segments must correspond to a circular arc with radius R.	EVALUATION: (1) Determine value of h based on nominal r; (2) Calculate r' (actual radius) based on h' = h + s.

TABLE 12.7.   EXAMPLE OF MULTIPLE-AXIS MEASURING-MACHINE APPLICATION
(Machine equipped with polar chart recorder)

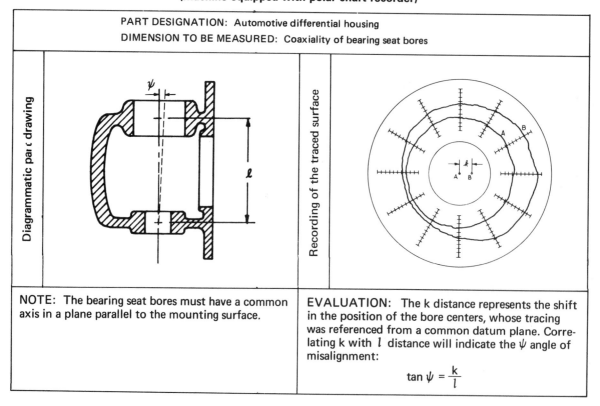

PART DESIGNATION: Automotive differential housing

DIMENSION TO BE MEASURED: Coaxiality of bearing seat bores

NOTE: The bearing seat bores must have a common axis in a plane parallel to the mounting surface.	EVALUATION: The k distance represents the shift in the position of the bore centers, whose tracing was referenced from a common datum plane. Correlating k with l distance will indicate the ψ angle of misalignment: $$\tan \psi = \frac{k}{l}$$

OPTICAL MEASURING MACHINES

Optical measuring machines—a designation assigned to multiple-axis self-contained measuring instruments—are not generally known as a strictly defined category. Optical devices occasionally are also used on other types of measuring machines, either for distance measurement or for the function of centering. There exists also a great similarity of operational purpose to the engineering microscope, although the application objectives and the dominant design concept of these two instrument categories are distinctly different. The essential purpose of the engineering microscope is related to its optical observation capabilities, which were discussed in detail in Chapter 8. The means for measuring the displacement of the microscope stage provide only a secondary, although often essential, function that complements the versatile observational potentials of the engineering microscope.

The purpose of defining the instrument category of optical measuring machines—without disregarding the interlacing of systems, functions and application techniques—is to assist the surveying of preferable uses of comparable measuring instrument types. These instruments, regardless of the similarity of purpose and of operational systems, are intended for different primary applications. Because it is the dominant kind of part configuration and dimensions that should direct the choice of the most suitable type of instrument, the optical measuring machine is discussed as a separate instrument category.

Optical measuring machines are distinguishable from other types of multiple-axis instruments discussed in this chapter by the exclusive use of line graduated master scales and optical scale reading devices, as means for measuring linear distances and angular spacings.

The distinguishing characteristic in relation to engineering microscopes is the method of targeting, mechanical contact being primarily used for this function in the measuring machines. Mechanical contact is the preferable method of targeting when clearly defined boundary surfaces of regular form are present on the part, and the only practicable means when the boundary points of the dimensions being measured are not accessible to optical observation.

Because of the greater depth control and reach of mechanical contact members, optical measuring machines are capable of making measurement over an extended vertical range. This condition makes them adaptable to measurements on much bulkier parts than are manageable on a microscope. For that rea-son, optical measuring machines are usually built with large staging surfaces that are dimensioned to carry relatively heavy objects without affecting the positioning and tracking accuracy of the translational movement. While on the engineering microscope the vertical adjustment of the head serves, in general, the purpose of focusing only, most of the optical measuring machines are designed for measurements along three axes, including the vertical. A typical optical measuring machine is shown in Fig. 12-23.

Obviously, the described characteristics of the optical measuring machines, as compared to those of engineering microscopes, involve higher cost that must be justified by the need for these additional capabilities. It should also be realized that there is a great variety of part configurations, geometric features, dimensional categories and bounding elements for which optical targeting by the highly adaptable means of modern measuring microscopes offer the best means of inspection. Optical targeting by visual observation of the part surface also may be used optionally on measuring machines, and is recommended when the overall size or weight of the part requires powerful staging members, yet the feature to be measured calls for microscope viewing.

The mechanical contact members of optical measuring machines commonly belong to one of the following basic types, which are listed, together with a

American SIP Corporation

Fig. 12-23. Optical measuring machine (SIP Trioptic), with optical scale readers by projected display for the translational movements of the table, cross-slide and head stock along three axes. A mechanically contacting, feeler microscope is used for targeting.

brief discussion of their characteristic advantages and limitations:

a. *Mechanical indicator.* Its use is limited to conditions where sufficient space is available for mounting the indicator in a position permitting parallax-free observation of the dial and the application calls for gaging displacements in one direction only. The mechanical indicator provides extensive measuring range with incremental values on both sides of the reference (zero) position.

b. *Electronic indicator.* It is distinguished by great sensitivity, high amplification potential, light contact force and readout on a remote meter with accurate incremental values over a bilateral range of limited extent. Electronic indicators are also preferred for centering type targeting when a precisely rotating spindle is provided on the measuring head of the machine for holding the transducer of the indicating instrument. They also are limited to single-directional sensing.

c. *Feeler microscope.* The feeler microscope is used as a fiducial indicator only, to signal coincidence of the contacted surface element with the reference position of the gaging. Functionally, it contains a tipping mirror rigidly connected with the arm of the mechanical contact point. Displacement of the contact element causes tipping of the mirror, which reflects a projected reference mark. In a specific tilt position of the mirror, the reflected reference-mark image appears in the field of view of the microscope and is observed in relation to the bifilar target lines of a reticle. Available in various designs, one (called the zero-indicator) was described in Chapter 8 as an optional accessory for measuring microscopes. The feeler microscope represents the basic sensing member for optical measuring machines, due to its great sensitivity, exceptional versatility assured by interchangeable contact points, light-gaging force and a unique characteristic: freedom from backlash or shift, even when reversing the contact direction while retaining the same line of sensing.

13.
Profile Measurements

Profile, in the aspects considered in the following discussion, is that part of the contour of a technical part or feature that is observed as its boundary line in a functionally significant cross-sectional plane. Such planes for regular geometric figures are the axial plane of a body of revolution, or the central plane of other types of symmetrical parts. From a functional point of view the profile may be considered in a plane perpendicular to the operational displacement movement, such as the cross-sectional plane of a guideway normal to the direction of translation of the guided member. Or, it could be the tangential plane in relation to a rotational path, as in a selected section of a propeller blade.

The basic profiles of engineering parts can have regular geometric forms, such as straight lines or circles (to be discussed in Chapter 14 under "Circular Contour Measurements"); a combination of straight lines and circular arcs in specific angular relationship to each other, such as thread forms; generated forms, such as the involute used in gears; or irregular forms developed to accommodate particular operational conditions (such as the camber or crown of a member subjected to heavy loads). Finally, the functional role of a part may require a form that varies over the surface, and although the changes are gradual and blend into each other, for practical reasons such shapes are often specified and measured by the profiles of precisely defined gaging planes, typical parts in this category being turbine blades and television screens.

The dimensioning of profiles in engineering drawings is commonly related to their basic form, or to the combined application of several basic forms that, in certain cases, must be blended in a manner expressly specified in the design drawing. The tolerancing is generally specified as deviations from the basic form, although for irregular profiles in particular, it is often practical to apply boundary zone tolerancing, requiring that the actual profile be contained between two essentially parallel lines, which are at a specific distance apart. Tolerance zones for profiles also may have other forms, such as gradually broadening areas to reflect the decreasing criticalness of adjacent sections. Finally, the tolerance zones may be located by straddling symmetrically the basic contour or may be shifted to one side only (permitting increased or reduced material conditions) in a manner comparable to the unilateral tolerancing of size.

Functionally, the profiles of the technical parts can be significant to various degrees and, in particular applications, a profile could represent the most important single condition of the part. The assigned significance is usually reflected by the specified tolerances, which could be considered as a guide in selecting the appropriate method of inspection and the most suitable type of profile-measuring instrument.

The inspection and measurement of the profile can be accomplished in many different ways, depending on the size of the part, the required functional properties and level of accuracy of the profile, its form characteristics and various others.

Frequently, the inspection is limited to checking the general shape of the profile, either disregarding the part's size, or in combination with the profile's effect on size. The specified tolerance may be large enough to permit visual inspection by comparison to a reverse contour master, or by gaging the coordinate positions of a few selected check points along the contour of the part.

The tolerancing of the profile may permit the reliance on one of the traditional toolroom methods, such as producing a light gap and estimating its width, bluing the surface and assessing the untouched areas, or depending on feel in the case of a full-contour limit gage.

For greater accuracy, enlarged viewing will be needed such as offered by a microscope or by an optical projector. The profile may even be critical to an extent requiring very high resolution for inspection in the significant sense, a kind of capability that is provided by the continuous scanning of the surface in a process known as profile tracing.

The dimensions of the profile tolerances are often more specific, requiring the profile to be contained within a defined tolerance zone, which can be displayed graphically on a magnified contour chart prepared for use on an optical projector screen.

Another way of specifying the profile tolerances is by calling out the sense and the dimension of the permissible deviations from a base line, most frequently either a straight line or a circular arc. Charts produced by profile tracing, either linear or rotary, are eminently adaptable for the assessment of the profile conditions in accordance with the latter kind of tolerancing.

Finally, the profile and its position with respect to a datum can be a meaningful element for *measuring* the linear or angular dimensions of a technical part that is examined by determining the amount of deviation from the nominal condition of its profile. For this kind of measurement very high sensitivity methods are generally used, such as provided by measuring microscopes and, particularly, by profile-tracing instruments.

Several processes used for inspecting and measuring the profile characteristics of various parts are discussed in conjunction with diverse measuring methods and equipment. However, in order to provide a comprehensive review of the applicable processes, and as an aid for selecting the best suited method and equipment, a survey of profile inspecting and measuring systems is presented in Table 13-1.

While not claiming completeness, this survey indicates the wide variety of methods by which the profiles of technical parts and features are measured.

Some of these methods may be considered as specific applications of generally used measuring systems and processes, which are discussed in other chapters of this book, as indicated by the process references of the table. According to the table listings, profiles having the basic form of a straight line can be inspected, with very high sensitivity, by means of *profile tracing*, a method of relatively recent origin.

Profile tracing is a particular system of profile measurement that has not been discussed to any

length in other chapters of this book, although occasional references were made to it, such as in Chapter 12 on measuring machines. Profile tracing, as a system of dimensional measurements, is considered so unique in its capabilities, and so versatile in its potential applications, that a separate discussion appears warranted. For that reason the subsequent sections of this chapter will concentrate on the methods, the instruments and the applications of profile tracing, including some principles of the staging techniques, as well as the evaluation of the gaging results.

PROFILE TRACING

In this context, profile tracing designates an inspection process in which a selected element of the work surface is scanned in a continuous translational movement by the stylus attached to an electronic sensing member, frequently called the pickup. Deflections of the stylus in a direction normal to its translational path originate electrical signals that are proportional to the surface variations causing the stylus to deflect. The stylus's basic translational path is commonly a straight line to which the traced surface element is compared. (Considering the capacity of the stylus to deflect, in order to intimately follow the traced surface, the term "stylus path" refers to an imaginary surface element that will retain the stylus in its median position, represented by zero indications on the meter).

The basic straight line path is established in a position to coincide with the theoretical contour of the part, or with a conceptual straight line to which the design contour is related. In practice, that positioning is accomplished by the mutual adjustment of the stylus path and of the specimen.

The straight line path is produced by the relative displacement of the object and of the sensor. That relative displacement along a precisely controlled track will result either from the movement of the sensor in relation to a stationary object or from the translational displacement of the object stage in front of a fixed position sensor.

Any adjustment needed to assure the proper location and orientation of the stylus path is carried out prior to the actual scanning operation, and then locked. The path, once aligned, will thereafter be followed by the relative translational movement along a very precisely maintained track whose maximum digressions from the basic straight line are usually guaranteed by the instrument manufacturer.

Another characteristic of profile tracing, as an inspection process, is the preparation of chart tracings

TABLE 13-1.   METHODS OF PROFILE MEASUREMENT — 1

SYSTEM	APPLICATION EXAMPLE AND DIAGRAM	DISCUSSION
Direct comparison to a master having the inverse form of the nominal contour	Contour gage for checking the profile form of a barrel roller	Digressions of the specimen contour from that of the master causes gaps which are observable by light transmission directly or viewed through a microscope tube. When applying a dye on the inspected surface the gaps will be observable as untouched sections. The operational element of the gage is often featheredged for improved contact definition.
Limit gages made for the entire contour of the part with dimensions corresponding to the tolerance limits of the design	Gap gage for the inspection of an extruded shape	Such gages are usually made in pairs, representing the GO and NOT GO sizes, which are determined by the maximum and minimum material condition resulting from the compounded effects of the form and size tolerances.
Indicators mounted side by side in a plane which contains the profile to be inspected	Inspecting the profile of a part with irregularly-curved contour	The judiciously located indicators (mechanical or other systems) must be set with the aid of a master representing the nominal profile. By substituting air gage sensing cartridges for the individual indicators, the deviations of the inspected points from the basic positions can be displayed remotely on closely mounted indicator columns.
Measuring the coordinate positions of characteristic contour points	The profile of a dovetail gage	The size and form of features bounded by essentially straight lines can be inspected by checking the coordinate positions of characteristic points along the operational profile or around the entire contour.

TABLE 13-1.   METHODS OF PROFILE MEASUREMENT — 2

SYSTEM	APPLICATION EXAMPLE AND DIAGRAM	DISCUSSION
Contour defined by polar coordinates.	Contour of a screw machine cam	The functionally essential contour characteristics of a cam can be specified, laid out and inspected by individual points whose positions are defined by means of dimensioning in a system of polar coordinates.
Magnified image of the contour observed conjointly with a graticule carrying reference lines of normal form and sizes.	The radius of a fillet measured by comparison to a circle graticule	The silhouette image of the inspected feature appears in the focal plane of a microscope, where a transparent graticule with concentric and individually dimensioned circles is mounted. The staged part is moved to coincide with the best fitting graticule line to show the size and the degree of conformity of the curvatures.
The magnified image of the contour is projected on a screen for comparison with the outlines of a screen chart.	Inspection of the total contour of a turned part	Special screen charts with single or double contour lines, the latter bounding tolerance zones, permit the inspection of size and form, either of an entire part or of a feature, with complex shapes, as long as the observable profile represents the meaningful parameters.
Comparing selected surface elements to straight lines established by the translational movement of the work stage in relation to a stationary gage tip.	The straightness and parallelism of sections of a stepped shaft	Once the part is properly located the change in the position of the contacting indicator instrument will not affect the relative parallelism of the displacement paths produced by the essentially straight line surface elements of the part, whose translational movement is assured by precisely guided stage slide.

TABLE 13-1.   METHODS OF PROFILE MEASUREMENT — 3

SYSTEM	APPLICATION EXAMPLE AND DIAGRAM	DISCUSSION
Comparing selected surface elements of a part to a straight line created as the path of a sensing point in its translational movement along a stationary part.	Checking the straightness of a bore surface element and the uniformity of the wall thickness	A moving sensor, when mounted on a slender arm, can penetrate into internal or other hard-to-reach areas for scanning the profile of specific surface sections in relation to a straight line path along which the sensing tip is being moved.
Circular form of a surface inspected by comparison to the rotational path of an indicator probe.	Checking the roundness of a bore in alignment with the axis of a machine-tool spindle	A common application is selected to illustrate the principles of the conceptually correct method of circularity measurement by comparing the contour of a feature, in a plane normal to its axis, to a precisely circular path described by the tip of the sensing instrument.
Generated form, such as an involute, created as the reference path of a sensor by concurrently acting constituent motions.	Involute form checker for gear-tooth profiles	Many gear profile checking instruments operate by having a straight edge, on which an indicator is mounted, rolling on the perimeter of a disk, whose diameter is equal to that of the base circle of the gear being tested. The indicator in contact with the gear profile, will move along a true involute path and show any deviation of the scanned surface.
Optical profile comparison by light sectioning.	Light section microscope for turbine blades	A narrow band of light is projected obliquely around the edge of the specimen blade, to illuminate a section which, by means of optical correction, appears as a normal cross section in the field of view of the special microscope, to be observed concurrently with a master drawing of the profile.

for displaying the magnified representation of the traced surface element. In order to dependably represent the investigated contour section, the tracing line on the chart must be proportional to the actual part profile, yet displayed at scales of magnification that are selected individually for each of the two basic directions: the vertical and the horizontal.

Reviewing the indicated functional elements of the process, it may be stated that in order to be adaptable for profile tracing the instrument must possess means to carry out the following functions:

a. To produce a relative displacement of the part and of the sensing member along a path that precisely follows a straight line.

b. To provide means for sensitively adjusting the mutual positions of the part and of the scanning path.

c. To possess a sensing device (pickup) with a contact member (stylus) capable of tracing a selected element of the work surface and to produce electrical signals proportional to the deviations of the surface from the basic scanning path. This member is supported by a remotely located amplifier unit.

d. To be equipped with a recorder that, by utilizing the signals of the sensing head, can produce a chart tracing line accurately representing the scanned physical surface at the selected vertical and horizontal magnifications.

Although this is a rather demanding list of requirements, it is possible, in principle, to adapt many types of machines and instruments, especially machine tools with translational movements, to profile tracing, by equipping them with the needed supplementary accessories. In practice, however, such adaptations are rather rare, for economic and also for technical reasons.

Economically, it is generally unwarranted to tie down for a measuring process machine tools that were dimensioned and built for chip removal operations and are required as production equipment. A sensitive measuring process by profile tracing usually involves only insignificant forces, yet extensive setup and adjustment time is required in addition to the actual tracing operation. Furthermore, such basic machines must be equipped with instrumentations whose value may represent the greater part of a single-purpose instrument's cost.

From a technical point of view, adaptations are limited by the fact that most profile-tracing operations require a degree of tracking accuracy of the translational motion that exceeds the level commonly needed for machine tools.

However, adaptations for profile tracing might still be justified under particular conditions and for this reason the potential feasibility was brought up. Actually, there are cases in which such adaptations extend the original capabilities of the basic equipment, which was not designed originally to operate as a profile-tracing instrument. The resulting extended capabilities permit measuring with profile tracing, critical features of a part that, due to its size, weight or shape, do not fit the staging capacities of profile-tracing instruments. Examples of such uses were mentioned in Chapter 12, discussing the operational potentials of jig borer type measuring machines.

Disregarding these exceptions, the majority of profile-tracing operations are being performed on special measuring instruments. For this reason, the following detailed discussions of the process and of its equipment are based essentially on instruments designed expressly for profile-tracing operations.

The Systems and Design of Profile-Tracing Instruments

As indicated earlier, the translational movement needed for the continuous scanning of the selected surface element is produced by the relative displacement of the part and of the sensing member. Both basic systems, namely, (a) the stationary sensor contacting a part in translational movement, or (b) a moving sensor scanning the surface of a stationary part, are equally capable of producing such a relative displacement. However, the manner in which the relative displacement between the part and the sensing member of the indicator instrument is accomplished affects many characteristics of the equipment's mechanical elements. It reflects particularly on the arrangement of the various adjustment movements that are needed for the positioning and alignment of the selected surface element of the part with respect to the path of the stylus.

Essentially, these adjustments are related to five axes, three of these being the basic axes of a system of rectangular coordinates (the X, Y and Z axes) and two are rotational axes, both normal to the scanning path and mutually perpendicular (see Fig. 13-1). The required range of these adjustments is rather small, particularly for the rotational axes that extend over only a few degrees in either direction from the basic.

While adjustments along and around these axes are generally needed for the proper positioning and alignment of the tracing path along a specific surface

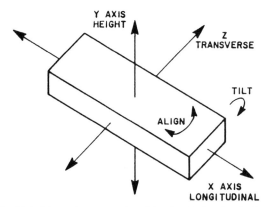

Fig. 13-1. The major axes of positional and directional adjustment needed for assuring the coincidence of the stylus path with a particular basic line with which the traced actual surface element will be compared.

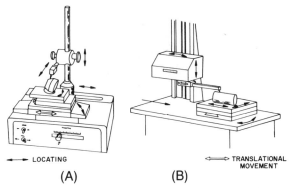

Fig. 13-2. Basic types of profile-tracing instruments: (A) with translational movement of the object stage; (B) with traversing sensing head. Arrows indicate the directions of the locational adjustments and of the tracing movement.

Giddings & Lewis Measurement Systems

Fig. 13-3. Profile-tracing instrument with translational movement of the work stage. The displaceable stand holds the extendable pickup arm, which can be adjusted in height and to different tilt positions.

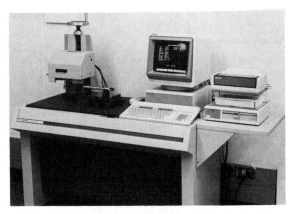

Giddings & Lewis Measurement Systems

Fig. 13-4. (Left) Linear "Proficorder" instrument with translational movement of the sensing head, which is mounted on the slide of the actuating member, termed the "Pilotor."
(Center) The workpiece is held on an adjustable stage.
(Right) The computer with printer for strip chart recording completes the system.

element of the object, not all profile-tracing instruments provide all five adjustment modes. Some instruments are designed to have a plane substituting for two adjustment axes; others offer optional accessories that complement the inherent adjustment potentials of the basic instrument. In still other cases, the missing adjustment modes must be compensated for by special workholding fixtures or by hand adjustment. Unless said adjustment potentials are provided in one way or another, the adaptability of the instrument is limited to a certain type of figure, usually of regular geometry and closely controlled size, or to parts mounted on special fixtures. Otherwise, the inadequate alignment could detract from the reliability of the resulting measurements.

The design principles of the two basic types of profile-tracing instruments, namely, those operating by the translational movement of the object stage and those operating by the translation of the sensor, are shown diagrammatically in Figs. 13-2A and 13-2B. These illustrations also indicate the possible arrangements for the five axes. However, the outlined arrangements are not unique; any other convenient alternative capable of producing comparable results may be satisfactory, and several varieties are actually applied for different types of profile-tracing instruments.

Figures 13-3, 13-4 and 13-5 show the photographs of a few, rather widely used instruments that were expressly designed for profile-tracing applications.

At this stage of the discussion involving the design principles of different types of instruments, a comparison of the translational systems on which these designs are based, seems appropriate. The comparison considers principles only, and lists *the*

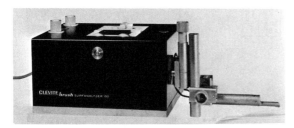

Brush Instrument Div./Clevite Corp.

Fig. 13-5. Straight line drive for profile tracing equipped with a vertically adjustable pickup arm. The drive also has a tilt adjustment for the leveling of the tracing path. The longitudinal and traverse adjustment, as well as the swivel movement for alignment, must be provided by the work stage.

particular advantages of the two basic systems of translational movement from an applicational point of view.

The purpose of reviewing these relative advantages is to indicate the diverse aspects that affect the suitability for specific measuring tasks of a particular instrument system, and also to point out the potential capabilities of these systems. Most of the cited advantages of one system in comparison with the other result from inherent characteristics that, however, are not necessarily present in all instrument models, even though, by the principles of their operations, such may be classified in a particular system. On the other hand, it is conceivable that by means of additional features or by other improvements, the capabilities of either basic system can be expanded, thus overcoming certain seemingly inherent limitations.

Advantages of Systems Using Traversing Work Stage

Some of the advantages of systems using a traversing work stage are:

a. The length of the traced surface section is not affected by the limitations of a cantilever tracer arm; it is practically equal to the travel length of the slide.

b. The size, weight and configuration of the part that is supported on an open slide are less limited than in the case of a confining instrument structure.

c. The location of the surface section to be traced must only be accessible from one side of the slide; otherwise, it may be in any position within the adjustment range of the gage stand carrying the sensor.

d. By having established the track of the slide in the horizontal, all tracing paths will then be contained in parallel horizontal planes, a condition per-

mitting the use of sensitive spirit levels for locating the part in relation to a datum surface that is also associated with the horizontal.

e. The parallelism of several tracing paths in different locations on the part, which is retained in an undisturbed staging position, is not affected by the repeated resetting of the sensor stand for establishing stylus contact with the subsequently traced surface elements.

f. Cross slides of adequate tracking accuracy and mutual perpendicularity may be used for inspecting the squareness of functionally interrelated, yet not necessarily adjacent, surfaces.

g. The system permits great flexibility of instrumentation because of its detached principal elements, which may be interchanged with units of different sizes and capacities, for example, a slide with longer travel, or a stand with increased extension.

h. The system is adaptable to machine tools when needed by utilizing the available translational movements for checking straightness, squareness, parallelism and so forth of machined surfaces.

Advantages of Systems Using Traversing Sensors

Some of the advantages of systems using traversing sensors are:

a. The light weight of the tracer arm carrying the sensing head (pickup) permits the use of optical flats for guide elements. A plane, accurate on the order of a microinch, can thus be assured for the stylus path.

b. The accuracy of the stylus path is not affected by the weight or shape of the specimen part.

c. The positive guidance and the sensitive adjustment of the pickup arm permit the use of fine diamond points for the stylus, such as needed for exploring arrow or partially obstructed surface sections.

d. The fine point stylus also makes the profile-tracing instrument adaptable for surface roughness measurements, provided the amplifier and the recorder assure the necessary response speed.

e. The guide surface of the stylus arm can also have other forms than the basic flat, such as a circular arc of a specific radius.

f. The angular setting (tipping) of the stylus path to accommodate a surface section with a small incline can be carried out conveniently and with great sensitivity, when the angular adjustment of the tracing plane is one of the design features of the instrument.

g. The position and the traversing section of the pickup can be precisely adjusted to trace a very narrow or only indirectly accessible surface element.

h. The surfaces of internal features, when open in the direction of the pickup arm, can be traced to substantial depth, even behind flanges, by using the appropriate stylus adaptors.

i. The fixed position of the stylus path makes the instrument adaptable to repetitive measurements without the need for individual alignment adjustments, when essentially identical parts that are held in staging fixtures, are to be traced.

Regarding the *guidance along a straight line path,* the methods applied be various instrument manufacturers differ in two major respects: basically, depending whether the sensor or the stage is the moving member, and also in design details, according to the intended operational characteristics and capacities.

For the translation of the stage, a design must be provided that, in addition to supporting the movable slide itself, is also capable of accepting loads that the object and its holding fixture represent, without causing sensible deflections or loss of guidance accuracy. For evaluating the accuracy of guidance along the useful range of travel of the slide, it is customary to distinguish between lateral tracking accuracy and rise-and-fall displacements. Digressions from the straight line path in the plane of the pickup action—commonly, a vertical plane containing the traced surface element—are more critical than those normal to that plane. The design and workmanship of the supporting elements must also assure a movement with very low friction, in order to avoid any stick-and-slip condition, or even variations of the lubricant film thickness to a degree that could affect the required accuracy of the slide positions.

Various design systems are found in instruments built exclusively, or used incidentally, for profile tracing. Examples are: ball slides, other types of rolling element slides, plain hand-scraped guideways, and also hydrostatic air bearings. The suitability of a particular system is governed by application of such factors as size, weight, length of travel, the required guidance accuracy, as well as the operational and environmental conditions.

The displacement of the pickup along a straight line path is not affected by the varying weight of the

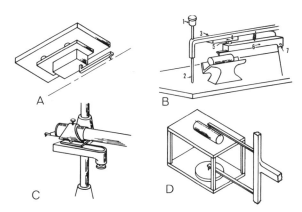

Fig. 13-6. Diagrams showing the principles of straight line guidance systems used in different models of profile-tracing instruments with traversing pickup.
(A) The pickup slide is forced against an optical flat by springs.
(B) A single contact point on the pickup arm is pulled against a narrow optical flat: *(1)* height-adjusting screw head for the datum beam stabilizer; *(2)* adjustable stabilizer-spindle for the datum beam; *(3)* datum beam; *(4)* optical flat datum; *(5)* single contact point; *(6)* stylus arm; and *(7)* hinge.
(C) The pickup arm is supported by an optical flat through single-point contact.
(D) Tandem arrangement of two pickups of a differential gage. One pickup traces the work; the other contacts an optical flat to provide the reference plane of the process.

inspected parts and this system is usually selected when measurements in the microinch discrimination range are required. Whereas the designs that are actually applied vary, the fundamentals are common for most models of this instrument category: a member with a surface of excellent flatness, for example, an optical flat, is used for a guide element in the critical sense (the rise-and-fall movement), and some conventional mechanical members confine the track laterally. A few examples of the design principles applied in different models of profile-tracing instruments for guiding the translational movement of the pickup arm are shown diagrammatically in Fig. 13-6.

Actual applications of these design principles are illustrated in the following photos. Figure 13-4 shows an instrument using an optical flat (not visible in the photo) to guide the pickup arm slide, bearing against the flat by contact through three ball segments. In Fig. 13-7, a bracket is illustrated that is steadied by an external support. This bracket contains a special datum element on which a skid in the top of the pickup arm is sliding; a spring pulls the hinged arm against the guide plane. Figure 13-8 shows an underslung arrangement in which a stirrup fitted over the hinged pickup arm carries the skid in

Rank-Taylor-Hobson/Engis Equipment Co.

Fig. 13-7. Straight line guide arranged over the pickup arm, which is pulled by spring force against the datum element mounted inside the attachment. In this device the reference master has the form of a cylinder segment whose straight line element can be aligned with the translational path of the pickup.

Rank-Taylor-Hobson/Engis Equipment Co.

Fig. 13-8. Straight line guide beneath the pickup arm whose hinged retainment permits gravity to assure contact between the traversing arm and the stationary datum element.

sliding contact with the optical flat of an adjustable datum block, with gravity holding the skid against the guide member.

Functionally effective straight line guidance of the pickup arm can also be accomplished by using a reference plane of excellent flatness to be traced concurrently with the specimen surface by means of a second pickup. The two pickups must be mounted on a common beam, at different levels in the vertical plane of the arm movement. The signals originating from the individual pickups are channeled into a differential circuit, whose output indicates the deviations of the specimen surface from the straightness embodied in the reference flat. In this case the rise-and-fall accuracy of the pickup-arm guidance is not critical. The principles of such an arrangement are shown in sketch (D) of Fig. 13-6.

The accuracy of guidance that can be accomplished with any particular system varies over a range of several orders of magnitude. Whereas some models are designed to provide an accuracy not much better than 0.001 inch and are intended for shop use only, the most accurate guidance systems used for profile tracing can maintain a stylus path digressing by a few microinches maximum from a theoretical straight line.

Considering that each guidance system has its characteristic applications, the wide range of available accuracy levels bespeaks the varied uses of profile tracing in inspection processes.

For the *actuation* of the translational movement, gear motors with reduced speed output driving a lead screw or a rack-and-pinion mechanism are generally used. Emphasis is put on a uniform speed rate. Most instruments have a single speed only for the translational movement. That speed is selected by the manufacturer, with due regard to the speed of response of the sensing and recording system, and also considering the degree of required resolution. Consequently, the speed rates may vary, depending on the particular model; as examples: 0.005 inch per second is the tracing speed of an instrument designed to provide excellent resolution, whereas over ten times this speed rate is found on another instrument type intended for shop applications where the tracing time is an important factor and high resolution might be more confusing than informative.

The Sensing Head and Its Stylus

Electronic indicator type instruments are used exclusively for profile-tracing processes, this being the only dimensional indicating system that offers the following properties; relatively small outside dimensions for the sensing member; light contact force; high sensitivity; very rapid response; remote location of the display members (meter, recorder); a wide range of alternative magnifications up to very high rates; and generation of signals that are reproducible, graphically, by means of a recorder.

All these properties, whose listing above does not follow an order of relative importance, are needed for the successful performance of profile-tracing operations; this wide range of requirements explains the exclusive use of electronic indicator devices for profile tracing.

The pickup for sensing the deflections of the stylus operates generally either by an LVDT (linear variable differential transformer) or by an inductive transducer. The movable member of the pickup is usually connected to one end of a lever, whose opposite end carries the actual scanning element in contact with the traced work surface.

That contact element, commonly termed the *stylus*, is kept under a light spring pressure to assure its staying in intimate contact with the specimen surface during the tracing operation. The stylus force, a design characteristic, is determined by evaluating such factors as the form of the stylus tip, the traversing speed, the mechanical design of the pickup and various others. Stylus forces of 0.1 gram or smaller, up to 1 ounce (28.35 grams) or greater are found in different instrument models.

The form of the stylus tip is either an actual ball, or a cone with spherical vertex. In either case the radius of the sphere is the dimension that expresses the functionally meaningful size of the stylus. Styli with diamond tips have radii in the 0.0001- to 0.001-inch range, whereas ball tips are selected to provide radii in the $\frac{1}{64}$th (0.0156) inch to $\frac{3}{32}$nd (0.0938) inch range.

Styli with fine tips are preferred for tracing narrow, short or partially obstructed surface sections, and for operations requiring a high degree of discrimination, even combining the measurement of the profile with surface texture measurements (see Chapter 15).

Ball styli are less sensitive to shocks, often require a less meticulous preoperational positioning and alignment; for these and similar reasons, ball styli are preferred for profile-tracing operations on the shop floor. There are also certain kinds of profile-tracing operations for which the ball form of the stylus tip, with its larger radius, is of distinct functional advantage because of its property to bridge narrow

valleys and smooth the effect of abrupt changes in the specimen contour. The produced chart tracing line represents a kind of envelope profile, because the minute details of the actual profile have been suppressed. As a result, the chart line will visually emphasize the general character of the surface geometry.

One of the properties of electronic indicators, as mentioned earlier, is the capacity of producing high and interchangeable *rates of magnification*. High magnification permits the display of even minute variations of the inspected surface element in relation to a straight base line at a precisely liner rate, providing reliable means for the numerical assessment of the detected profile conditions. The interchangeability of several magnification rates substantially increases the adaptability of profile-tracing instruments. Surfaces of different degrees of conformity with the datum line can be inspected at optimum magnifications, whose rates are determined as a best compromise between two conflicting factors: the desirability of high magnification for improved discrimination, and the need for limiting the magnification in order to contain the representation of the surface variations within the available recording range. While three magnification rates are quite common for electronic indicators, a few instrument types provide as many as 6 or even 8 different rates.

The available magnification rates are selected by the instrument manufacturers in a manner that permits assigning round values to the graduations of the recorder chart. Such graduation lines, often referred to as divisions, are usually 0.100 or 0.250 inch apart. Frequently used rates of magnification are: $500\times$, $1000\times$, $2000\times$, $5000\times$, $10,000\times$, $20,000\times$, $50,000\times$ and $100,000\times$. The actual rates available in various types of instruments are often selected from the values mentioned above.

These magnifications refer to the digressions of the traced surface element from the basic line and are commonly known as vertical variations, because of the general arrangement of the tracer instruments designed to contact the topmost element of the investigated feature's surface.

The horizontal magnification, producing an enlarged representation of the length of the inspected section, is accomplished by the difference between the scanning speed and the recorder chart advance. The scanning speed, whose rates were mentioned above in conjunction with the actuation of the translational movement, is usually fixed, although a few instrument types provide more than a single speed for this movement. The chart speed of most types of recorders can be varied by fixed steps with the aid of gear boxes. Some instruments provide as many as 6 different chart speed rates, which produce an equal number of horizontal magnification rates. The range of the available rates differs according to the instrument type, the most frequently used horizontal magnifications being on the order of from about 10 times to 500 times.

Higher rates, while improving the resolution of the representation, require a substantial amount of chart paper. Excessively long chart recordings are also unwieldy in handling, difficult to reproduce with conventional equipment and can actually be of disadvantage in the course of chart evaluation. For these reasons, the availability of several rates of magnification can be a valuable feature of instruments used for diverse types of inspection operations.

At this stage of the discussion, it is interesting to note the widely different order of the magnifications in the vertical and in the horizontal directions. This unique and important characteristic of the profile-tracing process, in general, will be discussed later in greater detail.

THE VERIFICATION AND CALIBRATION OF PROFILE-TRACING INSTRUMENTS

In order to obtain dependable profile-tracing charts, various geometric conditions and indicating characteristics of the instruments must be assured to a degree commensurate with the expected accuracy of measurement. While the design and workmanship of the instruments capable of providing the required or guaranteed accuracy belong to the responsibilities of the equipment manufacturer, the testing and periodic verification of the instrument performance must be carried out by the user. For this reason the basic accuracy requirements of profile tracing are discussed in the following, conjointly with the principles and the commonly used methods of testing and verification.

1. The *guidance of the scanning movement* whose purpose is the establishment of a straight line path for the stylus along the inspected surface element. The requirements regarding the accuracy of this movement can best be analyzed by its constituent factors, considering the conditions that may affect the straightness of the resulting translational movement.

　　a. The *rise-and-fall* error refers to the deviations of the stylus path from a datum plane that

is normal to the plane of the stylus deflections during the tracing of a surface. The extent of that error, when present, can be measured by tracing the surface of an optical flat of known excellent flatness, for example, on the order of 2 to 3 microinches or better, and recording the deflections of the stylus at a high rate of magnification, such as $50,000\times$ (see Fig. 13-9A).

　　b. *Tracking* error refers to lateral digressions of the stylus path from a vertical plane that also contains the sensing deflections of the operating stylus. A practical and reliable method of verification consists in tracing the top element of a gage cylinder of known, excellent straightness. The gage pin or cylinder must be precisely aligned with the stylus path, at least at the two end portions of the traced section. It is advisable to use a pin with the smallest diameter compatible with the required degree of straightness, in order to increase the indicating sensitivity to lateral digressions (see Fig. 13-9B).

2. The *chart indications* must possess two important properties to assure the reliability of the measurements, namely, linearity and calibration accuracy.

　　a. *Linearity* refers to that property of an amplifying system by which variations in the "height" (representing material condition) of the traced surface element cause proportional changes in the vertical location of the pen marks on the chart. Perfect linearity is accomplished when the magnification retains constant rate over the entire recording range in any of the available magnification steps. Usually, linearity within a maximum deviation of 0.5% or 1% of the indicating range can be expected from commercially available instruments. The linearity of the representation may also be affected by too rapid changes in the traced specimen surface; limitations of the linearity in the indications that may result from such circumstances are explained in the discussion of the response speed.

　　b. *Calibration* expresses a condition of assured correlation between the scale values of the recorded line on the chart and the actual height variations of the inspected surface, which the chart line represents. The scale values of the chart tracing line are determined in relation to the chart rulings (graduations or divisions, which have assigned dimensions expressed in standard units of length, commonly, decimal fractions of the inch).

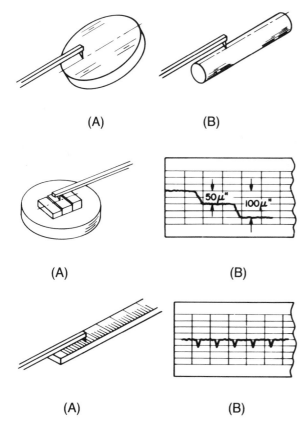

(A)   (B)

(A)   (B)

(A)   (B)

Fig. 13-9. (Top) Verification methods for determinating the precision of the guidance in profile-tracing instruments. (A) Rise-and-fall errors detected by tracing the surface of an optical flat. (B) Tracking errors evaluated by tracing the top element of a well-aligned gage cylinder of small diameter.

Fig. 13-10. (Center) The principles of calibration procedures used to verify the vertical response accuracy of profile-tracing instruments.
(A) Gage blocks wrung on an optical flat to produce steps of known magnitudes. (B) Chart line produced by tracing across the calibration blocks.

Fig. 13-11. (Bottom) A method for verifying the horizontal magnification accuracy of profile-tracing instruments. (A) Master scale with etched graduation lines. (B) Chart tracing resulting from scanning along the graduated portion of the scale.

The chart indications may be affected by a further condition, known as the *response speed* of the instrument. In this connection, it should be recalled that profile tracing involves a series of process steps for converting the variations of the surface element that are contacted by the stylus into highly magnified chart lines. The mechanical deflections of the stylus are transduced into electrical signals, these

then are amplified, occasionally also filtered, and made available to actuate another type of transducer device, which reconverts the electrical variations into the mechanical motions of the pen arm. Lagging response to the sensed variations of the material condition of the traced surface element can result from electrical causes (such as those related to the carrier frequency) and from mechanical sources (due to the play, hysteresis or deflections in the linkages).

While the high response speed may not be critical in many profile-tracing processes (actually, for certain uses it will even be deliberately suppressed by electrical or mechanical means), in many other cases insufficient speed of response can detract from the reliability of the surface portrayal as produced by profile tracing. When the recorder pen does not respond to variations of the traced surface instantaneously, a distorted representation of the inspected surface will result. The character and the extent of such distortions will be affected by a number of factors, such as the discriminating capacity of the electronic indicator system; the lag of the recorder action; the tracing speed; the stylus geometry; and various other factors. Distortions of the chart line may be due to a belated reproduction of the surface variations (horizontal shift) or to the display in an incomplete scale, occurring particularly when the inspected surface has abrupt variations; all these causing a "smoothing" of the chart line.

Sudden deflections of the stylus—due to the combined effect of surface variations and tracing speed—might, in some cases, cause an "overshooting" of the recorder arm; this is another condition that should be considered in the course of evaluating the indications of the instrument.

Other aspects of the response speed of instruments used for surface tracing processes and the evaluation of its effect on the true representation of the inspected surface elements are considered in Chapter 14, dealing with the measurement of circular contours.

In discussing linearity and calibration, reference was made only to height variations that are sensed by the stylus and cause vertical displacements of pen marks on the chart. Because the so-called vertical represents the more critical sense in profile tracing, indications in that direction were selected for reference purposes in order to clearly visualize the concepts. It must be pointed out, however, that the horizontal magnification, although needed at a lower scale only, is also an important constituent of the reliable surface representation. For that reason, the principles of linearity and calibration also apply to indications representing the length of the traced surface.

Diverse test methods can be devised and are actually used for the *verification* of both the linearity and the calibration accuracy of the profile-tracing instrument. It is well to remember that while linearity, that is, the proportionality of indications, is usually an inherent characteristic of a specific instrument, the calibration (resulting from a particular "gain" setting of the electronic amplifier) can be adjusted. The calibration, after it has been adjusted for one of the magnification rates, should be effective at all the other available rates as well; nevertheless, it is advisable to check that correlation occasionally in order to avoid measurement errors due to undetected instrument deficiencies.

Although the concepts of linearity and calibration accuracy were discussed separately, the actual testing can be consolidated into a common procedure. Of course, the applicable testing process differs for the vertical indications (resulting from the stylus deflections caused by surface variations) and for the horizontal indications (the ratio between the length of the traced surface element and the length of the resulting chart line).

The *vertical calibration accuracy* can be checked with the aid of a master prepared by wringing adjacently placed calibrated gage blocks of different sizes to the surface of an optical flat (see Fig. 13-10). The size difference between the adjacent blocks should create steps that are well within the recording range of the instrument at the rate of magnification chosen for the calibration test. It is practical to establish different steps in a testing master that then can be used for verifying the calibration accuracy at more than a single rate of magnification. The resulting steps in a continuous chart line must represent the scale values of the actual height differences of the traced gage blocks. The accuracy of the representation must be within the guaranteed limits, which are usually specified either as the maximum deviation in the representation of adjacent levels, for example, ±1% maximum height difference or as an error not exceeding a certain percentage of the entire indicating range, for example, ±0.5%.

The *horizontal calibration accuracy* can be verified by means of a line-graduated master rule (see Fig. 13-11). Certain conditions should be met and precautions observed for the successful accomplishment of this type of test. It is preferable to select a scale with graduation spans corresponding to the scale value of the horizontal chart graduations. The

master scale must be precisely aligned and parallel with the stylus path in order to avoid cosine type errors. A sharp stylus is needed that will sensitively respond to the very shallow and narrow graduation grooves found on master scales. By this method, verification tracings can be prepared that supply dependable information on both, on the station-to-station, as well as on cumulative errors in the horizontal indicating performance of the instrument. Horizontal magnifications with a maximum calibration error of 1%, or even 0.5%, can be expected from instruments designed for metrological applications.

Applications of the Profile-Tracing Process

The primary and the most obvious purpose of the profile-tracing process is to inspect a selected section of the part's contour in relation to a straight line or, exceptionally, another type of datum, for determining the degree of congruence. However, this should not be considered the only uses of the process in dimensional metrology. There are several types of dimensions and geometric conditions whose measurement can be carried out under uncommon circumstances and to a high degree of accuracy by means of profile tracing. Actually, for certain locational conditions, profile tracing represents the most suitable and, frequently, the only method by which significant dimensions can be measured.

In substance, these measurements consist of comparing the actual position, direction, length or form of the selected surface element of a part, in a precisely defined staging location to a fixed straight line that is produced by the path of the stylus tip.

The following characteristic properties of the profile-tracing process are utilized in these varied applications:

1. The tracing path produced by the tip of the stylus is a line that is straight to a very high degree of accuracy.

2. The location and direction of that straight line is defined with respect to two mutually normal datum planes, to which the staging of the part can be related.

3. The deflections of the sensing member, the stylus, are confined to a plane with which the imaginary cross-sectional plane of the part can be brought into coincidence.

4. The deviations of the scanned surface element from the basic straight line are detected with very

great sensitivity and measured by highly magnified indications. As mentioned earlier, in most instrument arrangements these variations of the inspected surface are measured in a vertical plane, hence, are commonly designated as vertical magnifications.

5. The longitudinal travel of the scanning movement—commonly in or close to the horizontal plane—is also indicated in true proportion and at magnifications that are usually smaller than those in the vertical direction. As the result of that difference, the contour of the traced surface is displayed at a selective magnification, intentionally distorted in a precisely controlled manner for the purpose of clarity.

6. Distinct from dimensional measurements in general, which compare a specimen and a master held in a mutually stationary state and produce a single value, profile tracing has a dynamic component, resulting in a continuous series of indications that join to produce a line as the representation of an infinite number of individual measurements.

The capability to produce indications at a *nonuniform and selective measurement* is one of the unique characteristics of the profile-tracing process, and for that reason it deserves a more detailed analysis.

For parts requiring sensitive measurements such as are provided by profile-tracing instruments, the deviations of the contour from a basic condition are usually of very small magnitude and therefore high magnification indicator instruments can be used. However, when instead of a single point on the surface, a section of the contour is being explored, the length of that section is always many times larger than the expected range of dimensional variations.

When both the deviations from a theoretical straight line and the length of the explored surface section are magnified at the same scale, such as by an optical projector, the rate of the applicable magnification is limited by the field of view or screen size of the instrument. Taking, for example, a surface section of one-inch length and an optical projector screen of 20-inches diameter, the maximum magnification showing the complete section is 20 times, a rate obviously too small for the measurement of deviations on the order for which profile tracing is primarily used.

When assuming a screen of about one hundred feet in diameter—an obvious absurdity—which would permit the projection of a one-inch section at,

for example 2000 × magnification, the sheer size of the projected image would make impractical even the detection and, particularly, the measurement of small digressions of the contour from a base line.

In order to make the chart tracing line truly informative with regard to the conditions being investigated, the magnification of the critical deviations must be much higher than that of the length of the inspected surface element. That requirement can be satisfied by producing *nonuniform magnifications*, meaning a higher rate in the critical sense, and a lower rate along the base line.

Most profile-tracing instruments provide several different magnifications for indications in the critical measuring sense, but in many models more than a single magnification is offered in the tracing direction as well. By changing the rate of magnification in either or both senses, the *ratio of magnification* can be varied. For each type of operation, the most informative ratio of magnification may be chosen from the available alternatives. For that reason the *selective magnification* is one of the important characteristics of most types of profile-tracing instruments, contributing to the versatile applications of the process.

The versatile applications of the profile-tracing process make a complete listing of potential uses impractical. Nevertheless, a survey of geometric conditions and dimensions that can be advantageously inspected or measured with the aid of profile tracing can serve as a guide for prospective users of the process in general.

Such a survey is contained in Table 13-2, to which the following complementary remarks apply:

1. Principles only are presented, without discussing actual measuring problems.

2. The conceptual role of a straight line in controlled position is considered the distinguishing characteristic of the different dimensional categories.

3. In these examples only the straight line is shown to represent the reference element for the process; this should not exclude the use of other regular forms that can be established as stylus paths.

4. No specific reference is made to the system of the profile-tracing instrument, whether operating by the translation of the work or of the stylus. Most processes can be implemented with either system, although due to conditions related primarily to the

**TABLE 13-2.   APPLICATIONS OF PROFILE TRACING FOR THE MEASUREMENT OF
GEOMETRIC CONDITIONS—BASED ON THE CONCEPTUAL ROLE OF THE STRAIGHT LINE – 1**

EXAMPLE OF THE DESIGN ELEMENT CONCEIVED AS A STRAIGHT LINE	THE GEOMETRIC CONDITION ASSESSED IN RELATION TO THE STRAIGHT LINE	
	DIAGRAM	DESIGNATION (AND QUALIFYING CONDITION)
The generatrix of a cylinder or of a cylindrical bore		The straightness of a cylinder
The generatrix of a cone or of a cone frustum		The regularity of form (of one basic parameter) of a conical body
The common, or the parallel elements of coaxial and straight features		The alignment (directional) of two or more bores (of substantially equal diameters)
Several straight line elements contained in a single plane indicate flatness of the surface		The flatness of a surface inspected by tracing it along several paths referenced from a common datum plane
A line element of the surface equidistant from another straight line or plane		Parallelism of the inspected surface element with the opposite (locating) surface of the part

**TABLE 13-2.   APPLICATIONS OF PROFILE TRACING FOR THE MEASUREMENT OF
GEOMETRIC CONDITIONS—BASED ON THE CONCEPTUAL ROLE OF THE STRAIGHT LINE – 2**

EXAMPLE OF THE DESIGN ELEMENT CONCEIVED AS A STRAIGHT LINE	THE GEOMETRIC CONDITION ASSESSED IN RELATION TO THE STRAIGHT LINE	
	DIAGRAM	DESIGNATION (AND QUALIFYING CONDITION)
A straight line surface element at right angles to a plane		The squareness of a surface element to another surface of the part which is located in a datum plane normal to the tracing path
A straight line surface element of a tapered figure presented in a position theoretically parallel with the datum plane of the tracing path		The actual angle of a wedge presented as the deviation from the theoretical condition; the wedge is supported on an incline having a slope angle equal to the nominal taper
The tangent to a curved surface element		The curved contour of a substantially straight but deliberately (or unintentionally) curved surface element, such as the profile of a crowned gear tooth
The distance between two boundary points along a straight line represents a specific linear dimension		The length of a section not accessible to direct measurement, assessed as the uninterrupted straight portion of the traced surface element
The datum line of a nominally flat surface, or of a body of revolution with nominally straight side elements		The waviness or other imperfections of an essentially regular surface assessed as contour digressions from a straight line datum (the envelope or the median line)

specimen characteristics or to the location of the surface element to be inspected, one of the systems might be preferable, or even exclusively adequate.

5. The examples are based on the potentials of the *process* and not on the capabilities of any particular model of *instrument*. The applicability of any process with the aid of a particular type of instrument must be evaluated by considering such circumstances as the instrument capabilities; the dimensional characteristics of the specimen in general and of the surface element's location; the required rate of magnification and measuring accuracy; and so forth.

The examples in the table also comprise processes for dimensional measurements, both linear and angular, by means of profile tracing. Such applications of the profile-tracing process are neither obvious nor used with a frequency comparable to the inspection of the profile form. The uncommon aspects of dimensional measurements by profile tracing warrant a more detailed discussion of the involved principles, as well as of the reasons for the application of profile tracing for such purposes.

Dimensional Measurements of Technical Parts With the Aid of Profile Tracing

Length measurements are commonly accomplished by determining the separation of two points that are considered to represent the two surfaces or elements whose distance from each other is the sought dimension. The angular measurement of parts supported on a sine plate is carried out in a similar manner with respect to determining the positions of discrete points, in this case their relative distance from a common datum plane.

No technical surface possesses a perfect regularity; variations in both the texture of the surface and the geometric form of the part must be expected. Under general conditions, however, such variations are considered small enough to avoid interference with the required degree of accuracy of the dimensional measurement. Expressed in other terms: the "deviation of repeat" due to surface irregularities is regarded as being kept within permissible limits.

In tightly toleranced parts, however, the deviation of repeat due to surface imperfections can exceed the limits of meaningless variations. In such cases it is often possible to derive valuable information regarding the causes of poor repeatability of the dimensional measurements by exploring the geometric form of the surface elements that contains the gaging points. Such information, supplied by profile trac-

ing, can prove of great value in devising the course of corrective actions.

In some cases it may be even warranted to measure the critical dimensions of the part with the aid of profile tracing. Such a process can supply information on the actual position of the surface element in relation to a datum, in contrast to the conventional measuring processes in which only a single point along that element is being contacted. The charts produced by the tracing process can also serve to establish the location of an assigned boundary element, as a basis for determining the functionally effective size of the part.

Such dimensional measurements might be needed for assemblies with extremely critical clearances, or for angle measurements where the predominant or characteristic incline of the surface element is considered the representative dimension.

Examples of such measurement by profile tracing are given in Fig. 13-12, which shows the diameter measurement of a plunger, and in Fig. 13-13, which shows the angle measurement of the rib face of a tapered bearing inner ring.

There is a special field for linear measurements by means of profile tracing that, although limited in the scope of its applications, deserves mention. Very small linear dimensions, such as the thickness of thin film deposits, can be measured by establishing

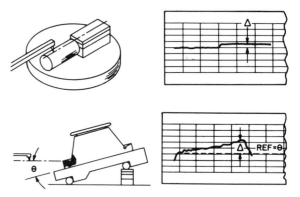

Fig. 13-12. (Upper) Comparative measurement of size by means of profile tracing. The part and a gage block stack, the latter representing the reference dimension, are staged on a common datum plane that is parallel with the stylus path. In the process the stylus will successively trace the part and the reference surface, producing a chart line with two sections that display the amount of dimensional differences.

Fig. 13-13. (Lower) Comparative angle measurement by profile tracing—the rib face angle of a tapered roller bearing inner ring. The part is located to present the surface being inspected basically parallel with the stylus path. Deviations of the actual angle of the traced surface element with respect to the basic angle can be measured by the slope of the produced chart line.

Rank-Taylor-Hobson/Engis Equipment Co.

Fig. 13-14. "Talystep" surface-tracing instrument with very high magnifications (up to 1,000,000×), designed for measuring height differences on the order of a microinch, or even smaller, on successively traced surface sections.

the traced section across the boundary of that layer. The stylus will then contact, successively, the surface of the base material and that of the deposit layer, the thickness of which is represented as a step in the resulting chart line. The height of the step, that is, the difference in the levels of the adjoining chart line sections, is the representation in the chart scale of the measured layer thickness.

Considering that the thickness of vapor deposit layers can be on the order of less than a microinch, and therefore often are measured in Angstrom units (Angstroms), a very special type of equipment is needed for the linear measurement of such extremely small magnitudes.

The photograph in Fig. 13-14 shows a unique example of a very sensitive type of profile-tracing instrument designed expressly for linear measurements at an extremely high rate of magnification. The following performance characteristics of this instrument illustrates the potentials of profile-tracing processes in the area of micrometrology:

Stylus force is adjustable from 1 to 30 milligrams

Vertical magnifications, in eight steps, from 5000 times to 1,000,000 times

Horizontal magnifications, up to 2000 times

The vertical resolution of this instrument is stated to be better than 0.000,000,02 inch (5 Angstroms).

PROCESS TECHNIQUES OF PROFILE TRACING

When profile tracings are made with the purpose of checking the regularity of form (generally the straightness) of the inspected surface element, the staging of the part must satisfy only a limited number of requirements. It is necessary to align the traced surface element with the stylus path in a manner that

produces a pen line well within the boundaries of the boundaries of the chart paper, and also in an essentially horizontal direction, the latter for the purpose of a more convenient evaluation.

Such alignment is usually accomplished by considering, mainly, the two ends of the section to be inspected, making positional adjustments to the effect that the selected reference points, when successively contacted with the stylus, cause substantially equal indications. Because the staging position needed for the inspection of the profile form is determined by the mutual positions of the stylus path and of the surface element being traced, it is irrelevant whether the displacement path, the specimen, or both are adjusted to arrive at the desired condition.

The alignment guided by two reference points near the ends of the inspected surface section, although generally practical, does not always assure the true representation of the selected surface element. The "lateral alignment error," which is discussed later in this section, can be the source of a partially distorted representation when surfaces curved normal to the scanning path are being inspected. However, the resulting digressions of the pen line on the chart have a characteristic form that differs from waviness and other periodic variations of the profile.

The staging and alignment requirements will be more rigorous when profile-tracing processes are used for dimensional measurements. In such applications, in order to make the recorded chart line sufficiently informative and reliable, the part must be presented for tracing in a manner that assures the coincidence with the stylus path of either the theoretical contour of the part or of a design element that is conceptually related to that contour, such as a tangent line or the chord of an arc. Deflections of the stylus caused by deviations of the actual part contour from the theoretical contour generate those digressions of the pen line on which the evaluation of the measurement is based.

Most technical parts are bounded by geometrically defined surfaces whose positions are interrelated. It is generally feasible to select a particular surface on the part that, when located properly, presents the surface element to be inspected in a position theoretically coincident with the stylus path. Here are two examples: a side of a cube located in a plane perpendicular to the stylus path presents an adjacent side parallel to that path; one side of a wedge located on a slope equal to the taper angle of the part presents the opposite surface in a plane par-

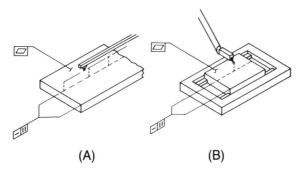

Fig. 13-15. Essential datum elements for locating the staged part in profile tracing: 1. flat tabletop in a horizontal plane, parallel with the path of the translational movement of the tracing process; 2. aligning datum element, essentially a vertical plane parallel with the pickup path.
(A) Traversing pickup type.
(B) Traversing stage type.

allel to that containing the displacement path of the stylus.

To accomplish such controlled staging locations of the specimen in relation to the stylus path, the following auxiliary elements, which are shown diagrammatically in Fig. 13-15, must be provided:

1. A datum plane, conveniently the flat top of a staging table, in a position dependably parallel with the displacement path of the stylus.

2. A datum line representing the projection of the stylus path on the datum plane. For practical reasons a physical surface, such as an edge mounted on the datum plane, is substituted for that datum element in a position shifted by a known distance from the theoretical projection line, yet strictly parallel with it.

In addition to these two fundamental datum elements, referencing along the tracing path may then be needed in measurements for which the exact location of the inspected section in the direction of the scanning displacement is also critical. For the purpose of such referencing, a specific surface element of the part itself is commonly used, either by direct contact with the stylus or through an attached auxiliary member that serves to transfer the reference point into the path of the scanning contact.

The role of the reference elements analyzed above in the staging of technical parts for dimensional measurements by means of profile tracing is not easy to visualize on the basis of principles only. For that reason, various examples for correctly located staging are presented in Table 13-3. Although the drawings of this table are diagrammatic and illustrate the

principles only, the accompanying discussion elaborates the concepts in such a way as to assist in designing fixtures for the practical application of the staging principles.

In the process of dimensional measurements by profile tracing, like most other systems of linear and angular measuring processes, the staging of the specimen directly affects the accuracy of the results. An example from the field of conventional measurements, such as dust on the stage in comparator gaging, visualizes the importance of dependable staging. However, in dimensional measurements by profile tracing, the staging is usually more complex than in any of the conventional processes involving only a single point contact. In profile tracing the scanned and indicated element is a line, that is, a single-dimensional element in contrast to the conceptually dimensionless point.

The accuracy of the measurement is affected by staging errors to different degrees, depending on the direction in which the locational inaccuracy occurs. One may distinguish errors of first order and of second order. Errors of the first order are those that substantially coincide with the direction of the "vertical magnification," that is, the sense of the stylus deflections. Such staging errors affect the accuracy of the measurement directly and often to the full effect of their magnitude.

Errors of the second order are those contained in a plane essentially normal to the primary direction of sensing. These may be considered as lateral alignment errors. Their effect on the measuring accuracy depends on the general shape of the part being measured. Here are four examples:

1. Flat surfaces, when located in the horizontal plane of the stylus path, are generally not sensitive to lateral alignment errors, except those affecting the representation of the traced section's length. However, even these are often considered insignificant, because the resulting deviations are of second order only, frequently designated as cosine errors;

2. Cylindrical surfaces generate hyperbolic digressions of the chart line when the tracing path is slanting across the axial plane of the specimen;

3. Conical parts, when traced aslant, introduce nonsymmetrical hyperbolic errors in the chart line; and

4. Lateral alignment errors in the staging of curved surfaces impart to the chart line convex digressions for external surface elements and concave digressions for internal surface elements.

TABLE 13-3. THE PRINCIPLES OF WORK STAGING FOR INSPECTION AND MEASUREMENT BY PROFILE TRACING – 1

MEASURED CONDITION	APPLICATION EXAMPLE		DISCUSSION OF THE LOCATING AND HOLDING REQUIREMENTS
	DESIGNATION	DIAGRAM	
Straightness	Straight edge by tracing the operating surface between two reference points	O = Reference point(s) for tracing path alignment. 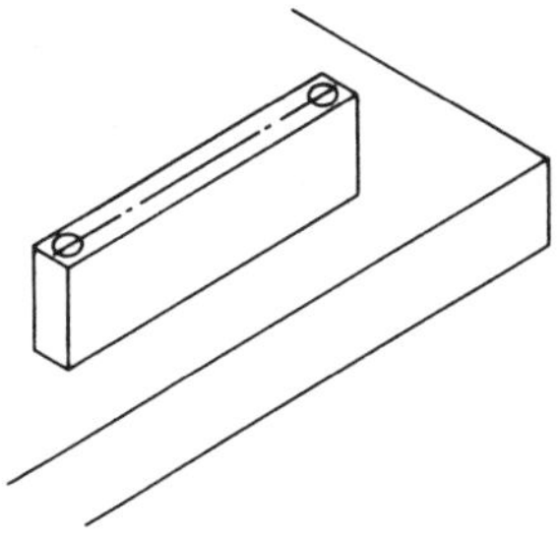	A straight surface, when scanned by a profile tracing instrument, will produce a straight line on the chart even in the case when the planes of the stylus path and of the surface are not perfectly parallel, unless a curvilinear recorder is used. For this latter, parallel adjustment is desirable, based on two reference points at the ends of the traced section, to avoid confusing distortions in the graphic representation.
Flatness	Toolmaker flat, inspected by tracing several randomly selected surface elements	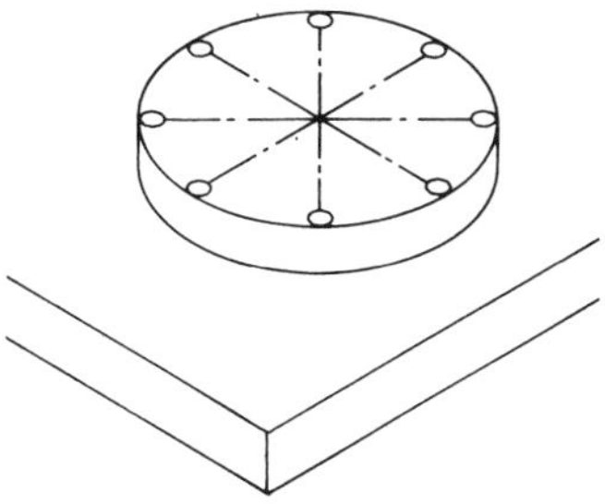	Considering a surface to be flat when several randomly selected elements, located in different directions, are essentially straight lines, the tracing of such elements will produce chart lines which are indicative of the flatness condition. In order to obtain significant values for deviations from the flat, regardless of minor variations in the staging plane due to the rotation of the specimen, the terminal reference points for each traced section must be adjusted for equal indications.
Parallelism	Cylindrical pin (e.g., a plug gage), —by comparing any slope of the traced surface element to a datum plane on which the part is located	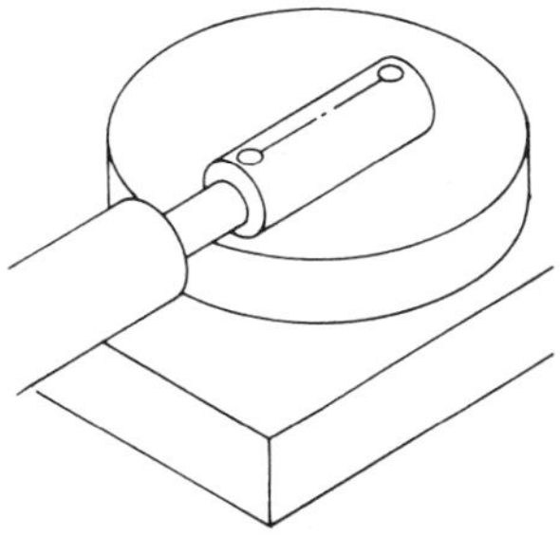	A surface element of the part in a position basically parallel with the plane tangent to the diametrically opposite surface element, is traced to determine the actual parallelism. To implement the concept, an optical flat is used to represent the tangent plane on which the part is supported. That plane also functions as the datum for establishing the horizontal displacement plane of the stylus.

TABLE 13-3.   THE PRINCIPLES OF WORK STAGING FOR INSPECTION AND MEASUREMENT BY PROFILE TRACING — 2

MEASURED CONDITION	APPLICATION EXAMPLE		DISCUSSION OF THE LOCATING AND HOLDING REQUIREMENTS
	DESIGNATION	DIAGRAM	
Angle—of a part with two mutually inclined flat surfaces	Wedge—such as an angle gage block, by locating the side opposite the inspected one on a slope of controlled angle		In order to assure the true representation by profile tracing of the wedge angle, the scanned element must be located in a plane which is at right angles to the vertex line of the part, i.e., the imaginary line in which the extensions of the mutually inclined sides of the part intersect. The support incline must also be so located that its imaginary vertex line is parallel with that of the part.
Angle—taper of a body of revolution	Taper plug gage—body angle determined by measuring the slope angle		For a conical part, the locating plane must be tangent to a surface element and the tracing path should coincide with the diametrically opposite element of the part's surface. In view of alignment and locating problems it may be easier to hold the part between centers which are resting on an incline with a slope angle equal to one half of the part's body angle. Imperfections of the centers and center bores will detract from the accuracy of the measurements by this simplified locating method.
Profile and geometric relation measured with respect to an auxiliary datum	Tapered roller bearing cup—pathway profile and angle		When a particular surface of the part, which is not a direct datum, has a definite geometric relation to the basic conditions of the surface to be inspected, it may be used for locating the part in the profile tracing process. In the example the face of the bearing cup is considered to be perpendicular to the part axis to a sufficient degree of accuracy permitting its use as an auxiliary datum when exploring the profile and the general slope of the pathway.

TABLE 13-3. THE PRINCIPLES OF WORK STAGING FOR INSPECTION AND MEASUREMENT BY PROFILE TRACING — 3

MEASURED CONDITION	APPLICATION EXAMPLE		DISCUSSION OF THE LOCATING AND HOLDING REQUIREMENTS
	DESIGNATION	DIAGRAM	
Substantially straight profile with deliberate camber	Crowned bearing roller—profile measured in relation to a basic straight line element of the design		The cambered profile, termed also the contour of a crowned surface, is an important design characteristic of certain gear teeth and bearing rollers, with the purpose of promoting uniform stress distribution. For determining the form, location and dimensions of the crowned part profile, the chart line should be related to a basic design element of the part, e.g., its axis or unrelieved side contour, a condition which must be assured by the proper staging for profile tracing.
Circular arc displayed in relation to a tangent line	Tapered bearing roller—the cross sectional contour of the end face		The combined rotational and planetary movement of a bearing roller with its axis tilted toward an apex, requires the land at the face of the roller to constitute a spherical zone. The general cross sectional contour of that zone, basically a circular arc, can be inspected by profile tracing, referencing from an imaginary line which is normal to and intersects the axis of the roller. It is advisable to select a low ratio for the vertical and horizontal magnifications.
Linear distance of very small magnitude	Thin film deposition for electronic circuits—the measurement of deposited layer's thickness		In order to obtain a dependable representation of the film thickness in the form of a distinct step of the recorded chart line, the tracing must traverse a section of the surface where the boundary between the base material and the layer to be measured is well defined. The tracing path and the base surface should be aligned to provide a reference line parallel with the horizontal ruling of the chart paper.

TABLE 13-3.　THE PRINCIPLES OF WORK STAGING FOR INSPECTION AND MEASUREMENT BY PROFILE TRACING – 4

MEASURED CONDITION	APPLICATION EXAMPLE		DISCUSSION OF THE LOCATING AND HOLDING REQUIREMENTS
	DESIGNATION	DIAGRAM	
Length of partially obstructed surface sections	The length of the relieved section of a bore surface— the pathway of a roller bearing ring		The location of the boundary between a straight surface element and of its relieved section, particularly when the approach angle is very small, can be determined with unequalled accuracy by means of profile tracing, even in difficult to reach areas. For referencing the section length from the face of the part, even though the edge is rounded, the plane of the face can be transferred into the tracing path with the aid of a knife edge blade in contact with the significant part surface.

These examples of lateral alignment errors and their effects on the resulting chart tracings are illustrated diagrammatically in Fig. 13-16.

When, for reasons of practicality, or in order to obtain information on certain geometric conditions, the part is staged for profile tracing on an auxiliary surface, the choice of that locating portion of the part is guided by various pertinent considerations. The following are a few examples of conditions that the selected locating surface must assure:

1. It should provide an area of sufficient regularity to permit positive location;

2. It should be adaptable for being solidly held on the supporting surface of the fixture, either by direct action, such as a magnet or expansion, or with the aid of clamping for acting on an adjoining surface; and

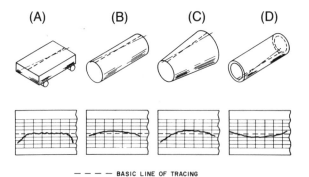

Fig. 13-16. Examples of lateral alignment errors and their distorting effect on the resulting chart lines. (A) Flat surface—checking parallelism. (B, C and D) Cylinder, cone, and cylindrical bore—checking the straightness of a surface element in a basically axial plane of the part. The slanting path of the tracing shown in the perspective sketches of the parts produces the indicated digressions of the chart line.

Rank-Taylor-Hobson/Engis Equipment Co.

Fig. 13-17. Setup for tracing the pathway element of a tapered roller bearing outer ring, which is held in a special fixture. The basic tilt is produced by a sine plate; the micrometer head permits additional leveling in relation to the stylus path.

Fig. 13-19. Setup with a tapered roller bearing inner ring supported on a sine plate during tracing of the rib face, for obtaining the dimensional information on both the angle and the profile of the scanned surface element.

3. It must be in a functionally defined relation to the surface actually being traced.

To satisfy the above and other similar requirements, special work-locating and workholding fixtures are often needed. Figures 13-17, 13-18 and 13-19 show examples of such special workholding devices designed for profile-tracing operations that supply information concurrently on the dimensional or geometric conditions of the traced element, as well as on the profile of the inspected surface.

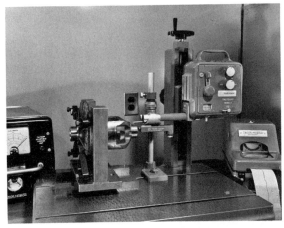

Rank-Taylor-Hobson/Engis Equipment Co.

Fig. 13-18. Setup with a gear housing held in a special fixture to assure the axial profile tracing of the hub surface in a plane normal to the mounting flange.

RECORDER CHARTS AND THE INTERPRETATION OF CHART TRACINGS

As profile tracing is a continuously sensing and indicating process, it is not practical to rely, for obtaining the measured values, on the observation of the constantly changing pointer position of a meter. For this reason, the meters with which various types of profile-tracing instruments are equipped, serve only for guidance in the adjustment phase of the process, preceding the actual tracing of the profile.

The generally employed indicating means of the profile-tracing process are the recorders. Either of two basic systems are used, namely, the $X-Y$ recorder and the strip chart recorder. Each of these systems has characteristic properties, some of which have direct bearing on the selection of either type for a particular set of applicational conditions.

$X-Y$ recorders, termed with reference to the two indicating axes, operate by two mutually independent inputs, one originating from the pickup, which registers the deflections of the stylus, the other related to the distance of the translational movement during the recording period. The stylus deflections are recorded in the vertical direction, along the Y axis, whereas the distance traveled is represented horizontally, along the X axis. Each is recorded at a rate of magnification that is set independently in either direction. Because the displacements of the pen along the X axis originates from electrical signals that are related to the momentary distance of the moving instrument member (pickup or table) from a starting point, variations in the tracing speed do not affect the proportionality of the chart recording (as long as the tracing advance requires pen tracking speeds not exceeding the capacity of the recorder). The substantial width of most types of X–Y recorder charts offers a wider range for recording the stylus deflections than strip charts in general.

Strip chart recorders require a drive mechanism that is reliably synchronized with the tracing speed in any of the selected magnifications. The length of the recording is almost unlimited; therefore, very high horizontal magnifications may be used when needed, and the recording capacity does not put limits on the length of the inspected profile section. Strip chart papers, supplied in rolls of substantial length, are more convenient to handle than individually mounted chart sheets, and generally are more economical when required for continuous tracing operations. Strip chart recorders are made in two basic types:

1. Those using chart papers with curvilinear ruling to record the arcuate movements of the pen; and

2. Those using chart papers with rectilinear ruling. This type requires a straight line movement of the pen, which is usually accomplished with the aid of a linkage mechanism for converting the rotational movement of the actuating element into a linear one.

Figure 13-20 illustrates the blank charts used in the above-mentioned systems of recorders. These are the X–Y type using a rectangular chart of fixed boundary dimensions, and the two types of strip

charts, the curvilinear and the rectilinear, which are used in the continuously recording systems of instruments.

The rectilinear ruling of the chart papers supplies a representation of the traced surface that is generally more conveniently assessed than the curvilinear type, because it can be readily associated with the general form of the part contour. As an example, the circular-arc contour of a part appears as a symmetrical curve on a rectilinear chart, in contrast to the lopsided tracing line drawn on a curvilinear chart. Supplying a faithful visual impression of the traced contour is a desirable characteristic of a chart representation, even though the exact evaluation of the chart must be guided by the scale values of the graduation lines. An exception in this respect is the chart evaluation with the aid of transparent templates, which may be prepared to contain the boundary lines of form tolerance zones. The application of this method is predicated on the undistorted graphical representation of the profile at the selected rates of vertical and horizontal magnifications.

The reason for the wide use of curvilinear chart rulings is the basically curvilinear track of the pen mounted at the end of the recorder arm, which swings around a fixed pivot. For transforming this arcuate track of the pen into a straight line motion, without affecting the dependable proportionality of the pen excursions, special linkages are needed. Several designs of these linkages can be found on rectilinear strip chart recorders of different makes, most of which assure a very good proportionality, approaching that of the direct, yet curvilinear, recording. The insertion of a mechanical linkage, however, reduces the response speed, a condition that might be undesirable for certain types of inspection processes.

A design that combines the advantages of the rapidly responding, direct pen actuation with the readout convenience of the rectilinear recording is found in the recorders of the Rank Corp. In this design the pen arm describes an arcuate movement, and the chart paper is presented on a curved backing plate. The combined effect of these two curvatures, a mutual canceling, generates a close-to-perfect rectilinear recording. The produced charts, however, have an effective width of only 2 inches, a condition that limits the display range of the stylus deflections.

Either of two types of marking media is used for producing chart recordings:

1. Ink permits the use of translucent chart papers that are adaptable to reproduction by conventional drafting duplicating machines; and

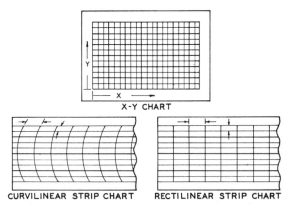

Fig. 13-20. Basic types of charts used in profile-tracing operations for recording both the deflections of the stylus and the traced distance, each at independently selected magnifications. *X–Y* chart—recorder operated by two separate input circuits. Curvilinear strip chart, and rectilinear strip chart—recorder has time-related drive and the input signals operate the vertical displacements of the pen.

TABLE 13-4.   THE INTERPRETATION OF CHART LINES PRODUCED BY PROFILE TRACING — 1

CHARACTERISTICS OF THE CHART LINE		INDICATED PROFILE CONDITION AND EXAMPLE OF OBJECT	DISCUSSION
DESIGNATION	DIAGRAM		
Straight line essentially horizontal, or—conditionally—regardless of direction	Dotted lines - - - - - - - are auxiliary marks for chart evaluation.	The straightness of the edge of a ruler, or of a gage pin, the latter assessed by a surface element contained in an axial plane of the cylinder	While a pen line parallel with the horizontal chart ruling is the obvious display of the scanned surface element, it requires sensitive staging adjustment. A sloping chart line, which results from a slightly out-of-parallel staging, is acceptable only when rectilinear charts are used. In such cases, a straight reference line drawn manually as a median of the tracing line, will assist the interpretation.
Straight line basically parallel with the horizontal chart rulings		Parallelism (or squareness) to the part's locating surface, such as the relation of sides in a rectangular block	Since the horizontal plane of the stylus path is being used for the datum, it is essential that the base surface of the part to which the inspected surface is related be located in a plane precisely parallel (or perpendicular) with this datum plane. Any slope of the chart line will then indicate deviations of the actual surface from the basic geometric relationship.
Straight line inclined to the horizontal chart ruling	$\frac{h}{l} = \tan \gamma$	Deviations of a surface element from a set angular position which corresponds to the theoretical angle of the part—such as of a taper gage plug or ring	The degree of dependability of the chart indications obtained by this method of angle inspection is determined by the accuracy with which the incline of the locating plane in relation to the datum is established. The slope of the chart line will indicate the part's angular variations which can be expressed in angular units, by substituting the actual values of the chart line slope into a tangent formula.
A pen line straight in its central section yet with symmetrical dropoffs at both ends	ψ = approach angle l = length of the straight section	Surface element straight in the center and gradually declining toward the extremities—the profile of an end-relieved bearing element	For easier evaluation of the recorded profile it is advisable to align the straight central section of the part with the stylus path. The design of such profiles frequently specifies the maximum approach angle for the transition between adjacent sections. This may also be assessed from the chart by calculating the angle of the pertinent tangent line.

TABLE 13-4.   THE INTERPRETATION OF CHART LINES PRODUCED BY PROFILE TRACING — 2

CHARACTERISTICS OF THE CHART LINE		INDICATED PROFILE CONDITION AND EXAMPLE OF OBJECT	DISCUSSION
DESIGNATION	DIAGRAM		
Line continuously curving at specific rates, and symmetrically distributed on both sides of a central point.	T = tolerance zone	Profile designed to assure uniform stress distribution under specific load, —e.g., the cross sectional contour of bearing rollers and gear teeth.	A convenient method of evaluating the agreement of the traced profile with the design specifications consists in using transparent overlay templates which contain, in addition to the basic profile, its tolerance zones as well. The template should be used in the "best fitting" position for appraising the functional adequacy of the inspected part.
Curved line with continuously increasing drop in both directions from the center.	$r = \dfrac{c^2 + 4h^2}{8h}$	Profile of circular arc form recorded at nonuniform rates of magnification—e.g., the contact surface of a rotating part in translational movement along a planetary track.	The profile represented by the chart line can be precisely inspected by laying out several mutually parallel chords and calculating the radius of the circular arc pertaining to the measured chordal length and chordal height of each section, using the assigned values in the scales of the chart. Identical radii are indicative of a regular circular arc profile.
Essentially straight line bounded at both ends by abrupt changes from the general direction.		The length of a distinctly bounded surface section in the traced direction is shown by the chart line against the background of a scale permitting direct assessment,— e.g., a seal seat section inside of a bore.	The surface section to be inspected must permit the unobstructed travel of the stylus and must be a plateau of a section distinctly bounded by grinding reliefs. The scanned surface should be aligned with the stylus path in order to avoid cosine errors and a sharp stylus is preferred for the precise definition of the boundary points.
A single pen line with two substantially parallel sections separated by a distinct shift in their levels.		The height difference between two adjacent and basically parallel surfaces can be determined from the chart line—thickness of a layer applied on an essentially flat surface.	The surface sections whose relative heights are compared should be distinctly separated from each other. Although an abrupt change of height is not required, the transition must lie within a reasonable chart length and the two surface sections being compared must be clearly defined.

TABLE 13-4.   THE INTERPRETATION OF CHART LINES PRODUCED BY PROFILE TRACING — 3

CHARACTERISTICS OF THE CHART LINE		INDICATED PROFILE CONDITION AND EXAMPLE OF OBJECT	DISCUSSION
DESIGNATION	**DIAGRAM**		
Line with alternating departures from an imaginary median represented by an inscribed reference line.	a = amplitude f = frequency	The amplitude and frequency, as well as the degree of regularity of the inspected part profile are clearly shown by the chart line— an essentially flat contact surface with limits on acceptable waviness.	In case the analysis of the surface conditions is not concerned with roughness, the closely spaced profile variations should not be shown on the chart. This requirement can be met either by mechanical means, using a large radius stylus (e.g., a ball tip), or by electronic means such as wave filters.
Line with irregular, although occasionally repetitive pattern, usually characterized by closely spaced departures from a general direction	w = roughness band width	The nonuniformly magnified profile shows the minute digressions from the basic contour highly exaggerated for clarity—required information for functional surfaces sensitive to asperities not revealed by average roughness.	Isolated peaks and valleys of the scanned surface element are represented in a manner permitting the evaluation of such surface texture characteristics as roughness band width, number and height of peaks, etc., either superimposed or isolated (by filtering) from waviness. The applicable interpretation is governed by the significant profile characteristics as specified on the pertinent product drawing.

2. Electrical marking requires a special type of opaque chart material, but needs no ink replenishment of the reservoir and is free of clogging. In addition, the thickness of the marking line can be varied by adjusting the current input.

The indicating means of the profile-tracing process, the line traced on a chart paper, is uniquely informative; it displays in a permanent manner, measurement results consisting of a series of individual values and it is capable of supplying information on more than a single parameter. However, due to the combination of several values that the recorded chart line may represent, it becomes meaningful only when properly interpreted on the basis of the pertinent process variables. Examples of the latter are given here:

1. The rates of magnification in each of the two fundamental directions;

2. The locating principles implemented in the staging of the part; and

3. The inspected geometric and dimensional conditions with regard to which the chart tracing is expected to supply information.

It would hardly be feasible to set general rules for the interpretation of profile-tracing charts. Instead of an attempt to formulate such rules, typical examples of profile-tracing processes were selected and the resulting charts are illustrated diagrammatically in Table 13-4. In order to avoid variables that could detract from the clear representation of the essential aspects, all the examples are illustrated by means of rectilinear strip charts. Most of these tracings can be displayed on other types of charts, as well, and require interpretations similar to those discussed in Table 13-4.

14.
The Measurement of Roundness and Circular Contours

Roundness expresses a particular geometric form of a three-dimensional figure, most frequently that of a body of revolution. Circular contour is the characteristic form of the entire or partial periphery of a plane figure.

In practical metrology these two concepts are closely associated by virtue of a generally accepted technique of roundness measurement, which consists in determining the circularity of the contour of the cross-sectional planes of a basically round body. This technological association of the two concepts is the reason for this chapter's title and for the consolidated discussion of the subjects it covers.

THE CONCEPT OF ROUNDNESS

Roundness, although designating, perhaps, the most frequently recurring geometric form of manufactured parts, states a condition that is not measurable by methods comparable to those used for linear or angular dimensions. Quantitatively, the degree of roundness is expressed by the dimension of out-of-roundness, that is, the deviation from the ideal form. That dimension, however, is not defined in terms that permit dependably comparable measurements by any of the various methods used for determining roundness. The reasons for the lack of uniformity in the systems of measurement and in the dimensions they actually determine as the representation of roundness may be found in two interacting aspects: (a) the purpose that the condition of roundness of a technical part is intended to serve; and (b) the methods of measurement that are preferable because of technological or economical circumstances.

In principle it is simple to describe a perfectly round part as having all points of its perimeter equidistant from the axis. Or, expressed in a different manner, the imaginary cross-section of a round part in a plane perpendicular to its axis will have for its contour, a perfect circle. In the terms of a recently issued standard (ANSI Y14.5M-1982 (R1988), Dimensioning and Tolerancing),

> Circularity (Roundness) is a condition of a surface of revolution where: (a) for a cylinder or cone, all points of the surface intersected by any plane perpendicular to a common axis are equidistant from that axis; (b) for a sphere, all points on the surface intersected by any plane passing through a common center are equidistant from that center.

However, it must be realized that a perfectly round part cannot be produced by any known means of manufacturing. Perfect roundness can be approached to varying degrees, but that is just the level where the responsibility of metrology takes over: to determine the extent to which the object's condition actually approaches perfect roundness. Or, expressed in more practical terms, to determine the character and the magnitudes of departures in the actual object form from perfect roundness.

The importance of the proper measurement of such errors is emphasized by the fact that the circular cross-section is perhaps the most frequently used basic shape in engineering design. There are many reasons for giving preference in design to bodies of revolution and to round internal features. Some of these reasons are recalled with the purpose of illustrating the role of roundness in engineering. The round form:

a. Is simple to specify, requiring a single dimension only—the diameter.

371

b. Is easiest to manufacture by a wide variety of machining methods, being essentially the only form that can readily be produced by the continuous path of the tool, whether in turning, drilling, grinding and so forth.

c. Can be measured in terms of the single characteristic dimension with less potential error and with simpler means than many other forms.

d. Has several functional advantages, such as easy assembly, smooth rolling and uniform strength in any direction symmetrical to the axis.

The various purposes that basically round objects must serve lead to a wide range of roundness requirements. The diversity of methods by which such parts may be produced results in different levels of accuracy with respect to both the dimension and the form. The latter aspect of basically round technical parts, the accuracy of the form and its measurement are the subjects of this chapter.

CHARACTERISTIC FORMS OF ROUNDNESS DEFICIENCY

The roundness of form, as well as its errors, originate in the manufacturing phases of technical parts. Because of the wide variety of manufacturing methods, equipment and conditions, the deficiencies in the roundness of manufactured parts may differ over an extensive range.

Certain characteristic types of roundness errors can frequently be associated with particular manufacturing methods and circumstances. It is well known that improperly aligned or unround center holes reflect on the form of the object that is turned or ground between centers. Clamping in three or four jaw chucks can distort parts to an extent where the essentially round form produced in the chucked condition will distort as soon as the clamping force has been relieved. Certain machining processes are prone to produce consistent roundness errors when not properly controlled. A typical example is centerless grinding, by which basically round parts with rather regularly distributed protuberances may be produced, such parts having 3, 5, 7 or some other odd number of lobes. Vibrations in the machine or workpiece can cause undulations on the machined surface that, when appearing with a high frequency, are considered chatter marks. Some of the conditions detrimental to the roundness of the finished part may occur simultaneously, resulting in superimposed forms of roundness deficiency.

The actual form of the finished part contour, when compared to the perfect circle, can differ in many ways, primarily depending on the manner in which the part has been produced. Although there are no rigid boundaries separating various types of roundness deficiency, some distinction can generally be made when the predominant characteristics of out-of-round forms are being considered. With the understanding that any classification of the types of out-of-roundness will be arbitrary to some extent, the assigning of the more frequently occurring types of roundness errors to categories has some usefulness for analyzing these deficiencies.

Such an analysis of various types of out-of-round conditions, in addition to being beneficial to the general appraisal of roundness deficiencies, will serve several practical purposes as well:

1. The characteristic form of deviations from the basic roundness is usually indicative of their origin, the manufacturing process by which the part was produced. When a certain type of out-of-round condition is considered objectionable because of its form or dimensions, the identification of the origin of these deficiencies leads to determining the course of action needed for avoiding or reducing the chance of a recurrence.

2. The currently used measuring methods differ in their capabilities to detect and to measure the diverse forms of out-of-roundness. The awareness of the category to which the expected roundness deficiency may pertain is helpful in selecting the appropriate measuring method and equipment. (For example, in case of roundness inspection in a Vee-block, the number of predominant lobes determines the included angle of the block that will be the most effective in detecting that particular type of roundness error.)

3. The functional effect of diverse forms of out-of-roundness differ, depending on the application or required service of the part. The category to which the predominant characteristics of the experienced out-of-roundness belong is a meaningful factor in the appraisal of the part's adequacy for the intended application.

Table 14-1 lists and illustrates schematically, against a background of concentric circles, some of the more frequently occurring basic types of out-of-roundness in manufactured parts. The designations used in the table are not based on a standard; they are intended only to distinguish certain types of out-of-round forms according to their predominant

TABLE 14-1.   CHARACTERISTIC IRREGULARITIES OF BASICALLY ROUND FORMS — 1

CROSS SECTIONAL DIAGRAM (EXAGGERATED)	DESIGNATION AND DESCRIPTION	POTENTIAL ORIGIN AND MEANINGFUL CHARACTERISTICS
	Oval with unequal axes approximately perpendicular to each other and of essentially symmetrical position.	Misalignment in the machine tool centers and/or the center holes in the part. Significant dimension is the difference between the major and the minor axes.
	Egg shaped, essentially oval yet the major and minor axes are not symmetrically located.	Inaccurate centers and/or worn-out center holes. The effect on the radial separation of envelope circles is usually indicative for the extent of functional consequences. (Envelope circles are mutually concentric and just contain the tracing line which represents the part's contour in the inspected plane.)
	Angularity characterized by odd number of undulations of essentially similar spacing. Typical numbers of undulations along the circumference are 3, 5, 7 and 9.	Centerless grinding, when not adequately controlled, tends to generate forms displaying characteristic departures from the ideal round. These irregularities have usually little effect on the consistency of the diameter, but can be detected in a Vee-block with appropriate included angle. Radial separation of enveloping circles and number of undulations are functionally meaningful.
	Contour undulations (medium frequency lobing). Departures of the contour from the basic round are essentially consistent in spacing and amplitude.	Vibrations in the machine tool or the set-up, and/or insufficient rigidity of the workpiece. Measurable either by the amplitude of undulations ascertained as the separation of the envelope circles, or by vibration analyzing instruments. The frequency of the undulations can also be of functional significance.

TABLE 14-1.   CHARACTERISTIC IRREGULARITIES OF BASICALLY ROUND FORMS – 2

CROSS SECTIONAL DIAGRAM (EXAGGERATED)	DESIGNATION AND DESCRIPTION	POTENTIAL ORIGIN AND MEANINGFUL CHARACTERISTICS
	High frequency undulations (frequently associated with surface roughness). Very closely spaced, low amplitude undulations of the contour, detectable only with sensitive tracer instruments or by optical means.	Usually correlated with the characteristics of the manufacturing process, like grinding wheel grain size, feed rates etc. Frequently intentionally disregarded or suppressed by electrical or mechanical filtering when preparing roundness tracings, because considered inconsequential in many applications.
	Roughness superimposed on angularity. The basically angular contour is not a smooth line, but displays closely spaced undulations (secondary form irregularities).	This combination of two different types of form irregularities originates from the simultaneous occurrence of conditions responsible for each of them separately. A rather frequent variety of roundness irregularities, whose precise analysis requires chart tracings made without the suppression of details.
	Random irregularities characterized by non-periodic occurrence of significant departures from the basic round form.	Could originate from various, not properly controlled conditions in the grinding process, frequently associated with inadequate stability in the work positioning and holding. Complete analysis is predicated on true roundness charts prepared by an instrument having sensitive response.
	Form irregularities of basically round surface sections whose designed cross sectional contours are regular circular arcs.	Departures from the basic form can be the combined effect of roundness irregularities and displaced arc centers. This latter contributory factor with error-causing potential requires the roundness measurement to be made with the rigorous control of the arc center and of the inspection plane locations.

TABLE 14-1.   CHARACTERISTIC IRREGULARITIES OF BASICALLY ROUND FORM – 3

CROSS SECTIONAL DIAGRAM (EXAGGERATED)	DESIGNATION AND DESCRIPTION	POTENTIAL ORIGIN AND MEANINGFUL CHARACTERISTICS
	Roundness deficiencies superimposed on coaxiality errors with conjugate effect on the functional adequacy of the part.	Digressions from a common datum during the manufacturing process, will cause errors of form which are superimposed on the form deficiencies of the individual surfaces. The adequacy of the part will be affected by the resultant irregularity. Measurements must be referenced from the functionally correct axis.

characteristics. The meaningful dimensions of the roundness errors, indicated in the exaggerated illustrations of the table, should point out those parameters that deserve primary attention in selecting the proper measuring method.

THE FUNCTIONAL NEED FOR ROUNDNESS

In order to function properly, or to meet other requirements related to the intended use of the product, most technical parts are designed with specific geometric forms that must be assured within the prescribed limits. When the basic form of a technical part is that of a body of revolution, one of its essential geometric conditions is roundness. The dependable measurement of roundness is, therefore, an important process for the assurance of the required functional characteristics of basically round parts.

However, there is no need to maintain the true form with equal accuracy for all types of parts that are, essentially, bodies of revolution. Actually, roundness requirements in engineering vary in two major respects:

a. The predominant characteristics of the roundness errors; these may be either objectionable or meaningless for a particular type of application.

b. The permissible limits of departures from the perfect roundness, in case the part displays the objectionable kind of roundness errors.

In principle, the definition of roundness requirements belongs to the design specifications and

should be spelled out on the engineering drawing. However, because deviations from the ideal roundness can have a wide range of different form characteristics, it is usually not possible to define them by a single dimension, such as is preferred for engineering specifications.

For that reason, some understanding by the metrologist of the functional consequences of roundness errors can contribute to making the measurements of roundness more informative with regard to the inspected part's adequacy for its intended application or service. Such understanding can be of assistance in two major respects:

1. In selecting the particular method of roundness measurement that, with the least effort and cost, can supply the needed information on the part's geometric conditions; and

2. In assessing the results of the roundness measurements from a functional aspect, by putting the emphasis on the dimensions of the significant form deviations.

As an introduction to the functional appraisal of roundness deficiencies, Table 14-2 lists, as examples, a few typical applications in which certain kinds of out-of-round conditions can have particularly harmful effects on the proper operation of the part. The diagrammatic sketches of this table are highly exaggerated in relative size, and also of distorted proportions, for the purpose of clarity.

Although for each of the randomly selected application categories of basically round parts, a single type of form irregularity was named, in the majority

TABLE 14-2.   APPLICATION EXAMPLES OF ENGINEERING PARTS WITH FUNCTIONAL NEED FOR ROUNDNESS — 1

DIAGRAMMATIC SKETCH	APPLICATION CATEGORY AND EXAMPLE OF HARMFUL FORM IRREGULARITY	TYPICAL FUNCTIONAL EFFECT OF DEFICIENT ROUNDNESS
	Running fits with clearance locally bridged by isolated lobes.	Protuberances which extend beyond the predominant surface, can interfere with the intended clearance of the fit, and by breaking up the hydrodynamic lubricant film, cause harmful metallic contacts.
	Press fits with contact concentrated on the narrow crests of the mating surfaces.	The limited area crests can collapse under high specific loads caused by stress concentrations much in excess of the designed values. The loosening of the press fit will result.
	Sealing fits with depressions and crevices penetrating beneath the predominant contact surface.	Gaps in the contact area between mating surfaces (e.g., a plunger in the cylinder) due to interruptions in the continuous contour can cause leakage of fluids which ought to be restrained by tightly fitting members.
	Measurement over the diameter for mating and segregation when the measured part diameter differs from the minimum diameter of the hole capable of containing the part.	In current shop practices mostly, and in automatic gaging almost exclusively, the diameter of the basically round part is considered the representative dimension of its size. Differences between the measured diameter and the diameter of the hole into which the shaft will actually fit, can be the cause of functional deficiency or difficulty during the assembly.

**TABLE 14-2.   APPLICATION EXAMPLES OF ENGINEERING PARTS WITH
FUNCTIONAL NEED FOR ROUNDNESS — 2**

DIAGRAMMATIC SKETCH	APPLICATION CATEGORY AND EXAMPLE OF HARMFUL FORM IRREGULARITY	TYPICAL FUNCTIONAL EFFECT OF DEFICIENT ROUNDNESS
	Axle supported in plain bearing when the journal of the axle has out-of-round form.	Out-of-roundness with widely spaced lobes, such as oval, egg shaped, triangular, etc. is detrimental to the locational stability of the rotating shaft, although the wrap-around design of the plain bearing tends to damp the harmful effect of roundness irregularities.
	Antifriction bearing roller with nonuniform diameter (diametrical out-of-roundness).	An important function of the rolling elements in the antifriction bearing is to consistently maintain a uniform distance between the outer and inner ring of the bearing. Varying diameters of the individual rolling elements interfere with this function.
	Bearing ball with surface defects affecting the ideal sphericity.	The ball bearing's smooth operation is contingent on the precise sphericity of each of its balls. Interruptions in the basic form, whether depressions or protuberances, will unfailingly reflect on the bearing performance and can cause audible noise.
	Antifriction bearing ring pathway with undulations of critical frequency and amplitude.	Undulations of the basically round surface have an amplitude measurable as radius variations, and a frequency associated with the number of lobes on the circumference. Vibrations in the operating bearing which are caused by these form irregularities become particularly detrimental when related to the natural frequency of the supporting members.

of cases this would be an oversimplification of the actual conditions. The reason for this kind of presentation is again, clarity, to permit a distinct analysis by isolating a predominant condition.

Cylindricity and Coaxial Roundness of Interrelated Surfaces

When the roundness of a technical part is a required geometric condition, for the purpose of its inspection it is customary to consider the part in one of its cross-sectional planes. This method of mental observation is encouraged by the wording of standard definitions, and is also applied in practice by most methods of roundness measurement.

The commonly used definitions of roundness consider the axis of the object as the reference (datum) element, from which all points of the surface must be at equal distance. Realizing that the axis of most engineering parts is a geometric concept only, the physically measurable elements are the surfaces of the figures of revolution to which these specifications refer.

For that reason it is practical and also customary to spell out form requirements that, by definition, are related to the axis, as a *runout condition*, using the wording "These surfaces must be concentric within . . . total indicator reading (TIR) or full indicator reading (FIR)." Actually, this specification requires that the designated surfaces should be individually round and in relation to a common axis.

The geometric conditions assessed by the measurable dimension of runout are frequently referred to as *concentricity*. This term, in general usage, is often applied to three-dimensional figures with respect to either a single surface, for example, that of a cylinder, or several interrelated surfaces. Regarding the usage of concentricity in engineering specification, it should be pointed out that although the term implies geometric conditions in a two-dimensional plane, such as the imaginary measuring plane of a body, it is also used to designate conditions related to a common axis, the term *coaxiality* being less current.

Runout specifications can also apply to other than basically round surfaces, as long as they are geometrically interrelated by means of a common rotational axis. Distinction is made between surfaces constructed around the axis or at right angles to the axis, although the fundamental principles of measurement are similar for both types of runout. It is worthwhile to note that, depending on the datum used in the measuring setup, the indicated values of runout can be higher than, theoretically, up to a

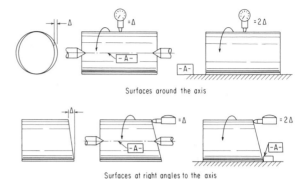

Surfaces around the axis

Surfaces at right angles to the axis

Fig. 14-1. The effect of datum selection on the indicated value of runout on surfaces around the axis and at right angles to the axis. To illustrate, basically cylindrical parts are shown with highly exaggerated imperfections. The datum is indicated by A, and Δ indicates the deflection of indicator pointer.

maximum of twice the dimension of the concentricity error. The possible effects of datum selection on the measurable runout are shown diagrammatically in Fig. 14-1.

THE DIFFERENT SYSTEMS OF ROUNDNESS MEASUREMENT

There are several essentially different methods of gaging in current use, for the purposes of determining the form trueness and of measuring the form irregularities of nominally round objects. These methods are outlined in the ASME specification B89.3.1-1972 (R1988), Measurement of Out-of-Roundness.

Although these methods are generally known to quality control and production engineers in metalworking, a survey is presented here with regard to one particular aspect, that is, the capability of any particular method to measure the functionally meaningful roundness irregularities of the inspected object.

There are essentially two basic systems of roundness measurement, which differ in several significant respects:

1. One is known as the conventional system in which the object is positively supported on selected points of its surface, and another surface point is contacted by the probe of an indicator gage of appropriate sensitivity. When the part is rotated, either into specific indexing positions or continuously, but resting on the same support elements, variations in the location of the momentarily contacted points of the surface are signaled by the indicator. The amplitude of these variations indicates the dimensions of the roundness irregularities.

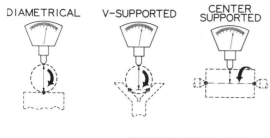

DIAMETRICAL   V–SUPPORTED   CENTER SUPPORTED

⊕ DATUM POINTS ON THE OBJECT SURFACE.

↕ DIMENSION WHOSE VARIATIONS BETWEEN DIFFERENT ROTATIONAL POINTS OF THE OBJECT ARE MEASURED.

Fig. 14-2. Principles of roundness measuring systems with intrinsic datum, implying referencing from the surface of the object.

There are several different methods that fall into this broad category, yet they all have a common characteristic: the referencing for the measurement originates from one or more points on the surface of the object. For this system of gaging the term *roundness measurement with intrinsic datum* was coined and is used in this discussion (see Fig. 14-2).

2. The other basic system of roundness measurement obviates using any part of the physical object as a datum. The referencing is accomplished from an external member, an ultraprecise spindle with almost perfectly runout-free rotation. The term *roundness measurement extrinsic datum* is used for this system (see Fig. 14-3).

Both systems have their particular merits, and limitations as well. The conventional method of roundness checking—termed the intrinsic datum system—is more widely used, whereas the extrinsic datum system provides information that, for most

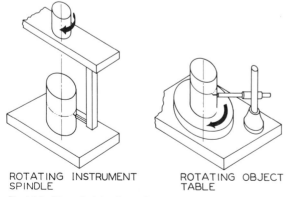

ROTATING INSTRUMENT SPINDLE

ROTATING OBJECT TABLE

Fig. 14-3. The principle of roundness measuring systems with extrinsic datum, implying referencing by means of a detached instrument member.

applications, is more pertinent to the functionally proper form of the object.

In industrial uses, the aspect of economics must, of course, be considered when seeking the appropriate approach to any problem: adequacy will usually prevail over perfection.

Some of the characteristic advantages of each of the two basic systems of roundness measurement are pointed out below. Considerations such as those listed, and perhaps complemented with others, should be weighed when choosing one or the other system for any particular application.

Advantages of the Intrinsic Datum Roundness Measuring Systems

The following are the advantages of the intrinsic datum roundness measuring system.

1. Inexpensive tooling, requiring a minimum of fixturing.

2. Relative rugged equipment may be used under adverse environmental conditions.

3. Quick to operate, the positively supported object needs no centering or squaring adjustments.

4. Sufficiently informative for many applications, when properly selected with due regard to predominant object conditions (may need to be supplemented with the occasional use of the tracer method, using an extrinsic datum).

5. Supplies values that are directly meaningful for several specific applications, or can be correlated with other requirements.

6. Adaptable for continuous in-process measurements and for monitoring automated production.

Advantages of the Extrinsic Datum Roundness Measuring System

The following are the advantages of the extrinsic datum roundness measuring system.

1. Supplies a true image of the geometric conditions of the object by selective magnification.

2. The rates of magnification and the suppression of inconsequential features can be varied to enhance the meaningful aspects of the roundness condition.

3. The image of the object supplied by the instrument can be evaluated by different methods of interpretation, choosing the one best suited to the functional requirements of the product.

4. Continuous tracing around the entire surface in the selected plane obviates the possibility of disregarding errors that can be missed by point-to-point measurements.

5. Graphic representation of the measured roundness conditions (charts) are valuable for both thorough analysis and permanent record.

ROUNDNESS MEASUREMENTS BY REFERENCING FROM THE SURFACE OF THE OBJECT

As an introductory note, applicable to all systems of roundness measurement that reference the instrument's indications from the surface of the object (direct referencing), it must be pointed out that none of these methods will supply information in complete agreement with the standard specifications of roundness. The degree of disagreement may vary, depending on the general conditions of the part surfaces being used for referencing, and also on the system and setup of the measuring process. However, in all cases there will be differences, since the principles of the methods differ from the fundamental concept of roundness measurement by comparison to an extraneous datum circle.

A detailed discussion of these methods of roundness measurement is warranted by their extensive usage in manufacturing practice. As mentioned in the preceding section, there are various circumstances in which the inspection of roundness conditions by direct referencing is an acceptable substitute, or may even be preferable to inspection by roundness tracing.

The three basic systems of roundness inspection by direct referencing are considered separately in the following discussion.

Diametrical Measurements of Roundness

The diametrical measurement is the most convenient method of inspecting roundness when considered solely from the viewpoint of gaging technique. The measurements can be carried out with standard measuring tools, including those designed for measuring linear distances, such as the diameters of round parts. The following considerations seem to justify the use of diametrical measurements for checking roundness:

When assuming that in a substantially round object: (1) any two radii aligned to each other to form a diameter have equal lengths, and that (2) the variations in the dimensions of several diameters mea-

sured in the same plane represent twice the differences in the length of the constituent radii, the resulting values comply with the standard definition of the concept, according to which the radius variations represent the characteristic dimension of out-of-roundness.

Although a rather convenient interpretation, it is far from faultless. The standard definition, of course, also specifies the *axis* of the object, which implies a single common axis for all measurements. However, in the case of out-of-round figures, the *centers* (middle points) of several diameters contained in the same plane are not coincident. Consequently, the requirement of a single common axis is not satisfied, and herein lies the weakness of roundness measurements based on the comparison of several diameters.

The lack of coincidence of the midpoints of several diameters becomes the most obvious when isodiametrical forms are considered. Figure 14-4 is a conventional illustration of a few common forms of isodiametrical out-of-roundness. This diagrammatic drawing shows figures that, although of essentially constant diameter, have perimeters with very pronounced departures from a true circle. In these or comparable figures, several diametrical measurements confined to the same radial plane would indicate either no errors at all or errors of such small magnitude as would not reveal the actual condition of out-of-roundness.

Consequently, for any application where roundness is a critical requirement, and nonsymmetrically distributed departures from the basic form are to be expected, roundness inspection by diametrical measurements can lead to erroneous appraisal and should not be used. However, there are applications in which, due to the functional role of the part, the constant diameter is the primary form requirement;

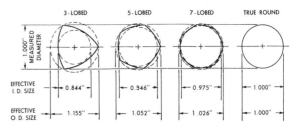

Fig. 14-4. Conventional although exaggerated representation of isodiametrical out-of-roundness and its effect on the diameters measured between two parallel planes in contact with the part surface. *Note:* The drawing shows conditions that are accentuated by assuming equal spacing of the lobes, and regular circular-arc form of the connecting perimeter sections.

in such cases diametrical roundness inspection is positively warranted.

For checking roundness by diametrical measurements, the sizes of several diameters must be determined. These measurements must be confined to the same plane and normal to the axis of the part. The greater the number of these measurements and the better they encompass the entire perimeter of the object, the more reliable are the inspection results to be expected.

When the inspection process is limited to determining the roundness condition of the part, only the dimensional variations are the significant values; the absolute size of the diameter has no direct meaning. Two basic methods of diametrical roundness measurement are used:

1. Making absolute measurements of the diameter and then comparing their values; and

2. Directly comparing the various diameters by using an appropriate indicator gage that shows only variations from a preset dimension.

Roundness inspection by comparing the actual lengths of several diameters. Depending on the sensitivity required from the particular roundness checking, standard types of length measuring tools may be used for either external or internal measurements. Such measuring tools rank from vernier calipers to different types of outside and inside micrometers. The drawbacks of this process are the limited sensitivity; the need for skill when measuring to fine increments; and the time-consuming nature of the gaging, which necessitates the adjustment and reading of the measuring tool for each gaging position. Furthermore, the readings that are obtained in absolute dimensions must be memorized or noted in writing, so that at the end of the measuring series the differences between the extreme values can be determined.

This method was mentioned for the sake of completeness of the listing only. Its actual applications are limited to very large parts, or those requiring only a rather low level of form trueness.

Roundness gaging by indicated diameter variations. These measurements consist in selecting any of the object diameters in a gaging plane as a reference dimension, and then, while bringing several other diameters successively into the measuring position, observing the range of the instrument indications.

The requirement of bringing several object diameters into measuring position can be met either by selecting discrete points on the object surface or by continuously rotating the part during the measuring process. Continuous measurement reveals the existing roundness errors in a more dependable manner, but may require some fixturing, special skill, or even mechanical equipment for rotating the object. Sequential gaging of individual diameters is easier to accomplish, and by aiming at the "high points," that is, at the maximum reading in each position, errors due to measurements shifted from the significant axial plane are avoided. On the other hand, in addition to exploring, only partially, the perimeter of the object, single-point measurements can be more time-consuming.

Figure 14-5 shows two types of gages commonly used for roundness inspection by direct diameter comparison. For external surfaces, either portable or stationary gages may be used, the latter being preferable for small parts that can easily be placed on the comparator stage. For internal surfaces, portable gages are more frequently used, although parts having dependable locating surfaces normal to the bore axis may be checked on an internal comparator, with the benefit of higher gaging precision. Although the portable gages are generally equipped with mechanical comparators, air or electronic indicating devices are often preferred for stationary comparators when used for roundness inspection.

For external measurements, flat or large-radius contact points are preferred, the part resting on a flat stage; thereby the probability of errors due to not measuring across the diameter can be reduced. For internal measurements, spherical contact tips, with radii definitely smaller than that of the bore, must be

Mahr Gage Co.

Fig. 14-5. Examples of standard types of indicator gages that may be used for measuring diametrical variations of external and internal cylindrical features.
(Left) Indicator snap gage.
(Right) Indicating internal gage.

used on both the anvil (fixed) and the moving member of the gage. When using bench comparators, back stops on the gage table, adjusted to retain the part's central plane coincident with the axis of the gage spindle, are a useful accessory of the setup.

Differential gaging of diametrical out-of-roundness. Substantial refinements of diametrical out-of-roundness measurements can be accomplished with the aid of differential gages. Both air and electronic gages are adaptable for differential measurements by using special circuitry and amplifiers.

Figure 14-6A shows the principles of a differential electronic gaging by means of a comparator gage setup for checking diametrical variations of basically round parts. The gage operates with two gaging heads facing each other in well-aligned positions and at a gap distance that is set to produce zero indication for a selected part diameter. The signals originating from both heads are channeled into a balancing system, which compensates one against the other. The signals are caused by the opposingly directed displacements of the sensing spindles. Only when the gap between the two contact points varies as a consequence of differences in the part diameter will deviations be indicated.

When the size and other characteristics of the part permit, essentially similar gaging conditions can be accomplished by means of reed suspension for a frame holding a single gage head and an opposing anvil (see Fig. 14-6B). The resulting instrument is functionally similar to an indicator snap gage, however, in a stationary version. When independent staging elements are used for supporting the part, it must be assured that during the gaging process both the anvil and the tip of the sensing head are in contact with the part's surface.

Both systems, the differential gaging and the reed-suspended gaging fork, permit the supporting of the part in a Vee-block. This type of staging assures pos-

itive transverse location either for stationary or sensitively rotated part, without interfering with the actual measurement that is carried out with the independent referencing of the described gage systems.

Roundness Inspection by Referencing From Vee-Block Supported Surface Elements

The commonly used term is *Vee-block roundness measurements*, designating a method of form inspection in which a basically round part is supported in a Vee-formed trough. Because of the similarity of the referencing principles, roundness inspection with portable indicating instruments having two supporting legs resting on the part's surface are also included in this category.

The commonly used Vee-blocks usually have flat walls, or flanks, symmetrically located in relation to a vertical central plane and the included angle of the walls is a defined dimension. The walls need not be actually flat surfaces; elements producing point contacts with the surface of the basically cylindrical part, such as round pins or balls held on stems, may also be used. However, the effective contact points of such elements must be contained in the imaginary planes of the walls that were replaced by contact elements of different forms.

Holding the object in a Vee-block for roundness inspection by means of an indicator instrument is a common process, widely used in general shop practice. There are various functional reasons and aspects of convenience for the application of this method of form inspection:

1. From a functional point of view, supporting the part in a Vee-block of appropriate angle can be an effective method of inspection when nonsymmetrical out-of-roundness must be detected and assessed.

2. It offers a convenient way for holding round objects by gravity, also assisted by the gaging pressure of the indicator acting in the direction of the gravitational pull.

3. It is also a simple, dependable process for locating cylindrical objects in such a manner that the part's axis is confined to the bisecting plane of the Vee-block, thus producing automatically controlled staging locations.

The roundness checking of parts supported in a Vee-block is always a comparative kind of measurement, using any optional orientation of the object for establishing the reference dimension. The locating

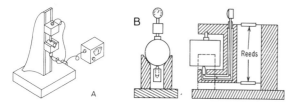

Fig. 14-6. (A) Inspection of diametrical variations by differential gaging, whose operation is, in principle, unaffected by the part's contact with the supporting Vee-block type stage.
(B) The operating principles of a reed-supported, stationary snap gage with indicator for the roundness inspection of parts held, but not functionally located, in a Vee-block.

elements on the object surface that are in contact with the Vee-block, are kept at a constant level in relation to the reference position of the gage point. The gage point contacts a third point on the object surface, in a position symmetrical to the locating points. In the measuring plane, these three points may be regarded as the vertices of an isosceles triangle, whose height variations constitute the actually measured dimension $H_I - H_{II}$, as shown in Fig. 14-7.

However, the measured variations are not equivalent to those of either the diameter or the radius of the object, and are proportional only when the surface undulations are of uniform height and spacing, and the Vee-block used for support has the appropriate angle. The significance of the latter requirement will be explained later in conjunction with the geometric analysis of this method of form inspection.

Applications of Vee-block supported roundness inspection. The principles of commonly used methods and of various techniques are discussed in Table 14-3. These methods differ with regard to the comprehensiveness of information they provide, as determined by the extent of surface contour that participates in the observable indications, and the number of positions in which comparative measurements are made. The latter may vary from continuous scanning around the entire surface to a limited number of checks at points that are assumed to represent the predominant characteristics of the inspected form. The table also contains information that can be of assistance in the choice of the applicable method of inspection.

Geometric analysis of information obtained in Vee-block-supported roundness inspection. The information supplied by Vee-block-supported roundness inspection in alternate positions of the object is the combined value of two categories of radius lengths:

a. That radius of the object that coincides with the bisector plane of the Vee-block; this is indicated directly as the variations of its actual length; and

b. The two radii pertaining to the contact points of the object with the supporting Vee-block; these are indicated indirectly as their effect on the level of the object axis.

A convenient way to visualize how these variations will affect the Vee-block type, roundness measurement is to consider the specimen as consisting of two half cylinders joined at the horizontal plane of their common axis. The radius of the upper half is measured, whereas the lower half, or supported portion, establishes the level of the imaginary dividing plane.

A diagrammatic illustration of these conditions is shown in the top part of Fig. 14-8, which is based on a theoretical part having lobes of equal height and of uniform spacing. For the purpose of clarity, two concentric circles, representing the circumscribed r_1 and the inscribed r_2 circles of the cross-sectional perim-

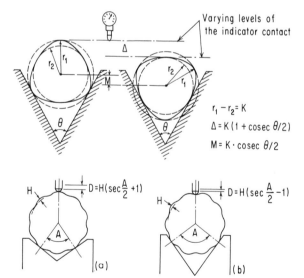

$$r_1 - r_2 = K$$
$$\Delta = K(1 + \operatorname{cosec} \theta/2)$$
$$M = K \cdot \operatorname{cosec} \theta/2$$

$D = H\left(\sec \dfrac{A}{2} + 1\right)$

$D = H\left(\sec \dfrac{A}{2} - 1\right)$

(a)   (b)

Fig. 14-8. The geometric analysis of Vee-block-supported roundness inspection.
(Top) Interrelations of the lobe spacing and of the Vee-block angle.
(Bottom) Effect of the Vee-block's included angle (180°, A) on the indicator readings in relation to the radial range H of the circumferential undulations.
The examples are based on equally spaced undulations and show the applicable formulas for (a) the most sensitive Vee-block angle; and (b) the least sensitive Vee-block angle.

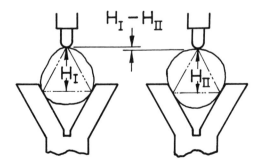

$$H_I - H_{II}$$

Fig. 14-7. Schematic illustration of the form conditions measured when supporting an imperfectly round part in an appropriate Vee-block.

TABLE 14-3.   BASIC VARIETIES OF VEE-BLOCK SUPPORTED ROUNDNESS MEASUREMENTS

MUTUAL POSITION OF WORK STAGE AND GAGE	DIAGRAM	CHARACTERISTIC MOVE-MENTS IN THE PROCESS	TYPICAL APPLICATION AND EVALUATION
Both members are in stationary positions		Object axis and gage axis are continuously aligned during the process. Part is re-oriented by rotation while in the gaging position.	Where the rotation of the object can be carried out in the staged position, it permits making all measurements in the axial plane of the object, without the need for repeated adjustments.
Sliding one of the members into gaging position (The "highest reading" in each position supplies the meaningful dimension)		The object is brought repeatedly into gaging position by sliding its holding fixture on the gage table. Part re-oriented between successive moves to gaging position.	Where changes in object orientation by means of rotation is not feasible while the part is positioned under the gage spindle.
		The gage stand is brought into the gaging position by sliding it on a surface plate which also supports the Vee-block. Part re-oriented between successive moves of the stand to gaging position.	Where the configuration or the bulk of the object make it preferable to move the gage and to leave the object in a stationary position. Can also be used with double Vee-block support for long shafts.

eter, have been added to the drawing. The part is shown in two alternate orientations in the supporting Vee-block, and the resulting variations in the level of the surface element contacted by the indicator, as well as of the axis, are pointed out. The drawing also shows the applicable formulas that, however, are valid only for a part of *theoretically regular* out-of-round form.

The combined effect on the resulting indications of both the spacing of the circumferential undulations and the included angle of the supporting Vee-block is illustrated schematically in the bottom part of Fig. 14-8, based again on the theoretical condition where both the heights and the spacings of the undulations are uniform. This illustration shows the central angle of the specimen, bounded by the radii that are normal to the walls of the Vee-block. The central angle of the part is the supplementary angle of the Vee-block. The two extreme indication ranges shown for the same part stem from the different Vee-block angles.

If the objective of roundness inspection in a Vee-block is to detect and clearly indicate the out-of-roundness condition of the part, that particular block angle is sought that will best expose the departures from the ideal roundness. The term "appropriate angle" is often used to designate that form of Vee-block that causes the widest range of indications to result from a specific type of round condition.

It is convenient to consider for the purpose of this analysis undulations of uniform spacing and amplitude, the variable being the number of undulations around the part perimeter. The appropriate angle of the Vee-block can be found by the simple formula

$$\theta = 180 - \frac{360}{n}$$

where

θ = "appropriate" included angle of the Vee-block in degrees;

n = number of uniformly spaced undulations around the part surface.

A Vee-block with an appropriate angle will actually exaggerate the extent of roundness, when the crest-to-valley distance (H in the bottom part of Fig. 14-8) is considered as the characteristic value of this condition. The following table shows, for a few typical numbers of undulations, the appropriate Vee-block angles, as well as the factors by which the indications on a part of theoretically regular form are increased over the actual out-of-roundness value.

Number of Uniformly Spaced Undulations on Object Surface	"Appropriate" Included Angle of Vee-Block	Factor by which Indicator Readings Are Increased over Actual Radial Out-of-Roundness
3	60°	3.00
5	108°	2.24
7	128° 34'	2.11
9	140°	2.06

The appraisal of the Vee-block roundness inspection process. The following conclusions may be made regarding the practical applications of roundness inspection by locating the specimen in a Vee-block:

a. When the primary objective of the inspection is to detect the presence of out-of-roundness due to odd-numbered lobing, the supporting of the part in a Vee-block is usually more effective than diametrical measurements.

b. The obtained out-of-roundness indications depend on the relationship between the included angle of the Vee-block and the number of undulations or predominant lobes on the object surface. When a series of parts whose characteristic lobing is essentially uniform is to be inspected, it is possible to select a best suited type of Vee-block. In practical applications this can be done by referring to an analysis of roundness charts made of a few parts considered typical by means of roundness measuring machines (to be discussed in the subsequent section) or by trying Vee-block with different angles to obtain the highest indications, either by alternating rigid blocks of different angles or with the aid of adjustable Vee-blocks.

c. The dimensions of out-of-roundness obtained in Vee-block measurements should be evaluated with circumspection, due to the tendency of this method to indicate a cumulative value of errors, arising from the contour characteristics at *both* the gaging and the locating points. Generally, but not necessarily, this combination results in higher indications than the radial out-of-roundness of the part.

d. The often-used factors for calculating the excess indications due to supporting the part in a Vee-block are based on theoretical figures, assuming uniformly spaced undulations; therefore, these formulas should not be applied unconditionally for evaluating the ratio between indicated variations and the radial out-of-roundness.

e. In the case of symmetrical out-of-roundness (even-numbered lobing), supporting the part in a Vee-block frequently tends to conceal the roundness errors, which could have a significant effect on the functional adequacy of the part.

Roundness Measurements Between Centers (Bench Center Radial Measurements)

This measuring process should, in principle, supply direct information on the most essential single parameter of roundness: the variations of the radius length around the entire perimeter of the object. The dimensions measured in this process are supposed to be based on a common datum, the axis of the part, which is theoretically correct and complies with the general interpretation of the concept termed roundness.

In practical applications, however, these theoretical conditions are only partially assured. The obvious reason for this discrepancy is the fact that the axis of the object, being only an imaginary element, is not actually present on the physical part. In practice, it is the inferred position of the axis, such as a connecting line between the vertices of center holes or centers, that is considered to represent the axis of the part being inspected.

The reliability of roundness measurements between centers can be affected by several factors. Any one of these in itself, or in combination with others, can result in indications that are generally in excess of the actual out-of-round condition of the inspected surface. These factors represent the major drawback of a system of roundness measurement that, in its principles, is correct and relatively simple to implement. Examples of such potential sources of error are discussed in Table 14-4.

Practical applications of the process. Roundness measurements between centers are commonly carried out by holding the object in bench centers, such as those shown in Fig. 14-9. After the part has been clamped between the centers of the bench, the probe of an indicating instrument is brought into contact with the selected element of the surface to be inspected. Then the part is rotated, usually by hand, and the range of indicator reading is registered as the dimension of radial out-of-roundness.

The following precautions in specimen preparation and gaging setup contribute to the reliability of this system of roundness inspection:

1. The center holes of the object should be made in close agreement with the design angle (commonly 60 degrees included), round and well finished. The lapping of the center holes is frequently applied for removing the scale and the soft skin when the part has been heat treated after the drilling of the center holes. Special center-hole grinders with planetary spindle motion are particularly useful for improving the center-hole geometry of heat treated parts. The principles of operation of planetary center-hole grinding are shown in Fig. 14-10.

2. The bench centers used for roundness inspection between centers are available in various grades of accuracy. The operative surfaces of the beds can either be ground (flat within 0.0005 inch) or scraped (flat within 0.00025 inch), and the center axes of head- and tailstocks are often held alike for height within 0.0005 inch. The alignment of the centers is usually assured by means of T-slots in the bed. In use, the bench centers should be set up with the bed in level position; for this purpose most types of bench centers are equipped with leveling screws. For high precision measurements, bench centers with accuracy values exceeding those of the generally used types are also available.

3. For holding the part in position between the centers, as well as to counterbalance forces due to the weight of the part and those applied by rotating it, an axial spring pressure is exerted through the

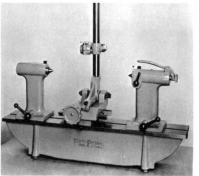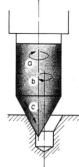

The Taft-Peirce Mfg. Co.
Bryant Grinder Corp.

Fig. 14-9. (Left) Bench centers with indicator stand for inspecting the radial runout of shafts and similar parts mounted between the centers.

Fig. 14-10. (Right) Center-hole grinder operational principles with planetary motion of the spindle. (a) Rotation of the grinding cone around its own axis. (b) Planetary motion of the machine spindle around the axis of the center hole being ground. (c) Reciprocating motion along the generatrices of the tapered center hole.

TABLE 14-4.   POTENTIAL SOURCES OF ERROR IN ROUNDNESS INSPECTION BETWEEN CENTERS

SOURCE OF ERROR	DIAGRAM (EXAGGERATED)	EXPLANATION
Misalignment of centers and center holes		The axes of all four contacting elements— the two center holes and two centers— must perfectly coincide, to prevent errors from this source affecting measurements.
Angles of centers and center holes differ	Edge contact Wobble	Whenever the contact between the centers and the center holes does not extend over an adequate area of the mating surfaces, instability of location can result.
Roundness errors in centers and in center holes	Out-of-Round Center Center hole	Roundness errors of either the centers or of the center holes will affect the gaging results in a manner similar to the actual roundness errors in the plane of measurement.
Displaced location of center holes in the object		When the otherwise well-aligned center holes are not concentric in relation to the object axis, twice the shift will be added to the indicator reading caused by actual out-of-roundness.
Inadequate surface condition of centers and center holes	Corrosion, nicks, etc.	Scale, corrosion, rust, nicks, even surface roughness of either the centers or the center holes, can reflect on the result of the roundness measurements.
Defective straightness of object		A bow in the object which has to be rotated for roundness measurement, will result in runout, which is twice the straightness error; it will combine with the runout indications of the out-of-round condition.

The Taft-Peirce Mfg. Co.

Fig. 14-11. Instrument grade bench centers used with an electronic test indicator for measuring radial variations (runout) of the hand-rotated specimen.

sleeve of the tailstock. This pressure should be adequate to maintain the object in its aligned position, yet not excessive to an extent where it could cause distortions in slender parts.

To reduce the potential errors due to axial runout in the roundness gaging of tapered parts, the inspection between centers can conveniently be carried out by supporting the bench center as a sine plate, tilted by half of the included angle of the tapered object (Fig. 14-11). Although the indicator will sense radial variations perpendicular to the surface and not to the object axis, the possible error usually will be less than in the scanning of the object surface normal to the axis, where even minute displacements in relation to the assumed radial plane can considerably affect the indicator readings, particularly when steep angles are inspected.

ROUNDNESS MEASUREMENTS BY COMPARISON TO A REFERENCE CIRCLE—THE CIRCULAR TRACING PROCESS

The roundness of a solid is considered to be represented by the circular contour form of its cross-sectional planes normal to the axis. For measuring the circularity of that contour, a comparative method may be used that has a similarity of principles with the linear profile-measuring process. Both methods operate by comparing the specimen's actual profile with that of a master of known degree of exactness.

That comparison is accomplished by the continuous scanning of a representative surface element of the specimen, and indicating or recording its departures from the ideal contour line. In roundness measurements, the actual scanning is carried out by tracing the surface with the stylus of a pickup, in a continuous, relative displacement around the object surface along a precisely circular path, hence the frequently used designation of the process: *circular tracing*.

Like the methods of linear profile measurement, the master is not presented in a physical form to the specimen whose contour geometry is being inspected. The master is represented by the rigorously controlled path of the displacement movement that, in the case of linear profile tracing, is a straight line; and in the case of circular tracing, an almost perfect circle.

The resemblance between the linear and circular profile tracing exists in still another significant aspect; both require a relative displacement of the pickup with respect to the specimen. Consequently, for the circular tracing process, either of two systems may be used, namely: (a) the rotational movement of the specimen against a stationary pickup, or (b) the circular displacement of the pickup carrying the stylus around a stationary part.

The Operating Principles of the Circular Tracing Process

The principles of operation of the two basic systems of circular tracing instruments are illustrated diagrammatically in Fig. 14-12. The essential member in the operation of either of these instruments is a precision spindle of very high rotational accuracy. The spindle serves either to carry the pickup attached to a cross beam or to support a rotary staging table.

The stylus of the pickup, which is functionally comparable to the probe of an electronic test indicator, is brought into contact with a point along the selected surface element of the part and adjusted to a position of zero indication (pointer in the center of the scale). From this randomly chosen reference position along the surface element to be inspected, the rotary displacement movement of the instrument is started, and the stylus remains in contact with the part surface. An uninterrupted succession of an infinite number of contact points is thus created, describing a complete circle around the surface of the part.

Variations in the distance between the axis of rotation and the contacted points along the surface ele-

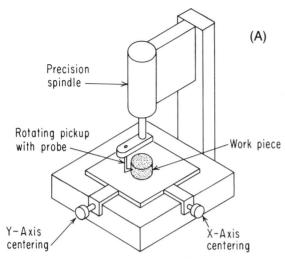

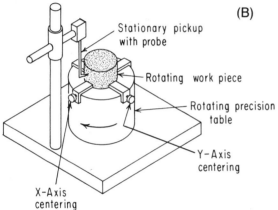

Fig. 14-12. Diagrammatic illustration of the two basic systems of circular tracing, roundness measuring instruments.
(A) Rotating table type.
(B) Rotating spindle type.

as represented by the axis of rotation of the instrument spindle, and (b) the axis of the part represented by an imaginary straight line at equal (or medium) distance from the surface of a basically round specimen part or feature. The coincidence of these two axes must be assured in both respects, namely, in alignment (directionally), and in location (laterally).

The staging of the specimen in agreement with this requirement of axial coincidence is one of the critical phases of roundness measurement by the circular tracing process. This operation, preceding the actual tracing, is generally termed centering. The circular tracing instruments are designed to permit the necessary alignment (tilt) and lateral adjustment of the specimen's staging position, to accomplish a degree of centering that is commensurate with the intended reliability of the applied roundness measurement.

The circular tracing instrument does not measure the actual length of the radius, but does display its variations. The dimension measured is, consequently, the form trueness, specifically, roundness of the inspected surface, and not its size.

Comparative Evaluation of the Two Basic Systems of Circular Tracing Instruments

Considering that both systems of circular tracing instruments, namely, the rotating spindle and the rotating table type, are widely used in industry, and are available from several prominent manufacturers, the question may arise as to which of the systems is better suited for roundness measurements. There is no generally valid answer to that question, because the suitability is dependent on the applicational conditions and the functional purpose of the instrument's operation.

In order to assist in an evaluation for determining the suitability of either system for a specific set of uses, examples of the relative advantages of each, the rotating spindle and the rotating table systems, are listed in the following:

Relative advantages of the rotating spindle system.

1. The centering and the tilt adjustment of a stationary object in relation to the planetary path of the stylus mounted on the rotating spindle is a more convenient and faster process than in systems where the part itself is rotating. The stationary staging table of the rotating spindle machine lends itself to a motorized jogging type of adjustment operated by pushbutton control, or even to automatic centering by a servo-system, thus substantially decreasing the most

ment cause deflections of the stylus. These deflections produce electrical signals in the pickup, which are electronically amplified at a preset rate, and then displayed on a meter or, more commonly, by a recorder. The distance between the axis of rotation and any contact point on the part surface is considered to represent the radius pertaining to the momentarily contacted surface point. The variations of the consecutive radii are the measure of departures from perfect roundness, as represented by the displacement path of the stylus.

The operation of roundness inspection by means of circular tracing is predicated on the essential coincidence of the axes pertaining to the two figures being compared. These are: (a) the axis of the master, time-consuming part of the roundness measuring operation.

2. The force acting on the precision spindle remains constant and its rotational accuracy is not affected by the weight of heavy objects.

3. When the basically round feature to be inspected is eccentrically located on the part, the roundness measurement can be carried out without the dynamic imbalance affecting the accuracy of the sensitive process.

4. Tracing along a circular arc of fixed length can be better controlled by limit switches than in the case of a bulkier rotary table.

Relative advantages of the rotating table system.
1. Tracing contacts can be established, without changing the original setup, at different levels and on sections of different diameters of the same part. Such measurements are valuable for the inspection of coaxiality conditions.

2. It is less sensitive to sudden temperature variations, particularly to drafts (not infrequent under shop conditions), which could affect the operational accuracy of the comparable system, requiring an upright to carry the spindle head.

3. Interrelated surfaces, parallel, perpendicular or inclined to a common axis, can be traced in the same setup, or even differential measurements can be carried out using two styli concurrently.

4. There are practically no limitations regarding the location of the traced section on the outside surface of parts mounted on a rotary table. Inspection behind shoulders or other protruding features can also be accomplished.

The examples above of relative advantages are, of course, not intended as a rigid delineation of instrument system capabilities. For the majority of roundness measuring tasks both systems are well adaptable, and accessories are being developed continuously to overcome many of the original limitations of either system.

The Application Potentials of the Circular Tracing Process

Analyzed in its geometrical aspects, the circular tracing process, by the relative movement of the stylus tip, establishes in space, a nearly perfect circle around an imaginary axis, and in a plane perpendicular to it. Although developed for the primary purpose of inspecting roundness, the process lends itself to the measurement of a great variety of geometric conditions. These can be visualized by considering the relationship between a true circle in strictly controlled position and specific surfaces of technical parts.

As a consequence of the importance attributed to the form accuracy of basically round parts, the performance characteristics of roundness measuring machines have been constantly improved in recent years. Certain models of roundness measuring machines may be considered to represent perhaps the highest degree of accuracy currently accomplished by commercially available rotating devices.

The application potentials, which tracer type measuring instruments in that accuracy category offer, deserve to be considered from two major aspects, namely:

1. As a means of accomplishing measurements related to diverse geometric conditions, in addition to departures from roundness; and

2. To find extended applications for roundness measuring machines, thereby assuring the versatile usage of this type of rather expensive measuring instruments.

To assist in appraising the varied metrological uses that, in principle, can be derived from the roundness tracing process and of its instruments, Table 14-5 presents a survey of various geometric conditions and interrelations. The measurement of these can be carried out advantageously, with the aid of roundness measuring machines of the appropriate type and model. Actually, for many of these inspection processes, roundness tracing machines are preeminently adapted, or even represent the only type of instrument capable of carrying out the measurements with the required degree of accuracy and reliability.

In these operations the departures of the traced surface element from the ideal circle are measured radially, or axially, or in both directions.

The Assessment of Roundness Condition

The circular tracing system of roundness inspection supplies information on the roundness condition of the specimen in a manner that is in agreement with the basic concept of roundness. For that reason, the conventional or standardized methods of roundness assessment are predicated on the application of the circular tracing process as the accepted method of roundness measurement. Consequently, all other systems of roundness inspection, such as those dis-

**TABLE 14-5.   EXAMPLES OF GEOMETRIC CONDITIONS EXAMINED BY
CIRCULAR CONTOUR TRACING – 1**

THE CONSIDERED GEOMETRIC RELATION TO THE CIRCLE	DIAGRAM	DESIGNATION OF THE GEOMETRIC CONDITION	THE APPLIED INSPECTION PROCESS
The circumferential contour of a body of revolution in a plane normal to its axis		Circularity Roundness	Measuring departures from roundness of cylindrical, conical, barrel-shaped, spherical, toroidal, etc., parts in selected planes normal to the part's axis (for the sphere in any central plane). The circularity of the traced perimeter is considered to represent the contour of the part.
The concentric contours of a body of revolution in several planes normal to its axis		Form regularity of a body of revolution	The true round form of a part designed to constitute a regular body of revolution can be inspected with a higher degree of reliability by tracing the contours of several cross sectional planes, all normal to the axis.
The concentric contours of a ring of uniform wall thickness in a plane normal to the axis		Concentricity Uniform wall thickness	Tracing successively and in the same plane the external and internal surfaces of a ring shaped part; the uniformity of wall thickness, as represented by the concentricity of the recorded tracing lines, can be determined.
The concentric contours of the several sections of a body of revolution regardless of the differences in basic sizes		Coaxiality	The roundness in relation to a common axis of the several sections of the part is measured, to display at high magnification any coaxiality error regardless of the different diameters of the inspected sections.

TABLE 14-5.   EXAMPLES OF GEOMETRIC CONDITIONS EXAMINED BY CIRCULAR CONTOUR TRACING – 2

THE CONSIDERED GEOMETRIC RELATION TO THE CIRCLE	DIAGRAM	DESIGNATION OF THE GEOMETRIC CONDITION	THE APPLIED INSPECTION PROCESS
A circle confined to a plane		Flatness	Sensing the traced surface with a stylus acting in a direction parallel with the axis, deviations from the ideal flatness will be displayed as proportional departures from roundness. *Note:* Concentric deviations from flatness, e.g., wide-angle cone, large-radius sphere, can be detected by tracing along a path nonconcentric to the axis of the part.
The identical departures of circles recorded in uniform orientation		Parallelism	Basically parallel surfaces at different axial levels, yet in planes normal to the axis, are inspected for parallelism by successively tracing the individual surfaces with recording on the same polar chart, and then comparing the chart lines for congruity of form.
The radial and axial departures from the ideal form of concentric circles established on mutually perpendicular surfaces		Perpendicularity Squareness	The perpendicularity of a flat surface in relation to an axis represented by the surface of a body of revolution, is measured in two steps: first staging the part's axis coincident with the rotational axis of the instrument, and then tracing the axial runout of the basically flat surface.
The circle considered as a curvature of constant radius		Circular form of a curve	Circular arc forms can be inspected for both (a) the rate of curvature, i.e., whether of constant radius, and (b) dimension, i.e., the size of the radius. However, the dimensional measurement of size requires the controlled positioning of the stylus tip in relation to the instrument's axis of rotation.

TABLE 14-5.   EXAMPLES OF GEOMETRIC CONDITIONS EXAMINED BY CIRCULAR CONTOUR TRACING – 3

THE CONSIDERED GEOMETRIC RELATION TO THE CIRCLE	DIAGRAM	DESIGNATION OF THE GEOMETRIC CONDITION	THE APPLIED INSPECTION PROCESS
The ideal contour of a perfectly round and smooth surface.		Microsurface in relation to basically round form	By establishing a circular base line—the path of the stylus—the circumferential surface texture of an essentially round part can be recorded as a continuous line at a high radial magnification.

cussed in the section on methods referencing from the surface of the object, produce values that (1) are only indirectly related to the true radial variations, or (2) are variations of parameters other than the lengths of radii related to a common axis.

In addition to the ANSI/ASME specification B89.3.1-1972 (R1988), Measurement of Out-of-Roundness, there are widely accepted concepts of roundness evaluation that make use of the following terminology:

Ideal roundness designates the geometric form of a body or feature whose profile constitutes a perfect circle, all parts of the profile being equidistant from a common point, the center.

The profile of a basically round part refers to the contour of a cross-sectional plane normal to the axis. For a surface of revolution the axis is a straight line equidistant from every point of the ideal profile.

The measured profile is the representation of the actual cross-sectional contour, which is usually determined by instrumentation.

Departures from roundness are generally expressed as a range and represent the deviations of the measured profile from the ideal profile. The range's dimension is stated in linear units, representing the radial span of the actual contour line's departures from the appropriately located reference circle. In the practical assessment of departures from roundness, the reference circle is often represented by its center, around which are laid out the two concentric boundary circles of the actual profile.

Out-of-roundness is another term for the condition designated as departures from roundness. Frequently, however, distinction is made between the meaning of these two terms. Out-of-roundness is used to indicate a generally unround form, such as an ellipse, while periodic variations of the actual profile about a reference circle are considered departures.

Lobing refers to a contour form that tends to lend a polygonal character to the highly magnified graphic representation of the actual profile. A typical origin for such conditions is an improperly controlled centerless grinding process, which can produce basically round parts with form defects represented as lobes. These contour traits are rather uniformly distributed around the perimeter and are, most commonly, of odd number, such as 3-, 5-, 7- or 9-lobed contours.

Undulations designate circumferential waviness, and are characterized, differentially to lobing, by a larger number of departures, distributed somewhat symmetrically on both sides of the reference mean circle. There is, however, no generally agreed distinction between the concepts of lobing and undulations.

Roughness can appear as very closely spaced undulations, usually as a secondary surface characteristic that is superimposed on wider spaced undulations or lobes. Several types of roundness measuring instruments provide means, by electronic filtering, for eliminating or substantially reducing the appearance of roughness indications on chart tracings, when such are prepared for the purpose of form inspection.

The *dimension of departures from roundness* is the radial distance between two concentric circles that just contain all the elements of the graphically displayed contour line.

Magnification of the graphic representation refers to the radial magnification of the actual profile's departures from ideal roundness. The diameter of the chart tracing has no relation to the size of the part: it may be larger or smaller than the actual part diameter. Because the chart tracing displays the highly magnified representation of a single form characteristic only, its proportions can differ substantially from those of the actual part.

THE INSTRUMENTS AND TECHNOLOGY OF ROUNDNESS MEASUREMENT BY CIRCULAR TRACING

The application of the circular tracing process as the means of accurate roundness measurement is a relatively recent development. Without detracting from the merits of the originators of this metrological system, the development was predicated on several concurrent technological advances:

a. The need for better form accuracy of manufactured parts of basically round form;

b. New design principles permitting substantial improvements in the rotation accuracy of special machine tool spindles; and

c. The availability of various electronic devices on which the operation of these instruments relies, such as displacement transducers for the pickup; the elements of the amplifier units; and accurate recorder instruments.

Continued progress in industry emphasized the demand for accurate and dependable roundness measuring instruments. Further advances, both in machine tool building and in electronic technology broadened the availability of the essential elements of this kind of instrumentation. As a result of this progress a rather wide choice of tracer type roundness measuring instruments is available to date. These instruments represent a substantial range of different systems, sizes, accuracy grades and degrees of adaptability.

Roundness Measuring Machines— Characteristics of Representative Models

A few representative models of roundness measuring machines are discussed briefly, particularly for illustrating the range of capabilities that currently available machines offer. This discussion is not intended to provide a complete listing of even the more common models of roundness measuring machines. In any case, that would be a futile attempt, in view of the continuous development of new types

of circular tracing instruments designed for the primary purpose of roundness inspection.

Figure 14-13 illustrates a more recent version of the pioneer in this field, the "Talyrond" Model 2. This is a rotating spindle type instrument of particularly solid design (machine weight about one ton), built for accepting heavy and bulky parts, up to a maximum weight of 1000 pounds. In contrast to its sturdiness, the machine permits very sensitive adjustments and has an exceptionally accurate spindle whose maximum radial departure from a true mean circle is not more than one millionth of an inch. The spindle rotates at 3 rpm for the graphing operation and also has a second, substantially higher speed (35 rpm) for the centering of the specimen.

For accommodating small- and medium-size specimens, a lighter type of Talyrond machine (Model 51) is also available. This model has essentially identical electronic sensing and amplifying, as well as recording equipment to that of the larger machine. While the maximum measurable diameter is the same for both models, that is, 14 inches for either external or internal diameters, Model 51 has a useful height adjustment of 16 inches and the maximum recommended load on the work table is 150 pounds.

Both Talyrond models can be equipped with various optional accessories, most of which serve either of two purposes.

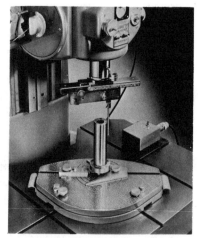

Rank-Taylor-Hobson/Engis Equipment Co.

Fig. 14-13. (Left) Rotating spindle type roundness measuring machine, designed for accepting long spindles and heavy parts without affecting the instrument's high operating accuracy.

Fig. 14-14. (Right) Close-up view of the major operating elements of the machine at left, when also equipped with a combined centering and leveling table.

a. To extend the applicability of the instrument by providing various stylus arms, styli and a variable gain preamplifier, needed for tracing and recording surfaces that are not readily accessible, or are located in a plane normal to the axis of spindle rotation; and

b. To facilitate the mounting, locating and centering of parts whose basically round surfaces must be inspected. This category of accessories comprises staging tables and various workholding fixtures. Figure 14-14 is a close-up view of the major operating elements of the Talyrond, showing also the use of the combined centering and leveling table, which is mounted on the instrument work table. The controlled location of the table's pivot center, one-half inch above the table surface, facilitates the leveling adjustments.

Figure 14-15 illustrates another type of roundness measuring machine, also of the rotating spindle category. This is a smaller, table type machine that was designed for providing a high degree of application versatility, accomplished partly by the variable speed–spindle rotation. This type of roundness measuring machine permits adjustment of the tracing speed to the diameter of the part being inspected, in order to obtain an approximately uniform circumferential tracing speed for an extended range of specimen diameters.

The uniform tracing speed is needed for producing chart recordings that display, with proper resolution, the surface texture characteristics of the part

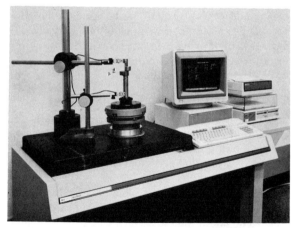

Giddings & Lewis Measurement Systems

Fig. 14-15. Instrument group for the comprehensive measurement of essentially round parts. The instrument is capable of supplying information in the form of either computer-generated circular (polar) or continuous (strip) charts. On the left is a rotating spindle type roundness measuring machine. In the center is a personal computer and dot-matrix printer.

around its perimeter. Because the texture of the surface comprises very closely spaced irregularities, the advance of the stylus along the surface must be contained within a speed range that permits it to follow and faithfully record all the meaningful departures from a mean profile line. This requirement is accomplished by conforming the rotational speed of the spindle to the diameter of the part. The spindle speed of this machine is continuously variable in two ranges from 0.1 to 1.2 rpm and from 1.0 to 12.0 rpm. At any of these speeds the polar chart recorder is rotating in synchronism with the tracer spindle.

The illustrated machine offers five different radial magnifications, from $250\times$ to $50,000\times$, which can be selected by means of push buttons. Also by using push-button controls, any of the following three sensitivity controls can be operated:

a. Filtered signals for surface-texture measurements, eliminating from the recorded chart line most of the major variations of the form;

b. Filtered signals for geometric form (roundness) measurement, suppressing the reproduction of the closely spaced departures of the surface profile; and

c. Total sensitivity, displaying the actual circumferential profile of the part at high magnification, showing the form variations on which the representation of the surface texture is superimposed.

The described machine can produce either polar charts or strip charts. The latter are of advantage when high circumferential magnification is needed, such as for the discriminating inspection of circular-arc profiles, for example, across the race of a ball bearing ring.

The following examples were selected to represent the *rotating table system* of roundness measuring machines.

Figure 14-16 shows an arrangement and application of the system's principles, implemented by means of units that are only electrically connected. The four major units of this instrument system are the following:

a. The rotary table supported on a pneumatic bearing;

b. The gage stand and adjustable arm carrying a lever type gage head with pickup stylus;

c. A "386" personal computer with 14-inch VGA monitor, including full-color graphics and touch screen control via circular and linear geometry analysis software; and

Federal Products Corp.

Fig. 14-16. Rotating table type of roundness-measuring instrument with detached yet electrically connected elements.

d. A high speed dot-matrix printer to generate a permanent, hard-copy record of measurement conditions, polar or linear traces and their measured values.

Several manufacturers supply the necessary elements for the measurement of roundness fully assembled as an integrated instrument.

An example of such a compact and self-contained roundness measuring machine, representing a modular type construction, is shown in Figure 14-17. A granite plate is the mounting base for the rotary table and also provides the datum plane for the height

Standard Gage Co.

Fig. 14-17. Rotating table type, roundness measuring instruments, assembled from modular units.

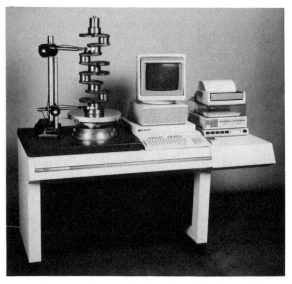

Giddings & Lewis Measurement Systems

Fig. 14-18. Rotating table type radial measurement system for roundness control of large or heavy parts.

gage type pickup holder. The instrument console in this particular version of the flexible modular design, contains an amplifier for roughness average reading, a gaging amplifier for roundness measurements, a polar chart recorder and a control panel with the necessary switches and adjustment knobs.

Figure 14-18 illustrates a roundness measuring machine that is built into a desklike console to assure the combination of compactness and operating convenience. This system can be used to measure various geometric interrelations of basically coaxial surfaces, such as wall thickness variations, cylindricity, squareness of a round surface in relation to a flat one, and several others.

Circular-Arc Curvature Measurements

There is a certain group of technical parts for which the circular tracing process is used to inspect the geometric form of a limited section of the part's perimeter, when that particular section has a circular arc for basic contour.

Perhaps the best known examples in this application group are the races (ball tracks) of ball bearing rings. The conditions to be inspected in these and similar applications differ from those of the regular roundness measurements, because in addition to the basically circular form, another parameter, namely, the rate of curvature, must also be measured. The curvature and its tolerances are usually expressed in the product specifications as radii of the circular arc and the dimension measured in the inspection process is the radial deviation of the actual profile from the circular stylus path at nominal radius.

While it is possible to carry out such measurements with certain types of roundness measuring machines, special instruments, designed expressly for the described particular application, are also available. Figure 14-19 illustrates a race curvature gage in which a stylus is moved over a circular path by means of a hand-operated, very precisely guided, swing arm. This swing arm carries an adjustable shaft, to which the pickup with the stylus is attached. The radius of the precisely circular path of the stylus can be set in a mastering process with the aid of gage blocks. When scanning the race profile with the stylus by swinging the arm in both directions from the central setting plane, any deviation of the race curvature or departures from the basic circular form appear as highly magnified indications on the meter of the instrument.

For recording the measurements in the form of a continuous trace line the same instrument can be equipped with a rotary potentiometer mounted on the hub of the swing arm. By channeling the outputs of the electrical signal generating devices into an X–Y recorder, namely, the potentiometer into the X axis, and the amplifier of the pickup into the Y axis, a chart can be produced. The X–Y chart shows the arc length (central angle) of the traced surface and the deflections of the stylus, which represent the departures of the actual profile from the ideal one. Curvature measurements with recording sensitivity on the order of 10 microinches, or even finer, can be made by this process, although for general applications a lesser degree of sensitivity, embracing a wider deviation range is often preferred.

Recorder Charts—Appraisal of Various Types and Characteristics

Circular contour tracing is essentially a graphically recording type of inspection process and the measurements are supplied in the form of a chart line with scaled departures from a basic circle (see Table 14-6 for examples of different types of tracings). Considering the role of recorder charts in the diverse applications of the circular tracing process, inclusive of roundness measurements, a brief appraisal of the various types of recorder charts used for different models of instruments seems warranted. The following appraisal is concerned with systems only, and not with specific makes. The relative advantages of the various basic types are considered from the viewpoint of their application in the circular tracing process.

a. The general form may be circular (polar) or linear (strip). Polar charts are more common and have the basic advantage of presenting an essentially circular tracing line that can be readily correlated with the actual object contour. Form irregularities shown on the chart tracing can be associated mentally with the corresponding sections of the physical part.

Strip charts are of advantage in applications where increased circumferential amplification of the traced contour section is desirable. This may be the case particularly for tracings that represent only a part of the circular contour, such as the tracks of ball and barrel roller bearing rings. When traced sections comprising less than a complete circle are recorded on a strip chart, some auxiliary means may be needed to determine the actual central angle that is subtended by the traced arc.

b. The size of polar charts differs according to the make and model. The 4½-inch diameter charts are perhaps the most common, offering an annular ruled track of 1-inch radial width; however, some instrument models operate with charts of 9- or 10-inch diameters. The smaller size has lower initial (instrument) and operating (chart paper) costs, and may be found more convenient for filing purposes. The larger size permits a more precise analysis and shows less pictorial distortions for certain types of roundness irregularities. Factors, such as response

Giddings & Lewis Measurement Systems

Fig. 14-19. Circular tracing instrument sweeping over an arc of about 120 degrees, designed primarily for the inspection of race curvatures of ball bearing inner and outer rings.

TABLE 14-6.   EXAMPLES OF THE CHART REPRESENTATION OF GEOMETRIC CONDITIONS INSPECTED BY CIRCULAR TRACING

DIAGRAM	CHARACTERISTIC FORM OF CHART TRACING	DESCRIPTION
		Roundness of External and Internal Surfaces Can be applied to any figure of revolution with overall size within the machine capacity, when surface to be explored is accessible to instrument stylus and can be traced, without obstruction, around an entire circle.
		Concentricity and Roundness of External and Internal Surfaces When measurements of parallel surfaces are made in the same plane, normal to the axis, the chart tracings also supply information on wall thickness variations.
		Coaxiality and Roundness of Parallel Surfaces with Different Diameters Chart with two tracing lines supplies information on both the roundness of the individual surfaces and the mutual alignment of the axes. Tracings are insensitive to differences in the basic sizes of the diameters.
		Coaxiality and Roundness of Nonparallel Surfaces on Figures of Revolution Because the tracing lines are contained in specific planes of measurement all of which are normal to a common axis, the surfaces traced need not be parallel to the axis. However, the position of the surfaces affects the magnification rate of error indication.
Axis of rotation	s−t = The dimension of error	**Condition of Flat Surfaces that are Normal to the Axis of the Object** Positions of the tracing lines in the measuring planes indicated by I, II, and III. Flatness—tracing lines I and II parallel. Parallelism—tracing lines II and III parallel. Squareness (perpendicularity)—tracing lines II or III concentric to the axis on chart.
Barrel roller contour	L = Effective roller length	**True Form of Bodies of Revolution with Circular Arc Surface Elements** When the measured surface element is only a section of a complete circle, high resolution of the chart tracing is essential for precise analysis. Strip charts, with extended tracing lengths are particularly informative.

speed, pen overshooting, and so forth, should also be considered when comparing the suitability for specific applications.

c. The pen system may be electrical (Teledeltos type) or it may use ink. The former needs no refilling of the inkwell, but requires more expensive chart blanks. Reproduction considerations could favor the ink pen when printing from transparent paper is preferred.

d. Curvilinear or rectilinear pen movement. The Talyrond instrument uses a rectilinear type of chart ruling; although the pen travels along an arcuate path, the curved surface of the chart support has a compensating effect on the resulting line form. Other systems have an arcuate pen travel and use a corresponding type of curvilinear chart paper in order to produce a true dimensional representation of the sensed contour departures even in the case of large diameter charts. Several other systems have also an arcuate pen movement but record on a rectilinear chart paper, considering the resulting distortions insignificant for certain types of inspection. Large, 10-inch diameter polar chart recorders, using rectilinear radial chart ruling, are also available; in this case the pen acts through a linkage, resulting in an essentially straight-line path for the radial pen excursions. When evaluating any particular design, the response speed affected by linkage drag, or by inertial overshooting, must be considered.

The Staging of Specimens for Roundness Inspection by Circular Tracing

In order to obtain a realistic representation of the traced specimen contour, the center of the part and the center of the rotational movement producing the reference circle must be coincident. It is not practical to meet this requirement perfectly, yet the eccentricity of the two circles must not exceed certain limits that depend on two factors:

a. The desired reliability of the measurements; and

b. The applied magnification.

While detailed mathematical analysis has been developed for determining the effect of improper part centering, as a rule of thumb, the following values may be considered.

The eccentricity between the chart ruling and the traced curve representing the part contour should not exceed 0.3 inch, as measured on the chart. For exceptionally accurate inspection that amount of ec-

centricity must be reduced by means of more precise centering of the part, whereas for the purpose of general information a somewhat greater separation of the two centers may be tolerated. However, it must be realized that the contour representation starts to be severely distorted when the eccentricity exceeds 0.5 inch. Considering that these values refer to distances between centers on the chart, the actual displacement of the part center from the center of rotation of the instrument is only a small fraction of the shift appearing on the chart. As an example: at 10,000 magnification, the 0.3 inch on the chart represents a physical displacement of 30 microinches.

For that reason the staging tables of roundness measuring instruments must be equipped with very sensitively acting adjustment means, usually fine-threaded regular or differential screws. A few types of instruments can be supplied with automatic centering devices, which operate by servo-motors.

The representation of the part contour in roundness tracing can also be distorted by angular misalignment between the axis of the specimen and the axis of the instrument spindle. In order to correct such a condition when present, most types of roundness measuring machines are equipped with means for the tilt adjustment of the staging table.

The Stylus of Roundness Measuring Instruments

The stylus is the member through which the departures of the traced surface from the base circle are sensed by the pickup. The tip of the stylus is the contacting element and its form affect the degree of discrimination with which the variations of the scanned surface profile are transmitted to the pickup.

For usual applications, a ball tip of $\frac{1}{8}$-inch diameter may be used. This will provide a reasonable amount of sensitivity even in the case of closely spaced undulations. For particular purposes, stylus tips with balls of other diameters are selected, such as a $\frac{1}{16}$-inch diameter for increased discrimination. In other cases, balls of more than $\frac{1}{8}$-inch diameter are preferred to make the stylus insensitive to closely spaced undulations that may be considered to border on surface roughness, and may interfere with the distinct representation of the form variations.

Stylus tips with smaller radii, on the order of 0.001 inch, are often made of sapphire. When the tracing of the circular contour is carried out with the objective of obtaining a total representation of the profile, including the very closely spaced departures with magnitudes on the order of surface roughness, diamond tips having a 0.0005-inch or smaller diameter are generally used.

The static force with which the stylus is held against the part surface should be selected according to the tip radius, also taking into account the hardness and the finish of the workpiece. As a general rule, that force should not exceed 1 gram for the smallest, and 20 grams for the largest tip radius.

The Frequency Response of Roundness Measuring Instruments

The relative rotation of the specimen and of the pickup presents continuously changing points around the specimen's surface to the contacting stylus. The rate at which the adjacent points are presented is a function of the circumferential speed of the scanning movement. A time lapse is needed for converting the mechanical deflections of the stylus into the proportional movements of the recorder pen arm. Consequently, the circumferential speed of the scanned surface relative to the pickup influences the completeness with which the departures of the surface from the ideal form is represented by the chart tracing.

The other factor affecting the graphic representation of the traced surface is the frequency response of the electronic system in combination with the speed of the recorder pen's sweep movement.

In surface tracing practice it is common, although not spelled out in any standard, to consider a less than 100% representation of the surface variations as true for most practical purposes. The rate of actual completeness is on the order of 85% to 90% near the borderline of the rated frequency response. As an example: a 20-cps frequency response, at 90% transmission, expresses the capability of an instrument to display on the chart, with an error not exceeding 10% of the full scale, all variations of the surface that were sensed in intervals of a 20th part of a second, or longer. Assuming a traverse rate of 0.005 inch per second, surface variations spaced not closer than

$$\frac{0.005}{20} = 0.00025 \text{ inch}$$

are represented with an accuracy within 10% of the total scale.

For the purpose of roundness measurements, however, the detailed display of very closely spaced surface variations is only seldomly required. Often, it may even be undesirable from the viewpoint of the distinct representation of the general geometric con-

ditions to cram the tracing line with too many non-pertinent details.

In the general roundness measuring practice, the departures from the ideal form are not judged by the circumferential spacing of the undulations, but by the frequency of their occurrence around the entire surface. Consequently, the response frequency of roundness measuring instruments is commonly expressed by the number of equally spaced undulations around the entire surface, regardless of the part's diameter. Generally accepted response rates are 5, 15, 50, 150 and 500 frequencies per revolution. The instruments are usually designed to produce the highest frequency response stated in the pertinent specifications, and the lower response rates are obtained with the aid of wave filters. Figure 14-20 illustrates a roundness chart with three traced curves, each representing the same specimen perimeter but recorded at different response frequencies.

It is also possible, and often advantageous, to adapt the frequency level of the recording to the desired information by means other than wave filters. As an example, the frequency response of the roundness measurement can be reduced by increasing the scanning speed and/or by using styli with relatively large radii.

Giddings & Lewis Measurement Systems

Fig. 14-20. Polar chart with three roundness tracing lines of different degrees of smoothness. All tracing lines represent the same surface, yet were recorded with different frequency cutoffs.

THE ASSESSMENT OF ROUNDNESS TRACINGS

The line traced by the recorder pen on a polar chart is a single curve of essentially circular shape, with ends meeting to form a closed loop. The digressions of that line from the true circular form represent, at a high radial magnification, the departures of the traced perimeter of the specimen from the ideal round form.

The magnitude of these departures, as a group should be expressed, according to the earlier cited standard definition of roundness, by the radial separation of two concentric circles that just contain the traced curve.

The implementation of this concept for the practical assessment of the chart tracing requires the laying out of two concentric circles that satisfy the specified conditions. This can be accomplished by either of two ways:

a. By actually striking with a drawing compass two circles around a common center; and

b. By using a transparent (glass or plastic) template with concentric circles at scaled distances apart, and shifting that chart reader on the surface of the recorded chart to a position where the two ruled circles containing the traced curve show the least separation. This procedure is also known as the *floating chart reader method.*

It is also practical to combine these two techniques, using the floating template to determine the location of the center around which the two concentric circles will be drawn with the compass.

In neither of these readout systems is the center of the circles in a defined location; actually, it can have several different positions. Depending on the center's location, the radial distance between the two concentric bounding circles can vary, as shown in Fig. 14-21, which is based on a typical roundness tracing line. It follows that the condition expressed as the separation of concentric bounding circles does not establish a unique value; it must be complemented with specifications regarding the location of the reference center.

There are four commonly used methods for establishing the center of the bounding circles:

a. By trying to find that particular center location with respect to which the radial separation of the bounding circles will be the least. In practice, that goal is usually accomplished by experimenting with different positions of the floating chart reader, aiming at the "best fitting circles." This method of assessment, designated as the MZC (Minimum Zone Center) system, is frequently used because it produces the smallest numerical value for the measured departures from roundness.

b. Considering the tracing of an outside surface, it can be assumed that from a functional point of view the external protuberances of the tracing line represent those areas of the surface where mechanical contact with mating parts will eventually result. For the evaluation of the roundness, a circle may be

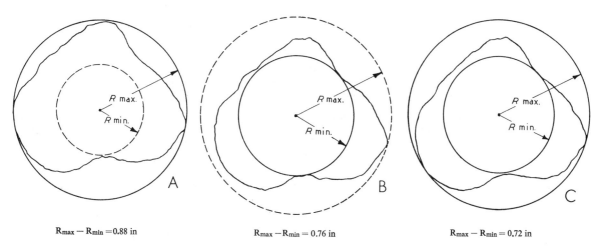

$R_{max} - R_{min} = 0.88$ in $R_{max} - R_{min} = 0.76$ in $R_{max} - R_{min} = 0.72$ in

Fig. 14-21. The assessment of the same roundness tracing curve by three different methods to illustrate the variations in the resulting annular zone width, which is considered the significant dimension of the departures from roundness. (A) Ring gage center method. (B) Plug gage center method. (C) Minimum zone center method. (British Standard 3730: 1964.)

sought that just contains these external protuberances, and the other bounding circles will be concentric with that guiding one, independently of the resulting radial separation. This separation could be substantially greater than the one obtained by establishing the best fitting, or minimum zone, concentric circles. In the described method where the outside bounding circle is the reference element, its function can be compared to that of a ring gage, and for that reason that method is known as the RGC (Ring Gage Center) system.

c. The third commonly used method of assessment is the reverse of the preceding one and is applicable to tracings representing a bore contour. This method is based on the PGC (Plug Gage Center) system.

d. A fourth method of assessment has been developed that has the advantage of uniqueness, that is, it provides a single numerical value and is independent of the variables that could result from experimentation connected with the use of the floating chart reader. This method, known as the LSC (least squares center) system, establishes a "least squares circle" as a mean line of the traced curve, from which the sum of a sufficient number of equally spaced radial ordinates has a minimum value.

Roundness tracing assessment in accordance with the principles of the LSC system can be accomplished by either of two procedures:

1. It can be calculated using approximation by laying out a sufficient number of equally spaced radial ordinates from the center of the chart. The distances of the points of intersection of the polar graph with these radial ordinates are measured from axes XX and YY. The values obtained, taking into account the plus and minus signs, are then used, by means of approximate formulas, to determine the position of the least squares center (see Fig. 14-22). By drawing around that center the circumscribing and inscribing circles of the traced curve the radial width of the resulting zone can be measured.

2. The same value can be established much more quickly than with the graphic method, and also with greater accuracy, with the aid of a computer that is available for the Talyrond type roundness measuring instruments. The computer can supply the following information on its four meters:

a. The maximum peak-to-valley height, which is the sum of the following items (b) and (c).

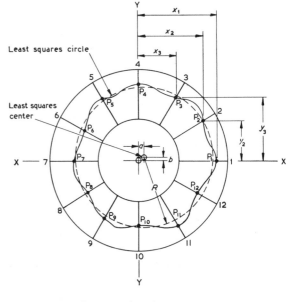

$$a = \frac{2 \times \text{sum of } x \text{ values}}{\text{number of ordinates}} = \frac{2\,\Sigma\,x}{n}$$

$$b = \frac{2 \times \text{sum of } y \text{ values}}{\text{number of ordinates}} = \frac{2\,\Sigma\,y}{n}$$

$$R = \frac{\text{sum of radial values}}{\text{number of ordinates}} = \frac{\Sigma\,r}{n}$$

Fig. 14-22. Layout for the determination of the least squares center and circle by computation, where R = the radius of the least squares circle.

b. The largest outward departure.

c. The largest inward departure.

d. The integrated average departure from the reference line.

The computer can also direct the pen of the recorder to draw in the chart containing the circular profile curve of the part, the least squares circle pertaining to that curve. The reference circle is drawn by the recorder pen during the next succeeding revolution of the machine spindle.

Distortions in the Graphic Representation of the Roundness Conditions

By reason of the dimensional limitations of polar charts and to enhance the significant characteristics of the inspected conditions, the chart representation of the scanned part perimeter is drawn to a nonuniform scale. The diameter of the drawn curve is governed by the available chart space and is in no relation to the actual part diameter. However, the departures of the part perimeter from the ideal round form are displayed at magnifications that may vary from about 500 × to as high as 100,000 ×, depending on the basic geometric conditions of the part and on the sought power of resolution of the measurement.

It follows that an exaggerated and distorted graphic representation of the part contour appears on the chart. That must be evaluated by means of the earlier described methods of assessment, and not judged by visual impression.

While this is a matter of importance in evaluating roundness conditions with the aid of circular tracing, the scope of this chapter does not permit a detailed discussion of the geometry of polar chart distortions. Therefore, only two examples of typical circumstances causing graphic distortions in the chart representation are shown diagrammatically, namely:

1. Chart line distortions due to the eccentric staging of the part (Fig. 14-23); and

2. Chart line distortions emphasized by increased magnification (Fig. 14-24).

The Verification of Roundness Measuring Machine Performance

For the users of roundness measuring machines, it is important to know the degree of confidence that can be assigned to the results of the inspection process as represented by the chart tracing.

There are three major aspects that have direct effect on the reliability of roundness measuring machine performance:

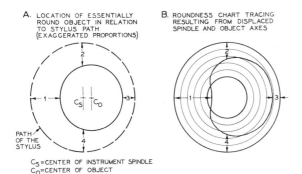

Fig. 14-23. Distorted geometric figure on roundness chart due to misalignment of object axis in relation to the instrument spindle.

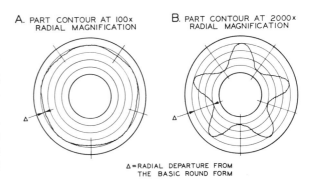

Fig. 14-24. Distorted geometric figure on roundness chart due to various degrees of nonuniform rate of magnification.

a. The rotational accuracy of the machine spindle or table, expressed as the maximum deviation from a perfect mean circle.

b. The axial movements of the spindle that, in combination with the rotation, can produce a wobble or axial shift. Such deficiencies are often assessed as *coning errors* of rotating tables, with reference to an imaginary inverted cone representing the zone of uncertainty in the rotational accuracy, at different levels from the table top.

c. The accuracy of the indications that represent the departures from roundness of the traced surface perimeter.

The principles and the commonly used methods for verifying the accuracy of roundness measuring machines in the above-mentioned respects are reviewed briefly in the following:

a. For inspecting the rotational accuracy of the machine spindle, a ball of excellent accuracy is used as a master, and a tracing of its perimeter is made at the highest magnification the machine offers. Allowing for the known inaccuracies of the master ball, usually on the order of a few microinches, the amount of the chart line's departures from a true circle are indicative of the rotational accuracy of the machine spindle.

b. The axial runout, or wobble, of the instrument spindle can be checked by tracing the surface of an optical flat staged in a position precisely normal to the axis of the spindle or table rotation.

c. A practical master for verifying the indicating accuracy, and also the response speed of the instrument, can be prepared from a gage cylinder of excellent roundness. A short flat, parallel to the axis, is ground into the side of the cylinder in the area through which the stylus path will pass. By successive measurements across an untouched diameter and along a diameter normal to the flat, the chordal height of the removed segment can be accurately determined. A roundless tracing of the perimeter passing through the flat shows on the chart line a departure from the basic circle. The radial depth of this departure must represent, in the scale of the chart, the dimension of the previously measured chordal height.

15.
Surface-Texture Measurements

Considering surfaces as the boundary areas of solids, such as technical parts, the characteristics of a surface can affect various conditions that contribute to determining a part's functional properties. It is the surface of the part that is in contact with the adjacent surfaces of other parts to which it is operationally related. The form and the size of the part are perceived and measured by sensing its bounding surface.

Most technical parts are conceived as regular geometric bodies, or as the combination of such three-dimensional figures. Theoretically, theses figures are bounded by surfaces of defined forms, the cross-sectional contours of which are commonly straight or circular lines. However, it is well known that such a theoretical concept cannot be accomplished in manufactured parts, although it is being approached to various degrees. The extent of permissible variations from the ideal is commonly specified on the product drawings as tolerances of size, of form and of surface texture. The inspection and measurement of size and form have been considered in the preceding chapters; the concept of surface texture and its metrology are the subjects of the following discussions and evaluations.

As an analogy, we might compare a technical part with our planet. Thus, the surface of the latter may be considered at three different levels of observations. The basic form is associated with the geometric concept of a sphere or, more precisely, of an oblate spheroid, having specific dimensions and form. The major variations of the earth's surface from a theoretically regular boundary to that geometric figure are designated as topography, often represented by relief globes, illustrated by means of contour maps, and identified as altitudes of selected points or areas. The comparable form variations of technical parts are often referred to as *waviness*, when displaying a distinct orientation in contrast to the random character of nature's topography. Waviness may therefore be measured by means of linear or circular contour tracing carried out along the profiles that are associated with functionally meaningful cross-sectional planes.

Finally, there are very small irregularities of the earth's surface that are mapped exceptionally only, but still constitute sensible digressions from a relative smoothness, such as might be represented by the surface of a road or of an airport runway. These variations from smoothness are often distinguished by terms indicative of the general condition, although for special purposes the individual digressions may also be measured. The conceptually corresponding digressions from smoothness of technical surfaces are designated as *roughness*.

While there are no exactly distinguishable boundaries between waviness and roughness of technical surface, an example of these two kinds of digressions from the ideal form may be helpful for visualizing the distinction between said concepts. Thus, in a grinding operation the vibrations of the grinding wheel spindle, chatter resulting from insufficient rigidity of the machine slides or tooling and similar circumstances may cause periodically appearing marks on the surface, which may be associated with waves. The grains of the grinding wheel, the short and irregularly distributed arcs of their contact with the part's surface and the dislodging of individual crystals from the work material result in randomly located minute tear-outs, fusions and so forth of varying depths or heights. The resulting departures from the theoretically smooth or regularly wavy surface are designated roughness.

For reasons connected with their technological origins, which are indicated diagrammatically in Fig. 15-1, such conditions usually appear concurrently, roughness being superimposed on waviness. The term *surface texture* designates the entirety of departures from the ideally smooth surface, inclusive of occasional flaws or other types of locally limited irregularities.

It is possible, although only exceptionally required in manufacturing practice, to penetrate even deeper in the analysis of the surface. Such investigations can comprise, as a next level, the crystalline formations of the surface, and finally, the lattice structure of the material as indicative, for example, of stress conditions. However, this kind of analysis of the "total surface" is generally considered to be beyond the scope of metrology and belongs to the domains of micrometallurgy and physics.

In order to indicate the diversity of functional requirements that determine the adequacy of certain types of technical surfaces for specific uses, and also to point out the need for the dependable assessment of critical surface conditions, various examples have been selected. These, representing typical applications of technical parts, are presented in Table 15-1, with the objective of illustrating two major aspects:

1. The importance of the appropriate surface texture for the intended service of technical parts; and

2. The different characteristics of the surface texture that may have a critical effect on the adequacy of a technical surface for specific uses.

From the point of view of industrial measurements, the surface-texture assessment, in order to be practical, must be limited to the evaluation of a single or of a few significant characteristics. The currently used methods of surface-texture measurement, which are discussed in the following, represent such approaches.

DEFINITION OF CONCEPTS AND TERMS

In the production processes of technical parts, various factors are interacting in a manner that causes digressions of the resulting surface from the ideal conditions. By disregarding in this study the variations of the general form and size of the part, the inspection of any portion of the surface that was generated in the process will reveal departures from the theoretical. Such digressions from the ideal surface were termed in the preceding section as the *texture*, comprising topographical deviations generally associated with more or less regular waveforms, on which the closely spaced random irregularities, termed roughness, appear superimposed.

Figure 15-2 shows the conventional diagrammatic representation of said surface characteristics, applying exaggerated scales for illustrating the concepts. Whereas the terms used in this illustration originate from the USA Standard ASME B46.1-1985, various foreign standards use similar diagrams and equivalent designations. The terms used in the diagram require supplementary explanation:

Waviness is the characteristic form of topographical variations that are measurable as the profile of the part in an actual or imaginary cross-section. The term waviness implies a repetitive and essentially regular occurrence of topographical features, an assumption that is based on the typical surface topography of machined surfaces.

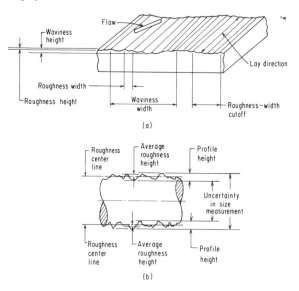

(a)

(b)

Fig. 15-2. Characteristic components of technical surfaces illustrated diagrammatically: (A) in perspective; and (B) in cross-section to represent the concepts on which the standard assessment of surface texture is based (ASME B46.1-1985).

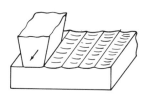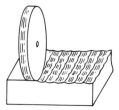

Fig. 15-1. Diagrammatic representation of the origins of waviness and roughness in basic metalworking processes.

**TABLE 15-1. EXAMPLES OF APPLICATION CATEGORIES REQUIRING
CONTROLLED SURFACE TEXTURE — 1**

FUNCTIONAL OBJECTIVE	APPLICATION		DISCUSSION— CRITICAL CHARACTERISTICS OF THE SURFACE TEXTURE
	EXAMPLE	DIAGRAM OR SYMBOL	
Resistance to wear	Machine tool guideways		A ragged surface exposes only a small portion of its total area to contact by a mating member of complying general shape. Surface roughness, by limiting the area available to carry the load, causes increased wear rate, or may require run-in before operation at maximum capability is feasible.
Fit clearance or interference	Bearing mounted on a shaft with interference fit		Protruding crests of a fit surface may display during gaging, a greater material condition (larger external or smaller internal size) than will actually develop in the interference fit when the members are assembled or after being subjected to operational conditions.
Preservation of an uninterrupted lubricant film	The track of a ball bearing ring		The continuity of the lubricant film which should prevent metal-to-metal contact, may be impeded by the peaks of a rough surface when such protrude beyond the general level on which the lubricant film develops.
Load carrying capacity	Gear flanks		Plastic deformation of a surface of originally adequate shape might take place when localized stresses in excess of the calculated values occur due to reduced contact area caused by excessive surface roughness.

**TABLE 15-1. EXAMPLES OF APPLICATION CATEGORIES REQUIRING
CONTROLLED SURFACE TEXTURE – 2**

FUNCTIONAL OBJECTIVE	APPLICATION		DISCUSSION— CRITICAL CHARACTERISTICS OF THE SURFACE TEXTURE
	EXAMPLE	DIAGRAM OR SYMBOL	
Resistance to chipping	The edge of a cutting tool, e.g. a milling cutter		Discontinuities due to surface roughness in the exposed thin edge of a cutting tool may subject the remaining area to high specific stresses which exceed the strength of the tool material and result in chipping with subsequent failure of the entire edge.
Resistance to corrosion	The plunger of a hydraulic cylinder		The roughness of the surface can contribute to reduce the capacity of withstanding corrosive effects, the pits and valleys of a ragged surface functioning as nucleating points of incipient corrosion.
Low coefficient of sliding friction	Plain bearing or bushing		Common experience points out the need for good surface finish on surfaces in sliding contact. Laboratory tests support these empirical findings and can also establish measurable relationships between surface roughness and the coefficient of sliding friction.
Smooth rolling contact	Antifriction bearing ball		For a surface in rolling contact, such as that of a bearing ball, excellent smoothness, measurable by its degree of roughness, is a prerequisite of quiet running, low operating torque and extended service life.

**TABLE 15-1.　EXAMPLES OF APPLICATION CATEGORIES REQUIRING
CONTROLLED SURFACE TEXTURE – 3**

FUNCTIONAL OBJECTIVE	APPLICATION		DISCUSSION— CRITICAL CHARACTERISTICS OF THE SURFACE TEXTURE
	EXAMPLE	DIAGRAM OR SYMBOL	
Reduced vibration and noise	Antifriction bearing pathway in the direction of rolling		High frequency vibrations, often in the audible range, can originate from closely spaced lobing which, by the standard terminology are classified as components of surface roughness when occurring within the selected cutoff width.
Avoidance of abrasive effects	Seal lip seats		Seal support surfaces on rotating or traversing members, against which the lips of rubber or other types of seals have to bear, require controlled roughness and texture pattern in order to assure the effectiveness of sealing without undue wear on the contacting lip surfaces.
Prolonged service life through increased fatigue strength	Pathways of roller bearings		Surface texture characteristics such as tool or grinding marks, particularly when normal to the direction of stressing, can substantially affect the fatigue resistance of continuously or periodically stressed surfaces. Empirical relationships between the depths of surface irregularities and the fatigue strength can often be established.
Assurance of structural strength	Fillet area of a shaft		To avoid excessive stress concentration in structural transition areas which, due to the general shape of the member are particularly exposed to such harmful conditions, a controlled surface texture is needed even when no contact with other members is intended.

**TABLE 15-1.   EXAMPLES OF APPLICATION CATEGORIES REQUIRING
CONTROLLED SURFACE TEXTURE — 4**

FUNCTIONAL OBJECTIVE	APPLICATION		DISCUSSION— CRITICAL CHARACTERISTICS OF THE SURFACE TEXTURE
	EXAMPLE	DIAGRAM OR SYMBOL	
Dependable dimensional measurements	Fixed gages— plug and ring		Excessive surface roughness affects the reliability of dimensional measurements for several reasons, such as: (a) the measured size will be that of an envelope based on protruding peaks which, however, will deform under load; (b) the rough surface of a fixed gage is conducive to wear to an extent causing significant size changes.
Smooth fluid flow	Clearance gap of an air bearing		Where the maintenance of a smooth, fluid flow is an essential functional requirement, such as in the clearance gap of an air bearing, an excellent finish is necessary to assure optimum operating conditions, particularly when the design specifies a very narrow space between the mating surfaces.
Base for developing ultra thin films	Gage blocks		Close to gapless contact between mating surfaces, accomplished by creating a fluid film of uniform thickness in the sub-microinch order, such as needed for wringing gage blocks together, or for assuring adequate sealing properties of a well-seated plug or plunger against high pressure gases, are predicated on surfaces of extremely high finish.
Non-functional (aesthetic) considerations	Sheet metal for automobile bodies		The assessment of surface conditions offering a pleasing appearance is often based on other parameters than those defining functional properties. The frequency and height of peaks in the base of painted surfaces, the lustre resulting from the pattern of the texture and the slope angle of the departures from the ideal surface, are typical examples.

Waviness width expresses the distance between adjacent crests of the essentially wavy profile.

Waviness height is the distance, in a direction normal to the general surface, between the crests and the valleys of the waves.

Roughness expresses the closely spaced digressions of the actual surface from its ideal form. These digressions are usually less regular in profile form and spacing than those termed waviness. Due to the technological circumstances from which roughness originates, the measurable digressions from a basic profile line are of higher frequency (closer spacing), yet of lower average amplitude (less height) than the waviness on which roughness is usually superimposed.

The boundary between waviness and roughness is not distinct although, based on the investigation of typical machined surfaces, distinguishing characteristics are presumed in the standard.

The diagram in Fig. 15-2A also shows a characteristic pattern: the essentially parallel ridges and valleys having a common direction that is termed the *lay* of the texture. That lay is illustrated to be normal to the cross-sectional plane in which the profile is observed. An oblique plane of observation (or oblique cross-sectional plane) causes the measurable distance between the consecutive profile features, the waviness width, to increase. A plane of observation parallel with the lay generally displays a substantially lesser amount of waviness, or no wave pattern at all. In Fig. 15-2 the cross-sectional plane is also shown as being perpendicular to the general plane of the investigated surface area. A cross-sectional plane at an incline would increase the measurable distance between the levels of the crests and of the valleys.

The surface texture of cylindrical parts, as shown in Fig. 15-2B, is usually examined in an axial plane, that being approximately perpendicular to the lay produced by most of the conventional manufacturing processes.

A few aspects of these interrelations between the lay and the orientation of the plane of observation are shown diagrammatically in Fig. 15-3.

It must be kept in mind that the conditions of the surface texture are illustrated here in an idealized manner in order to assist the understanding of the concepts on which most of the currently used methods of surface-texture measurement and assessment are based.

Actually, the pattern of the surface texture is not always as regular in repetitiveness and lay orientation as shown in the diagrams. Nor is it always possible to select a plane for measurement that is normal

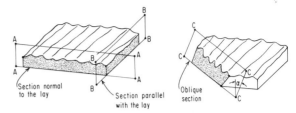

Fig. 15-3. Diagrams of surface sections to visualize the interrelations between the direction of the lay and the orientation of the plane of observation.

to both the lay and the general plane of the surface. Finally, there are cases in which the direction normal to the lay is not the meaningful one with regard to the functional role of a particular surface.

The effect of the direction and pattern of the lay on certain application requirements of technical surfaces is recognized in the American and several foreign surface-texture standards, by providing definitions and symbols for the specification of such conditions on product drawings. These specifications refer to the orientation of the predominant surface topography, as represented by the distinguishable grooves produced in the applied technological process. Some foreign standards also provide means to specify the profile of the machining marks or groove forms, because of the potential functional effects of these surface-texture characteristics.

THE METHODS OF SURFACE-TEXTURE MEASUREMENT

The general concept of measurement implies specific dimensions, the extent or size of which must be determined and expressed in numerical values based on standard units of measurement.

The commonly measured linear and angular dimensions are usually delineated by specific elements on the part's surface, and the dimension to be measured is the distance between these boundaries. In the case of surface-texture measurements, however, such readily perceivable and exactly defined dimensions are not available. In the inspection of surface texture, the dimension to be measured is the calculated or qualified extent of a multiplicity of departures of the actual surface from the theoretical.

The number of such individual departures from the theoretical is usually very large, even when a relatively small portion of the surface is being examined. Not all these departures are equally significant and, regardless, it would not be practical to measure all the variations of the surface. Neither is it practicable to examine the entire surface of a technical part for measuring its texture; therefore, only a relatively small area or a short element of the surface is usually inspected and the results are considered to be representative of the larger surface.

For assessing the functional effects of different types of surface variations, not all imperfections are considered equally significant, or even requiring informative measurements. Accordingly, the applicable methods of surface-texture measurement are often chosen to supply reliable information on the meaningful aspects only, neglecting, partially or totally, the other types of surface irregularities.

Examples of different categories of surface-texture parameters are listed in the following, with implied reference to the pertinent methods of measurement:

1. The average range of departures from a mean or the total span between the high and low points of the surface;

2. The height of the most protruding elements (peaks) of the surface and/or the frequency of their occurrence;

3. The interrelations between the widely spaced (waves) and closely spaced (roughness) departures from the theoretical surface;

4. The pattern of the surface and its orientation;

5. The cross-sectional form of the repetitive form irregularities (ridges); and

6. The presence of random discontinuities, and various others.

Depending on the effect of these variables on the operational adequacy or performance of the part, inspection methods may be needed that are capable of measuring primarily, or exclusively, the critical type of surface-texture conditions.

Other factors that may have to be considered in the choice of the appropriate method of measurement are:

1. The need for expressing the measured dimension by a single or a limited number of values;

2. The assurance of repeatability, in order to obtain essentially equal values independently of the inspecting individual or location; and

3. A method that can be implemented by means that are relatively simple to operate and will supply direct values without the need for interpretation.

The purpose of the preceding survey of various surface-texture conditions and inspection requirements is to point out the need for different methods

of assessment. Because the present USA Standard ASME B46.1-1985 considers a single dimension in its listing of preferred values for surface texture, namely, the value of the "average roughness," it is often assumed that said dimension will always provide a dependable guide in evaluating the adequacy of a technical surface for a particular kind of application.

Actually, there is a rapidly growing understanding of the fact that the interrelations between surface texture and operational performance of technical parts often require the inspection of other parameters in addition to, or instead of, the average roughness value. This trend is reflected by the diversity of inspection methods in use, and the availability of appropriate instrumentation for the measurement of various other parameters of the surface texture, which may or may not permit direct correlation with that specific representative value, known as average roughness.

Table 15-2 surveys the more frequently used methods of surface-texture measurement, the applied systems of instruments and the kind of information that their application will provide.

SURFACE-TEXTURE MEASUREMENT WITH STYLUS TYPE INSTRUMENTS

The principles of this general category of surface-texture measuring processes may be described in the following manner. On the surface to be inspected, an element is selected in a location and orientation that is considered to provide a dependable representation of the texture of that surface. A needle type instrument member, the stylus, is moved along the selected surface element. The stylus is connected with a pickup that is maintained in a controlled level, yet that permits the stylus to follow intimately the physical surface of the part. Variations of the surface in relation to the level of the pickup translation cause the stylus to deflect from a reference position. The mechanical deflections of the stylus during its scanning movement along the part's surface are translated by the pickup into electrical signals. These signals either reflect the velocity of the stylus deflections or, more frequently, produce a voltage or current proportional to the amplitude of those deflections. The electrical signals are electronically amplified and either displayed as an average value on a meter or, in the case of displacement sensing pickups, are channeled into a recorder to produce a magnified chart tracing of the scanned surface element. The level in which the pickup is held during

TABLE 15-2. METHODS OF SURFACE-TEXTURE MEASUREMENT – 1

DESIGNATION OF METHOD	EXAMPLE OF INSTRUMENT		DISCUSSION AND EVALUATION
	TYPE	DIAGRAM	
Visual or tactile comparison	Specimen blocks		Unaided sensory perception by eye or by fingertip can be rather discriminating in evaluating the roughness of technical surfaces when guided by samples of comparable pattern, from which the best matching duplicate of the inspected surface must be selected. While not equal in accuracy to instrument measurements, the method is inexpensive, quick and flexible in application.
Visual comparison aided by optical instrumentation	Microscope with two simultaneously observed stages		Similar in principle to the preceding method, yet with substantially upgraded discriminating capacity due to the magnification accomplished with the microscope. The shape and the accessibility of the surface to be inspected can impede the application of this method.
Surface tracing with average indications	Stylus type roughness measuring instrument with averaging amplifier		The most widely used and generally accepted method of surface roughness measurement providing numerical values in agreement with current standards. The extended application throughout the industry is accomplished by virtue of the variety and flexibility of the available instrumentation.
Surface profile tracing with recording	Profile tracing instrument with amplifier adapted to surface texture analyses		For the inspection of surfaces which require a more detailed examination than provided by a single average value, recorded profile tracings are often needed. Such tracings are made with instruments offering high sensitivity and, in special cases, electronic wave filters to eliminate the effect of major profile variations.

TABLE 15-2.   METHODS OF SURFACE-TEXTURE MEASUREMENT – 2

DESIGNATION OF METHOD	EXAMPLE OF INSTRUMENT		DISCUSSION AND EVALUATION
	TYPE	DIAGRAM	
Viewing under high magnification	Regular light microscope		For the general evaluation of surface texture, in distinction to the limited scope of roughness height measurement, the observation at high magnification is often a very informative method. Oblique lighting, or projecting light from two opposed sources of different color can enhance the topography. Cross hair reticule with micrometer slide permits scaling of features.
Light sectioning	Light section microscope		Surfaces whose degree of roughness and raggedness makes the stylus type measurement difficult or impractical, can be inspected by non-destructive "sectioning" using an obliquely applied thin light band and observing the illuminated contour line by a microscope. Being a non-contacting method it is also adaptable to soft materials.
Interferometric measurements	Reflected light interference microscope		By producing several interference bands in the light reflected from a small surface area under observation at high magnification (e.g. 100x to 400x), the boundary lines and the general pattern of the fringes will reveal the significant surface texture variations of that area with a resolution permitting assessment with microinch discrimination.
Electron microscopy	Electron microscope		The examination of minute details of ultra-fine surfaces, with a discriminating capacity in the sub-microinch range, may only be feasible with the unique resolving power of an electron microscope. This is a purely laboratory method, which requires elaborate replica preparation, including vapor deposition at a controlled angle of incidence.

TABLE 15-2.   METHODS OF SURFACE-TEXTURE MEASUREMENT – 3

DESIGNATION OF METHOD	EXAMPLE OF INSTRUMENT		DISCUSSION AND EVALUATION
	TYPE	DIAGRAM	
Replica techniques	Various replica materials		Although not an independent method it is an auxiliary process which can substantially extend the application areas of several methods for surfaces whose location impedes direct access by the basic instrument. For example, lacquer replicas are used for light section microscopes and stylus instruments, mylar films for light interference measurements and air gaging.
Successive probing along a surface element	Oscillating probe instrument with optical magnification (Leitz-Forster system)		A unique method of scanning the roughness of selected surface elements is accomplished by an instrument using an oscillating sharp probe for contacting in rapid succession the surface of a slowly moving object. The variations in the amplitude of the alternating probe movements is optically magnified, photographically recorded and displayed as the contour line of a silhouette image.
Taper sectioning	Special grinding equipment and magnifier microscope		By removing the top layer of a machined surface along a plane slightly inclined to the general surface, those irregularities which are substantially perpendicular to the direction of the incline can be made to appear enlarged. A common, yet not exclusively used, angle of incline is 2°17½′, which produces 25x magnification.
Reflectance	Reflectometer		This method is based on the relationship between the roughness of a surface and the amount of incident light which, instead of being reflected, will be diffracted. In the case of surfaces of fine finish with roughness smaller than the wavelength of the used light, that relationship becomes reliable enough for providing numerical roughness values of a limited area or, by scanning and signal averaging, of an extended portion of the surface.

its translational movement can be either an envelope surface of the part, or may be provided by an extraneous datum that has been established essentially parallel with the general direction of the inspected surface element.

Considering the above-mentioned variables, namely, the selection of the reference plane to which the departures of the surface are related, and the indications of the sensed surface irregularities, the use of a stylus type instrument does not exclusively define the process. Actually, there are several systems of stylus instruments used for surface-texture measurements that differ in operating principles and also in the information they supply.

In many cases it is possible for the values indicated by different types of instruments to be correlated, and with the concepts of the ASME B46.1-1985 standard. However, even under optimum conditions, as provided by a workpiece with essentially uniform texture, such correlations are only approximations and by no means comparable with, say, the excellent agreement between values of repeated length measurements. The correspondence of indications originating from different types of instruments becomes poorer when inconsistently occurring departures characterize the surface being examined.

The complexity of surface conditions, which are designated collectively as texture, and the diversity of their effects on specific operational requirements, warrant the efforts of instrument manufacturers toward devising and building different systems of surface-texture measuring instruments. Although a complete listing of all the available systems and methods of surface-texture measurement by stylus type instruments would be difficult to accomplish in view of continued progress in this field, the major systems and the basic characteristics of the information that they supply are reviewed in Table 15-3.

Subsequent sections will review: (a) the major elements and functions of the instrumentation, as well as their various forms of execution; (b) characteristic examples of diverse types of stylus instruments; and (c) the assessment of indications obtained by means of different, currently used stylus-tracing processes.

Elements of Stylus Type Surface-Roughness Measuring Instrument

For explaining the operation of various types of surface-tracing instruments and the differences in the representation of the surface texture they supply,

the principal instrument elements and their functions will be reviewed.

The *stylus* is the element through which the instrument senses the departures of the traced surface element from a reference line. The stylus must have a fine point permitting it to follow the textural variations of the surface, yet be of a form and dimension that will avoid marring the surface being traced. The point of the stylus is made of diamond with the basic form of a cone, having a 90-degree included angle and a spherical vertex. For average roughness measurements that spherical surface of the stylus point has a radius of 0.0005 inch; however, for tracing finely finished work surfaces, several models of instruments can be equipped with styli having 0.0001-inch or, exceptionally, 0.00005-inch radii.

The stylus is held against the surface being traced with a force that should be heavy enough to assure continued contact, yet not in excess of 2.5 grams. Some manufacturers, particularly when using smaller stylus tip radii, prefer a lesser contact force, frequently 1 gram, or even much less, exceptionally, as low as 0.1 gram.

The *pickup* may operate by any of several systems, such as piezoelectrics, electromagnetics, using a linear variable differential transformer (LVDT) and so forth, according to the choice of the instrument manufacturer. Originally, the pickups used for surface roughness measuring instruments were frequently of the velocity sensing type that, however, could not portray the actual profile of the surface, but only certain characteristics that could be correlated with the concept of average roughness. Most of the currently available instruments have displacement sensing pickups that produce electrical signals proportional to the excursions of the stylus.

Skids and shoes. When the measurement of surface roughness is the only objective, it is a convenient and general practice to traverse the pickup along an envelope element of the surface. That imaginary envelope element is established as the path of a supporting member mounted on the pickup head and in continued contact with the object surface along which it is sliding. That contact member, termed the skid or, when of long flat form, the shoe, has a radius substantially larger than the stylus radius. Consequently, the skid will not penetrate into the closely spaced valleys of the profile, but will rest on the most protruding components of the surface; these latter are often termed the peaks or the crests.

Regardless of the greatly different order of magnitude of the stylus and of the skid radii, respectively, variations in the form and location of the skid

TABLE 15.3   OPERATIONAL CHARACTERISTICS OF DIFFERENT SYSTEMS OF STYLUS TYPE SURFACE-TEXTURE MEASURING INSTRUMENTS

SYSTEM CHARACTERISTICS AND OPERATIONAL OBJECTIVES	DIAGRAM OF PRINCIPLES	BRIEF SURVEY OF THE SUPPLIED INFORMATION
Pickup head with skid or shoe for average roughness indications		This is the most generally used system, because it is easy to operate, requiring little setup and adjustment for supplying directly numerical information which is in agreement with the specifications of the standard. The great variety of available pickup heads and skids (see Table 15-4) contributes to this system's extended applications.
Pickup head with support roll to indicate average deviations from an envelope surface ("E" system)		The objective of this method in actual use in several overseas countries, is to assess the total deviations of the surface from a reference line developed to represent the contour of an imaginary envelope which is spatially bounding the part.
Skidless pickup head extraneously supported for average indications in restricted areas		This system, whose mechanical arrangement resembles that for surface profile measurements, has average indications for an objective, and relies on an external skid support only as a substitute for the insufficient area of the surface being inspected. In the case of very narrow surfaces (less than five times the minimum cutoff length) the resulting indications may be lower than the actual roughness value of the surface.
Extraneously guided (datum supported) pickup head for recorded indications of the entire surface texture		Where the operational conditions of a surface require the control of the total texture, including the characteristics which are not revealed by an index value, such as the Arithmetical Average, stylus type instruments with extraneous datum are used to produce a recorded tracing, at high magnification, of all the surface deviations from a reference line which is established by the datum element.

can affect the result of the surface-texture measurement. For reasons usually related to the form characteristics and space limitations of the surface to be inspected, most instrument manufacturers are making available different types of skids, designed to accommodate different part configurations and dimensions.

A few characteristic examples, comprising various skid forms and arrangements, are shown in Table 15-4, with the double purpose of:

a. Helping the reader to visualize the potential effects that these skid variables can have on the establishment of the envelope line to which the excursions of the stylus are related; and

b. Indicating the flexibility of design, which is reflected by some of the currently used skid types.

Extraneous datum for pickup guidance. Although skid support is used most commonly for stylus type surface-roughness measurements, there are various applications in which the envelope of the surface, as represented by the path of the skid traverse, cannot be established, or does not supply the appropriate reference element.

Thus, space limitations may prevent the use of skids for supporting the pickup, particularly on extremely narrow or short surfaces, inside very small holes, and in recessed areas.

For reasons defined by the sought-after information, a datum of known geometry may be preferable to the undetermined, commonly wavy form of an envelope element. Such cases will include: (a) the exploration of the entire profile comprising both waviness and roughness; and (b) short surfaces where the available traverse length is not sufficient to supply dependable average values.

In such cases it may be desirable, or even indispensable, to substitute an extraneous datum for the envelope of the surface being inspected. With the use of an extraneous datum, stylus type measurement can be made of both the average roughness and the entire profile, the latter displayed in the form of chart tracings.

Extraneous datum type guidance devices are supplied by several manufacturers of surface-roughness measuring instruments. The design principles of such devices differ, reflecting the primary purpose any particular type of guidance system is intended to serve. The principal systems are reviewed, and their relative merits briefly discussed, in Table 15-5.

The traversing drive. In order to accomplish the scanning of the surface being inspected along a specific element, the pickup must be traversed in a plane parallel with the surface. The traverse movement must be carried out at a speed whose rate will be a factor in the resulting measuring characteristics of the instrument.

It is possible to approximate the requirements of controlled rate and uniform speed by manual traversing, when carried out by a skillful operator. A few types of surface-roughness measuring instruments can still be supplied for shop applications with handheld pickup heads. However, the majority of surface-roughness measurements by stylus type instruments are made, at present, by applying a mechanical traverse motion for the pickup displacement.

Such mechanical traversing devices, examples of which are shown in the subsequent section, must satisfy several requirements. A few of these are the following:

1. Uniform traversing speed.

2. Controlled rate of advance, usually 0.100 or 0.125 inch per second for general use, and a lower speed, for example, 0.040, or even as low as 0.005 inch per second when very high resolution is needed, such as during the tracing of very fine finish surfaces, with a small radius stylus, or for confined short sections. An example of the latter application is shown in Fig. 15-4.

3. Adjustable stroke length, usually in the range from $\frac{1}{16}$ inch to 1 inch or more. However, there are drive devices that operate even in lower length ranges, to explore very short sections. An example of such a device with a minimum stroke length of 0.010 inch is shown in Fig. 15-5.

Fig. 15-4. Surface-texture measuring instrument (Talysurf) with special accessories for the tracing of a narrow surface section in a confined area.

TABLE 15-4.   SELECTED DESIGN CHARACTERISTICS OF PICKUP HEADS, STYLI AND SKIDS – 1

DESIGN ASPECT	EXAMPLE		DISCUSSION
	DESIGNATION	DIAGRAM	
Arrangement of the skids in the pickup head	Aligned		The skid is located either (a) ahead of the stylus, or (b) behind it in the direction of travel. The aligned arrangement is the most universally applicable, although it does not provide lateral guidance to the pickup head. Aligned single skids are also used for tracing inside small bores.
	Straddling		The straddling skids provide contact with the surface being inspected at both sides of the stylus, and assure guidance along the axial plane of parts with curved surfaces. The straddling skid may consist of two buttons (a) or the symmetrical contact is obtained from a Vee-formed support member (b).
Stylus tip forms and dimensions	Cone with regular tip radius	90° r = 0.0005 in.	This type of stylus is specified in the standard, because it is adequate for the majority of surface conditions without being extremely sensitive in usage. Commonly used contact force: about 2 grams.
	Cone with fine tip radius	r r ≦ 0.0001 in.	Selected for the examination of very fine finishes by means of recording instruments. Requires pickups with extremely light contact force, e.g., 0.1 gram.
	Chisel form	r = 0.0005 in. $l = \frac{1}{16}$ in. r l	Developed for the measurement of very narrow surfaces, such as wires, blade edges, etc., where the lateral guidance of straddling skids is not sufficient to assure tracing along the crest.
Skid forms (a limited number of examples)	Flat (shoe)	l l ≈ 0.040 in.	Designed for the examination of very rough surfaces, such as produced by cutting tools, which require long support members to bridge the widely spaced crests of the texture.

TABLE 15-4.   SELECTED DESIGN CHARACTERISTICS OF PICKUP HEADS, STYLI AND SKIDS – 2

DESIGN ASPECT	EXAMPLE		DISCUSSION
	DESIGNATION	DIAGRAM	
Skid forms (continued)	Large radius	r = 2.5 in.	A design preferred for certain types of rough surfaces where smooth gliding is considered more important than the theoretically correct tangential contact of the shoes.
	Common form for aligned arrangement	r = 0.012 in.	Used for finer surfaces which are measured with cutoff lengths of 0.030 inch or less. The "m" dimension may be critical when tracing narrow surfaces.
	Recess type	d ≈ 0.350 in.	For the examination of surfaces which are recessed behind shoulders. Made in different sizes, to accommodate the required depth ("d" dimension on the diagram).
	Hole bottom type	d ≈ 1.0 in	With shaft at right angles to the common arrangement and with 0.25 in. radius skid on the face of the shaft, it permits tracing deeply-lying surfaces, such as the bottom of holes and grooves.
	Straddling, for small diameter cylindrical parts, also for flat surfaces		The straddling skids are located close to each other to avoid excessive chordal height variation for different diameters. May also be used for flat surfaces.
	Straddling, for internal surfaces within specific size range		The bore diameter must be cleared by both the body of the pickup and the curvature of the skids. Provides lateral guidance but not adaptable for very small bores.
	Straddling, for large diameter internal surfaces and for flat surfaces		The straddling skids for large bores consist of two relatively widely-spaced buttons whose radii exceed that of the largest bore for which the pickup head is designed.

TABLE 15-5.   DESIGN PRINCIPLES OF EXTRANEOUS GUIDANCE SYSTEMS FOR SURFACE-TEXTURE MEASURING INSTRUMENTS

GENERAL CATEGORY OF REFERENCE PATH	FORM OF EXECUTION	FIGURE	DISCUSSION
STRAIGHT LINE	Rigid pickup beam with cantilever support, solidly attached to a ram which is guided inside the actuator (motor drive)	Fig. 15-5	This is the most commonly used system due to several advantages, such as easy setting, uncluttered approach, etc. Extended beam length causes sag which can detract from the expected accuracy of the guidance.
	Pickup beam hinged on a pivot or on flexure strips, with one degree of freedom in the direction of the stylus excursions, where a detached datum element is exerting the required constraining effect on the pickup head	Fig. 15-4	Permits a dependably accurate guidance close to the effective working area of the stylus, thereby avoiding the effect of sag. However, the leveling and alignment requires more skill than in the case of the rigid beam system.
CURVED	(A) A pivoting arrangement that causes the pickup at the end of the beam to follow a curved path. (B) Another design uses a hinged pickup beam whose extraneous datum is an element with curvature corresponding to the ideal form of the object surface.	(A) Fig. 15-13 (B) Fig. 15-14	For small radii the pivoting system can provide dependable guidance. When the section of a cylinder is used as an extraneous datum element, the radius of the generated stylus path can be varied by swivelling the datum. The orientation of the stylus does not follow the part contour, therefore both systems are intended for short arcs only.
CIRCULAR	The pickup beam is rotatingly supported on an axis which coincides with the center of the circular arc to be traced. The cross-slide of the staging table permits the adjustment of the object into the properly centered position. The stylus is always directed toward the center of the arc to assure that sensing is consistently normal to the surface.	Fig. 15-17	Extended lengths of circular arc contours, such as are to be followed in the across-track scanning of ball bearing races, are best assured with crank type pickup beams rotated with uniform angular velocity whose rate can be adjusted to provide the circumferential speed needed for the tracing. The signals originated by the stylus excursions can be used either for average reading or chart recording.

Giddings & Lewis Measurement Systems

Fig. 15-5. Special skidless Profilometer surface-texture measurement system for the inspection of surface sections requiring reciprocating stylus traverse with very short strokes to produce average-roughness indications.

4. Shockless reversals at the ends of the strokes when reciprocating transfer motion is required. The latter is a rather widely practiced procedure for the momentary average indicating meters, used in the majority of industrial surface-roughness measuring instruments. Reciprocating motion prolongs the time available for the observation of the meter, and thus permits determining the scale value associated with the predominant position of pointer. Integrating type instruments do not require a reciprocation of the trace direction.

5. Adjustment means for height setting to accommodate work surfaces at different levels relative to the base of the instrument.

6. Quick acting couplings or easily changed connections at the end of the drive ram, for accepting the beams of the selected types of interchangeable pickups.

Examples of actual instrument drives that satisfy the requirements listed above may be seen in the illustrations of the subsequent section.

The amplifier. This is the general designation for the entire electronic instrumentation, except for the remotely located pickup. The amplifier is connected with the pickup through an electric cable, and is operated by electric current either from the shop power line (115 volts) or from rechargeable nickel–cadmium batteries.

Most types of surface-roughness measuring amplifiers are equipped with several multiple-position switches, for selecting the suitable factors of surface-texture assessment. The two major selection categories are:

a. Cutoff selection for the desired roughness-width cutoff value. Although the standard lists a rather wide range of different values, most instruments offer, in adherence to general industrial practice, a selection of three values, namely, 0.010-, 0.030- and 0.100-inch cutoff lengths. The effective cutoff values will be reduced, proportionately, when lower than regular trace speed is applied. The respective switch operates the low frequency cutoff of the amplifier in a manner such that signals received at frequencies below the selected value are substantially attenuated, even to the degree of complete suppression, and do not participate in the indicated average value.

The relationship between the tracing speed and the low frequency (long wave) cutoff is mathematically expressed by the following formula:

$$\text{Instrument long wave cutoff in cycles} = \frac{\text{speed of trace (in./sec)}}{\text{selected roughness-width cutoff (in.)}}$$

b. Amplification selection that establishes different ratios between the average value of the stylus excursions and the length of the scale. Depending on the type of the instrument there might be up to seven scale selections, ranging, for example, from 3 to 3000 microinches, although shop type instruments usually cover a narrower amplification range only, and offer a lesser number of rates.

Additional selections and connections are also available on certain types of amplifiers, such as a changeover from arithmetical average (AA) to root mean square (RMS) values and a recorder connection.

The meter is usually part of the amplifier, with the dial located on the front face of the instrument. Most meters indicate continuously, the momentary average-roughness value of the traced surface. Integrating type meters are also available with certain types of instruments, to show a single average value at the end of the trace covering a controlled traverse length. This is a fixed value that does not have to be estimated as the predominantly appearing pointer position; consequently, it offers a higher degree of repeat accuracy of the readings than those from even a well-damped, continuously averaging meter. Integrating meters, however, require more sophisticated instrumentation, and the available straight line traverse length must be at least five times, preferably, seven times the selected cutoff length.

Calibration of Stylus Type Average Surface-Roughness Measuring Instruments

In order to assure the compliance of the instrument's indications with the concepts and dimensional relationships of the standard, periodic verifications are required. For that purpose a reference specimen is traced with the instrument at standard speed and in a direction normal to the lay. The average-roughness indications of the instrument should agree with the value marked on the reference specimen.

Essentially, such a calibration master (see Fig. 15-6) is a flat block, on one surface of which a field is covered with consecutive parallel grooves of specific shape and size. The basic cross-sectional shape of the grooves is that of an isosceles triangle with 150 degrees included angle, as specified in the standard. The peak-to-valley height of the triangles is four times the average roughness size they represent, that is, 0.0005 inch for 125-microinch average roughness.

The 125-microinch average-roughness value has been selected for a widely used type of reference master, as the scale value for calibrating average roughness measuring instrument. The same reference specimen also has a second field, with finer grooves, having a spacing that theoretically represents 20-microinch average roughness. However, the tracer instrument, when calibrated at the 125-microinch scale does not indicate the nominal value of a theoretically dimensioned array of closely spaced grooves. The stylus of the instrument has a finite radius on its tip, which cannot penetrate into the sharp bottom of the valleys, consequently, the span of the actual stylus excursions will be smaller than the distance between the peaks and valleys of the reference specimen.

The ratio between the actually traced and the theoretical values will be close to one for the wider spaced grooves, but it will decrease as the spacing becomes narrower. There is a geometric relationship between the stylus radius and the reduced indications due to incomplete penetration of the stylus tip into the bottom part of the groove valleys. This relationship is utilized for determining the effective radius of a particular stylus tip in the calibration procedure by:

1. First calibrating the instrument on the 125-microinch scale, and then

2. Tracing the finer, 20-microinch scale, and observing the difference on the meter, between the actually indicated and the nominal scale values.

These principles of stylus type instrument calibration are illustrated in Fig. 15-7.

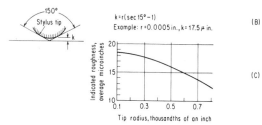

Fig. 15-7. Concepts related to the calibration of stylus type, average-roughness measuring instruments.
(A) The basic profile of the calibration specimen in the cross-sectional plane of the stylus traverse, and the development of the original peak-to-valley height *h*, through the unilateral height *m*, into *Y* height representing the assigned AA value of the surface.
(B) The geometric conditions limiting the penetration of the stylus of finite radius to the full depth of the valley, thereby reducing the actual range of stylus excursion in relation to the peak-to-valley height of the theoretical profile.
(C) Curve representing the relationship between stylus radius and the indicated average roughness on the 20-microinch scale; the decrease resulting from the reduced stylus excursions.

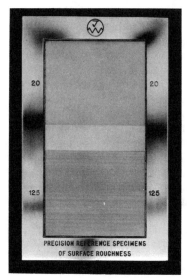

Ringler-Dorin, Inc.

Fig. 15-6. Cali-Block reference specimen for the calibration of stylus type, surface-roughness measuring instruments with average indications. The block is made of nickel by an electroforming process, from a gold master, produced with a very accurate ruling machine. The block has two fields representing 20- and 125-microinch average roughness, respectively.

REPRESENTATIVE MODELS OF STYLUS TYPE SURFACE-TEXTURE MEASURING INSTRUMENTS

The manufacturers of stylus type surface-texture measuring instruments, with due consideration of the widely varying conditions of technical parts, are offering instrumentation that is often quite versatile and can be adapted to different methods of surface-texture inspection. Although a high degree of versatility is not characteristic of all types of instruments, a rigid categorization is not feasible and overlapping capabilities must be recognized.

With these reservations in mind, an attempt will now be made to discuss the characteristic types of stylus instruments by categories. These have been established for the purpose of this survey by considering that kind of information that specific types of instruments are primarily capable of supplying.

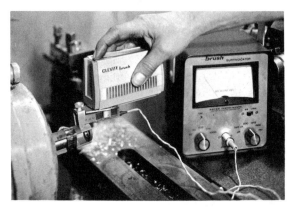

Brush Instrument Div./Clevite Corp.

Fig. 15-8. Handheld motor drive of a portable, surface-roughness measuring instrument in use on a lathe. Inside the supporting legs is the pickup with the stylus that is given a reciprocating movement by a small electric motor. The cord is the only connection with the remotely located amplifier/meter.

Instruments with Handheld Pickup Heads

The simplest way to trace a surface is to move the pickup by hand, along an essentially straight line path. A few of the original, average-roughness measuring instruments were actually operated in this manner. These offered the advantages of low cost and the elimination of setting up and adjustment of workpiece and instrument in mutually controlled positions.

Advantages of manually operated tracing are still present in specific application areas, such as for rapid checks in the shop; the inspection of parts mounted on a machine tool, or those that are too bulky to be moved to the instrument; and surfaces located in vertical positions; to name but a few.

For this reason, some up-to-date instruments have auxiliary equipment for manual tracing as an optional accessory. Such equipment consists, essentially, of a rugged pickup head that can be held directly by hand, or mounted on a handle; the latter contributes to steady guidance and permits the reaching of remote surfaces as well.

It must be realized, however, that manual tracing, even when performed by a skillful operator, can only approximate the reliability of measurements made by power-driven pickups. The tracing speed is a factor of the roughness-width cutoff; consequently, variations in the speed of the trace, unavoidable in manual traverse, affects the resulting roughness indications. For this reason the use of manual tracing should be limited to surfaces that offer a straight line sampling length of at least $\frac{1}{2}$ inch, and whose roughness is not less than 10 microinches.

A combination of the handheld and the mechanically operated pickup traverse is offered by a relatively recent development, the handheld motor drive, shown in Fig. 15-8, in use on a workpiece held in a lathe.

This drive is a self-contained pickup traversing device, using a standard battery for the power supply of its motor that imparts the uniform $\frac{1}{8}$ of an inch per second tracing speed to the reciprocating ram movement. The pickup with its integral skid is mounted on that ram. The drive can be operated in any position: horizontal; vertical or inverted; held by hand; resting on the work surface; or mounted on an adjustable extension arm. The pickup mount can be rotated by 90 degrees in either direction, permitting the substitution of lateral movement for the regular axial traverse, when such orientation is required due to the configuration of the work.

The amplifier used in conjunction with this drive is also battery operated and portable. It offers five magnification ranges, representing 10 to 1000 microinches over the meter scale, and a selection of three different cutoff widths, 0.003, 0.010 and 0.030 inch.

Average Indicating Instrument with Interchangeable Pickups and Alternate Drive Units

The instrument selected to typify this group (Fig. 15-9) represents the repeatedly improved model, with substantially extended capabilities, of an original that may be considered the prototype of stylus

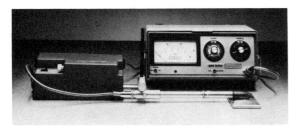

Giddings & Lewis Measurement Systems

Fig. 15-9. A general type of surface-roughness measuring instrument comprising an amplifier unit with meter, a motor drive for reciprocating movement of an adjustable beam that holds the pickup head with stylus, shown in contact with the workpiece.

using average-roughness measuring instruments. This historical background, although barely recognizable in the exterior appearance and engineering characteristics of the present-day instrument, still persists in two significant respects: (a) The basic principle of measurement by the conversion of surface-roughness-caused stylus excursions into proportional electrical signals, and (b) the trade name Profilometer that, in this country's technical language, is widely associated with the general method of stylus type surface-roughness measurement.

The illustrated group of units is one of several alternative combinations, yet it is typical enough to represent the principles of this category of instruments. Three principle members can be distinguished:

a. The *amplifier*, powered by rechargeable batteries, is an average-roughness indicating instrument that offers seven different amplifications. The entire span of the meter represents 3000 microinches in the lowest, and 3 microinches in the highest scale. Any one of three cutoff widths, that is, 0.010, 0.030 or 0.100 inch can be selected. These cutoff lengths are based on the 0.3 inch per second tracing speed of the illustrated type of drive. The reduction of the tracing speed proportionately decreases the cutoff length; as an example, for a tracing speed of 0.1 inch per second, the resulting cutoff values are 0.003, 0.010 and 0.030 inch.

b. The *drive unit* is also battery powered and provides a uniform traverse speed of 0.3 inch per second, continuously reciprocating over the selected sampling length, with quick reversals at the terminal position. The length of the stroke can be adjusted from 0.050 to 2.50 inches, keeping in mind that for obtaining dependable average-roughness values the stroke should be not less than five times the selected cutoff length.

The ram of this drive unit ends in a vertical post along which the position of the swiveling extension arm can be adjusted to the height corresponding to that of the surface being traced. The linkage provided by this extension element compensates for minor alignment errors between the ram axis and the traced surface.

In addition to the described basic drive unit, other types are available, such as one for exceptionally short surface sections (see Fig. 15-5), for balls, for circular surface elements and so forth.

c. The *tracer head* is interchangeable, although the illustrated type is highly versatile and satisfies the majority of generally occurring inspection tasks. Different tracer heads are used, for example, for small, deep holes and also for handheld uses. The versatility of the basic tracer head is assured by a variety of quickly exchangeable skid mounts, offering the choice of skid locations and sizes best suited for a particular surface configuration.

In addition to the described principal members, various accessory units are also available for this instrument. Two of these are discussed.

The peak counter. In various industrial uses, such as the production of sheet steel that has been prepared to serve as a base for high quality finishes, surfaces subjected to frictional loads as in brake drums, and many others; the individual peaks (protruding components) of the general surface, particularly the height and the relative number of such irregularities, represent functionally significant surface characteristics. While peaks participate in establishing the average-roughness index of the surface, information regarding the above-mentioned parameters of the individual peaks must be obtained by other means.

Figure 15-10 illustrates such an instrument grouping in which the principal members discussed

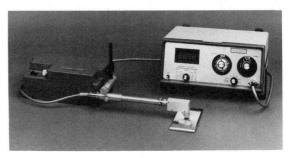

Giddings & Lewis Measurement Systems

Fig. 15-10. Surface-roughness measuring instrument similar to that shown in Fig. 15-9, and complemented with a numerical display and digital printer.

above are supplemented by an electronic peak counter. The information supplied by this instrument is the digital display of the number of peaks counted over a trace length of one inch. The term peak, in the context of this instrument's indications, refers to specific surface components that protrude over a preselected average-roughness height. By varying the base height setting, an analysis of peak incidence and characteristics can be carried out quickly, using the distinctly displayed numerical values.

The *reading recorder* supplies a chart tracing that graphically freezes the momentary indications of the average meter in the form of a continuous pen line. The imprinting of the chart paper is a duplication of the meter scale and the ordinate position of the pen line corresponds to the pointer position on the average meter. Due to the continued variations of that position on a nonintegrating type of average meter, the resulting pen mark is not a straight line, but still, it lends itself to a more dependable determination of the predominant level than the visual observation of the fluctuating pointer position. Such recordings, which originate from a skid-supported stylus trace, should be clearly distinguished from datum referenced profile tracings that represent the magnified image of the actual profile.

Surface Analyzing Instrument with Computerized Recorder for Multiple Characteristics

The instrument group shown in Fig. 15-11 comprises members that were designed with particular

emphasis on the analyzing, visual display and recording capabilities of the system. The probe pickup, which is of a breakaway design to eliminate damage, has full measurement capability including skid support for average-roughness sensing, or it can be externally referenced by being supported on a cantilever arm that is precisely guided inside the drive unit. The linear drive stroke length can be adjusted form 0.020 inch to 4.00 inches.

A personal computer utilizing an advanced surface metrology program and touch screen technology eliminates the need for keyboards, knobs, switches and push-button control panels. The computer display allows the operator to visualize the surface measurements. Hard-copy records of measurement data can be outputted via either a standard dot-matrix printer or the system's own data acquisition controller unit printer/recorder. This system offers complete analysis of all forms of profiles, roughness and waviness, including all combinations thereof.

Integrated Averaging and Profile-Recording Instrument with Different Datum Accessories

Integrated averaging and profile-recording instruments are represented by the instrument shown in Fig. 15-12. Perhaps the most remarkable characteristic of this instrument is versatility, making it one of the preferred types of equipment for research and development laboratories, as well as for various industrial applications where the texture of the surface is of critical importance.

For industrial uses in which the average surface roughness (AA) values are sought, the integrating capability of the instrument assures a repetitiveness of the measurements that is independent of the operator's judgment. The value indicated by the meter at

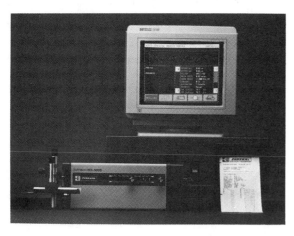

Federal Products Corp.

Fig. 15-11. Surface-analyzing instrument designed for comprehensive analysis, visualization and hard-copy recording of surface texture by exploring various parameters, such as actual and average roughness, waviness and so forth.

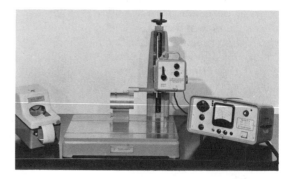

Rank-Taylor-Hobson/Engis Equipment Co.

Fig. 15-12. Universal surface-texture measuring instrument for integrated average indications, as well as for profile recording, using various types of datum devices to establish reference lines corresponding to the ideal surface contour.

the end of the pickup traverse movement is the integrated, average surface roughness of the entire traced section. The traversing length, established as the function of the selected cutoff value, is set prior to starting the drive for the translational movement.

Recorded tracings can be prepared with any of eight magnifications, ranging from 500× to 100,000×. The overall recording accuracy stated to be better than 3% of full scale for any magnification, in combination with the rectilinear type of graphs, permits a very dependable evaluation of the inspected surface characteristics. The discriminating capability of the instrument is enhanced by the small (0.0001-inch) tip radius of the stylus that bears on the specimen surface with a force of about 0.1 gram.

A particularly large variety of pickup beams and heads with interchangeable nosepieces, comprising different skids and shoes, can be supplied for adapting the trace movement to diverse external and internal surfaces. The latter may be directly accessible or recessed, including those requiring the mounting of the pickup beam at right angles to the traverse of the drive ram.

As optional accessories, different types of datum attachments can be obtained for both straight and curved surfaces. A few of these accessories deserve special mention because of the unusual design and uncommon application potentials.

The *pivoting type curved datum device* shown in Fig. 15-13 can be adapted to both convex and concave surfaces. For guiding the flexibly mounted pickup beam, a ball-ended radius bar of adjustable length is held between two spherical sockets, whose

Rank-Taylor-Hobson/Engis Equipment Co.

Fig. 15-13. Pivoting type datum-generating device for guiding the pickup beam in a manner causing the stylus to follow a curved path of convex (illustrated) or concave form, whose radius is adjustable within specified limits

relative locations differ for the convex and the concave curves. Radii from $\frac{1}{8}$ of an inch to $3\frac{1}{2}$ inches can be traced, although, for obtaining meaningful average readings, a minimum of $\frac{3}{4}$ of an inch radius is needed.

The *variable curvature datum attachment* shown in Fig. 15-14 is mounted on an arm above the hinged pickup beam. Spring tension causes this beam to bear, through its contact button, against the overhead datum. The datum itself, made of optical glass, has a plane section for straight-line paths, as well as separate positive and negative cylinders for curved surfaces, with radii ranging from two inches to infinity. The cylindrical datum sections are held in a rotatable mount for varying their orientation over a range of 90 degrees. Graduations on the periphery of the mount indicate the approximate radius corresponding to any particular rotational position of the datum cylinder.

The described, as well as other datum attachments, that guide the basically straight displacement movement of the pickup have certain functional limitations, namely:

a. The length of trace is, at most, equal to but frequently less than the traverse of the instrument's drive mechanism; and

b. The excursions of the stylus are confined to a path perpendicular to that of the basic traverse movement and not normal to the inspected surface, a condition that causes distortions in the scanning of curvatures with very small radii.

These limitations can be eliminated, while retaining the potentials of the basic electronic instrument system, by either traversing the specimen in front of a stationary pickup whose stylus stays in contact with the surface being inspected or by using a separate traverse device to accomplish a controlled pickup travel. Several of these devices, designed to operate as accessories of the basic Talysurf instrument, are reviewed, indicating in the pertinent subtitles the essential functional purpose.

Tracing the circumferential elements of a sphere. The device, designated as the ball unit (see Fig. 15-15) has interchangeable seating plates for balls from 0.04-inch to 1-inch diameter, and contains a pickup whose stylus can be adjusted to contact, from beneath, the seated ball. The drive unit of the basic instrument carries a hinged arm that is equipped with a friction pad resting on the specimen ball. The drive movement of that arm imparts to the ball, a constant peripheral speed of rotation, which agrees with the tracing speed at which the average indication as well

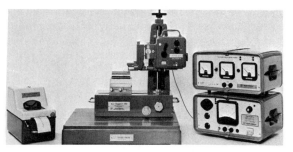

F A G Kugelfischer/Engis Equipment Co.

Fig. 15-16. Air-supported traversing table used in conjunction with a Talysurf surface-texture measuring instrument, whose motor drive stays inactive while the object mounted on the table slide can be traversed along a straight-line path to a maximum length of 6 inches.

Rank-Taylor-Hobson/Engis Equipment Co.

Fig. 15-14. (Left) Curved datum element in a rotatable mount for varying the radius of the resulting stylus path by changing the orientation of the datum.

Fig. 15-15. (Right) Special surface-roughness inspection device originally developed for balls. In reverse of the common arrangement, the stylus is located inside the device. The specimen, in this case a needle roller, is rotated by the translation drive of the basic instrument.

as the chart tracings of the basic instrument are produced.

Tracing a straight line element over an extended length. Motorized traversing tables can be supplied for measurements over a maximum straight-line length of 6 inches. Two different types of such tables are available:

1. For average-roughness measurements and for inspecting profile errors on the order of more than 0.0001 inch, a table with accurately machined guideways that assure the retention of the traverse path with 0.0001-inch maximum departure from straightness is used.

2. For very accurate profile analysis, an air table (see Fig. 15-16) is supplied whose traversing accuracy is within 4-microinch error in relation to a straight-line reference.

Tracing circular arcs with stylus excursions aligned with the momentary specimen radius. The drawing in Fig. 15-17 shows a rotary device for carrying the pickup around a circular path that can be either convex or concave, that is, the stylus type is directed toward or away from the center. Perhaps the most important, although not the only application of this device is the across-track tracing of ball bearing ring raceways. The pickup is held in a rotatable crank arm with power drive. The angular velocity of that rotational movement can be adjusted to produce the required peripheral trace speed, and the maxi-

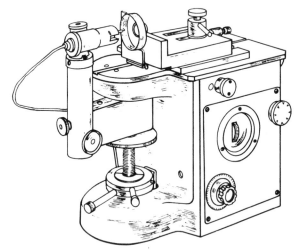

Rank-Taylor-Hobson/Engis Equipment Co.

Fig. 15-17. Rotary tracing device with stylus for scanning external and internal surface elements of convex and concave circular-arc form.

mum sweep of the arm is 200 degrees. The radius length can also be adjusted from 0 to $\frac{5}{8}$ of an inch for measuring ball bearing rings across track, and over a larger range for surface locations that are accessible to a vertically positioned pickup.

THE ASSESSMENT OF SURFACE TEXTURE WITH STYLUS TYPE INSTRUMENTS

The assessment of such complex conditions that characterize the texture of technical surfaces is usually guided by the essential requirements of the inspection process. These requirements might be conflicting regarding the feasibility of their implementation when considering the following aspects of the applicable measuring process:

a. The assessment should be based on a single characteristic dimension susceptible of being expressed as a value that is comparable, reproducible and easily established; and

b. The dimensions selected and measured in the process should define the functionally significant characteristics of the surface.

For the majority of industrial applications, the concept of average roughness as determined by means of stylus instruments represents a fair compromise of these requirements, with emphasis on the

Standard Evaluation of Technical Surfaces

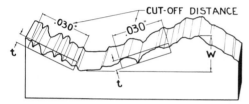

Technical surfaces are never perfectly smooth. They usually reflect the marks of the final machining process, resulting in a predominant lay (roughness measurements must be made in a plane normal to that lay). The actual surface texture commonly displays wide-spaced waves and closely spaced ridges and furrows. Surface roughness measurements disregard the waves (*W*) and consider only asperities (*t*) over a shorter distance (cut-off).

Standard Measurement of Average Surface Roughness

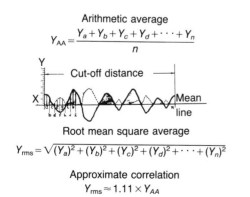

Arithmetic average
$$Y_{AA} = \frac{Y_a + Y_b + Y_c + Y_d + \cdots + Y_n}{n}$$

Root mean square average
$$Y_{rms} = \sqrt{(Y_a)^2 + (Y_b)^2 + (Y_c)^2 + (Y_d)^2 + \cdots + (Y_n)^2}$$

Approximate correlation
$$Y_{rms} \approx 1.11 \times Y_{AA}$$

Fig. 15-18. Diagram illustrating the basic concepts of assessing the conditions of a surface by means of average-roughness values.

practicality of the measuring process. However, most technical surfaces do not conform with the ideal regularity of a repetitive pattern on which the standard evaluation of average roughness is based. To visualize the probable relationship between ideal and actual conditions, Fig. 15-18 illustrates some of the assumptions and concepts on which the determination of the average-roughness value is based.

In contrast, a few typical conditions that occur on technical surfaces are shown in the diagram in Fig. 15-19. These examples illustrate the limitations of average roughness as a representative value for functionally significant surface conditions. Because of these limitations of the single-average value, the inspection of critical surfaces is often carried out with the aid of stylus type instruments by producing chart tracings, also termed graphs, as the magnified portrayal of the traced surface.

The representation of an inspected surface element by means of a chart tracing, however, requires interpretation for providing any specific type of information. In addition to the basic variables of all recorded tracings, namely, the different scales of vertical and horizontal magnifications, there are also other variables, which are related to characteristics of the applied methods and instruments. Examples of these types of variable factors are:

a. The datum for the translational movement of the pickup;

b. The degree of fidelity of representation, as determined by the relationship between the speed of response of the instrument system and the speed of the trace (scanning) movement; and

c. The type of electronic filtering used, if any, in processing the signals supplied by the pickup.

The existence of these variables was pointed out in the preceding discussions; at this point, their effect on the produced chart tracing and on the representation of the surface conditions will be considered.

For this purpose, Table 15-6 reviews several currently used systems of portraying, graphically, the traced surface element, together with various methods applied for the assessment of the resulting chart tracings.

Only a few of the listed methods are defined in some foreign standards, and none in the current American Standard. However, the valuable information such tracings can supply on certain critical surface characteristics explains the expanding application of surface-texture inspection by means of

**TABLE 15-6.   SURFACE-TEXTURE ANALYSIS BY CHART RECORDING—
EXAMPLES OF CHART TYPES AND ASPECTS OF ASSESSMENT**

CATEGORY	TYPE	DIAGRAM OF CHART	DISCUSSION
CHART TRACING TYPES	Average roughness level		The level of an essentially straight pen line on the chart is the graphical representation of the pointer position on the average roughness meter. Digressions of the line, if any, indicate variations of the average roughness value.
	Actual roughness; short cycle asperities		The chart line is the highly magnified representation of the actual stylus excursions which occurred while tracing the surface during a period corresponding to that of the stylus traverse over the selected cutoff length.
	Actual waviness; long cycle departures		The chart line represents the deviations of the traced surface element from a reference line, displayed with the exclusion of closely spaced digressions, which are eliminated by means of wave filters.
	Total surface digressions; short and long cycle combined		The actual profile as it appears in relation to a reference line representing the ideal contour of the surface, recorded without signal processing and by guiding the stylus path with an extraneous datum.
SYSTEMS OF ASSESSMENT	Peak-to-valley roughness		The maximum separation between the highest point (peak) and lowest point (valley) of the chart line displaying short cycle asperities, is considered the representative roughness value h of the surface.
	Total profile, including roughness and waviness		Commonly assessed by determining two values: (1) waviness, expressing the amplitude of the roughness mean line; and (2) roughness, the maximum distance between the bounding lines of the tracing.
	Qualified total— by referencing from an envelope line		A line parallel with the path described by the center of a circle rolling over the specimen surface (actually a large radius skid) establishes a datum which rests on the exposed peaks of the profile. The area between the datum and the actual contour line A is the index of roughness.
	Qualified total— by selective elimination of excursions	$$\sum P = \frac{L}{20} \qquad \sum V = \frac{L}{10}$$	Protruding peaks and valleys are disregarded in a manner such as to keep the aggregate horizontal length of the stubs within specified limits (e.g., 5% for peaks and 10% for valleys). The thus resulting bandwidth w is the representative value.

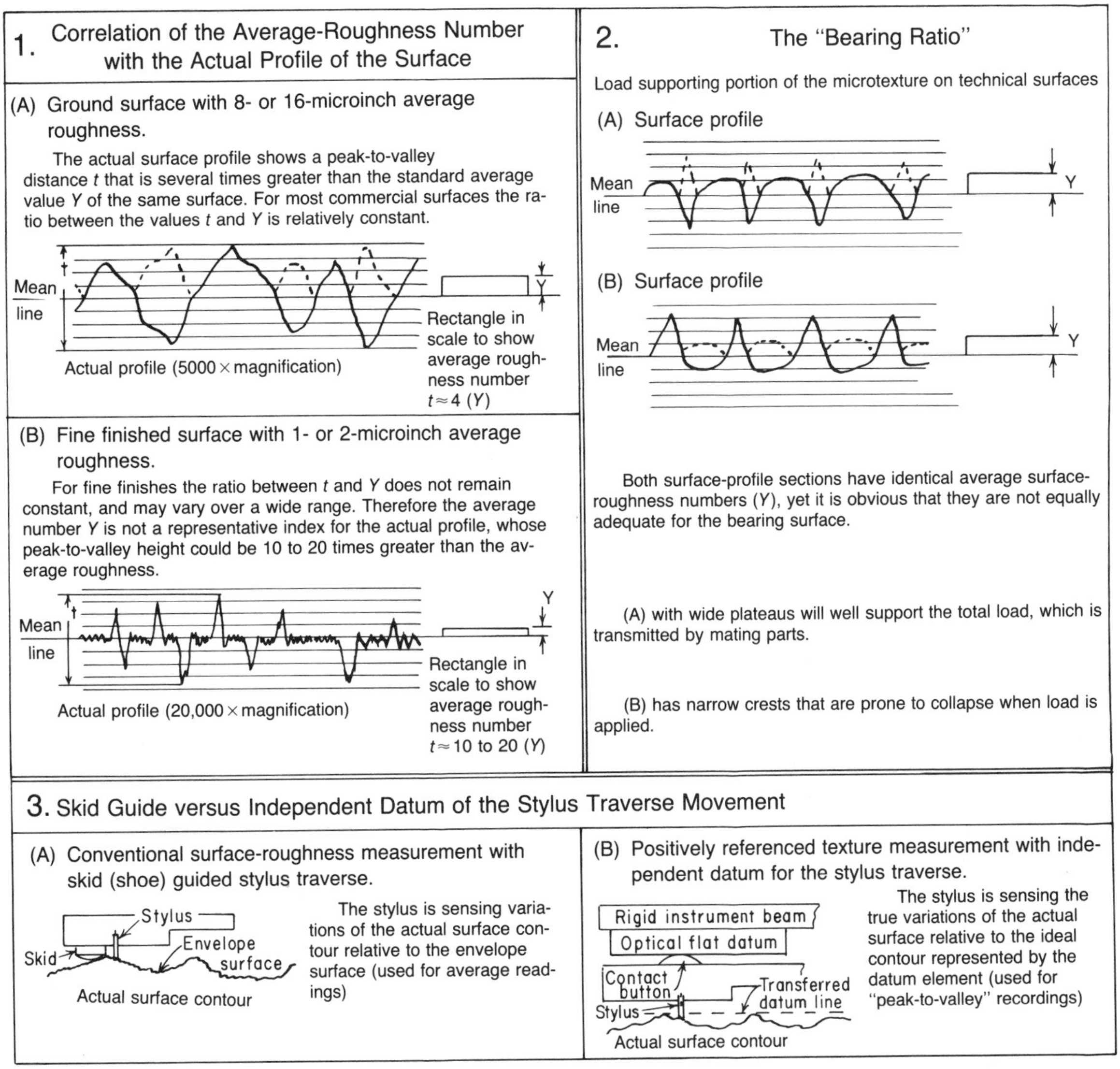

1. Correlation of the Average-Roughness Number with the Actual Profile of the Surface

(A) Ground surface with 8- or 16-microinch average roughness.

The actual surface profile shows a peak-to-valley distance t that is several times greater than the standard average value Y of the same surface. For most commercial surfaces the ratio between the values t and Y is relatively constant.

Mean line

Actual profile (5000 × magnification)

Rectangle in scale to show average roughness number $t \approx 4$ (Y)

(B) Fine finished surface with 1- or 2-microinch average roughness.

For fine finishes the ratio between t and Y does not remain constant, and may vary over a wide range. Therefore the average number Y is not a representative index for the actual profile, whose peak-to-valley height could be 10 to 20 times greater than the average roughness.

Mean line

Actual profile (20,000 × magnification)

Rectangle in scale to show average roughness number $t \approx 10$ to 20 (Y)

2. The "Bearing Ratio"

Load supporting portion of the microtexture on technical surfaces

(A) Surface profile

Mean line

Y

(B) Surface profile

Mean line

Y

Both surface-profile sections have identical average surface-roughness numbers (Y), yet it is obvious that they are not equally adequate for the bearing surface.

(A) with wide plateaus will well support the total load, which is transmitted by mating parts.

(B) has narrow crests that are prone to collapse when load is applied.

3. Skid Guide versus Independent Datum of the Stylus Traverse Movement

(A) Conventional surface-roughness measurement with skid (shoe) guided stylus traverse.

Stylus
Skid
Envelope surface
Actual surface contour

The stylus is sensing variations of the actual surface contour relative to the envelope surface (used for average readings)

(B) Positively referenced texture measurement with independent datum for the stylus traverse.

Rigid instrument beam
Optical flat datum
Contact button
Stylus
Transferred datum line
Actual surface contour

The stylus is sensing the true variations of the actual surface relative to the ideal contour represented by the datum element (used for "peak-to-valley" recordings)

Fig. 15-19. Diagrams showing certain types of limitations of the stylus type, average-roughness measuring method.
(1) The relationship between the actual digressions from the theoretical profile and the indicated average values.
(2) A functionally important condition of the surface texture that is not revealed by average-roughness indications.
(3) The selective assessment of surface variations by skid-guided tracing as compared to the datum-guided, surface-texture tracing method.

chart tracings. The survey in Table 15-6 should be helpful in the selection of a suitable system for any particular set of conditions and functional requirements.

SURFACE-TEXTURE INSPECTION BY AREA SAMPLING

The term *area sampling* is used here to distinguish the methods of surface-texture inspection to be discussed in the following, from those that examine an element of the surface by scanning it with a contacting stylus. This latter system, commonly known as the stylus method, may be considered for the purpose of this comparison as *surface-element sampling*.

The distinction between the two general systems of surface-texture measurement may be better visualized by comparing the basic concepts on which each of these systems rests.

The surface-element sampling method is based on the assumption that a bounding line of a judiciously selected cross-section through the specimen is sufficiently representative of the pertinent surface. Actually, the cross-sectioning is imaginary, a tracing of the bounding line only being effected by a stylus. However, the traverse and the excursions of the stylus are confined to said imaginary plane, and are assessed by referencing from a line contained in that plane.

The area sampling method considers the surface in relation to a two-dimensional reference element, commonly a plane, which is coincident with the ideal form of the inspected area. While conceptually superior, this method has limitations in the implementation techniques, which usually constitute a compromise between conflicting objectives, such as:

1. Comprehensive surveying versus assessment of the conditions by one or a few dimensions, for example, viewing through a microscope versus the stylus method;

2. The extent of the area of observation versus the degree of resolution, for example, low and high magnification; and

3. Direct appraisal versus conversion of actual conditions into measurable magnitudes, for example, regular and light interference microscopes.

It follows that there is no single system of surface-texture measurement that may be considered generally superior to all others. The choice should be based on these characteristics and parameters of the surface that must primarily be determined, in combination with several other considerations related to the functional purpose of the part, as well as the technical feasibility of the measurements.

Most of the commonly used methods of surface-texture measurement by area sampling, together with the processes based on stylus type instruments, were listed in Table 15-1. In the following additional details are reviewed for a few characteristic methods of the area sampling category, extending over a wide gamut of sensitivity levels, from shop to laboratory instruments.

Sensory Assessment of Surface Texture

Visual sensing by eye and tactual sensing with fingernail are direct methods of perception. These may be used for surface-roughness evaluation when guided by similarly sensed reference surfaces, whereby the process becomes a comparison type measurement.

In order to assign to the thus inspected surfaces numerical values that are meaningful in the light of current practices, such numbers should preferably express equivalent arithmetical average (AA) values.

However, it is very difficult, and often not feasible at all to recognize and to measure on surfaces having different patterns the correlations with the dimensional conditions on which the standard's index numbers are based. Whereas the average-roughness value of several inspected surfaces may be essentially identical, the perceivable impression they convey can differ substantially. Such differences usually arise from the manufacturing process by which any particular surface was produced.

The applied production process is usually known in plant practice, or is easy to recognize, and that information can help in bridging the lack of direct correlation between sensory impressions and the concepts of the standard. The actual links in establishing such correlations are reference surfaces that have patterns similar to the object being inspected and, at the same time, have assigned AA values as well.

Figure 15-20 shows some typical surface-roughness specimens for visual and tactile comparison with surfaces under inspection. On the specimens, various degrees of roughness are shown, originating from different machining processes, each producing a distinctive pattern. The individual specimen fields are marked to indicate the applied machining process and to state the equivalent AA roughness values. Other specimens, containing different samples associated with particular processes and producing specific AA roughness values, are also available.

For assisting the visual comparison of the workpiece with the specimen, a magnifying viewer can be used. That device has its own light source and a holder for the retainment of the specimen in the appropriate position. In the split-vision field of view of

GAR Div./Heli-Coil Corp.

Fig. 15-20. Surface-roughness reference specimens for visual and tactile comparison.

the magnifier, both the workpiece and the pertinent specimen section can be observed concurrently.

Surface-roughness specimens made for sensory comparison should never be used for calibrating stylus type surface-roughness measuring instruments.

Surface-Texture Examination by Light Interference

The development of light-interference bands on a reflective surface illuminated by out-of-phase light beams is explained by the wave theory of light. In practical applications a single source of light is used, whose directed and collimated rays are split and made to follow two different paths before striking the object surface. By varying the length of one of the optical paths, the phase shift needed to produce distinctly separated interference bands, often termed fringes, can be adjusted.

According to the wave theory of light, the fringe pattern, consisting of dark and bright bands, is alternating at a rate, which is the product of two factors: the length of the light waves and the difference in the lengths of the paths over which the component beams have to travel. By using monochromatic light of known wavelength, one of the factors generating the fringe pattern will be exactly defined.

The physical phenomenon of light interference is utilized in dimensional metrology for different purposes, the major applications being the following:

a. *Length measurement.* This should be considered the first in importance because our present system of dimensional measurements is based on units originating from this application of light interference. In length measurements the correlation between constant wavelength associated with a specific source of light, and the distance separating the two wavefronts is utilized for determining the linear distance between two concurrently observed surfaces.

b. *Form trueness measurements,* by displaying as fringe pattern variations the lack of parallelism between a reference surface and the specimen surface. Because of the technical difficulties of producing reference surfaces that are geometrically true in the microinch order, this application of light interference is limited to a few basically regular forms, such as flat, cylindrical and spherical.

c. *Surface-texture measurements.* This application of light interference is discussed more specifically in the following. Assuming a surface area of essentially regular form, either flat or symmetrically

receding from a tangent plane, such as a cylinder or a sphere, the reflection of out-of-phase light rays from that surface produce a fringe pattern consisting of parallel bright and dark bands at equal distances (in the case of a flat surface) or with regularly decreasing separations, whose rate corresponds to that at which the observed surface area is receding from the tangent reference plane.

The texture of a surface, that is, its departures from ideal smoothness, alters the distance between the reference plane and any particular portion of the specimen surface. Such distance variations affect the fringe pattern originating from reflection of that particular area of the surface. Consequently, digressions of the fringe contour are representative of the height variations of the observed surface, which can be measured in fringe units or fractions thereof. The width of the fringe is equal to one half of the wavelength of the light that produces it, either by direct emission or through a monochromatic filter. As an example, a thallium bulb emits monochromatic light with a 21.2-microinch wavelength, therefore, one fringe represents a height separation of 10.6 microinch. Because fringe pattern variations can be evaluated with a discrimination of better than $\frac{1}{10}$ part of a bandwidth, the sensitivity and accuracy of the light-interference method of surface-texture measurement can be considered to be in the microinch order.

A simplified way of demonstrating the working principles of light interference for surface-texture analysis is shown in Fig. 15-21. It depicts the fringes that appear on the specimen surface as originating from parallel bright and dark planes, which are inclined and half a wavelength apart. The fringes will be deflected by variations of the surface from the ideal smoothness. The amount of deflections, di-

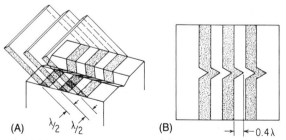

Fig. 15-21. Diagrams illustrating the principles of surface-texture examination by light-interference microscopy.
(A) A grooved plate intercepted by dark and bright planes.
(B) As seen in the microscope.
(λ = the wavelength of the monochromatic light used.)

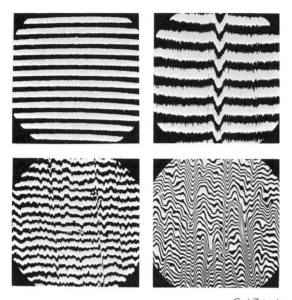

Carl Zeiss, Inc.

Fig. 15-22. Interference micrographs of flat surfaces with different degrees of surface roughness and flaws.

rectly measurable in fringe spacing units, can be expressed as a microinch value.

Figure 15-22 shows a microphotograph of an essentially flat surface with maximum peak-to-valley roughness on the order of 8 microinches as this condition appears in the field of view of an interference microscope. Figure 15-23 illustrates the interference image of a bearing ball of excellent general geometry and with a surface roughness on the 4-microinch peak-to-valley order, yet with two exceptions: (a) there are a few isolated digressions in excess of the

Carl Zeiss, Inc.

Fig. 15-23. Interference micrograph of a ball with excellent sphericity, as revealed by the parallelism of the circular interference fringes.

general roughness, and (b) the presence of a scratch is displayed by the defection of the fringes by about two bandwidths, equivalent to about 20 microinches. This picture shows clearly the gradual narrowing of the spacing between the substantially parallel fringes, as the effect of the receding spherical surface in relation to an imaginary tangent plane.

There are several optical systems by which the light beam can be split and directed to follow different paths for the purpose of generating light-interference bands when reconverged on the object's surface. One of these systems is shown diagramatically in Fig. 15-24. The interference microscope that operates by this system is illustrated in Fig. 15-25. This particular instrument is equipped with a turret containing three different objectives to provide 80 $\times$, 200 $\times$ or 480 $\times$ magnification with an 8 $\times$ eyepiece. The latter is interchangeable, thus offering an even wider range of different magnifications. The illustration shows the microscope equipped with a camera; this is an optional accessory, particularly useful for the detailed analysis of the interference image observed through the eyepiece.

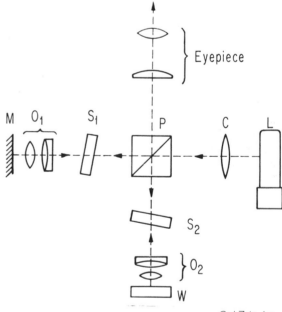

Carl Zeiss, Inc.

Fig. 15-24. Diagram of the operating principles of a modified Michelson type, interference microscope. Light-source (thallium vapor lamp) L; objective C; and beam-splitting prism P. One part of the light passes through the plane plate S_2, and objective O_2, to workpiece W, and hence back to P, where the two beams are superimposed upon each other and are reflected into the eyepiece.

Carl Zeiss, Inc.

Fig. 15-25. Light-interference microscope for the examination of surface texture, operating by the principles of the diagram in Figure 15-24.

Carl Zeiss, Inc.

Fig. 15-26. (Top) Interference micrograph of a generally flat surface, interrupted by a deep groove in the center of the observed area. The abrupt change in level causes the compression of the fringes beyond the limit of resolution.

Fig. 15-27. (Bottom) The specimen area shown in Figure 15-26, as it appears when prepared by immersion technique. A replica made from the surface is immersed into a medium with high refractive index, the value of which is considered as a factor when computing the depth of the now distinctly appearing groove.

The total depth-measuring range of this particular instrument is from 1.2 microinches to about 80 microinches, although its preferred use is the examination of surfaces in the higher order of finish.

Various other types of instruments are also available for surface-texture inspection by light interference. Several of these instruments are of simpler design, offering a single, usually lower, magnification, although the objectives and/or eyepieces may be interchangeable. Although less power of magnification reduces the discriminating capacity, it permits measurements to greater depths, that is, on rougher surfaces, and the field of view will also be increased.

A further increase in depth penetration by light-interference microscopy can be accomplished with the immersion method. This consists of observing the specimen surface through a medium with a high refractive index, such as oil, into which the object, or its replica is being immersed. With the immersion method, the depth of penetration of light-interference microscopy can be increased to about 0.001 inch. Figures 15-26 and 15-27 illustrate the advantages of the immersion technique when inspecting surfaces on which the departures of certain features exceed the discriminating capacity of regular light-interference microscopes.

While monochromatic light is preferred to obtain a reliable scale in the form of known fringe separation, most interference microscopes offer the alternative use of white light, as well, for producing a single dark fringe, often termed zero fringe. This latter is easier to follow in the assessment of the interference image when variations of the object surface are causing fringe shifts that amount to several bandwidths.

Multiple-beam interference microscopes are sometimes preferred for the examination of very fine surfaces where the clear definition of the fringe boundaries offered by the multiple-beam system is required. In this system, the mutually interfering light beams are forced to pass the interference space several times, resulting in particularly sharp fringe lines, permitting measurements in the submicroinch range. The multiple-beam system, however, requires focusing very close to the objective, which usually limits its application to flat surfaces of excellent geometry, unless replicas of the test piece surface are used.

In conclusion, it can be stated that the interference microscopy, although subjected to certain practical limitations, offers many technical advantages for the analysis and measurement of surface texture. A few characteristic properties are mentioned in the following, to permit an evaluation of the suitability of the interference method for the surface examination of specific workpieces.

The following are the *limitations* of interference microscopy:

a. The depth range that can be covered by direct measurements excludes its use for rough surfaces, unless an immersion technique is applied;

b. It is adaptable only to external surfaces, which are accessible by the microscope, and of substantially regular form, unless replicas are made and inspected; and

c. The assessment of surface conditions, as presented by the interference image, is not directly correlated with the average-roughness values specified by current standards; however, good correlations exist with the stylus type, peak-to-valley measurements.

The following are the *advantages* of interference microscopy:

a. It is an area sampling method, which can be visualized as simultaneously presenting a series of closely spaced cross-sections of the inspected area;

b. The direction of these imaginary cross-sectional planes, as well as their spacing, can be selected to provide the most informative fringe pattern;

c. It offers its own standard of measurement of the highest accuracy and never requires recalibration;

d. Being a noncontact method, it is nondestructive and applicable to materials that, because of their softness, could not be inspected by the stylus method; and

e. Even minute variations of the surface that defeat faithful scanning with a stylus of finite dimensions can be clearly shown.

16.
Screw Thread Gaging and Measurement

GENERAL CONSIDERATIONS

Screw threads are important elements of mechanical design with wide and varied applications, particularly for controlled translational motion and for fasteners providing disengageable connections.

Screw threads are composite design elements that, distinguished from, for example, a cylindrical shaft or hole, are defined by specific form and dimensional parameters, several of these being considered as principal and the others as complementary. The principal parameters determine the basic functional properties of the screw thread, such as whether it can be assembled with a mating part, whereas the complementary parameters affect the service performance of the threaded machine components.

Threaded components operate in conjunction with a mating member, which is also a threaded machine element, such as a nut fitting a screw. Occasionally, in particular mechanical designs, the mating member may contain only skeleton elements of the complete screw thread deemed sufficient to support and guide the fastening or translating functions of the completely threaded member. In other designs the mating members of the threaded couple are not in direct contact with each other, but operate by means of intermediate transfer elements, such as in the case of ball screws.

The basic shapes of threaded mechanical elements are those of a cylindrical shaft and a cylindrical hole, except in the case of taper threads, such as used for pipes and fittings. Consequently, screw threads are produced equally on external and internal surfaces, a characteristic that affects both their production and measurement.

As in all other mechanical elements, the actual sizes of screw threads on manufactured parts are not exactly identical to the pertinent design sizes. Such deviations may be within acceptable— the *toleranced*—limits or exceed the applicable tolerances. Components with out-of-tolerance screw thread dimensions are considered defective products.

The accuracy of screw threads is mandatory for assuring several significant functional characteristics of threaded mechanical components, such as the dependable assembly of threaded mating components, also termed assemble-ability; the interchangeability of the corresponding threaded parts; the consistent proportional relationship between the imparted rotational and the resulting translational movements; and the mechanical strength of the threaded connection, which is dependent on the proper distribution of stress over most of the designed contact area.

The objective of screw thread measurement is to determine whether the different parameters of the thread agree, within the toleranced limits, with the pertinent design dimensions. Because of the composite design of screw threads, comprising angular dimensions and contour forms in addition to linear dimensions, some of the latter being defined by reference points that are not directly accessible, the measurement of screw threads requires its own particular methods and instruments. These processes and means of screw thread measurement usually differ for external and internal threads.

The extent of the required screw thread measurement varies according to the number and types of parameters that it involves: it may be limited to a single significant dimension or it may be applied to most or all of the parameters by which a critical screw thread is defined.

The methods and instruments also vary depending on the required accuracy of the screw thread, expressed as the applicable tolerance grade. The accuracy of the measurement must be, as a rule, several times higher than the allowed tolerance of the inspected part's dimension.

It is also obvious that different methods and instruments may be required depending upon the size category of the threaded component and of the screw thread itself. Finally, the adequacy of the method and instrument depends on the number of threaded parts to be inspected, differing substantially when a single or a few pieces are measured in a gage laboratory and when production measurements are made. The latter, when only a limited number of parts or small lots are involved, are usually carried out manually; however, when a large number of threaded components are inspected in a continuous process, automated equipment may be required.

In view of the special methods and means of measurement that the general design of screw threads involve, compounded by the large variety of parameters that may be specified as important and assigned to different tolerance grades, the measurement of screw threads has developed into a distinct branch of dimensional measurement, warranting its discussion in a separate chapter.

BASIC CONCEPTS OF SCREW THREAD DESIGN

Before discussion the actual techniques of their measurement, some of the basic concepts of screw thread design are reviewed.

The *basic overall form* of a screw thread is the helical groove adjoining and mutually bounding the helical ridge on the surface of a generally cylindrical shaft or of a hole of corresponding shape. Most threads consist of a single helical groove and ridge, although multiple or multistart threads, with more than a single pair of grooves and ridges, equally spaced and following parallel helical paths, are also used.

The *profile of the screw thread,* when viewed in the axial cross-section of the threaded part, is most commonly a symmetrical Vee-shape, although screw threads with other cross-sectional shapes, and also some with nonsymmetrical profiles, are used for particular purposes.

In the majority of screw thread types the symmetrical Vee-shape of both the groove and the ridge

have a 60 degree included angle. That basic shape results in a sharp crest on the top of the ridge and a sharp root at the bottom of the groove. Such thread forms, however, are seldom specified and, for manufacturing reasons due to tool wear, are never produced perfectly. For replacing the uncertain effect of tool wear with more consistently producible form specifications, and also for such functional reasons as increased strength of the threaded part and interference-free assembly with the mating component, the specifications of most screw thread types define the thread form as a Vee with truncated (flattened) crest and root. Again, for manufacturing reasons, those specified truncations are not actually surfaces with a continuous straight line contour joining the adjacent sides of the Vee-shape in a manner forming sharp obtuse angles, but have a slightly curved profile. The remaining side portions of the groove–ridge boundary are designated as the thread flanks and represent those elements of the screw thread form that actually perform the desired functions of the thread.

Standard screw threads. Different types of screw threads are used throughout the industry; various types were designed to satisfy particular functional requirements, and others were introduced prior to the development of the present standards. For the purpose of this introduction to screw thread measurement, the most widely used system, the American National Standard Unified Screw Thread, is reviewed. The basic design form of that thread system is shown in Fig. 16-1 as a means of visualizing some of the parameters that will be discussed subsequently.

The drawings in this figure indicate the location of an important reference line in the design of screw threads, that of the *pitch line.* This line is the boundary of an imaginary cylinder, coaxial with the screw thread; on the surface of that cylinder, the widths of the grooves and of the ridges are theoretically equal. (In the case of a taper thread that cylinder is replaced by an imaginary cone.) This conceptual line is an important reference element of screw thread measurement because it represents the location where the pitch and the pitch diameter are measured.

The standard system of screw threads comprises, in addition to a wide range of sizes, different groups or classes distinguished by the pitch of the thread (the number of grooves and ridges along a specific length) and the sizes affecting the fit conditions of the assembled threaded parts.

Thread series. The discussed thread system comprises several series, which are groups of diameter–

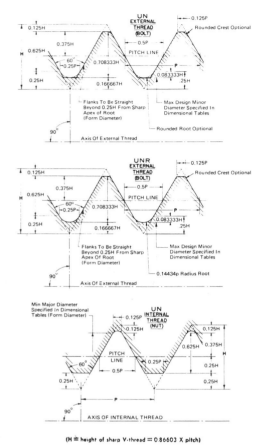

Fig. 16-1. Standard design form of the Unified external and internal threads. (The UNR external thread form has a design with rounded root, a detail that is optional only for the UN external threads.)

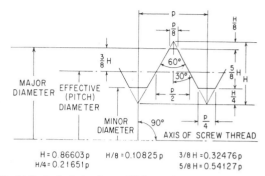

Fig. 16-2. Basic design form of ISO metric screw threads with 60 degrees included thread angle.

mating thread parts. The fit of the assembly also can be regulated by designs specifying different combinations of classes for the mating components.

Metric screw threads. The gradual transition to metric measurements in the United States makes it necessary to consider the metric screw threads as objects of current and future screw thread measurements.

That particular system of metric screw threads that is the equivalent of the American National Standard Unified Screw Thread, having a triangular thread profile with 60 degree included angle, is the ISO Metric Screw Thread, which has its dimensioning specified in millimeters. That dimensioning also applies to the pitch that, in metric threads, is specified directly in millimeters, or decimal fractions thereof, in distinction to the American system that states the number of threads per inch. The design forms of ISO metric threads, internal and external, are shown in Fig. 16-2.

The authority for the standardization of metric screw threads is the ISO (International Standards Organization). There is also a British Standard for ISO metric screw threads. (Comprehensive excerpts of this Standard are also contained in the latest edition of *Machinery's Handbook.*[1])

Several American Standards have been developed to define metric threads and their inspection. These standards are the following:

1. ANSI/ASME B1.3M–1986, Screw Thread Gaging Systems for Dimensional Acceptability—Inch and Metric Screw Threads.

2. ANSI/ASME B1.16M–1984, Gages and Gaging for Metric M Screw Threads.

pitch combinations, distinguished from each other by the number of threads per inch applied to a series of specific diameters. There are three groups with graded pitches, with the number of threads per inch changing as a function of the major diameter; these are the coarse-thread, the fine-thread and the extra-fine-thread series. In addition, there are eight other series with constant pitches in which the number of threads per inch (ranging from 4 to 32) remains constant, regardless of the major thread diameter.

Thread classes. The American National Standard ANSI B1.1-1989 specifies three classes each for external threads (designated as 1A, 2A, and 3A) and for internal threads (1B, 2B, and 3B). These thread classes are distinguished from each other by their amounts of tolerance and allowance, factors that affect both the gaging and fit of the threaded components and also control the ease of assembly of the

---

[1]*Machinery's Handbook*, 24th rev. ed., Industrial Press, New York, 1992.

3. ANSI/ASME B1.BM–1982 (R1987), Metric Screw Threads for Commercial Mechanical Fasteners—Boundary Profile Defined.

4. ANSI/ASME B1.19M–1984, Gages for Metric Screw Threads for Commercial Mechanical Fasteners—Boundary Profile Defined.

5. ANSI/ASME B1.21M–1978, Metric Screw Threads—MJ Profile.

6. ANSI/ASME B1.22M–1985, Gages and Gaging for MJ'' Series Metric Screw Threads.

Because all the dimensions of metric threads are specified in millimeters, the methods of screw thread measurement to be discussed subsequently, when applied to metric threads, must be implemented with metric instruments and metric auxiliary elements, such as thread measuring wires.

THE DIMENSIONING OF SCREW THREADS

The five most important dimensions of screw threads, which are also those generally measured for determining the nominal size and the dimensional conditions of a threaded component, are the following:

1. *Pitch* is the distance, measured parallel to the part axis, between corresponding points of adjacent thread flanks. The *lead* of the thread designates the distance by which one of two mating parts is moving axially when rotated by a full turn, while the other part is stationary. The pitch and the lead are identical in single-start threads. The term pitch is also used, incorrectly, to designate the number of threads per axial inch. (In Fig. 16-1 the pitch is designated by the letter *p*.)

2. *Pitch diameter* on a straight thread is the diameter of the pitch cylinder, which is an imaginary body that, in axial cross-section, is bounded by the pitch line (see Fig. 16-1). In a theoretically perfect thread, the pitch line, parallel to the part axis, is located where the widths of the thread ridge and the thread groove are equal. In taper threads that reference body is the pitch cone.

3. The *flank angle* expresses the incline of the thread flank (side) with respect to a reference plane perpendicular to the axis of the thread. In symmetrical threads the flank angle is one-half of the thread angle, hence the alternate term half-angle of thread ($60°/2 = 30°$, in Fig. 16-2).

4. The *major diameter* of a straight thread is the diameter of an imaginary cylinder that bounds the crests of an external thread or the roots of an internal thread.

5. The *minor diameter* of a straight thread is the diameter of an imaginary cylinder that bounds the roots of an external thread and the crests of an internal thread.

In addition to these five basic dimensions, there are several other parameters that define the geometrical elements of screw threads. Some of these parameters are interrelated with the basic dimensions listed above, and deviations in the actual size of any of these elements are reflected in the measured and the functional conditions of the basic screw thread dimensions, particularly the pitch diameter.

Conversely, size deviations of some design elements, such as lead and flank angle, can be detected, although not dependably identified, by their effect on the measured pitch diameter. In specific cases, however, such as in the inspection of thread gages or tools for controlling the setups of threading machines or for identifying sources of composite errors, several of the following thread dimensions are measured individually:

1. The *flank* is the side of the thread that, in an axial plane, is theoretically a straight line, inclined by the flank angle with respect to a plane perpendicular to the axis. Because the flank is the functional element of the screw thread that carries the load, its shape and length are significant quality factors.

2. The *crest* joins the flanks on opposite sides of a thread ridge and is the surface farthest from the axis of the thread. The crest would be a sharp line in the case of intersecting flanks; however, the design of most thread types specifies a truncated crest that, when viewed in an axial plane, is a straight line parallel to the axis of the thread.

3. The *root* joins the flanks of adjacent thread ridges and is the surface located closest to the thread axis. The root also is generally truncated, resulting in a joining surface farther away from the axis than the theoretical intersection of the flanks.

4. *Truncations* of crest and root are the radial distances between the actual joining surfaces and the theoretical intersections of the flanks. The amounts of the truncations are generally checked by their effect on the major and minor diameters, respectively. In practice the truncations do not join the adjacent

flanks by forming sharp angles, but because of tool wear, are rounded; a condition that is permissible as long as the rounded corners do not interfere with the required material limits and therefore do not cause interference with the mating threaded component.

5. The *helix* is the geometric term for the path of the pitch and lead around the axis of the thread. It is measured by the *helix angle,* which is the incline of the thread with respect to the axis, or by the *lead angle,* which is the complement of the helix angle. The lack of consistency of these angles along and around the thread results in helix variations, or helical path deviations, a potentially harmful condition commonly termed "drunken thread."

SIZE LIMITS OF SCREW THREAD ELEMENTS

In order to assure the intended functional properties of threaded components, such as easy assembly, unconditional interchangeability, or appropriate strength, the dimensions of the screw threads must be held within specific limits. These limits are established in the Standards for different classes in a manner to provide the allowances (if clearance between the mating members is intended) and a tolerance range. The diameters and related parameters involved in specifying the limits of screw thread sizes are shown in the diagrams of Fig. 16-3. These dimensioning diagrams, selected as examples, apply to screw thread classes designed to provide a minimum of allowance (Fig. 16-3A), or to avoid interference (Fig. 16-3B) even in the case where both members of the threaded couple, such as a bolt and a nut, have sizes at their maximum material limits. This condition is present when the pitch diameter of the bolt is at the maximum and that of the nut is at the minimum size limit.

As is clear from Fig. 16-3, the size limits of the thread are expressed as diameter values and these dimensions can be measured individually on a single thread. The size of the pitch diameter reflects the actual material limit of the thread, as long as the thread form is symmetrical, its flank angles correspond to the basic size, for example, 30 degrees for the Unified Screw Thread, and the thread has a perfect lead. Of course, such a screw thread with perfect angles and lead is a theoretical concept only; deviations will always occur, although their magnitudes can be held within the toleranced limits.

Deviations of the thread angle from the basic size cause the flank boundary to depart from its theoretical line and thus increase the material limits over

the design value. Increased material limits reduce the designed allowance and may result in unintended interference with the mating part. The effect of thread angle deviation on the pitch diameter for the Unified American form can be expressed by the formula

$$\delta E = 1.25p \tan \delta\alpha$$

where

δE = increase of pitch diameter resulting from the deviation of the flank angle;
p = pitch (equals $1/n$);
α = basic flank angle of the thread;
$\delta\alpha$ = deviation in the flank angle.

The deviations of the lead, by shifting the actual location of the thread flanks in an axial direction from the plane of the mating thread with basic lead size, also affect the effective size of the pitch diameter. The effect of such an error on the pitch diameter, designated as the diameter equivalent of lead deviation, can be expressed for the Unified American thread by the formula:

$$\delta E = 1.7321 \, \delta p,$$

where

δE = increase of pitch diameter caused by lead deviation;
δp = the maximum pitch deviation between any two of the threads engaged.

Because the effect of lead deviation absorbed by the pitch diameter, regardless of whether that deviation extends or reduces the actual lead, increases the effective material limits of the thread, the resulting diameter equivalent must be added to the individually measured pitch diameter for external threads and subtracted for internal threads.

The effect of lead variations on the effective pitch diameter is illustrated, in a deliberately exaggerated manner, in the diagrams of Fig. 16-4. A similar effect results from the flank angle deviation, both types of error increasing the effective material limits of the threaded component.

If only one of these variable dimensions, the flank angle or the lead, deviates significantly and the pitch diameter of the screw thread has nearly perfect size, then either substantial flank angle or lead deviations might occur without their effect exceeding the pitch diameter limits. Although the occurrence of such a lopsided composite error is rather rare, the Standard limits the permissible extent of deviations of individually measurable dimensions, such as flank angle

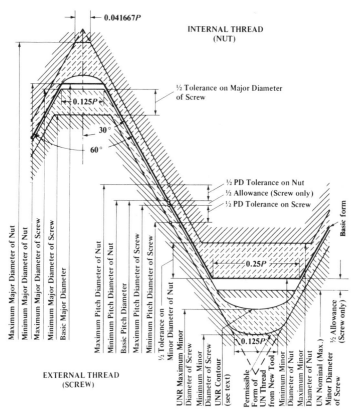

Fig. 16-3A. Relationship of the dimensional limits of tolerances, allowance (clearance space along the flanks) and crest clearance for Unified internal and external threads, Classes 1A, 2A, 1B and 2B.

and lead, which are absorbed by the pitch diameter. The diameter equivalent of the deviations of any of these dimensions is limited to a maximum of one half of the pertinent pitch diameter tolerance.

Although deviations of the flank angle and lead always increase the material limits, other variations such as out-of-roundness or taper may have either positive or negative diameter equivalents.

The length of engagement—its effect on the functional screw thread size. The toleranced dimensions

of the pitch diameter are established to include, in addition to the variations of the pitch diameter of a single thread (one pitch), the effect (the diameter equivalent) of flank angle, lead and pitch variations, as well as of other imperfections.

The assemble-ability of the threaded members is assured only when their pitch diameters along the entire length of engagement are within the specified maximum material limits. However, the pitch diameter of a single thread will not reveal the angle and lead deviations that are absorbed by the pitch diameters that participate in the engagement of the threaded members. Therefore, the dependable inspection of threaded components requires the simultaneous measurement of all the pitch diameters within the length of engagement.

THE PURPOSE OF SCREW THREAD MEASUREMENTS

Screw thread measurements may serve different purposes, which then define the parameters to be inspected and determine the applicable processes to be used. The following are examples of different pur-

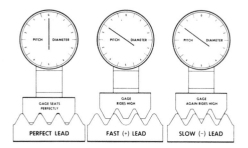

Standard Pressed Steel Co.

Fig. 16-4. Diagrams pointing out the effect of lead variations on the pitch diameter.

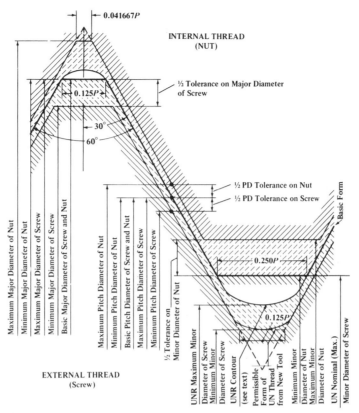

Fig. 16-3B. Relationship of the dimensional limits of tolerances and crest clearance for Unified internal and external thread, Classes 3A and 3B. Note that there is no allowance (clearance).

poses for screw thread measuring, as well as of the areas they serve in the field of industrial production. Screw thread measurements are used:

1. To verify by single-element measurements whether all significant parameters of the screw thread are within the specified limits—such as needed for thread gages.

2. To ascertain by gaging that the functionally significant dimensions are within the required limit values—such as for product acceptances tests.

3. To assure that the operating elements are of the proper form and size—such as for threading tool inspection.

4. To control the production process of threaded components by checking the size conditions of the variable product dimensions—such as on threading machines, including thread cutting, rolling and grinding.

5. To measure the accuracy of specific thread dimensions that are critical for particular applications—such as the consistency of the helix of lead screws used for machine tools and instruments.

6. To determine the nominal size of a threaded part for use in assembly or for providing the proper mating part—such as identification measurements in shop practice.

As an indication of the different objectives that determine the appropriate measuring processes, Table 16-1 provides a concise overview of the common purposes of screw thread measurements and of the generally applied methods.

While the pertinent standards define the limits of form and size of the screw threads to be inspected, the following must be selected by the quality-control engineer: (a) the required measuring processes, (b) the applicable gaging methods, and (c) the suitable measuring instruments.

SCREW THREAD INSPECTION BY ATTRIBUTES AND VARIABLES

The majority of screw thread measurements, with regard to the inspection purposes they serve, may be assigned to one of the following categories:

TABLE 16-1.   EXAMPLES OF PURPOSE-RELATED SCREW THREAT MEASUREMENTS

PURPOSE	PROCESS CHARACTERISTICS	DIMENSIONS MEASURED	REMARKS
Thread Gage Measurement	Comprehensive single-element measurements.	Most or all of the toleranced thread dimensions.	Because of the tight tolerances specified for thread gages and setting masters, very sensitive metrological instruments and rigorously followed procedures are generally applied.
Acceptance Inspection of Threaded Components	Gaging by attributes.	Inspection of one or a few parameters, primarily of the effective pitch diameter.	Either solid or indicating type gages for checking the functional (virtual) pitch diameter are used; the inspection gages being set and verified with master gages.
Threading Tool Inspection	Inspection of specific geometric characteristics.	Principally the complete thread contour and the pitch.	The thread contour elements and, in the case of multi-thread tools, the pitch are also measured for verifying the required accuracy of these parameters.
Manufacturing Process-Gaging	Gaging, preferably by indicating instruments, the variable product thread dimensions.	Commonly the single thread pitch diameter only.	Having verified earlier the correctness of tool form and of the machine setup, the variations of the product thread are generally limited to those of the pitch diameter.
Verification of Particular Function-Related Thread Dimensions	Analysis of the dimensional characteristics of the critical dimensions.	Typical examples are the helix accuracy of lead screws and of micrometer spindles.	The critical thread dimensions of screws serving translational movement, such as of machine tool lead screws, are the parameters of the helix, including the lead and the consistency of the lead angle.
Thread Identification Measurements	Processes using basic measuring tools generally provide adequate information.	Pitch (number of threads per inch) and the major diameter (external threads) or minor diameter (internal threads).	Applied particularly on repair jobs involving the replacement of threaded components whose specifications are unknown, but the measurement of a few characteristic dimensions permits type and size identification.

1. *Inspection by attributes* serves a functional purpose. The applied method and instruments are designed to determine whether the inspected thread is within specific size limits that are defined to assure a particular function, such as assembly with a mating part also having acceptable thread dimensions. Another dimensional condition generally checked in inspection by attributes is whether the size of the examined thread provides the specified minimum material limits, that is: (a) the external thread is not too small (LO size); and (b) the internal thread is not too large (HI size). The prime purpose

of requiring minimum material limits is to assure the proper strength of the thread or of the threaded connection.

Gaging by attributes is generally used for the acceptance inspection of threaded fasteners and other components, the most commonly gaged dimension being the functional or virtual diameter of the thread. The term *virtual thread diameter* designates the pitch diameter of an imaginary thread with perfect geometric dimensions that just envelopes the actual thread over a specified length of engagement and thus reflects the maximum material limits of the inspected thread (that is, the largest functional diameter of an external thread and the smallest functional diameter of an internal thread). The virtual diameter of the thread results from the combined effect of form and size deviations, the sum of their diameter equivalents being added to (for external threads) or subtracted from (for internal threads) the actual single pitch diameter.

2. *Inspection by variables* provides more specific information on the actual size of the measured thread element. It is used, for example: (a) to obtain detailed information on the dimensional conditions of thread gages and threading tools; (b) to collect data on the performance accuracy of threading equipment; (c) to analyze the origins of virtual diameter variations that reflect the compounded effect of deviations in various interrelated thread dimensions; and (d) to measure individually the actual size of several interrelated and independent dimensions, such as those needed in the metrology of thread gages.

SCREW THREAD GAGING AND MEASUREMENT

The *gaging of screw threads* is done to determine whether the inspected dimensions conform to the prescribed limits of size. The tools used for these inspection processes are, basically, the *solid screw thread gages*, which physically incorporate the limit sizes of specific dimensions by containing the inverse replica of the inspected product element.

These limit sizes represent the boundaries of the tolerance zones that separate the maximum and minimum material limits prescribed for a particular dimension. For acceptance of the product, gages that incorporate the maximum material limits of the inspected product element must permit assembly with the manufactured threaded component. Consequently, a thread ring gage at maximum size assembles with the corresponding external thread that

does not exceed the maximum size (effective diameter), or a thread plug gage at minimum size enters a corresponding threaded hole that has a size not less than the minimum limit.

Conversely, gages representing the minimum material limit of particular dimensions should not accept the manufactured product with a corresponding nominal thread. Assembly with a gage made to incorporate the minimum material limit of the inspected dimension reveals that the actual size of the pertinent part's dimension is at the boundary surface of less material than required as a minimum, that is, a plug is too small or a hole is too large.

These fundamental principles of gaging apply equally to a simple dimension, such as those inspected with plain cylindrical ring or plug gages, as well as to complex dimensions, such as those inspected on screw threads. The gages for screw threads differ, however, by requiring particular designs. In thread gages the boundary surfaces representing the maximum material limits may incorporate the compounded effect of all the contributing individual dimensions, as long as these effects can be absorbed by the actually measured dimension. Screw thread gages designed for checking the minimum material limit usually verify the actual size of one or a few individual dimensions that are deemed critical for the intended service of the threaded component.

Solid gages, both fixed and adjustable types, are designed to make the simple determination of whether the inspected dimension is within the prescribed limits of size, but not to supply information on such details as

How close is the actual size of the inspected dimension to the upper or lower limit of the specified size?

In the case of out-of-tolerance parts, by how much does the inspected dimension deviate from the limit size?

Such information may be obtained readily by means of differential gaging, which is carried out with *indicating screw thread gages*. Indicating screw thread gages have contact elements with contours similar to those of the solid gages. These contact elements, however, are mounted on gage members that permit their movement in the direction of the expected product size deviations, that is, generally in the radial direction with respect to the thread axis of the inspected component.

When the indicating screw thread gage is set to indicate zero at a specific size—generally the design size of the inspected dimension—deviations of the actual size of the inspected screw thread dimension cause deflections of the gage indicator, displaying in dimensional units the amount of that deviation. The indicated size deviation may be within the toleranced limits, usually shown on the face of the gage indicator by adjustable tolerance pointers, or outside those limits by the indicated amount.

In many inspection operations the indicating type screw thread gages are the preferred instruments for reasons that will be reviewed later.

The survey in Table 16-2 lists the most commonly used types of screw thread gages, together with the inspection purpose they serve. Each type of gage is manufactured for various screw thread types, sizes and classes.

The *measurement of screw thread elements* is the process by which the actual size of a particular dimension is determined or, in the case of interrelated angles or forms, the deviations from the design size or contour are measured in basic dimensional units.

Whereas gaging is the satisfactory method for the majority of screw thread inspections, the measurement of screw thread elements provides more detailed and potentially more accurate information on the actual dimensional conditions of the screw thread. Measurement is also the only means by which the actual sizes of various screw thread elements can be determined individually. It is by measurements that the sources of compound deviations are analyzed, and the tools of inspection by gaging— the screw thread gages—are also verified by the measurement of their individual elements.

Various methods and instruments of measurement, both of which are applicable to screw threads, as well as those designed exclusively for screw thread measurement, will be discussed later.

SOLID SCREW THREAD GAGES

Solid screw thread gages that physically incorporate the essential functional dimension of the mating part are, in principle, the best suited means for checking an important property of the product screw thread, namely, its assemble-ability.

A solid screw thread gage that corresponds to the product thread size at maximum material limit provides the assurance that any product assembled with that gage also assembles with the mating part having acceptable thread dimensions.

As a reminder:

A screw with thread at maximum material limit has the largest acceptable pitch diameter;

A nut with thread at the maximum material limit has the smallest acceptable pitch diameter.

For gaging purposes the maximum material limit is specified by the pitch diameter of a screw thread assumed to have perfect lead and flank angles, and also to be free of any other deviations that could affect the pitch diameter. The screw thread gages are manufactured to satisfy, within very close tolerances, these theoretical concepts of a screw thread with perfect form.

The actual tolerances of the product pitch diameter are tighter than those of the screw thread gage, because deviations of the lead, flank angles and other dimensions cannot be avoided in the manufacture of threaded parts. Several of these deviations are absorbed by the functional pitch diameter, in most cases increasing the material limits of the virtual contour of the screw thread (increasing the functional pitch diameter of the screw and decreasing the functional pitch diameter of the nut). Consequently, it is the compounded effect of several variables over a specific length of engagement that is determined by inspection with a solid screw thread gage.

In addition to their capacity for the dependable checking of assemble-ability, solid screw thread gages also offer other advantages, such as

The determination of that attribute, the assemble-ability, without reliance on human judgment and free of interpretational errors;

The incorporation of the critical material limit in a tamper-proof and drift-free manner.

Solid screw thread gages are subjected to wear, of course, and for that reason a systematic gage surveillance, with periodic verification of the size conditions, is mandatory. Screw thread gages are generally manufactured from materials and with processes that assure extended wear life. Thread plug gages are often made of high speed steel with a nitrided surface, and the thread ring gages, which are made of special tool steel, are designed to permit wear compensation by adjustment (see Figs. 3-19 and 3-20).

The full-thread profile type screw thread gages, both solid plug and adjustable ring types, are manufactured for different standard screw thread systems and classes, their general design resembling the corresponding size groups of the plain cylindrical gages, which are discussed in Chapter 3. Examples of a few typical screw thread plug gages are shown in Fig. 16-5.

The full-thread type screw thread gages are particularly suitable for use as *GO gages*,

TABLE 16-2.   SCREW THREAD DIMENSIONS INSPECTED BY GAGING

	GAGED DIMENSIONS	INSTRUMENTS USED	
		SOLID GAGES	INDICATING GAGES

For Product Internal Threads	Pitch diameter
	GO (acceptable) thread size
	(Gaging the functional thread pitch diameter)
		Solid thread plug (GO) gage	Indicating gage with
			    two segments or
			    three rolls
			having full thread profile

	HI (not acceptable) thread size
	Solid gage for functional diameter material limit
	Indicating gages also for single-pitch minimum material limit
		Solid thread plug (HI) gage
		(with gaging element shorter than that of the GO gage)
			Indicating gage with
			    two segments or
			    three rolls
			either with full thread profile or with cone and vee form single-pitch gaging elements.

	Minor Diameter for minimum and maximum material limits	Plain plug gages (GO and NOT GO)	Indicating internal gages with extended cylindrical contact elements (for larger diameters only).

	Single Pitch Diameter
	(also used for discovering and appraising the effects of variations in several other thread dimensions)
		Not used	Indicating gages with cone and vee contact elements. Gages with
			    two arms,
			    two segments,
			    three segments or rolls.

For Product External Threads	Pitch Diameter
	GO (acceptable) thread size
	(Gaging the functional thread pitch diameter)
	LO (not acceptable) thread size
	(preferably by gaging a single pitch diameter)
		Thread ring gage (adjustable)
		Roll thread snap gage
		Thread ring gage
		Roll thread snap gage
		  multi-ribbed elements
		  single pitch elements
			Indicating multi-ribbed gages with
			    two contact segments
			    two contact rolls
			    three contact rolls.
			Indicating single pitch diameter gages, with
			    cone and vee contact elements,
			    three contact rolls.

	Major Diameter GO and NOT GO for maximum and minimum material limits	Plain cylindrical rings Plain snap gages	Indicating snap gages with flat contact elements.
	Minor (Root) Diameter GO and NOT GO	Snap gages with acute angle contact elements (cone and vee or rolls)	Indicating snap gages with root diameter type (acute angle) rolls

	Single-Pitch Diameter
	(also as a means of discovering
	flank angle deviations,
	lead deviations,
	out-of-roundness, and taper)
		Not used	Indicating
			    snap gages,
			    two-roll gages, and
			    three-roll gages, with elements
			      contacting the thread
			      (a) along the flank or
			      (b) coincident with the pitch
			        line.

	Pitch Diameter Limit Sizes for setting the thread ring gages and indicating gages	GO thread setting plug gage LO thread setting plug gage (usually made with a truncated and a full thread portion)	Not used.

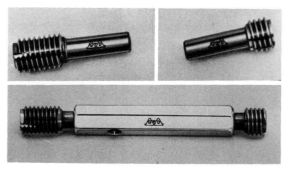

TRW Greenfield Tap & Die Div.

Fig. 16-5. Screw thread plug gages for checking the limit sizes of internal product threads, by determining whether the size is within the maximum material limits (GO gage, upper left), yet without having less than the minimum material required for adequate strength (HI gage, upper right). Both elements may be combined in a double end gage (bottom).

The thread plug gages made to the lowest limit of size,

The thread ring gages made to the highest limit of size,

specified for the functional pitch diameter; both types of gage serve for inspecting the acceptable maximum material condition of the product screw thread.

Solid screw thread gages with full-thread profile are also made in sizes specified for checking the NOT GO condition of threaded components,

HI thread plug gages at the highest limit of size,

LO thread ring gages at the lowest limit of size;

both types of gage serve for inspecting the pitch diameter equivalent of the acceptable minimum material condition of the product screw thread.

For checking the NOT GO condition of the product screw thread the full-thread type solid gages are not the best means. The assembly of such gages with the product thread, indicating inadequate material condition that calls for rejection, might be prevented by deviations of various dimensions that are absorbed by the functional pitch diameter.

As an example, excessive lead and/or flank angle deviations of the screw thread can increase significantly the material limit of the functional pitch diameter, thus preventing assembly with the NOT GO gage. In such cases, owing to the increased material limit of the functional pitch diameter, thread dimensions inspected by a full-thread type NOT GO gage

appear to be adequate and the product is accepted, although its single-pitch diameter, the dimension controlling the operational strength of the screw thread, may be less than the least material limit (the single-pitch diameter may be HI on an internal thread or LO on an external thread). Consequently, the NOT GO inspection with gages checking the minimum material limit of a single thread, or of a reduced number of threads, is the more dependable method for determining whether the actual size of the critical screw thread dimensions assure the required operational strength.

In order to check dependably the limit size of product internal threads in the direction of minimum material, the HI (NOT GO) thread plug gages have their flank contacts reduced by truncation, making them less sensitive to the effects of flank angle variations, and the interference of lead deviations is reduced by making the length of the gaging element shorter than that of the GO gage.

Consequently, the HI plug gages detect the less than minimum material limit of the product thread as long as that condition affects the pitch diameter size checked with the gaging element in contact with the flanks. Less than minimum material limit is not discovered, however, when the gage element meets sufficient product flank surface to prevent the gage from entering the product thread, irrespective of the area of the actual contact.

Thus even a substantial reduction of the flank contact area, resulting in insufficient load bearing surface, with potentially deleterious effects on the service performance of the threaded connection, may remain undetected. The incidence of such hidden defects is potentially more frequent in the case of internal threads, the flank forms and angles of which cannot be checked by inspection methods, such as optical examination, which are readily applicable to external threads.

Another gaging system has been developed to discover most of the potentially harmful form and angle defects of internal threads that, when checked with conventional HI plug gages, would pass inspection. That system, designated as *double NOT-GO gaging*, uses two HI plug gages with selectively truncated operating surfaces, one with elements contacting the product thread flank near the root, the other near the crest.

The operating principles of such double NOT GO plug gages, in comparison with the regular type HI plug gages, are pointed out in the diagrams of Fig. 16-6. A double NOT GO thread plug gage is also shown.

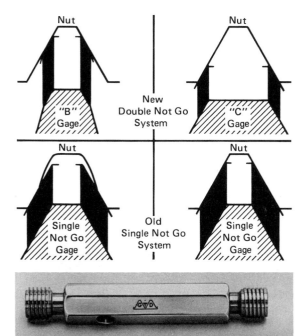

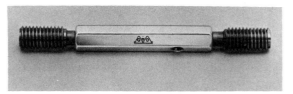

TRW Greenfield Tap & Die Div.

Fig. 16-7. Truncated type thread setting plug gage in double-end design with GO and LO elements for the setting and verification of adjustable thread ring gages. Each gage element has a full threaded and a truncated portion with identical pitch diameter. The full-form portion checks the pitch diameter and the major diameter clearance; the truncated portion (in front) checks the pitch diameter and the flank angle wear, which affects the flank angle.

TRW Greenfield Tap & Die Div.

Fig. 16-6. The double NOT GO gaging system, using sequentially two NOT GO (HI) plug gages, one of them (the B gage) contacting the flanks near the major diameter, and the other (the C gage) contacting the flanks near the minor diameter of the inspected internal thread, thus ascertaining the presence of the minimum flank boundaries needed for adequate shear strength. (Top) Diagrams comparing the gaging contact principles of the double HI gage (upper row) and the single HI gage (lower row). (Bottom) Double NOT GO gage with different flank profiles at each end.

Thread Setting Plug Gages

Thread setting plug gages are used primarily for the setting and checking of adjustable thread ring gages, but are also used for several other applications, and are made in design sizes in accordance with the intended use.

Setting gages for the control of thread ring gages are manufactured for both the GO and the LO (NOT GO) ring gages with pitch diameter and major diameter sizes corresponding to the material limits specified in the standards for different classes of screw threads.

The setting plug gages for thread ring gages, both the GO and the LO type, are designed to have about one half length truncated major diameters; the remaining length of the plug gage thread has full form, and the pitch diameter is constant along the entire threaded length of the gaging member. The full-form portion checks the pitch diameter and the major diameter clearance, and the truncated portion checks the pitch diameter and the flank angle wear. Thus a single gage checks several important thread dimensions of the thread ring gage.

Figure 16-7 illustrates an example of a partially truncated type thread setting plug gage.

Other uses of thread setting plug gages are for master gages for the adjustment, setting and checking of adjustable solid gages (roll thread snap gages for external threads) and for indicating gages of all types used for external threads. Thread setting plug gages for these applications, also referred to as thread setting masters, have screw thread dimensions that correspond to the basic size of the thread that is to be inspected.

The accuracy requirements of thread setting plug gages are more rigorous than those for working gages used for the direct inspection of product threads. The thread setting plug gages are usually made to W standard tolerances, which are applied unilaterally, allowing for the pitch diameter minus tolerance on the GO and plus tolerance on the LO gages. Typical values of the W tolerances used for thread setting plug gages are:

Nominal Size	Tolerances (inch)			
	Pitch Diameter	Major Diameter	Lead	Half Angle ($\pm$)
$\frac{1}{4}''$–20	0.0001	0.0005	0.00015	0° 8′
$\frac{1}{2}''$–13	0.00015	0.0006	0.0002	0° 6′
1″– 8	0.0002	0.0007	0.00025	0° 5′

Roll Thread Snap Gages

Screw thread inspection with snap gages having threaded rolls for gaging elements, as shown in Fig. 16-8, offers several advantages over the use of the regular thread ring gages for product external threads.

1. Two pairs of gaging elements for inspecting the GO and the NOT GO (LO) size of the product thread are combined in a single gage, permitting these two operations of the screw thread inspection to be performed in a single pass.

2. The roll gaging members, which rotate when contacting the product thread, are less prone to wear than the ring gages, which have sliding contact; the extended operating surface of the rolls further prolongs the wear life of these gages.

3. The inner LO member rolls contain only two ribs for checking the significant single-pitch minimum diameter of the product thread.

4. The length of the multiribbed GO rolls can be selected to be approximately equal to the applicational length of engagement of the product thread.

5. The same gage can be used for checking right-hand and left-hand threads of the same nominal size.

6. For contacting the work the roll thread gage is approached in a radial direction, in distinction to the axial assembly direction needed for thread ring gages; this approach direction permits one to use the

TRW Greenfield Tap & Die Div.

Fig. 16-8. Roll thread snap gage with annular profile on the rotatably mounted gaging rolls, which are adapted for checking both right- and left-hand product threads. The outer rolls have multiple ribs for checking the functional GO condition, the inner rolls have only two ribs for determining the LO condition by gaging the minimum pitch diameter.

gage on work that is mounted between centers on a thread cutting machine.

The roll gaging elements have annular ribs and are mounted on eccentric pins for adjustment. Full-form thread setting plug gages with GO and LO members, and threads made in the size of the selected class, are used as masters for setting and checking the roll thread snap gages.

INDICATING SCREW THREAD GAGES

Indicating screw thread gages, using a dial or other method of display, show in standard units of measurement the amount by which the actual size of the gaged object dimension differs from the previously set basic size. The applied method is usually referred to as *differential gaging*, in distinction to the simple "yes or no" information obtained with solid gages when used for checking a limit size.

Differential gaging also determines the compliance with a limit size by having adjustable tolerance markers on the dial, occasionally complemented with signal lights or other means. The supplemental information on the size condition of the gaged dimension, indicating in small dimensional increments the actual amount of deviation from the basic size, whether inside or outside the specified limits, is quite valuable in many applications.

Screw thread dimensions checked with solid gages for determining the limit size condition can be inspected, as an alternate method, with indicating gages, because essentially equal contact elements are used in both systems of gages. The contact elements of indicating screw thread gages must be composed of, at least, two parts, one of these being installed on a movable member that is connected with the indicator instrument. The range of movement of this gage member permits free displacement in accordance with the deviations, in either the plus or minus direction, of the actual size from the set basic size, together with an additional retraction movement providing adequate clearance for the insertion of the object, or of the gage, into the gaging position.

In screw thread gaging the required distance of retraction is somewhat greater than that needed for plain surface objects because of the screw thread's particular form, on the surface of which the boundaries of the dimensions to be gaged are generally contained inside (or, for internal threads, outside) of a plain envelope cylinder. Apart from the design of the gaging elements, that need for additional retraction is the major difference between the indicating screw thread gages and other types of indicating gages used for plain cylindrical objects, which were discussed comprehensively in Chapters 5, 6 and 7.

Indicating screw thread gages are well adapted to the inspection of several significant product thread dimensions, including both those of single elements and of the composite effect of several elements that produce a functional size. The listing in Table 16-3 presents a comprehensive review of the more commonly inspected screw thread parameters, the size conditions of which can be determined with indicating gages, the applicable types also being pointed out in the table. Some of the gaging processes listed in the table are frequently used methods of screw thread inspection; others are mentioned to indicate the potential size information supplied by differential gaging, not excluding, however, still further methods of size determination, both actual and indicative, that can be obtained by means of differential screw thread gaging.

Indicating screw thread gages, similarly to other instruments of differential gaging, require setting by means of masters, in this case master thread gages, which are manufactured with very close tolerances to correspond to the nominal form and the selected basic size.

Differential screw thread gaging with indicating instruments offers many *potential advantages* over the use of solid gages, whether of the fixed limit type or of adjustable design. Some of these advantages are enumerated in the following; one should keep in mind, however, that not every screw thread inspection process can benefit from all or even most of the listed characteristics of indicating type screw thread gages.

1. *Indicating capability,* that is, pointing out in increments of standard units of measurement the deviation of the gaged dimension from its basic value to which the instrument is set. Numerical values of deviations from the basic dimension are required in various applications, such as

   a. For the adjustments of machines and tools used for producing products with screws threads;

   b. For supplying data needed in statistical quality control or for control charts serving other purposes.

2. *Speed of measurement,* both for supplying incremental values of deviations and for determining compliance with the specified limit size, particularly when using dials with tolerance-position hands or gages equipped with limit-position signaling lights.

The indicating screw thread gages generally are designed to establish gaging contact by a radial approach, in distinction to the axial assembly needed for inspection with solid screw thread gages.

3. *Selectivity* by equipping the indicating gage with the appropriate contact elements for the gaging of particular screw thread parameters, such as functional pitch diameter, single-pitch diameter and root diameter; or, by using consecutively two indicating gages with different types of contact elements, the gaging can comprise straightness, flank angle and helix conditions (angle and consistency).

4. *Dependability.* Indicating screw thread gages are designed to establish and to maintain contact with the product thread by means of spring pressure, which, once set, remains uniform, thus eliminating reliance on the "feel" of the inspector. Also the pointer position on the dial of the gage provides dependable information on the size condition of the gaged dimension, eliminating reliance on human judgment.

5. *Versatility* of the indicating screw thread gages, which are usually designed to be adaptable to a wide range of thread types, sizes and classes, simply by setting the gage to the required basic size and exchanging the measuring contacts (needed only when the pitch or form is different).

6. *Economy of use,* resulting from both the versatility of the instrument and the easy compensation for wear, unless the wear affects the form of the contact elements; in this case only the relatively inexpensive members have to be replaced.

7. *Adaptability to automation,* by generating signals either at limit positions or in convertible analog values, which can trigger such actions as segregation, rejection or even feedback signals for machine adjustment.

INDICATING SCREW THREAD GAGING INSTRUMENTS

The instruments to be discussed were selected as examples to comprise a wide range of systems, each designed for a particular field of application, implementing many of the potential advantages of screw thread inspection with indicating gages. The instruments discussed, although the products of prominent specialized manufacturers, generally are not unique; indicating screw thread gages of other designs and makes are also available for most of the gaging capacities offered by the selected examples.

**TABLE 16-3.   INSPECTION OF SCREW THREAD VARIABLES
WITH INDICATING GAGES—1**

	INSPECTED DIMENSION	GAGE AND CONTACT ELEMENTS	FIG. REF.	REMARKS
Product — External — Threads	Functional (Virtual) Diameter— Maximum Material Size	Full thread type gaging elements, engaging over several threads and contacting the flanks over ⅝H of the Unified series threads Element types: (a) Two thread segments (b) Two contact rolls (c) Three contact rolls (d) Snap gage with two pairs of contact rolls	 16-12 16-9, 16-11 16-13 to 16-17 16-9, 16-10	Gaging conditions approximate those produced by thread ring gages, with the added advantage resulting from differential gaging (a) Engage product thread over minimum of 50% of the periphery. (b) & (c) Assure extended wear life of the gaging elements. (d) Similar in function to the thread roll snap gage, with the addition of indicating capability.
	Pitch Diameter— Minimum Material Size	(a) Indicating thread snap gage with cone and vee. (b) Three thread rolls with ribs contacting on one pitch length ⅜H of the flank (LO gaging). (c) Three single-rib thread rolls with radii corresponding to that of the "best wire size" for contacting the flanks on the pitch line.	16-9, 16-10 16-13 to 16-17 16-16	(a) & (b) For gaging minimum material size of a single thread, but determination may be affected by flank angle deviations. (c) Minimum material size of a single pitch determined unaffected by flank angle deviations.
	Major Diameter (Crest Truncation)	Indicating snap gage with flat anvils.	5-24 5-25	May be used to discover insufficient (or excessive) crest truncation; also taper and ovality when these deviations affect the major diameter.
	Minor Diameter (Root Clearance)	(a) Indicating snap gage with acute angle cone and vee contact elements. (b) Thread roll type indicating gage with roll angles more acute and with smaller radii than those of the product thread.	16-9, 16-10 16-16	The purpose of these gagings is to determine whether adequate product thread root clearance is assured. (Inadequate root clearance is caused usually by wear of the threading tool, e.g., cutter, roll, or grinding wheel.)
	Flank Angle Deviations	Two indicating gages are used, each with single pitch thread rolls in the following designs: ribs contacting ⅜H of flanks, followed by gaging with ribs made to duplicate best wire size contacting on pitch line.	16-16	Differences in the indications obtained with these two types of contact elements (on gages set with the same master) reveal flank angle deviations and their approximate magnitude, but not whether the angle is wider or more acute than specified.

**TABLE 16-3.   INSPECTION OF SCREW THREAD VARIABLES
WITH INDICTING GAGES—2**

	INSPECTED DIMENSION	GAGE AND CONTACT ELEMENTS	FIG. REF	REMARKS
Product External Threads	Lead Deviation	Two consecutively used three-roll-type indicating gages with different sets of gaging elements: one with single-pitch contacting rolls; the other with full length thread rolls.	16-14	Successive contacts along the product thread with single-pitch elements will determine straightness. When satisfactory, then two consecutive gagings with the full thread elements, one engaging over the entire length, the other over the first full thread will indicate the diameter equivalent of lead deviation, if present.
	Helical Path Deviations (Drunkenness)	(a) Three-roll-type indicating thread gage with full length thread rolls. (b) Two-segment-type indicating thread gage, with one segment fixed, the other floating laterally.	16-14 16-13	Rotating the work while in contact with the functional diameter gaging elements might produce erratic indicator readings which reveal drunkenness condition (although may be caused by out-of-roundness too). Type (b) gage provides more distinct indications.
	Straightness deviations (Taper)	Three-roll-type indicating thread gage with single-rib rolls, which have radii duplicating best wire size.	16-14, 16-16	Gagings at various points along the product screw thread will reveal straightness deviations (taper) by the consistent change of indicator readings.
	Out of Roundness (a) Two-lobed (Oval) (b) Three-lobed	Indicating gages for functional diameter gaging; (a) with two contact elements (rolls or segments); (b) with three contact rolls, at 120° apart.	16-11 16-14 to 16-16	Gaging with both types of contact elements is needed for discovering the most common types of out-of-round condition. *Caution:* drunkenness can cause similar indications as out-of-roundness. In case of doubt, single-element measurement is advised.
Product Internal Threads	Functional (Virtual) Diameter— Maximum Material Size	Internal indicating gage with full thread gaging elements engaging several threads and contacting the flanks over ⅝H of the Unified series threads (a) Indicating plug gage with three thread segments; (b) Indicating internal gage with two thread segments; (c) Indicating internal gage with two or three thread rolls.	16-22, 16-25 16-23, 16-27 16-20, 16-25	For the gaging of internal thread diameters one of the gaging elements must be mounted on a retractable member to permit the insertion of the gaging elements. After releasing the retracting lever, spring force causes the movable member to bring the gaging element into contact with the product thread. Thread ring gage master is used for setting.

**TABLE 16-3.   INSPECTION OF SCREW THREAD VARIABLES
WITH INDICATING GAGES—3**

INSPECTED DIMENSION	GAGE AND CONTACT ELEMENTS	FIG. REF.	REMARKS
Pitch Diameter—Minimum Material Size	Indicating internal diameter gage with single-pitch contact elements of cone and vee form, respectively.	16-21, 16-25	Gaging elements designed to determine the largest (minimum material limit = LO) pitch diameter of a single thread or of several threads gaged individually.
Major Diameter (Root Clearance)	Indicating internal diameter gage with multiple or single thread form gaging elements having more acute thread form than the design size of the product thread.	16-25	For setting the indicating gage equipped for root clearance checking, a plain ring gage with corresponding bore size can be used.
Minor Diameter (Crest Truncation)	Indicating internal gage with plain cylinder segment gaging elements.	5-45	For setting the indicating gage used for internal thread truncation gaging, a plain ring gage is recommended, but any other rigid OD gage with flat parallel contacts may be used.
Flank Angle Deviations	Two indicating internal gages equipped with single pitch engaging elements one contacting ⅜H of the flank; the other with best wire size radius, contacting the pitch line.	16-25	Differential indications will reveal flank angle deviations. Seldom used method, except for large diameter internal threads for which roll type contact elements are applicable.
Lead Deviations / Helical Path Deviations (Drunkenness) / Straightness Deviation (Taper) / Out-of-Roundness	The principles of inspecting these conditions with indicating type internal gages are similar to those used for external product threads, however, their use is limited to large diameter internal threads.		Gaging these types of deviations of internal threads with the use of indicating gages, often requires skill and judgment for discovering and identifying the defective condition. The measurement of individual elements may be the preferable method of inspection.

(Internal Product Threads)

Indicating Thread Snap Gage (Fig. 16-9)

Indicating thread snap gages are used for inspecting the actual pitch diameter of any single pitch of the product thread by using the appropriate cone and Vee contact elements installed in the instrument spindles. The gaging elements made for the different pitch sizes have standardized shanks, are easily interchangeable and are dependably seated in the opposite ends of the gage spindles. One of the spindles is adjustable over a wide range for accommodating workpieces of different diameters. The opposite spindle, which is connected with the indicator, has spring actuation for producing a constant gaging force, and can be raised with a lever for the free positioning of the workpiece. The adjustable stop, which is attached to the frame, can be set to contact the outside diameter of the workpiece being inspected, thus assuring proper work-positioning for the gaging of consecutively inspected parts of equal nominal diameter. Setting plug thread gages are used

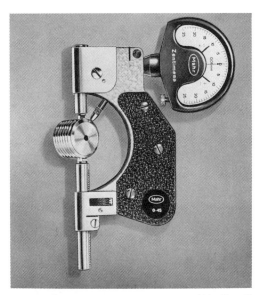

Mahr Gage Co. Inc.

Fig. 16-9. Indicating thread snap gage with interchangeable anvils, which are available for a wide range of thread forms and sizes. The lower spindle provides the coarse adjustment of the diameter, and the fine adjustment of the indicator produces the zero reading with the set-ring master. For introducing the work, the movable upper spindle is retracted by a lever (not visible), and the adjustable stop assures proper centering.

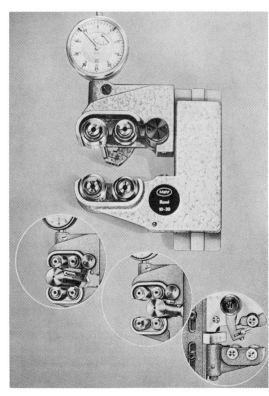

Mahr Gage Co. Inc.

Fig. 16-10. Indicating roller thread snap gage for GO and NOT GO checking of external threads. Both the full-thread GO rolls and the truncated single-thread contacting NOT GO rolls are interchangeable. Setting is with master plug gage or with gage blocks. The adjustable end stop is used for functional gaging and retracts for NOT GO checking. Insets (from left to right): GO gaging. NOT GO checking. Setting with gage block.

for guiding the setting of the gage, by means of the fine adjustment incorporated into the indicator, to the basic size of the product thread pitch diameter.

These portable thread snap gages are available in the following work diameter ranges: 0 to $1\frac{3}{4}$ inches; $1\frac{3}{4}$ to $3\frac{5}{8}$ inches; $3\frac{5}{8}$ to $5\frac{1}{2}$ inches; and $5\frac{1}{2}$ to $7\frac{1}{2}$ inches.

Similar indicating thread snap gages are also available for the inspection of taps with an odd number of flutes, specifically three- and five-flute taps. For such workpieces the gages have frames with a greater free space depth in order to permit the installation of a bridge type secondary spindle holder on the adjustable spindle of the gage. Two secondary spindles are retained at the two ends of the bridge type holder; these spindles are also individually adjustable and face each other in mutually inclined positions at 120° (for three-flute taps) or at 72° (for five-flute taps) apart.

Indicating Roll Thread Snap Gages (Fig. 16-10)

These gages are used for the rapid inspection of both the functional (effective) diameter and the single-pitch diameter of product thread within the nominal diameter ranges of $\frac{7}{16}$ to $1\frac{1}{16}$ inches and $1\frac{3}{16}$

to 2 inches listed as the individual capacities of two different models of this type of gage.

The front rolls have full thread for determining the GO condition (functional maximum size) and the inside rolls, with cone- and Vee-type single-pitch thread forms, serve the inspection of the LO condition (single-pitch thread diameter beneath the minimum size limit).

The distance between the roller holding brackets can be set by means of gage blocks placed on the parallel face buttons near the inside ends of the brackets. That method might be substituted for the setting with thread masters, particularly for repeat settings, after the required distance has been established by means of a thread plug master gage.

The positioning of the product to the proper depth and alignment in the functional (GO) gaging stage is facilitated by an adjustable end stop; for LO condition checking, that stop is retracted.

Indicating Thread Comparator with Full-Thread Contact Rolls (Fig. 16-11).

This bench type gage offers a wide diameter range capacity, accepting workpieces from 0.120 to 2 inches. The gage permits rapid setting to the functional pitch diameter of the product thread to be inspected by using the appropriate thread plug gage master. Full-thread type contact rolls are easily exchangeable on the shanks of the two brackets, the lower one of which is moved into a required position for coarse adjustment, where it stays fixed during the gaging. The upper bracket is double spring loaded and connected with the indicator, that instrument also providing the means for the fine setting. An adjustable stop pin assists the positioning of the workpiece at the proper level and alignment.

Indicating Thread Gages with Segment Type Gaging Elements (Fig. 16-12)

Indicating gages with full-thread type segments as gaging elements are used for the functional size inspection of screw threads, particularly in applications where the 50% minimum peripheral contact

The Johnson Gage Company

Fig. 16-12. Indicating thread gage with segment type contact elements, for checking the functional diameter of external threads. This portable gage is equipped with particularly long segments and is shown while being set with a master plug gage held, for convenience, in a simple stand.

with the product thread by the two opposite gaging elements is a desirable inspection condition.

Functional thread size comparators with segment elements are manufactured in different sizes, each accommodating a particular diameter group and used with segments corresponding to the nominal pitch of the inspected thread. The total nominal diameter size range, for which various models of the segment type thread comparator gages are available, extends from 0.060 to 20.000 inches. The indicating thread gages with segment elements are supplied as portable instruments or bench models. The latter type may be combined on a bench stand with a second indicating thread gage equipped with three single-pitch rolls for the consecutive inspection of (a) the functional size (maximum material limit) and (b) the single-pitch diameter (minimum material limit) of the product thread (Fig. 16-13).

Thread size comparator gages with segment type contact elements are available in other designs, such as in the system of the bench type internal thread gages with segments shown in Fig. 16-27. In these gages the work is staged with the axis in the vertical position, permitting the basic functional thread gage to be equipped with special attachments and supplemental indicators for checking the position of the thread functional diameter with respect to other surfaces of the workpiece. Such inspections are needed

Mahr Gage Co. Inc.

Fig. 16-11. Indicating thread comparator with full-thread rolls for rapidly checking the functional thread diameter (GO condition). The roll holders, one of which can be set for a wide range of work diameters, permit the easy interchangeability of the contact rolls.

for assuring the accuracy of particular geometric relationships between the thread and other surfaces when precise concentricity or perpendicularity (squareness) is specified. Indicating thread gages with segment elements used for external threads differ, however, from the later-illustrated internal thread gaging version in several respects, such as for loading of the work; the movable segment expands the space between the gaging elements.

Indicating Thread Gages with Three Contact Rolls (Figs. 16-13—16-15)

Indicating thread gages of this system are manufactured both as bench type instruments, up to 3 inches nominal thread diameter, and as portable models, covering the same nominal diameter range as the segment type portable gages previously discussed.

The contact rolls of these gages are supplied with both full threads for functional diameter size gaging and limited contact single-pitch profiles for inspecting the pitch diameter size of the thread. The full-thread type profile rolls contact the product thread flanks over the length specified for the functional thread size gaging, that is, $\frac{5}{8}$H for the Unified system threads. The single-pitch gaging contact rolls have profiles corresponding to cone and Vee, truncated to a reduced contact length of 0.09H, positioned half way along the flank height.

A special type of pitch diameter contact roll is designed with that 0.09H-long contact section in a location bounded by, but outside of, the pitch line of

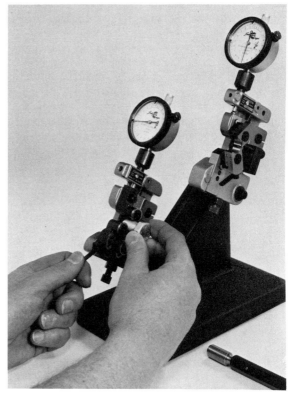

The Johnson Gage Company

Fig. 16-14. Indicating thread gages, each with three rolls, mounted on a common bench stand. The gage in the rear has multiple-thread functional diameter gaging rolls, and that in the front checks single-pitch diameter for LO size with single-thread contacting truncated profile rolls.

The Johnson Gage Company

Fig. 16-13. Indicating thread gages installed on a common bench stand, for inspecting both the functional (GO) and the single-pitch (LO) diameters of product external threads. The functional gage, in the rear, has segment elements for extended peripheral contact. The single-thread pitch diameter gage, used for LO inspection, has rolls with single-rib elements.

The Johnson Gage Company

Fig. 16-15. Two indicating thread gages for checking the GO and LO conditions with appropriate gaging rolls as in Fig. 6-14, but in portable design to be used on work held in stationary position. The pertinent setting plug gage is shown between the two instruments.

the product thread. Contact rolls with this type of truncation can be set with plain cylindrical gage plugs for threads having a helix less than one degree.

Efficient work thread inspection processes can be carried out by installing two of these gages on a common bench stand and checking consecutively the product functional thread diameter on the gage equipped with full profile rolls and then gaging the single-pitch diameter on the second instrument with single-pitch cone and Vee profile rolls.

Colt Industries—Pratt & Whitney

Fig. 16-16. Three-roll type thread comparator in bench-stand-mounted design, shown equipped with three full-thread rolls for functional thread diameter checking. The rolls are easily interchangeable with other types needed for different inspection processes.

Colt Industries—Pratt & Whitney

Fig. 16-17. Introducing a setting plug gage between the contact rolls of the three-roll gage shown in Fig. 16-16. By actuating the lever at the back of the gage, the front roll is retracted, permitting the specimen to be introduced in a radial direction, without causing wear of the rolls.

Three-Roll Type Thread Comparators (Fig. 16-16)

The three-roll type of indicating gage operates by thread gaging principles similar to those implemented by the instruments discussed in the preceding paragraphs; however, the design of the gage is different. Several frame sizes are available to cover a total range of 0.060 to 3.375 inches nominal thread diameter. The retraction of one of the three gaging rolls, in order to introduce the setting master or the threaded product into the gaging position, is carried out conveniently by depressing a lever in the rear of the instrument, as shown in Fig. 16-17. The instruments are supplied with a bench stand, but the gage can be removed from the stand and used for in-process inspection as a portable instrument in order to gage products mounted in threading machines. There is also a larger handheld model of these gages for thread diameters up to 6 inches.

Gage rolls of different lengths, approximating the commonly used lengths of engagement in the medium range of nominal thread diameters, are used as the gaging elements of these instruments. The gage rolls are supplied in four different types, each type interchangeable with others of the same length. These gage rolls, shown in the diagrams of Fig. 16-18, are designed to permit, in addition to the inspection of the thread diameters, a rather complete thread analysis, generally by examining the part consecutively on two gage frames equipped with different types of roll sets. Such inspections will determine, in terms of the equivalent diameter, deviations from the nominal conditions of straightness, flank angle, root clearance, lead angle and helix consistency or the lack of it, resulting in drunkenness.

The most significant size conditions, those of the functional (effective) and the single-pitch thread diameters, are determined by first inspecting the product thread with the multiple-rib gage rolls to check the maximum material limit (GO condition), and then with one of two types of single-rib gage rolls to check the minimum material limit (LO condition). For the LO condition checks, either the "cone and Vee" type or the "best wire size" type gage rolls may be used.

For preparing the instrument for the required inspection process, the set of gage rolls in the diameter—pitch combination corresponding to the product thread is mounted on the hardened pins of the gage. These pins are precisely aligned and located to assure engagement of the work by the rolls in positions 120 degrees apart. With the proper tightening of the

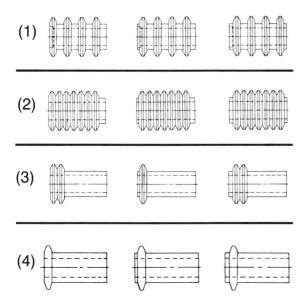

Colt Industries—Pratt & Whitney

Fig. 16-18. Gage roll types used for different inspection purposes on the three-roll indicating gage (thread comparator) shown in Fig. 16-16.
(1) Multiple-rib type, with alternate ribs removed, used for general-purpose functional inspection.
(2) Full multiple-rib type, used in nonfloating mounting for critical functional inspection.
(3) Single-pitch diameter checking rolls; the two outside rolls have two ribs; the one in middle has a single rib.
(4) Single-rib rolls with best size wire radius for accurate pitch diameter size gaging.

locking screws the gage rolls can be installed (a) freely turning and floating, (b) freely turning without lateral movement, or (c) locked. The locking of the gage rolls is applied for critical gaging operations; by turning the locked rolls occasionally, thus presenting different sections of their periphery to contact with the product thread, a rather uniform wear of the roll profile can be maintained, thereby extending the useful life of these gaging elements.

The described thread indicator gages are applicable to both right- and left-hand product threads by simply exchanging the positions of the gage rolls on the two lower pins.

INDICATING GAGES FOR PRODUCT INTERNAL THREADS

Indicating Internal Gage with Thread Form Contact Elements (Fig. 16-19)

The indicating internal gage with thread form contact elements is designed for the rapid differen-

Mahr Gage Co. Inc.

Fig. 16-19. Indicating internal gage adapted for the gaging of internal thread diameters when equipped with appropriate contact elements, as shown in the following figures. The instrument has the capacity for a wide range of diameters. When gaging internal thread diameters, thread ring masters are used for setting.

tial gaging of internal diameters or, by means of an integral reverse mechanism, for external diameters. One of the two arm-holding brackets can be moved along the hardened and ground column for the coarse adjustment of the diameter to be gaged. The smallest model of these gages has a work diameter range of 1 to 4 inches, but similar gages with capacities for larger diameter ranges are also available. The other arm-holding bracket of the gage is mounted on a high precision guide bushing, which transfers its movement without friction to the indicator.

Different type arms can be installed on the brackets. For internal thread gaging the arms can be equipped with the following:

(a) Multiple-rib thread form rolls for functional diameter gaging (see Fig. 16-20);

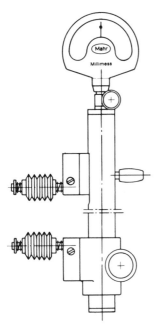

Fig. 16-20. Indicating internal gage, equipped with a high sensitivity type indicator, with multiple-rib contact rolls installed on the gaging arms for checking internal thread functional diameter.

(b) Cone and Vee gaging elements for checking the single-thread diameter of internal threads (see Fig. 16-21).

Different gaging elements are needed for various thread types and pitch sizes; however, these elements can be used for gaging product threads with corresponding forms regardless of their nominal diameter or whether they are right- or left-hand threads. Thread ring masters are used for the dependable setting of the gage to the basic pitch diameter of the product.

Indicating Plug Gages for Product Internal Threads (Fig. 16-22)

Indicating plug gages of this type when used for internal thread inspection can examine either the minor diameter or the functional pitch diameter of the product thread by installing the appropriate gaging elements.

The gaging element consists of a set of three segments, two fixed and the third, in a position between the fixed ones, attached to a sliding member of the instrument. The unobstructed edges of these seg-

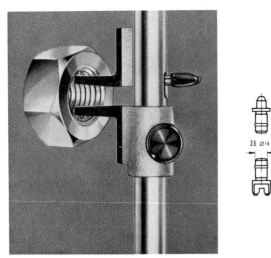

Mahr Gage Co., Inc.

Fig. 16-21. Use of the gage shown in Fig. 16-19 for checking the single-pitch diameter of an internal thread. Interchangeable spherical and Vee-contact elements, shown in outline, are used for different pitch ranges, for example, 48–32, 28–19, 18–14, 12–10 and 9–7 threads per inch, or equivalent metric sizes.

Mahr Gage Co. Inc.

Fig. 16-22. Indicating thread plug gage with three interchangeable thread-profile segments for checking internal thread functional diameter. Cylinder segment form gaging elements, used for smooth holes, can serve for the gaging of the minor diameter.

ments are the contact surfaces having the form corresponding to that of the work surface:

(a) Cylinder segments for contacting the envelope cylinder of the product thread minor diameter;

(b) Segments of a plug thread gage for contacting the product thread flanks over several pitches.

The sliding member of the instrument to which the centrally located gage segment is attached is spring loaded for producing constant gaging force, and its movement is transferred to the indicator. That movable segment can be retracted by lever action to permit the contact-free introduction of the gage segments into the product bore. By confining the contact between the gaging elements and the product to the actual gaging, the wear, which is common with fixed plug gages, is substantially reduced.

The gage bodies are made in five different sizes, covering a total diameter range of 0.080 to 4.80 inches; however, the individual segment sets have narrower ranges determined by the retraction movement of the slide member. For internal thread pitch diameter gaging, segments with pitch corresponding to that of the product must be used.

These portable gages are convenient to operate and permit rapid gaging, particularly for the inspection of parts in positions where portable instruments are needed.

Indicating Internal Thread Gages with Segments (Fig. 16-23)

The illustrated types of gages are designed as portable units, although stands are also available for mounting the gage when bench type instruments are better suited for the inspection process. A significant characteristic of these gages is the large arc over which the segments contact the product internal thread, as shown in Fig. 16-24. For functional size gaging, full-form segments are used, and the pitch diameter size is gaged with truncated cone and Vee-form segments.

Two-segment type internal thread gages are available in different models, covering a nominal diameter size range of 0.250 to 2.5 inches. Fixed, non-adjustable thread ring gages are recommended as masters for setting these gages.

Segment contact element type internal thread comparator gages for larger product diameters, specifically in models with a total range of 1.500 to 3.000 inches are manufactured with three contact segments and are also supplied with gaging elements for both functional size and pitch diameter size inspections.

The Johnson Gage Company

Fig. 16-23. Portable indicating gages for internal threads with segment gaging elements providing extended peripheral contact.
(Left) Gage for single-pitch diameter checking.
(Right) Gage for functional size inspection of an internal thread with one long length of engagement. Adjacent to the gages are the applicable gage setting rings.

The Johnson Gage Company

Fig. 16-24. Disposition of gaging elements inside a large diameter internal thread.
(Left) Three full-profile segment elements in contact with nearly the entire thread periphery.
(Right) Cone and Vee profile rolls for pitch diameter size gaging.

Indicating Internal Thread Gages with Three Rolls (Fig. 16-25)

Indicating internal thread comparator gages with three rolls as contact elements are manufactured in

The Johnson Gage Company

Fig. 16-25. Portable indicating gages with three contact rolls, designed for the inspection of internal threads above 3 inches in diameter.
(Left) Gage with full profile rolls for functional size gaging.
(Right) Gage with cone and Vee profile rolls for pitch diameter inspection.

many different models, the capacity of that series covering a nominal diameter range of 3.001 to 20 inches.

The contact rolls used for these gages are supplied for both functional size gaging with full-profile and pitch diameter size gaging with cone and Vee profile. The dimensional conditions determined by that type of indicating gage are shown in Fig. 16-26, which points out the imperfections whose diameter equivalents can be gaged with different combinations of these contact elements.

Special gage stands, holding the internal thread indicating gage and supplemented with an additional indicator instrument, are manufactured for measuring the functional product thread size, as well as the runout and concentricity of thread-related surfaces of the workpiece.

Comparators for Gaging Internal Thread Diameters and Thread-Related Geometric Conditions (Figs. 16-27 and 16-28)

The operating elements of these gages are pairs of interchangeable segments, each installed on one of the two principal members of the instrument. Ac-

cording to the objectives of the inspection, segment pairs to gage either one of two critical sizes are used, namely,

(a) The functional (effective) diameter of the thread—for determining the maximum material limit size, which controls the assemble-ability of the thread;

(b) The single-pitch diameter of the thread—for determining the minimum material limit size (HI limit), specified to assure the minimum design strength of the component thread.

The contact conditions that these two types of gaging elements establish with the product thread are shown in the diagrams of Fig. 16-29.

In order to practically eliminate the interference of flank angle deviations with the gaging of the minimum material limit of the pitch diameter, these single-pitch gaging elements are truncated for greatly reduced flank contact, thus simulating the conditions of three-wire measurement used on external threads.

The gaging segments used for these particular instruments have a "pilot diameter," which overrides the threaded section, thus protecting it during the insertion of the contact elements into gaging position inside the internal thread.

The instrument itself consists of two cast members. One member is stationary and is the stand of the bench type or the handle of the portable unit. The movable member is attached to the stationary part by means of steel spring reeds, a design assuring consistent alignment of these two principal instrument members. The movable member also carries the thumb lever for retracting this member during loading; by releasing the lever the gaging segment installed on this member is impelled by a constant force to establish contact with the surface of the product thread. Figure 16-30 illustrates, in a line diagram, the elements of the bench type internal thread gage; also shown is the way the internal thread gaging segments are mounted by means of locating studs and retaining screws. The indicator is installed on the stationary member in such a way that the sensing point of its spindle bears against the movable instrument member. The instrument also has an adjustable workrest for mounting the specimen part at the correct level, and an adjustable stop screw limiting the retraction of the movable gage member.

The gaging capacity of the described type internal thread comparator is from 3 to 5 inches. The bench instrument, equipped with the appropriate thread

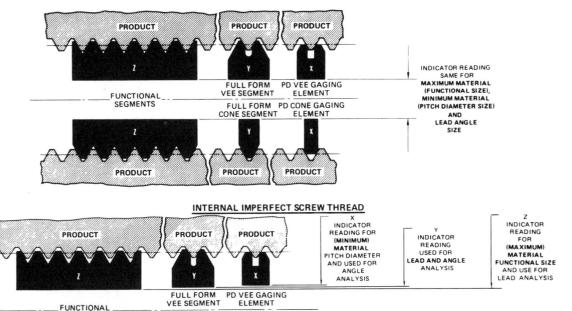

INTERNAL PERFECT SCREW THREAD

FUNCTIONAL SEGMENTS

FULL FORM VEE SEGMENT / PD VEE GAGING ELEMENT

FULL FORM CONE SEGMENT / PD CONE GAGING ELEMENT

INDICATOR READING SAME FOR MAXIMUM MATERIAL (FUNCTIONAL SIZE), MINIMUM MATERIAL (PITCH DIAMETER SIZE) AND LEAD ANGLE SIZE

INTERNAL IMPERFECT SCREW THREAD

X INDICATOR READING FOR (MINIMUM) MATERIAL PITCH DIAMETER AND USED FOR ANGLE ANALYSIS

Y INDICATOR READING USED FOR LEAD AND ANGLE ANALYSIS

Z INDICATOR READING FOR (MAXIMUM) MATERIAL FUNCTIONAL SIZE AND USE FOR LEAD ANALYSIS

FULL FORM VEE SEGMENT / PD VEE GAGING ELEMENT

FUNCTIONAL SEGMENTS

FULL FORM CONE SEGMENT / PD CONE GAGING ELEMENT

The Johnson Gage Company

Fig. 16-26. Diagrams showing the engagement with the component internal thread of the full-thread type and of different single-pitch type gaging elements. Different combinations of these elements permit the diameter equivalents of lead and thread angle deviations to be gaged, as well as the maximum material functional size of an imperfect internal screw thread.

$Z-X$ = combined effect of diameter, lead and angle deviations;
$Y-Z$ = diameter equivalent of lead deviations;
$X-Y$ = diameter equivalent of single-pitch angle deviations.

Ex-Cell-O Corp.—Micro-Precision Div.

Fig. 16-27. Bench type comparator gages equipped with different type gaging elements for a comprehensive inspection of product internal threads.
(From left to right) Gage with full-thread elements for functional diameter gaging; gage with single-thread elements for gaging the single-pitch diameter; gage with plain cylindrical elements for gaging the minor diameter. The applicable setting ring gages are shown in front of the instruments.

Ex-Cell-O Corp.—Micro-Precision Div.

Fig. 16-28. Portable type comparator gage for product internal thread; shown while being set for functional gaging by means of a thread ring gage. Beside the instrument is a pair of segments with gaging elements for single-pitch diameter inspection.

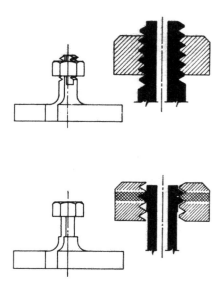

Ex-Cell-O Corp.—Micro-Precision Div.

Fig. 16-29. Diagrams showing the relative engagement with the product internal thread of:
(Top) The full thread gaging elements checking the functional size that determines the assemble-ability.
(Bottom) The single-pitch diameter gaging elements for detecting defective condition due to HI size.

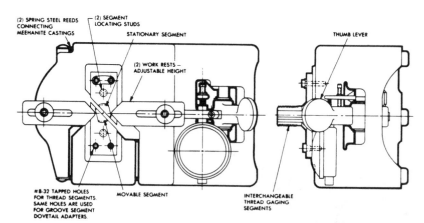

Ex-Cell-O Corp.—Micro-Precision Div.

Fig. 16-30. Diagrams showing the operating elements of the bench type comparator gage illustrated in Fig. 16-27. A member of the gage body is reed-supported and carries the movable segment gage element. The two gaging elements, the stationary and the movable, can be replaced easily with other segments corresponding to the product thread being inspected. Actuating the thumb lever causes the retraction of the member holding the movable segment; after releasing the lever, an adjustable spring pressure forces the gage elements into positive engagement with the product thread.

segments, can also be used for gaging external threads, after reversing the direction of the retracting motion produced by the depression of the thumb lever. That changeover is accomplished by a simple mechanical rearrangement, causing the roller assembled to the thumb lever to act on the opposite side of the coil-spring-loaded cam, which is attached to the movable instrument member.

The bench type instrument holds the workpiece in a positively controlled position by the spring action of the gage segments that engage the product thread. That retainment of the workpiece is a dependable staging for the gaging of other work surfaces, the locations of which must be in an accurate geometric relationship with the functional thread surface. For gaging the accuracy of such relationships the described bench instrument can be complemented with various accessories and attachments with indicators for signaling any deviations of such thread-related feature conditions as squareness of the work side away from or facing the instrument (of an over- or underface), the concentricity of an external or internal cylindrical surface, the coaxiality of a second thread and so forth. Specially designed attachments permit the continuous indicator sensing of deviations from the required squareness while the workpiece retained on its thread is rotated; an attachment arm raises or lowers the holder of the supplemental indicator to follow the level changes of the rotated workpiece.

These gages also can be used for inspecting the minor diameter of the product internal thread, using plain cylindrical segments for that purpose. For setting the described internal indicator gages, solid ring gages are used as masters.

HAND-MEASURING TOOLS FOR SCREW THREADS

In the production of component threads, as well as in the assembly of parts with screw threads, there are setting, inspection and identification operations for which specially designed hand measuring tools are used efficiently, particularly in job shops and tool rooms. A few of the most commonly used types of hand measuring tools for screw threads are pointed out.

Center Gages (Fig. 16-31)

Center gages have several uses in the shop, such as the following:

Rapid determination of the number of threads per inch on a component, the gage having graduations

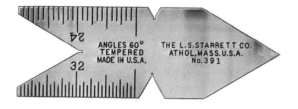

The L. S. Starrett Company

Fig. 16-31. Center gage—a simple hand-tool with multiple uses in shops producing screw threads, such as a template for thread cutter grinding and for the adjustment of the thread cutting tool in a lathe.

for the most commonly used thread types, namely, those with 14, 20, 24 and 32 threads per inch;

As a template for the grinding of screw cutting tools;

As a guide for setting the screw thread cutting tool in the lathe.

Of course, these are rather coarse methods of gaging, but often useful for work not requiring particular accuracy.

Screw Pitch Gages (Fig. 16-32)

Screw pitch gages serve for determining quickly the pitch of component screw threads by comparing the profile of the thread on the part to a template

The L. S. Starrett Company

Fig. 16-32. Screw pitch gage—one of many different types, available for all the common screw thread systems and sizes, used primarily for determining the pitch size (threads per inch) of screw threads on products used for assembly.

with identical pitch. The templates are in folding leaves arranged at both ends of a steel case, the leaves having teeth with forms in agreement with the flank angles of standard screw thread systems, and the number of teeth per inch on each leaf corresponds to a particular pitch. By choosing the leaf with teeth matching the thread of the component part, the corresponding pitch can be read from the marking on that particular leaf.

Screw pitch gages are made in a wide range of sizes, with different numbers of leaves and for various pitch ranges in the most common screw thread systems.

Screw Thread Micrometers (Fig. 16-33)

Screw thread micrometers are used for measuring the pitch diameter of product external screw threads. The end of the spindle is pointed to form a 60 degree cone, and the anvil, which is either a fixed or swivel type, has the form of a Vee. The two opposite gaging elements thus have corresponding forms and match completely when brought into contact by advancing the micrometer spindle. Consequently, the reading on the micrometer sleeve and thimble represents the pitch diameter of the product thread, assuming that it is free of angle deviations with diameter equivalents causing the pitch diameter to appear larger than its actual size.

Screw thread micrometers of the illustrated type, with solidly attached spindle cones and Vee anvils, are supplied in various sizes, each usable for a narrow range of product screw thread pitch sizes, such as for 14 to 18, 20 to 24, or 28 to 32 threads per inch. These micrometers are manufactured in two models with 0 to 1 inch and 1 to 2 inches respective gaging capacities, each tool available in several sizes with contact elements dimensioned for different ranges of threads per inch.

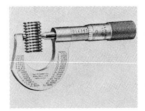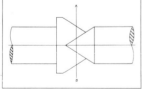

The L. S. Starrett Company

Fig. 16-33. Screw thread micrometer for measuring the pitch diameter of product external threads in directly displayed values, by using a spindle with cone point and a Vee anvil of matching form and size. These elements produce a mutual contact, as shown in the diagram, the *A–B* line corresponding to the zero reading of the micrometer.

Another system of screw thread micrometers is designed with interchangeable spindle inserts and anvils to accommodate various ranges of threads per inch. Such screw thread micrometers are manufactured in four models, with gaging capacities from 0 to 4 inches in one-inch increments. Each of these models can be used with a series of interchangeable contact elements. This design of screw thread micrometers is also available in the metric system, with the corresponding contact inserts dimensioned for pitch ranges expressed in millimeters.

The zero setting of the screw thread micrometer with 0 to 1 inch measuring range can be accomplished without the aid of setting masters because of the matching form of the spindle cone and anvil Vee. Micrometers with measuring capacities over one inch do not establish direct contact between their gaging elements and therefore appropriate setting masters, matching the form of the cone and Vee elements, are needed for the zero adjustment; such masters are commonly supplied with the larger models of screw thread micrometers.

SCREW THREAD SINGLE-ELEMENT MEASUREMENTS

The measurement of particular single elements of screw threads, with the exclusion of the effect of interfering variables, is needed for various purposes, such as indicated by the following examples:

1. *Screw thread gages*, particularly plug gages and, to a more limited extent, ring gages, when serving as setting masters for adjustable solid type or for indicating gages, have to be made to exacting specifications within very narrow tolerances. In the case of working gages the tolerances, although less rigorous than for setting masters, are still much tighter than those of the products for the inspection of which these gages are used. The prescribed tolerances apply to a greater number of dimensional parameters than those commonly gaged on product screw threads, thus necessitating the individual inspection of variables by dependable methods of single-element measurement.

2. *Gaging elements* of screw thread gages, such as thread form rolls and segments with various degrees of truncation, and cone and Vee inserts for screw micrometers, are examples of components that incorporate screw thread elements requiring individual inspection by discriminating and accurate methods of measurement.

3. *Threading tools,* including both the cutting tools and those used for the chipless forming (rolling) of screw threads. The category of thread cutting tools includes both the single-point and the multiple-point (chaser type) lathe cutters, threading machine chasers, taps, thread milling cutters, as well as the operating profiles of thread grinding wheels. The thread rolling tools may be flat or round dies, each requiring different methods for the measurement of significant variables. In this area, too, the measurement of the individual elements requires procedures and instruments commensurate with the specified tolerances, which for threading tools are more rigorous and discriminating than for the product manufactured with these tools.

4. *Manufacturing control,* including the performance analysis of threading machines, as well as the setup of machines and tooling used for producing threaded components. Although the size-holding performance of threading machines is generally monitored by using indicating screw thread gages, the setting up of such equipment requires more specific measurements with emphasis on particular parameters that are prone to be affected by improper machine and tool setup at the start, and by tool wear in the course of the process.

5. *Component parts with multiple-function screw threads,* one common example being the dryseal type product screw threads that, in addition to assuring easy assembly with the mating component, must also provide a pressure-tight joint, accomplished by thread truncation held within the specified special limits. The profile of the thread crest and root are critical elements that require individual inspection, complementing the gaging of the basic size conditions.

6. *Threaded machine and instrument elements,* particularly when required to perform critical functions, are often inspected by the accurate measurement of functionally important thread parameters. Characteristic examples are the lead screws of precision machine tools and the threaded spindles of measuring instruments, including the common screw micrometer. The accuracy and consistency of the lead and the freedom from helical path deviations (drunkenness) are conditions of great significance for the applicational adequacy of such components.

In the following a few of the more commonly used methods of screw thread single-element measurement are discussed, complemented with brief descriptions of applicable measuring instruments. Some of these instruments are adapted to measuring processes used for determining other dimensional conditions and were discussed in previous chapters of this book. In such cases, reference is made to the pertinent chapter in order to permit a review of the instrument from the aspect of screw thread single-element measurement.

Over-Wire Method of Screw Thread Pitch Diameter Measurement

The pitch diameter, although the most important single dimension of screw threads, is only a conceptual parameter; therefore, its accurate measurement is difficult. As was pointed out earlier, the pitch diameter of a screw thread represents the diameter of an imaginary cylinder with surface passing through the thread at points where the widths of ridges and grooves are equal.

It is theoretically possible to select a ball or wire of predetermined size that, when placed in a groove of a perfect thread, will make contact with the flanks at the points where the pitch cylinder intersects these flanks. If two or, preferably, three wires or balls of the correct size are placed in the thread grooves diametrically opposite each other and a measurement taken over the outside of the wires or balls, it is possible using this measurement to compute the pitch diameter of the thread.

The correct wire size, usually called the "best wire size" for a given thread, can be calculated by the formula

$$w = 0.5p \sec \alpha$$

where

w = diameter of wire or ball;
p = pitch of the thread;
α = one-half the included angle of the thread.

For a 60 degree thread this reduces to $w = 0.57735p$.

The formula for finding the pitch diameter of the thread for a given measurement over wires is

$$E = M_w + \frac{0.86603}{n} - 3w$$

where

E = pitch diameter of the thread;
M_w = measurement over-wires or balls;
w = mean diameter of wires or balls;
n = number of threads per inch = $1/p$.

It should be pointed out that the helix of the thread shifts the actual positions of the points where the wires touch the flanks away from the positions viewed in the axial plane of the screw. Although that shift in the actual positions of the points of contact affects the measured size, in most cases that difference is so small that in general practice it is uniformly neglected. The difference between the actual pitch diameter and the dimension determined by the simplified formulas is, however, considered significant when the position of the contact points affects the measured pitch diameter by more than 0.00015 inch; that is the case for multiple threads and single threads with exceptionally large lead angle. In such cases a formula with a correcting term, expressed as constants for different lead angles, must be used. A table with the constants applicable to 60 degree threads with large lead angle (λ = 5 degrees and larger) is contained in the standard (National Bureau of Standards, Handbook H28). Formulas also have been developed to provide direct and dependable solutions for the extremely accurate measurement of threads, by taking into account the effect of the lead angle. These are explained in *Machinery's Handbook*.[2]

Three-Wire Method of Measuring Pitch Diameter

In order to determine the pitch diameter of screw threads by measuring the corresponding over-wire size, the most practical procedure is the use of three wires, actually small hardened steel cylinders, placed in the thread groove, two on one side and one on the opposite side of the screw. That arrangement of the wires, as indicated in the diagram (Fig. 16-34), permits the opposite sensing elements of a length-measuring instrument to be brought into simultaneous contact with all three wires, thus providing a dependable measurement of the over-wire distance.

Measuring wires in sets of equal size within a certain diameter range may be used, as long as the wires have a minimum diameter that projects over the crest of the thread, when in measuring position, and a maximum diameter that permits the wires to touch the flanks just below the crest. The use of "best-size" wires is recommended, however, for the following principal reasons:

(a) When the wires touch the midslope of the thread flanks, the interference of flank angle variations with the measured pitch diameter is practically eliminated;

---

[2]*Machinery's Handbook*, 24th rev. ed., Industrial Press, New York, 1992, p. 1669.

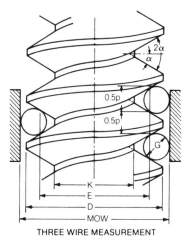

THREE WIRE MEASUREMENT

The Van Keuren Company

Fig. 16-34. The positions of the screw and of the gage elements in the over-wire measurement of the thread pitch diameter.

MOW = measurement over-wires (alternate symbol M_w);
D = major diameter (d);
E = pitch diameter (d_2);
K = minor diameter (d_1);
α = half-angle of thread;
G = wire diameter (w)

(b) Measurements made with "best-size" wires (and assuming zero lead) permit the use of the simplified formula quoted previously, whereas the pitch diameter has to be calculated when wires other than the "best size" are used.

The three-wire method of pitch diameter measurement is used primarily for determining accurately the size of thread plug gages and of critical product threads made with similar tight tolerances. This method is, however, not recommended for the size determination of components made with substantially lesser accuracy; the inspection of such threads is carried out by one of the gaging methods discussed previously.

Other Uses of Over-Wire or Ball Measurements for Screw Threads

A few other methods of measurement for determining the size of screw thread elements are pointed out in the following, all these methods having the common characteristic of using intermediate members (cylinders or balls), which can be readily located for accurate measurement in positions accessible to the sensing surfaces of the measuring instruments.

Over-wire measurement of the pitch diameter of taper thread gages. For these measurements the front face of the gage is used as a reference surface, from which the axial distance of the measuring plane is determined with the aid of gage blocks. On one side of the thread a wire, designated as the fixed wire, is placed in the groove coincident with the measuring plane; in the opposite side groove portions, at levels straddling the measuring plane, the complementary wire(s) is (are) placed. The element placed in this latter position may be either of the following:

One wire, used in two consecutive positions for over-wire measurements in conjunction with the fixed wire, producing two measurements, the average value of which is considered the pitch diameter in the measuring plane *(two-wire method)*; or

Two wires are used on the side opposite the position of the fixed wire; however, in this case a measuring instrument with sensing surfaces in self-aligning retention is needed in order to establish simultaneous contact with the wires placed side-by-side *(three-wire method)*.

In both types of measurement the pitch diameter must be calculated by the appropriate formulas, which are more complex for taper threads, and are different for each of the two methods listed. (Formulas for calculating the pitch diameter of taper screw threads measured over wires are quoted and explained in *Machinery's Handbook*.[3])

Measurement of pitch diameter of thread ring gages. Measuring the pitch diameter of thread ring gages by direct contact is difficult to accomplish, particularly for small sizes. For that reason a widely used indirect method of inspection is to fit the ring gage on a thread setting plug gage.

It is possible to measure the internal thread pitch diameter of a larger sized ring gages by using balls as contact elements, preferably balls with the same diameter as that of the best-size wire recommended for the mating external thread. For carrying out such direct measurements of the pitch diameter on an internal thread, measuring machines with special ball-holding gaging arms are needed, set up in an arrangement indicated in the sixth diagram of Table 12-2. Instruments designed to provide that measuring capability operate by applying the gaging force required for secure contact between the ball elements and the thread flank. The applied measuring method is a high-precision differential gaging in

which the reference position of the ball contacts is set with the aid of a ring gage.

Thread angle measurement with two sizes of wires. The diameters of the wires used for this method are selected to assure that both touch the flanks when placed consecutively into the groove of the thread; the levels of flank contacts are near the root and crest, respectively. It is recommended that wires be used having diameters that are approximately over and under that of the best-size wire by the same amount.

The principles of such Vee-groove angle measurement with different diameter pins are shown in the fifth diagram of Table 10-6. In the case of thread angle measurement it is often practical to use a sensitive depth gage with a thin or ground-down spindle diameter and to rest the beam of the depth gage on two large-sized wires, which are placed in thread grooves adjacent to the inspected one.

Wires and Processes of Over-Wire Screw Thread Measurement

The wires used for over-wire screw thread measurements are accurately finished hardened steel cylinders, with excellent roundness and straightness, assuring constant diameter with variation not exceeding 0.000020 inch. Most commonly these wires are used in sets of three, having the same diameters within 0.000020 inch. The actual diameter of the wires should not deviate by more than 0.0001 inch from that defined as the best size for the pitch to be measured.

Calibrated wires supplied in sets of three have the exact size, determined with an accuracy of 5 microinches, shown on the label of the case, together with the pertinent constant C, the value of which must be subtracted from the over-size wire M of the corresponding 60 degree thread for determining the pitch diameter PD. Usually such wires are applied in the manner shown in Fig. 16-34.

Some manufacturers supply measuring wires with auxiliary holding elements, which are useful in measuring arrangements where the specimen gage is mounted with its axis in the horizontal plane containing the axes of the spindle of a horizontal length-measuring machine.

Such holding elements attached to the wires are:

a. Suspension eyelets by which the wires are held on hooks over the gaging area, while not in actual engagement with the contacted thread and instrument face.

---

[3]*Machinery's Handbook*, 24th ed. rev., Industrial Press, New York, 1992, p. 1684.

b. Slip-on holders, which can be slipped over the spindles of the measuring instrument and retained in place by spring tension. Such holders either have a hole through which the instrument tip has direct contact with the wires, or the wires rest on the gage block type back of the holder, which stays interposed between the face of the instrument spindle and the wires; the known thickness of that gage block type back must be subtracted from the measured M value.

For the dependable measurement of its diameter in instruments with indicating accuracy of 10 microinches or better, the wire should be stated between a high-precision cylinder anvil (with its axis perpendicular to that of the specimen wire) and the flat face of the measuring machine spindle. Alternate arrangements, which assure that the diameter of the wire is actually being measured, are also acceptable.

The contact force applied in pitch diameter measurement with wires must be sufficient to assure the proper seating of these auxiliary elements, yet the force should not exceed the limit over which a permanent deformation of the wires or the specimen gage could occur. The following contact forces are recommended for various pitch ranges:

Thread per Inch	Equivalent Metric Pitch (approx., mm)	Measuring Force (± 10%)	
20 or less	1.25 or larger	2.5 lb	1100 g
20–40	0.6–1.25	1 lb	450 g
40–80	0.3–0.6	8 oz	225 g
80–140	0.2–0.3	4 oz	115 g
Above 140	Finer than 0.2	2 oz	55 g

Instruments of Over-Wire Screw Thread Measurement

Although screw micrometers with 0.0001-inch discrimination may also be used for over-wire measurement of screw thread pitch diameters, the expected accuracy of the measurement generally requires dependable instruments with higher resolution. Particularly well suited and frequently used for this type of high accuracy measurement are the single-axis length-measuring machines, examples of which are shown in Figs. 12-1–12-4. Several of these instruments provide a means for adjusting the contact force, thus satisfying in this respect the requirements of dependable over-wire pitch diameter measurements.

The Van Keuren Company

Fig. 16-35. Over-wire measurement of screw thread pitch diameter on a vertical spindle length-measuring machine having large contact faces.

On the vertical type length-measuring machines the proper staging of the work in combination with the three measuring wires is feasible, but might be difficult to accomplish without some auxiliary device, unless an instrument with large anvil surface, permitting staging as shown in Fig. 16-35, is used. The horizontal type, single-axis, length-measuring machines are much better suited for such measurements, particularly when the specimen part has a flat face in a plane dependably perpendicular to the thread axis. That is the case for most types of thread plug gages, which can rest on their flat face in a position permitting the convenient insertion of the measuring wires into the thread grooves at the level of the instrument spindles, and also observation of the proper seating of these complementary gaging elements. Figures 16-36A and 16-36B are close-up photographs showing over-wire pitch diameter measurement of thread gage plugs mounted on the vertically adjustable staging table of the horizontal measuring machine shown in Fig. 12-3 and described in the accompanying text.

A much less commonly applied method, but based on the same principles as the over-wire measurement of external threads, is the over-ball measurement of internal thread pitch diameters. The method is mentioned here because of its affinity with the over-wire measurement, and reference is made to the diagram in Table 12-2 relating to over-ball measurement of internal threads. That diagram indicates the principles of such measurements, which require measuring machines with special arms for

(A)

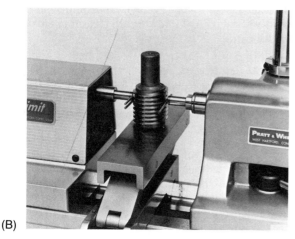

(B)

Colt Industries—Pratt & Whitney

Fig. 16-36. Over-wire measurement of thread plug gage pitch diameters on the horizontal length-measuring machine shown in Fig. 12-3, but equipped with vertically adjustable staging table.

internal diameter measurement, such as are available with the instrument shown in Fig. 12-4.

The third diagram on the same table page indicates the principles of the over-wire measurement of taper thread plug gage pitch diameters. Such measurements are carried out in two steps, using two wires only, one of which is held in a constant position, while the wire on the opposite side of the specimen is placed consecutively into two adjacent grooves in locations straddling the plane of the constant position wire.

It is also possible, and sometimes more convenient, to measure the pitch diameter of taper thread gages with three wires. In that case, however, the specimen must be staged in a position with inclined

axis, in order to assure simultaneous contact of the flat and parallel gage tips with all three wires.

The applicable formulas for calculating the actual pitch diameter from the distance measured over the wires are different, depending on whether the two-wire or the three-wire method has been used for taper thread gages.

Adjustable Thread Gage System

The hand-measuring of in-process and finished threads has recently been made possible by the marriage of threaded anvils and electronic calipers (see Fig. 16-37). Anvils that conform to the pitch and form of the threads being measured are required. Interchangeable thread anvil sets are available for both English units and the metric system. Digital calipers also offer the RS232 interface for outputting measurement information to most electronic data collectors for statistical analysis.

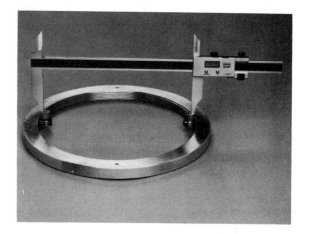

Flexbar Machine Corp.

Fig. 16-37. O.D/I.D digital caliper thread gaging system.

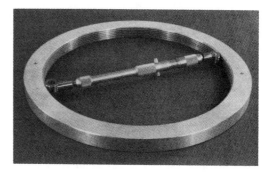

Flexbar Machine Corp.

Fig. 16-38. Internal micrometer thread measurement system.

Another type of adjustable thread gage is the internal micrometer (see Fig. 16-38). This system along with the electronic caliper thread gage can detect out-of-roundness and taper, which are impossible to check with threaded rings and plugs. Also, both tools provide accurate readings to 0.001 inch total and eliminate the difficulties typically encountered with over-pin or over-ball thread measuring (that is, calculations and loose details).

SCREW THREAD INSPECTION WITH OPTICAL INSTRUMENTS

Optical instruments, which present a magnified view of the specimen or a portion of the specimen that fits the available field of observation, are useful and commonly applied tools of screw thread inspection, specifically single-element measurement. The pertinent instruments are the engineering microscope and the optical projector: the design, operation, and applications of these measuring instruments are discussed in great detail in Chapters 8 and 9.

Although both systems of optical instruments can, in principle and in the case of many models actually, present either a magnified image of the profile or, with surface illumination, of the entire surface of the viewed specimen area, for the purpose of screw thread measurement the profile viewing of the workpiece is the predominantly applied process. The profile of an external thread presented by these instruments appears as a distinct contour line, representing the boundary of an axial cross-section of a basically round surface. This mode of observation is consistent with the specifications of the major screw thread elements, which are also defined with reference to an axial plane of the threaded component. The observable and measurable screw thread elements include all the diameters (pitch, major and minor), the profile features (flank angle and form, contours and dimensions of both the crest and the root), as well as pitch (either as a directly measured distance between adjacent threads or the number of threads per inch).

These capabilities apply to external threads only; internal threads are not accessible to direct optical viewing. The lack of optical access to internal threads may be compensated, to a limited extent, by auxiliary means, but these are exceptional uses of optical instruments, much less commonly applied than the inspection of external threads.

However it must be considered that optical instruments, while providing the means for measuring a great variety of screw thread elements, require appropriate work staging; the selection and installation of the applicable graticules or screen charts; specimen focusing; and a measuring operation that usually comprises two phases: the consecutive alignment of image contour lines with the pertinent graticule or screen chart lines and the measurement of the distance over which the work must be traversed to accomplish these interrelated contour alignments.

While the listing of these process elements might infer a time-consuming operation, the actual procedure of optical screw thread element measurement is greatly simplified and accelerated by various accessories and equipment characteristics, as well as techniques developed for such operations. A few of these means that, when available and applicable, can greatly assist optical screw thread measurements are the following:

a. *Staging of work* between retractable centers or in self-centering holding devices, when continuously inspecting parts of the same nominal size, greatly reduces or actually eliminates the need for repeated focusing.

b. *Graticules and screen charts,* which contain the contours of several size threads within the same system, permit the optical equipment to operate over extended or even indefinite periods without the need for installing different charts.

c. *Magnification rates* in conformance with the required degree of discrimination can be changed on most engineering microscopes and various types of optical projectors by installing the proper objective or, in certain types of optical projectors, simply by switching the lens selector. Microscopes designed with an intermediate image system can be operated with the same graticules, irrespective of the applied magnification; optical projectors require screen charts at scales corresponding to the magnification of the projected specimen image.

d. *Multiple dimensions* are incorporated into most types of graticules and screen charts, permitting one to inspect simultaneously the conditions of several elements, for example, flank profile, form, length and angle, root clearance and crest dimensions, where their inspection by contact gaging would require several individual operations.

e. *Tolerance zones* laid out on various types of screen charts assist in determining, at a single glance, whether the inspected work contour line,

**TABLE 16-4. SYSTEM-RELATED CHARACTERISTICS OF
ENGINEERING MICROSCOPES AND OPTICAL PROJECTORS—1**

CHARACTERISTICS		DISCUSSION	APPLICATIONAL ASPECTS
Engineering Microscope	Overall Size	Engineering microscopes are compact instruments, commonly mounted on a bench or table and easily transportable to the selected location, where they can be plugged into the electric lighting line as easily as a regular lamp.	The reduced space requirement permits the use on a bench adjacent to thread cutting or grinding machines in the shop, or in laboratories where floor space is limited. The actual optical part of the engineering microscope, which provides magnification and comparison with graticule lines, but without the work staging and traverse member, is particularly compact, adaptable to installation on machine tools, e.g., thread grinders, an application supported also by the safe low-voltage illuminating system.
	Flexibility of Application	The oculars of engineering microscopes have revolving templets or graticules, which usually contain the contours of most thread sizes within a system; such oculars are quickly interchangeable with other types, permitting the inspection and single-element measurement of a large array of screw thread types and sizes.	In the shop, e.g., in the thread grinding department, where one microscope serves the inspection needs of several machines producing different threaded components, or in an inspection laboratory where the type of thread to be measured is varied, the quick availability of the required profile templets within the same eyepiece, or in an easily installed exchange eyepiece, is a valuable characteristic. Most models of microscopes also have angle-measuring oculars, very convenient for the rapid, yet accurate measurement of the flank angle or of its deviations.
	Resolution of Viewing— Accuracy of Measurement	The clear view of the magnified object profile with crisp, well-defined boundaries which are easily aligned with the distinct and very fine lines of the graticule, permit the setting of the object to the pertinent reference positions with a repeat accuracy which is commensurate with small incremental measurements.	The critical link in the sequence of the measuring steps on optical instruments is usually the establishment of dependable reference positions, the precision of which defines the accuracy of the optical measurement. The other component of the process, the traverse distance of the staging table between the two reference positions (distance boundaries), can be determined by high-resolution length-measuring devices (micrometer screw, optically observed line-graduated master scale, or various electronic systems) with an accuracy which is commensurate with that of the reference positions.

which may be the composite of several individually specified dimensions, is within the limits of the allowed tolerances.

f. *Digital readouts* of engineering microscopes and optical projectors equipped with such devices, permit the inspector to determine by a simple comparison of the specified and displayed numerical values, whether the examined dimension, for example, the pitch, the directly measurable diameters, or the diameter equivalents of other variables, are in agreement with the pertinent specifications or, in the case of nonconformance, the amount of the determined deviation.

**TABLE 16-4.   SYSTEM-RELATED CHARACTERISTICS OF
ENGINEERING MICROSCOPES AND OPTICAL PROJECTORS—2**

	CHARACTERISTICS	DISCUSSION	APPLICATIONAL ASPECTS
Optical Projector	Convenience of Image Viewing	The generally large diameter screen can be observed from a comfortable distance, and permits the use of large scale magnifications which still embrace the portion of the work profile needed for screw thread measurements. Concurrent viewing by several persons is possible.	Preferred system for continuous inspection by optical means, because less tiring to the operator than viewing through an ocular. Multi-person observation desirable for joined examination, demonstration, etc. Screen image susceptible to measurements by directly applied graduated scale. Inspection of particular features by means of specially designed overlay charts is also applicable.
	Adaptability to Special Inspection Requirements	The generally large and open work staging area permits the use of special fixturing needed to hold workpieces of uncommon configuration or for presenting the observed specimen feature in a specific geometric relationship with another, not projected surface of the workpiece.	Components might be designed with threaded features in position defined in relation to another surface to assure, e.g., concentricity, squareness, alignment, etc., of the effective thread surface with functionally interrelated elements of the part. The special holding fixtures needed can usually be installed and properly located on most models of optical projectors.
	Specimen size Capacity	The system of optical projection can be applied to very large instruments, in the size order of machine tools, with work staging tables many times larger and stronger than those of even large sized engineering microscopes.	Large and heavy workpieces with screw threads over the entire length, or confined to specific sections, are often still within the work mounting and positioning capacity of appropriate models of optical projectors, thus adaptable to the optical examination of the thread elements, a process which for very large threads may be the only practical inspection method.
	Inspection by Attributes	The dimensional limits of screw threads specified for assuring assemble-ability at maximum material condition and the size limits of minimum strength, can be translated into envelope surfaces, the contours of which, when represented by substantially parallel chart lines, are the tolerance zone boundaries of the screw thread profile.	For screw thread inspection involving various dimensions, the optical method may replace several individual operations performed by diverse contact-type instruments. The advantage of optical inspection is further enhanced by screen charts having double contour lines representing the limit surfaces of essential thread elements, thus providing the means of single operation functional and minimum size gaging by attributes for determining the GO and NOT GO (HI) conditions.

The system-related advantages of engineering microscopes and optical projectors, particularly with regard to applications for screw thread element measurement, differ to a degree that warrants their review with reference to the intended uses of the instrument and its condition of operation.

For facilitating such a comparative review, on the results of which the selection of the applicable type of instrument should be based, a survey of the system-related advantages of both families of optical measuring instruments is presented in Table 16-4. The listed characteristics, however, must not be regarded as each benefiting exclusively one of the two systems; furthermore, advantages attributed to either of the compared types of instruments are those generally inherent in the systems, although not actually assured in all models of instruments designed in that system.

Depending on the intended applications of the instrument for screw thread measurements, some of the listed distinctions might have minor importance only, and other characteristics are regarded as decisive in selecting the instrument best suited for a particular use.

Optical Examination of Internal Thread Forms

The inspection of form characteristics, including flank angle, root and crest contour and so forth, of internal threads is not feasible by direct observation with an optical instrument. An indirect optical examination of such critical surface contours, however, can be accomplished by preparing and using an inverse replica of a section of that internal screw thread. Such replicas can be prepared as castings made of a sulfur–graphite mixture, plaster of paris, copper–amalgam, or one of the available proprietary compounds. Although the preparation of cast replicas involves an auxiliary process preceding the actual measurement, the profile of the thus produced inverse replica of the thread form is suitable for precise inspection by optical instruments, on an engineering microscope or an optical projector.

LEAD MEASUREMENT

The accuracy of the lead of component screw threads—the lead being, of course, identical with the pitch of single start threads—can be inspected by several different methods, some of which were pointed out earlier in this chapter. Such lead accuracy inspection methods are:

1. Gaging with indicating instruments the diameter equivalent of lead deviations, which produces different diameter readings when indicating gages with multiple-thread and single-thread elements, both set with the same master, are used consecutively for checking the diameter of the component thread; and

2. Inspection with optical projectors or engineering microscopes, in conjunction with the specimen stage traversing mechanism of the instrument.

These methods, although adequate for many applications, have various limitations. In differential

gaging the effect of lead deviations on the pitch diameter is determined, assuming only negligible interference by variables other than the lead. In optical gaging the repeat accuracy of the visual targeting, whether on the screen or through the ocular, is the limiting factor, also due to the applicable rate of magnification that must be consistent with the requirements of encompassing a full thread, at least, in the field of view of the instrument.

In applications where the lead is of critical importance, more discriminating and sensitive methods of lead measurement are needed; these are assured by using instruments specifically designed for lead measurement or even helix analysis.

The *electromechanical lead tester* shown in Fig. 16-39 is applicable to both straight and taper threads and has a potential measuring accuracy of 0.000025 inch. In its general design this instrument resembles the horizontal length-measuring machine shown in Fig. 12-3, with the addition of a cross-slide that carries the sensitive thread-locating head. That head has interchangeable ball point contact elements with different radii, which are selected according to the pitch of the thread being measured. The approach and retraction movements of the cross-slide are by hand, and an attached counterweight assures a constant contact pressure. The sensitive and accurate measuring head is used to produce the traverse movement of the carriage on which the specimen is mounted between centers and prevented from free rotation by precision type drive dogs. One of the centers is installed in a stock with index plate, which might be used for the manually controlled analysis of the thread, by checking the consistency of the helix at different intermediate positions of a full turn.

A microammeter signals when the thread-locating ball point is in the proper position inside the thread

Colt Industries—Pratt & Whitney

Fig. 16-39. Electromechanical lead tester with high accuracy measuring head (at right) operated manually for traversing the carriage with the center stocks. The thread-locating head is installed on the cross-slide and advanced manually; the microammeter with 3000 to 1 magnification signals the proper position in the thread of the ball point sensing element.

flanks; to reach that position the carriage has to be traversed by means of the measuring head. When the proper contact position of the thread-locating ball point has been established, the carriage position is read from the three adjacent scales of the measuring head. Differences, if any, between the distances of the consecutively inspected threads and the pertinent nominal pitch sizes, are the measured lead of pitch deviation.

The measuring head has a range of one inch, which can be extended to a total specimen thread length capacity of 4 inches by the use of square gage blocks interposed between the measuring head spindle anvil and the carriage face. Workpieces to a maximum diameter of 8 inches, installed on a maximum center distance of 12 inches, can be measured on the described instrument.

Instrument for inspecting the drunkenness of screw threads. In a constant helix screw thread the thread grooves on one side are always exactly midway between those of the opposite side, in any axial plane of the screw. In a drunken thread that relationship exists in a single axial plane only, whereas in other axial planes the relative positions of the opposite thread grooves vary. Instruments have been designed and are used for measuring and displaying as indicator readings the amount of such variations between opposite thread grooves.

HELICAL PATH ANALYZER

Deviations of the screw thread helix, which affect the accuracy of the lead, may be consistent (progressive), periodic (drunkenness), erratic, or any combination of these potential errors. Consequently, inspection processes directed at determining and gaging the amount of either one of these individual errors may not be sufficient for assuring the lead accuracy required for screw threads used in critical applications, such as for instrument traversing mechanisms.

The helical path analyzer shown in Fig. 16-40 is a special instrument designed for detecting, measuring and recording on charts all the potential errors of the specimen screw thread helix over any three revolutions of the threaded member.

Figure 16-41 is a close-up view of the operating area of the instrument, with a specimen installed between the centers and the cover removed for displaying the principal functional elements. The motor at left rotates the part through three revolutions in one setting, transmitting at the same time a cor-

Colt Industries—Pratt & Whitney

Fig. 16-40. Helical path analyzer for recording with a high rate of amplification chart tracings that represent the helical path conditions of the specimen thread. The instrument accepts specimens to 6 inches in diameter, to 8 inches in distance between centers, and to a maximum thread length of 5 inches.

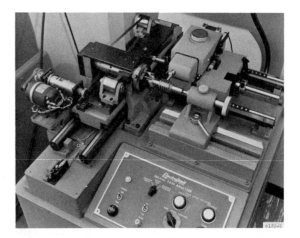

Colt Industries—Pratt & Whitney

Fig. 16-41. Close-up view of the operating elements of the helical path analyzer in Fig. 16-38, shown with the cover removed for displaying the principal operating elements.

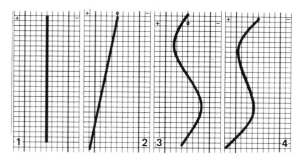

Colt Industries—Pratt & Whitney

Fig. 16-42. Examples of chart tracings produced by the helical path analyzer (Figs. 16-38 and 16-39). The tracings represent different conditions of the inspected helical path, such as correct *(1)* and with deviations, which are progressive *(2)*, periodic *(3)* or cumulative *(4)*. The instrument has dual magnification, producing chart tracings with each chart division corresponding to 0.000020 or 0.0001 inch.

responding motion to the chart drive of the recorder. Concurrently, a driving ribbon, running off the drum of the headstock, advances the sine bar, against which bears the follower of the gage head slide. The advance of the sine bar causes a lateral movement of the gage head at a rate corresponding to the nominal lead of the specimen screw thread. The sine bar is set to the proper angle prior to the start of the operation; pitch blocks corresponding to the number of threads per inch of the specimen are used.

The instrument's recorder produces tracings on chart paper with each division representing 0.000020 inch with the traced line having a 9-inch length for three revolutions of the specimen; these values result in a high rate of amplification. Figure 16-42 shows examples of four typical chart tracings, representing (1) a theoretically perfect helical path, (2) consistent (progressive) deviation of the lead, (3) periodic lead errors (drunkenness), and (4) progressive and periodic deviations superimposed. These are examples of the principal lead deviations, but similar informative chart tracings are also produced from more complex lead errors, for example, those involving erratic conditions.

17
The Measurement of Gears

THE APPLICATION AND OPERATIONAL REQUIREMENTS OF GEARS

Gears are widely used mechanical elements that transmit rotary motion, generally in conjunction with a modification of the speed and/or direction of rotation. While rotary motions are the most common, gears operating in engagement with a rack are used to convert rotational into translational motion. Gears are also used for indexing, that is, producing partial rotation over a precisely controlled angular distance, or as timing gears, listed as examples of the applications with particularly rigorous accuracy requirements.

Motion transmission by gears, although often carried out at high speeds and involving substantial force, must be performed at controlled ratios and uniform angular velocities, within particular limits of vibration and noise, and perhaps in compliance with other types of specifications.

Gears are usually designed to provide optimum operational conditions. Design concepts, however, cannot be incorporated perfectly into any manufactured product, including gears; consequently, tolerances must be established for the compliance of the produced gears with the theoretically correct dimensions. Since gears are complex elements, even in their basic forms as spur gears and more so in the cases of many other types of gears, tolerances are specified for the significant parameters. The ranges of these tolerances are established for different quality numbers, which are selected by the user according to the envisioned application.

Compliance of the produced gear with the applicable tolerances must be determined by dimensional measurements, which, for gears, require specific techniques and equipment, and represent a particular branch of metrology. This chapter will review the more commonly employed methods and instruments of gear measurement, with emphasis on the most widely used involute profile gears. The concepts of gear design, as well as data on dimensions, tolerances, and similar specifications, will not be part of the discussion. For that information the reader is referred to the pertinent standards[1] and to recognized mechanical engineering handbooks, for example, the *Machinery's Handbook*.[2]

THE PURPOSE OF GEAR MEASUREMENT

The main purposes of gear measurement are the following:

1. For process control, particularly in the setup of gear making machines, and for the inspection of gear cutting tools; often the errors of individual tooth elements must be determined in a series of *analytical measurements*. The discrete measurement of particular gear elements is also needed for the inspection of gears intended for critical applications; and

2. To determine how the gear will perform in continual engagement with a mating gear; instruments providing a close approximation of the actual

---

[1] *AGMA Standards*, available from American Gear Manufacturers Association, 1500 King Street Alexandria, VA 22314.
[2] *Machinery's Handbook* 24th rev. ed., Industrial Press, New York, 1992.

operating conditions are used, and the *composite action* of several potential tooth errors is measured by the *functional checking* of a work gear, commonly using for the mating member a "master gear" of known accuracy.

Both categories of gear measurement are used for process control and the final inspection of gears.

Measurements for process control in the manufacture of gears are applied for guiding the setup, monitoring the performance and identifying the sources of errors in the operation of gear cutting machines; furthermore, measurements are also made to determine the condition of gear cutting tools and to analyze the effect of heat treatment on gear tooth size and form, as a means of process planning for the assurance of the optimum quality of the produced gears.

Final inspection is needed, particularly for gears that are manufactured to specifications based on higher quality numbers. In the large volume production of gears requiring 100% inspection, semiautomatic or fully automatic gear inspection machines are frequently used.

In addition to the dimensional parameters of gears, other conditions are also inspected. As examples, semifinished gears may be inspected for the presence of size or form errors, excessive burrs and so forth, conditions that could be harmful in subsequent operations, such as damaging expensive finishing tools. Checking the performance characteristics of certain types of gears that have to comply with particular applicational requirements, such as not exceeding a specific noise level when running under the expected operational conditions, may also be considered a method of gear measurement. Finally, a method of gear measurement is used for inspecting the adequacy of gear cutting tools, such as hobs.

ANALYTICAL GEAR MEASUREMENTS

Analytical gear measurements are used to check for the compliance of particular gear tooth parameters with the pertinent design specifications or, for deviations, to measure the characteristics and extent of the error.

The analytical or discrete measurement of gear tooth dimensions may be needed for various reasons, which may be related to the expected functional performance of the gears or to gear manufacturing process control. A few examples of the purposes of analytical gear inspection are:

1. To determine the sources of improper gear operation, either experienced in use or discovered by functional checking.

2. To inspect the accuracy of specific gear tooth parameters, for example, involute profile, tooth spacing or blacklash, errors that may be mutually compensating or for other reasons not revealed by functional checking, but that are still critical in certain applications, such as where the gear has an indexing role, and is required to produce incremental rotational movements that are rigorously proportional to the input motion.

3. To provide guidance in the proper setup of gear manufacturing machines, thereby avoiding the production of faulty gears.

4. To check the appropriate execution or sharpening of gear cutting tools, such as hobs and gear cutters.

5. To determine the effects of the applied heat-treating processes so that appropriate modifications in work material or procedures can be made, when needed, or compensating adjustments can be applied to the tooth form design of the soft (nonhardened) gears.

The gear tooth parameters that are commonly inspected by analytical measurements and the instruments used for the individual types of inspection procedures are reviewed in Table 17-1. The listing of this table does not indicate an order of importance or frequency of use; the types of analytical gear tooth measurements to be carried out are determined according to the applicational or technological requirements of the gears.

The subsequent detailed discussion of each measurement type and examples of the instruments used will follow the order of that table.

RUNOUT

The runout of gears reflects the total variation of specific elements of the tooth surface from the axis of rotation of the gear. Runout is commonly measured in a direction perpendicular to the axis of rotation of the gear, the *radial runout*; but, in particular cases, runout is determined by measuring the variation of specific gear surface elements in a direction parallel to the axis of rotation, the *axial runout*.

The runout of the inspected gear may include the effects of various irregularities associated with the

TABLE 17-1.   PARAMETERS AND METHODS OF ANALYTICAL GEAR MEASUREMENTS

GEAR DIMENSION	INSPECTED CONDITION	METHODS OF MEASUREMENT	FIGURE REFER-ENCE
Concentricity	Gear tooth runout.	Single-probe runout check—in radial direction. Two-probe runout check—in tangential direction.	17-1
Pitch	Pitch variation. Spacing variation. Index variation.	Pitch measuring instrument checking variation of spacing. Automatic tooth space comparator. Index position measuring instrument.	17-2,-3, -4,-5,-6
Tooth profile	Deviations (deliberate or unintentional) from the true involute form.	Involute profile measuring instruments with interchangeable base circle discs. Involute profile measuring machines with variable base circle.	17-7,-8, -9,-10, -11,-12, -13,-14, -15,-16, -17,-18

Tooth thickness	Deviations from the calculated value and tooth-to-tooth variation.	Tooth thickness caliper.
		Addendum comparator.
		Span measurement.
		Measurement over pins.
		Gear center distance measurement
		   (a) in tight mesh
		   (b) at operating center distance (backlash).
			17-19,
			-20,-21,
			-22,-23

Gear lead	Deviations from the design value and variation.	Lead-measuring instruments.	17-24, -25,-26, -27,-28
Clearance between meshing teeth	Backlash of meshing gears.	Indicator method for assembled gears. Gear thickness measurement with caliper gage (indirect method).	
Gear tooth shape	Evaluation of the actually contacting gear tooth surface.	Coating the gear teeth with marking compound and running the mating gears under light load, followed by visual analysis of the resulting tooth bearing pattern.	

basic form of the gear (out-of-roundness); the locations of the teeth with respect to the gear axis (eccentricity); and/or errors in diverse tooth parameters, such as variations of the profile, tooth spacing, or thickness.

Runout may be measured as an individually inspected condition, using methods to be described, or that irregularity may be detected in conjunction with the measurement of specific gear tooth elements, the variation of which could be causing the runout. Gear rolling checks, used for inspecting composite errors that result in center distance variations, may also reveal runout.

A common method of runout inspection, called the single-probe check and shown in Fig. 17-1A, uses an indicator with a single ball probe whose di-

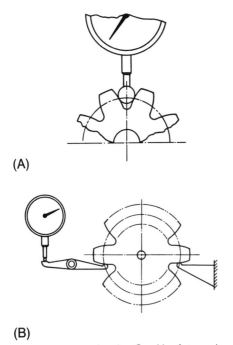

(A)

(B)

American Gear Manufacturers Association

Fig. 17-1. Principles of gear runout check by an indicator probe making consecutive contacts with identical surface elements of all the gear teeth.
(A) Single-probe runout check.
(B) Two-probe runout check.

ameter makes contact with the flanks of adjacent teeth in the area of the pitch circle.

The two-probe check is an alternate method, using a fixed and a free-moving probe, positioned on diametrically opposite sides of the gear and making contact with identically located elements of the tooth profile, as shown in Fig. 17-1B. The range of indications obtained with the two-probe check during a complete revolution of the work gear is twice the amount resulting from the single-probe check.

In the single-probe runout check of bevel and hypoid gears the variation is indicated in a direction perpendicular to the pitch cone.

PITCH VARIATION

Pitch is the theoretical distance between corresponding points on adjacent teeth. The actual distance between those points is the tooth-to-tooth *spacing*, a directly measurable dimension. The theoretical angular position about the axis of the gear is designated the *index*.

The pitch can be measured by simple, manually operated pitch-measuring instruments with indica-

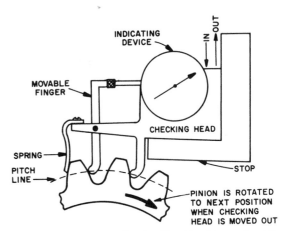

American Gear Manufacturers Association

Fig. 17-2. The operating principles of a pitch-measuring instrument with separate positioning and measuring probes.

tors; the schematic arrangement of such a device, designed for spur gears, is shown in Fig. 17-2.

Tooth spacing measurement by manually indexed instruments, however, is a rather slow process, and when the spacing variation of all the gear teeth is to be measured, the use of such simple indicating instruments also involves the manual recording of the individual indications. For the regular inspection of the tooth spacing in the production of gears with rigorous pitch tolerances, automatic measuring instruments, such as those described as examples in the following paragraphs, are generally used.

The *automatic tooth space comparator*, with two probes or fingers, inspects the pitch accuracy of the entire gear by comparing consecutively the spacings of all the teeth. One of these fingers is the solid driving finger, the movements of which cause the work gear to rotate continuously but at varying angular velocity, and the other is the checking finger, which freely follows the surfaces of the gear teeth. The instrument checks the tooth spacing while the gear rotates, indicating and recording in the form of a strip chart tracing the results of all the measurements. The work gear is mounted on the spindle of the machine, and an adjustable crank drive, the operating principles of which are shown in Fig. 17-3, produces the movement of the driving finger for advancing the gear in increments of one circular pitch. The design characteristics of that crank drive provide the varying angular velocities of the gear rotation, which result in a fast indexing cycle and a slow rotation for the measuring action of the checking finger. The movement of the checking finger acts on an electronic pickup head that sends its signals to a re-

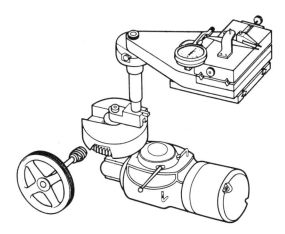

Illitron Div. of Illinois Tool Works Inc.

Fig. 17-3. The operating principles of the adjustable crank drive, which imparts by contact through a solid driving finger continuous rotation at varying angular velocity to the work gear mounted on a tooth space comparator instrument.

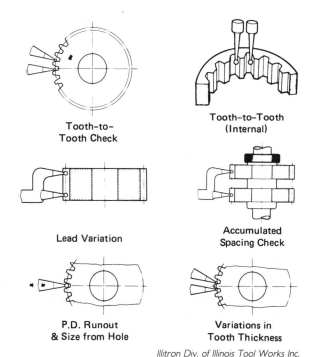

Illitron Div. of Illinois Tool Works Inc.

Fig. 17-4. Examples of gear tooth position checks that can be carried out, and the measurement results recorded, with the automatic tooth space comparator.

corder, which produces a continuous tracing line with narrowly spaced consecutive valleys and peaks. The peaks represent the actual tooth space recorded by the probe in the measuring position, and the comparison of the peak levels provides information on the measured tooth space error. Two tracings can be recorded consecutively on the same chart, one representing the left side and the other the right side of the teeth. The instrument is adapted to perform various types of tooth spacing checks, examples of which are shown in the diagrams of Fig. 17-4.

The *index position measuring instruments* shown in Fig. 17-5 operate with a single probe that intermittently advances into contact position with the specimen surface at selected and consistently repeated identical points. The probe contacts take place after the specimen has been indexed automatically by a preset angular distance. The instrument is well suited for tooth spacing checks on gears, and it is also adapted for index position checks on other types of mechanical components with critical index specifications. The machine spindle on which the gear or other workpiece is mounted is indexed by a crankshaft and connecting rod mechanism; the length of the crank throw can be adjusted. Attached to the connecting rod is a spider arm carrying two stop buttons, which contact adjustable anvil blocks as the connecting rod oscillates. The distance between the anvil blocks determines the degree of arc through which the work spindle rotates.

When in gaging position, the sensing probe feeds its signals into the amplifier of a recording system. Examples of the chart tracings produced are shown in Fig. 17-6. Each index measurement is represented by a short plateau of the tracing line, at levels displaying the index differences between adjacent elements, as well as the variation around the entire periphery. Index measurements of both sides of the gear teeth, or other surface features, can be recorded consecutively on the same chart, using identical chart positions for the starts of the two measurements. Such chart tracings permit the ready evaluation of the workpiece index accuracy.

TOOTH PROFILE MEASUREMENT

The profile of the gear tooth is the shape of its functional surface that during the operation of the gear is in contact with the corresponding tooth surface of the mating gear. The involute curve proved to be particularly well adapted for the profile of most types of commonly used gear teeth because it can assure a constant velocity transmission of the rotation in a slippage-free rolling contact. The functional profile of most types of spur, helical and herringbone

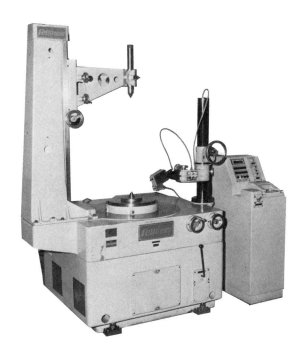

Fellows Corporation

Fig. 17-5. Index measuring instruments with single probe and precisely adjustable, self-contained mechanical indexing mechanism, complemented with recorder for rectangular charting at six different rates of error magnification.
(Left) Instrument for gears with $13\frac{5}{8}$-inch maximum pitch diameter capacity.
(Right) Large instrument with 36-inch maximum pitch diameter capacity.

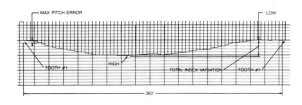

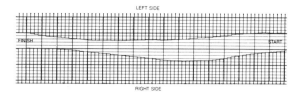

Fellows Corporation

Fig. 17-6. Chart tracings produced by the index-measuring instruments shown in Fig. 17-5.
(Top) Chart of one side of a 72-teeth gear, displaying the individual pitch errors and the total index variation.
(Bottom) Chart with two tracing lines, representing the opposite sides of the teeth of the same gear.

gear teeth is the involute curve, and the following discussion of tooth profile measurements applies to this category of gear tooth shapes.

The form of an involute curve can be visualized as the path of any specific point of a taut string while it is unrolled from a cylinder. That relationship between a cylinder and the unrolling taut string, which is always coincident with a straight line tangent to the cylinder at any momentary point of separation, is also the basis for the development of the involute gear profile in the manufacturing process, as well as in the commonly used methods of gear profile measurement.

Figure 17-7 shows the theoretical profile of involute gear teeth, together with the nomenclature commonly used in describing the measurement of the tooth profile. It is possible to inspect an involute gear profile by the coordinates of selected points along the profile using a gear tooth caliper, measuring them over balls or pins, inspecting the tooth profile on an optical projector, or, as a recent development, making an extended series of measurements

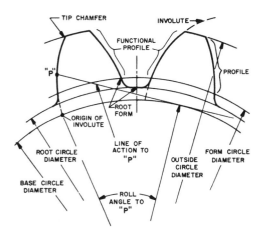

American Gear Manufacturers Association

Fig. 17-7. Nomenclature of the elements of the gear tooth profile.

on a computer-controlled measuring machine. The preferred and most widely used method, however, is by instruments generating a controlled involute path for a probe that is in continuous contact with the profile surface of the gear tooth being inspected.

Figure 17-8 shows a series of diagrams visualizing the manner in which the involute path of the inspection instrument is generated, in a modified application of the basic concept: the unrolling of a taut string from the surface of a cylinder. Referring to Fig. 17-7, the cylinder represents the base circle and the position of the taut string corresponds to that of a line of action for a given P point along the tooth profile.

When rolling the cylinder–straight-rail couple, the locus of consecutive P points will be an involute curve. Positioning the probe of a profile-measuring instrument coincident with that P point in relation to a properly mounted work gear, the path of that probe will correspond to the theoretical involute profile of the contacted gear tooth. Consequently, any deviations of the actual tooth profile from the true involute curve cause proportional deflections of the instrument probe. The amount of the probe deflections can be displayed by an indicator, but much more satisfactory information is obtained by contacting the tooth surface with the probe of an electronic gage, whose signals are fed into a recorder.

According to the design concept shown in Fig. 17-8, the cylinder used in conjunction with the straight rail must have a diameter equal to the theoretical base circle of the gear being inspected. In the simpler types of involute profile-measuring instruments, which may be preferred for use in the production

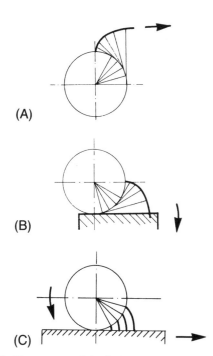

Fig. 17-8. The concept of developing an involute curve and its application in the design of involute gear tooth profile-measuring machines; the loci of the ends of the imaginary taut string are regarded as the consecutive portions of the point P.
(A) As a string is unrolled from a cylinder, any point of that tautly held string describes an involute curve. (*Note:* The fine tangent lines in the diagram represent the string in its consecutive stages of unrolling. The heavy line, coincident with a particular point along the unrolling string, is the generated involute curve.)
(B) Replacing the taut string by a tangent straight rail, pivoted around the stationary cylinder: any point along the rail describes an involute curve.
(C) Applying the relationship between a cylinder and a tangent rail in slippage-free rolling contact to an arrangement with a cylinder rotating around a stationary axis and the tangent rail moving along a straight line; here again, each point on the rail travels along an involute path in relation to the cylinder. This arrangement is practical for the purpose of involute profile testing.

shop, that requirement is satisfied by operating with interchangeable discs, installing the one with the diameter of the applicable base circle. Instruments of this type are designated as *involute profile-measuring machines with interchangeable base circle discs.*

The more versatile types of tooth profile-measuring instruments are built with special linkages of various designs, which permit adjustment of the effective diameter of a single base circle disc to cover a range of different base circle diameters. The latter are the *involute profile-measuring machines with adjustable base circle diameter.*

Examples of both types of profile-measuring machines are discussed.

Involute Profile-Measuring Machines with Interchangeable Base Circle Discs

For gear tooth profile measurements on the production line, where the same type of gear must be checked for a considerable period before the inspection instrument can be set up to accommodate gears with different base circle diameters, involute profile-measuring machines with interchangeable base circle discs offer several advantages. Such machines, because they are less complex in design, can be built sturdier, are lower in cost, and relieve the operator from sensitive instrument adjustment procedures that, when incorrectly performed, may be the source of inspection errors.

Involute profile-measuring machines with fixed base circle disc diameters are now generally built with designs permitting a limited amount of base circle diameter adjustment to compensate for small differences between the available base circle discs and the theoretical base circle diameter of the work gear. The range of such compensating adjustment may be on the order of 0.8 to 2 inches, depending on the particular make and model.

Figure 17-9 shows the work area of an involute profile-measuring machine with interchangeable base circle discs. The selected disc, installed under the work gear in a rigorously coaxial position, is vis-

Mahr Gage Co.

Fig. 17-9. The operating area of an involute profile-measuring machine with interchangeable base circle discs. The instrument has the capacity to inspect and record the tooth profile of gears within a base circle diameter range of 0.3 to 13 inches and is adapted for both spur and helical gears, also for checking helix angles over the range of 0 to 45 degrees.

ible in the figure, which shows the disc in contact with the straight edge or rail of the gage head slide; the relative positions of the work gear, the straight edge and the gage probe can also be distinguished.

The gage head of this instrument is mounted on a vertical slide installed on the horizontal base or tracer slide, which carries the straight edge. That vertical slide permits the height setting of the gage head and may also be used for measuring the tooth helix angle using a special drive of the instrument. This vertical movement represents an additional capability of various types of modern profile testers, including the illustrated instrument model.

The diagram Fig. 17-10 illustrates the operating principles of the mechanism that produces in the instrument shown in Fig. 17-9 the coordinated actions of (1) the relative movement between the work gear and the gage probe along the involute curve, which represents the theoretical profile of the work gear tooth; (2) the adjustment of the effective base circle diameter to compensate for minor differences between the diameters of the theoretical base circle and that of the employed disc; and (3) the production of a coordinated vertical movement of the gage head for testing the helix angle of the gear teeth. The mechanical elements shown in the diagram and their movements are the following, described in the order of the functions just listed.

1. The work gear (Z) and the base circle disc (G) are mounted coaxially on a common spindle, which is installed on the work slide (S), that being advanced against the straight edge (T); a clutch system assures the slippage-free rolling contact between the disc and the straight edge. The straight edge is mounted on a tracer slide (TS), the movement of which causes the rotation of both the disc and the coaxially mounted work gear. The point of contact between the gage probe (M) and the gear tooth must be located precisely over the active face of the straight edge in order to ensure the relative probe movement along the theoretically correct involute path. The micrometer screw (MS) serves for the fine setting of the gage problem.

2. The adjustable pivot point (D) is positioned on the setting device (E) to a dimension corresponding to the difference between the theoretical base circle diameter and the actual disc diameter. The mounting point (P) on the transmission lever (U) is firmly connected to the machine base; the linkage thus set imparts to the straight edge an additional movement in relation to that of the tracer slide, thereby producing the required compensating action.

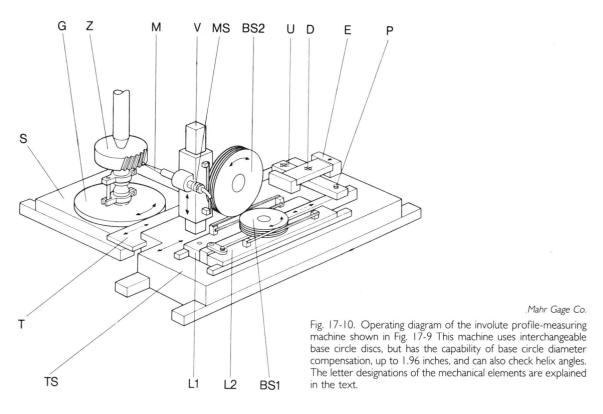

Mahr Gage Co.

Fig. 17-10. Operating diagram of the involute profile-measuring machine shown in Fig. 17-9 This machine uses interchangeable base circle discs, but has the capability of base circle diameter compensation, up to 1.96 inches, and can also check helix angles. The letter designations of the mechanical elements are explained in the text.

3. After the pivot point *(D)* has been set by means of the device *(E)* to a position corresponding to the gear helix angle, the movement of the bar (*L1* or *L2*, depending on the direction of the gear helix) acts on the steel tape pulleys *(BS1* and *BS2)*, which produce the vertical displacement of the gage head along the ways *(V)*. These movements take place concurrently with the rotation of the work gear caused by the action of the straight edge travel, resulting in a relative probe path that corresponds to the theoretical helix angel.

Involute Profile-Measuring Machines with Variable Base Circle

In inspection departments where a variety of gears must be checked for the correctness of several parameters, including the involute profile, instruments with variable base circles offer the advantages of reduced tooling and quick setup changes. The operating principles of such variable base circle profile-measuring instruments are essentially similar to those using interchangeable base circle discs, with the significant difference, however, that for a wide range of different base circle diameters a single disc can be used, because the effective base circle diameter of the movements produced by that disc is mod-

ified by an adjustable linkage that is inserted into the motion-transfer mechanism.

Different design principles can be applied to accomplish that modification of the motion generated by a single base circle. Figures 17-11, 17-12, and 17-14 are diagrams of different mechanisms that are used for base circle modification by several prominent manufacturers of involute profile-measuring machines. Various types of these instruments combine the capability of involute profile measurement with that of testing other gear parameters, as is pointed out in the following description of the illustrated systems.

Variable Involute Profile-Measuring Instrument with Master Base Circle Sector

The diagram in Fig. 17-11 shows the operating principles of a base circle modification system incorporated into instruments that use a single base circle sector banded to a generating carriage, both members rolling together in slippage-free contact. A ratio bar, linked to the generating carriage, converts the motion originating from the fixed radius master base circle sector into the movement of the contact finger along an involute path, which corresponds to the set base circle diameter. For setting the ratio bar

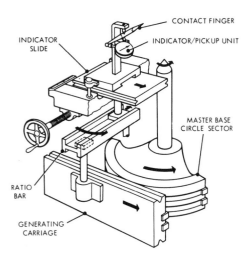

Fig. 17-11. Operation diagram of a variable involute profile-measuring instrument using a master base circle sector and a ratio bar for varying the effective base circle diameter.

Illitron Div. of Illinois Tool Works, Inc.

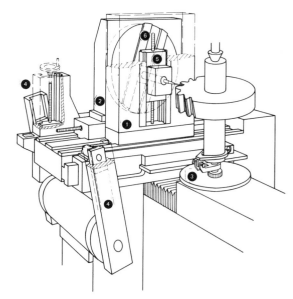

Dr.-Ing Hoefler GmbH

Fig. 17-12. Principles of the mechanical motion-transfer system used in a variable involute profile-measuring machine, with helix-angle-checking and gear contact line tracing capabilities. (The reference numbers *(1)–(6)* shown in the diagram are explained in the text.)

to the required base circle diameter, an optical indicator gage is used, reading with 0.0001-inch discrimination the graduations of an etched stainless steel scale. That method of base circle setting is fast and accurate, and is done without the use of gage blocks or similar auxiliary tools.

The described system of base circle variation is used on involute profile-measuring machines that are manufactured in two models, with rated capacities of 12-inch and 24-inch maximum gear diameter. These instruments measure profile deviations within 0.0001-inch accuracy and locate such errors within 0.1 degree of roll. The measured deviation of the actual tooth profile from the theoretical form can be read from the position of the indicator pointer, and the position can be read from an optionally supplied roll indicator. The preferred use of the instruments, however, is in conjunction with an electronic chart recorder that produces permanent records as the means of dependable tooth profile analysis.

Variable Involute Profile- and Helix-Checking Instrument with Single Base Circle Disc

The diagram in Fig. 17-12 shows the operating principles of an involute profile- and helix-testing machine that is adapted to measure any one of the following gear tooth parameters: *(A)* the tooth involute profile, *(B)* the lead, or *(C)* the contact line, by moving the stylus (probe) of the instrument along the lines indicated on the gear tooth diagram (Fig. 17-13).

The motion-transfer system shown in Fig. 17-12 operates with a single base circle disc *(3)*, in contact with a straight edge mounted on the main slide. The mechanism is designed to compensate for the difference between the diameter of that disc and the theoretical base circle diameter of the work gear by using a system of levers *(4)* that are equipped with optically aided adjustment. The following elements of that mechanism are brought into action for testing various gear tooth parameters, which are listed with reference to Fig. 17-13:

(A) *The involute profile.* The measuring slide *(1)* and the main slide *(2)* are rigidly locked together. The horizontal lead screw of the main slide is driven.

(B) *The lead of the teeth.* The measuring slide *(1)* and the main slide *(2)* are connected through the sliding block of the stylus slide *(5)*, and the measuring slide is locked in position. The vertical lead screw of the measuring slide is driven and the resulting helix path is controlled by the angle setting of the sliding block guideway in the main slide.

(C) *The contact line.* The contact line results in the form-generating cutting of the gear teeth with a rack-shaped tool, such as a hob, a rack cutter, or a

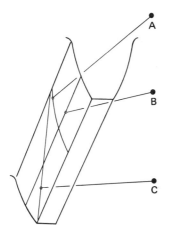

Dr.-Ing. Hoefler GmbH

Fig. 17-13. Gear tooth surface elements, the forms of which can be inspected, and deviations from the theoretical profile recorded by the involute profile measuring machine operating with the mechanism shown in Fig. 17-12.
(A) Involute profile.
(B) Lead (helix angle).
(C) Contact line (path of the tooth form generating tool).

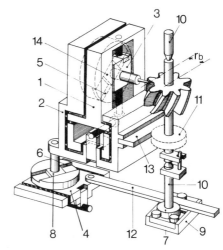

Klingelnberg Gear Technology, Inc.

Fig. 17-14. Operation diagram of an infinitely variable base circle type involute profile-measuring instrument, adapted for the measurement of several different gear tooth parameters, including the helix angle. The base circle radius is r_b. (The reference numbers (1)–(14) shown in the diagram are explained in the text.)

grinding wheel with straight flanks. The degree of straightness of the chart tracing produced by the contact of the stylus, which is moved along the theoretical path of that line, is a valuable guide for process control in the manufacture of gears. For contacting the tooth surface along the contact line, the main slide (2) is locked together with the straight edge and the work gear spindle. The stylus slide (5) is moved by the vertical lead screw, causing the movement of the measuring slide (1) in correspondence with the helix angle setting of the guideway (6).

This system of motion variation is used in a type of involute profile-measuring machine, which is built in two models, adapted for gears with 16-inch and 24.8 inch maximum tip circle diameter. The minimum work gear root circle diameter for both models is 0.2 inch, and helix angles within the range of 0 to 60 degrees (optionally 90 degrees) can be checked.

Variable Involute Profile- and Helix-Measuring Machines for Large Size Gears

The mechanism shown in the operational diagram (Fig. 17-14) incorporates adjustment means for both the infinite variation of the effective base circle diameter and of the helix angle, permitting the testing of different tooth flank lines, leads and lines of

action. Mounted on bridge type guides is the measuring slide (1) with its vertical guides and the rolling slide (2), which moves parallel to and partly envelopes the measuring slide. The bridge type guides also support the vertical superstructure with its large optics for the angular setting of the straight guide (5). The vertical slide (3), which carries the measuring head and moves inside the vertical guide of the measuring slide (1), is connected to the angularly adjustable straight guide (5) of the rolling slide (2) by means of a rotating sliding block (14). The movement of the block (8) attached to the rolling slide (2) engages the angularly adjustable straight guide (6), which functions as a sine bar of the control slide (4). That control slide can be brought into positive contact with the firmly fitted rolling disc (7) by means of a rolling straight edge (12). The clamping fixture (9), which can be engaged to prevent slack in the rolling motion, the rolling disc (7), and the test gear support bearing (10) are all attached to the work slide (not shown in the diagram). Power movement to slides (2), (3) and (4) are produced by an electric motor and are transmitted by feed screws; these movements can be engaged selectively, according to the test function being performed. A rolling straight edge (13) can be installed to permit the use of interchangeable base circle discs (11) for inspection processes in which the use of a fixed diameter base circle disc is preferred.

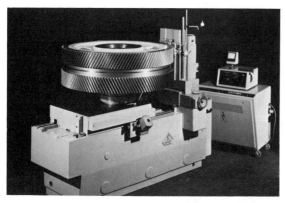

Klingelnberg Gear Technology, Inc.

Fig. 17-15. Large size, 63-inch maximum gear diameter capacity, involute profile-measuring machine, operating with infinitely variable base circle diameter and adapted for lead testing from a 0 to a 90 degree angle, shown in rear view during the measurement of a heavy herringbone gear.

Fellows Corporation

Fig. 17-16. Involute profile-measuring machines with variable base circle diameter, for external and internal spur or helical gears. (Right) The vertical axis model is designed for gears with 12-inch maximum pitch diameter. (Left) The horizontal axis model is designed for gears with 24-inch maximum pitch diameter.

Involute profile- and helix-testing machines using the described system of motion-modifying mechanism are manufactured in three models, for gears with maximum base circle diameters of 23.6, 45, and 59 inches, and are adapted for testing both external and internal gears. Figure 17-15 shows the rear view of the largest model in the process of measuring a heavy herringbone gear.

Tooth Profile-Measuring and Recording Instruments for External and Internal Spur and Helical Gears

Two widely used models of involute profile-measuring instruments with infinitely variable pitch circle diameter are shown in Fig. 17-16; these machines are designed for gears with 12-inch and 24-inch maximum pitch circle diameter. Both machines are adapted for measuring and recording the entire tooth profiles of external and internal spur or helical gears. The special gooseneck pointer shown in Fig. 17-17 provides a convenient means for checking internal gears.

These machines are operated by means of a handwheel mounted on a shaft that, through a friction safety clutch and worm, rotates the work spindle. Mounted on the work spindle is also the master involute former cam, which has a base circle approximately equal to the maximum capacity of the instrument. Two alternate methods are available for setting the desired base circle radius for the path of the probe that contacts the gear tooth being measured: (a) size blocks with micrometer and dial indicator or (b) direct reading optical measuring de-

Fellows Corporation

Fig. 17-17. The gooseneck pointer in contact with the tooth of an internal work gear, installed on the vertical axis involute profile-measuring machine shown in Fig. 17-16.

vice. An electronic meter, graduated in ten-thousandths of an inch, gives visual readings for the setup and calibration of the instrument. For the positive check of the instrument's proper calibration, a master involute is supplied; that master has three contour sections, corresponding to (1) a true involute, (2) a modified involute and (3) a plain offset flank.

The machines are equipped with a solid-state rectilinear recorder, producing chart tracings at any of six different magnifications, ranging from 100× to 5000×.

The dual ratio chart paper drive produces tracings with each horizontal graduation line of the chart paper representing a 3 degree or a half degree roll of the gear being tested. Examples of the chart recordings produced are shown in Fig. 17-18. In the illustration these charts are accompanied by the diagrams of the pertinent tooth profiles displaying, at an exaggerated scale for clarity, the contour modifications that are distinctly recorded by the actual chart tracings. The location of each profile modification is shown on the chart in a manner associated with its angular position, and the amount of departure from the theoretical involute profile is presented at a scale of 0.0002 inch for each chart paper graduation.

TOOTH THICKNESS MEASUREMENT

In the manufacture of gears the teeth are usually made thinner than the width of the tooth space along the pitch circle in order to assure the specified backlash. That additional space is commonly required to avoid tooth jamming as the consequence of tolerances in pitch, profile, concentricity and so forth, and also to provide for the increase in the diameter of the operating gear caused by thermal expansion. On the other hand, too much backlash caused by excessive tooth thinning results in noisy gears. The thickness of the gear teeth, therefore, is measured for process control, and tooth thickness measurements are also used in the inspection of finished gears.

Tooth thickness is the length of the pitch circle section bounded by the opposite flanks of the gear tooth. The length of the circle section is the arc thickness of the tooth, but for practical reasons the straight line distance between the end points of that arc, the chordal thickness of the tooth, is the distance actually measured.

The radial distance of the pitch circle from the periphery (outer diameter) of the tooth is the addendum, but for locating the plane of action with the commonly used types of tooth thickness measuring tools, the chordal addendum is used. The design principles of these gear tooth dimensions are shown in Fig. 17-19. The length of the distance termed chordal thickness is slightly less than that of the arc thickness. The number of gear teeth and the required degree of accuracy determine whether that difference can be ignored or whether the measured distance should be modified by a calculated correction factor.

A direct method of tooth thickness measurement is by means of a *gear tooth caliper*, such as the one shown in Fig. 17-20. The illustrated type of gear tooth vernier calipers are manufactured in two sizes,

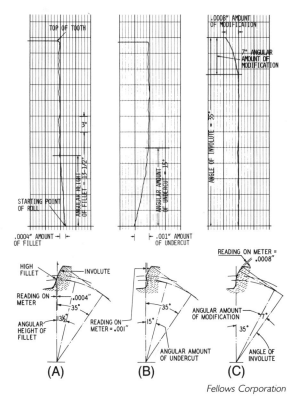

Fellows Corporation

Fig. 17-18. Examples of profile tracings on rectilinear charts, produced by the involute profile-measuring instruments shown in Fig. 17-16. The chart tracings distinctly record the amounts and locations of gear tooth profile errors and modifications, such as high fillet, undercut and tip modification. The tooth profile diagrams, drawn at an exaggerated scale and placed under the individual charts, visualize the contour deviations recorded by the chart tracings.

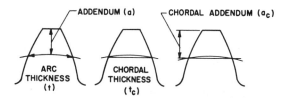

American Gear Manufacturers Association

Fig. 17-19. The relative positions of dimensions used in the design and measurement of gear tooth thickness.

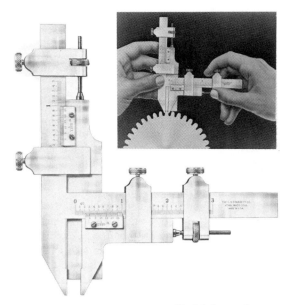

The L.S. Starrett Company

Fig. 17-20. Gear tooth caliper for measuring the chordal thickness of gear teeth. The inset illustrates the use of that tool.

with 20 to 2 and 10 to 1 diametral pitch capacity, or as metric tools with $1\frac{1}{4}$- to 12- and $2\frac{1}{2}$- to 25-millimeter module capacity. The gear tooth caliper has two adjustable members, the tongue for setting the (chordal) addendum and the jaws for measuring the chordal thickness of the tooth. Each member has a vernier for measuring the distance in increments of 0.001 inch, or 0.02 millimeter for the metric tools. Because the tooth thickness measurement with calipers is based on setting the contact plane to the theoretical chordal addendum, differences in the outside diameter of the gear, due, for example, to undersize or oversize blanks, will affect the measured tooth thickness. When such tooth outside diameter variations are intentional or otherwise known, their effect on the measured tooth thickness can be calculated and used as a correction value.

The *gear tooth addendum comparator*, Fig. 17-21, also utilizes the interrelation between the addendum and the tooth thickness, but reverses the approach applied with the gear tooth caliper; in this case the instrument is set to the calculated equivalent of the nominal tooth thickness, using locating elements that correspond to the pertinent basic rack, and the indicator is set to zero at a level representing the nominal addendum. A plus reading of the indicator results when the tooth is thinner than the standard size, a condition permitting the tooth to project farther toward the indicator probe.

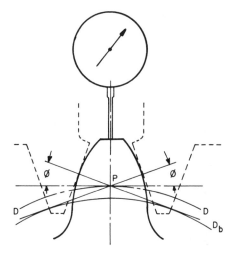

American Gear Manufacturers Association

Fig. 17-21. The operation of the gear tooth addendum comparator showing in outline the locating elements having the form of a rack with pressure angle equal to that of the gear tooth being checked:

ϕ = pressure angle;
D = pitch diameter;
D_b = base circle diameter;
P = pitch point.

The tooth thickness can be derived from the *measurement of the span chord* over a number of teeth sufficient to permit parallel probe faces of an instrument to contact two nonadjacent teeth on the opposite sides of their involute flanks, in the manner indicated in Fig. 17-22. The instrument used may be a vernier caliper or a disc-type micrometer. The distance measured is the sum of base pitches of the spanned teeth minus one $(n-1)$, adding the thickness of one tooth at the base circle. The tooth thick-

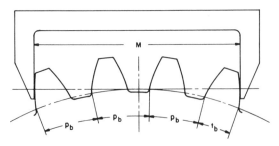

American Gear Manufacturers Association

Fig. 17-22. The principles of span measurement for the inspection of gear tooth thickness.

M = distance spanned by the measuring instrument;
P_b = base pitch;
t_b = tooth thickness at the base circle.

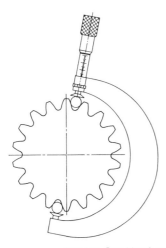

American Gear Manufacturers Association

Fig. 17-23. Principle of tooth thickness inspection by measurement over pins.

ness determined by span measurement is not affected by variations or runout of the outside diameter of the gear.

Inspection by *tooth thickness measurement over pin*, as indicated in Fig. 17-23, is an accurate method because the results are not influenced by the outside diameter or runout of the gear. The measurements are affected, however, by errors in tooth spacing and profile. The instrument used is either a disc-type micrometer of sufficient capacity or a length-measuring machine with flat probe faces. Data regarding the required pin diameters and the nominal distances over pins, listed for different pressure angles and numbers of teeth, are found in technical handbooks, for example, *Machinery's Handbook*.

The *composite method of gear inspection* for determining the functional properties of a gear by rolling it in tight mesh with a mating master gear is another method that signals variations in tooth thickness. Since gear rolling tests register center distance variations resulting from the composite effect of all tooth errors, that method supplies only indications of tooth thickness variation but does not provide a means to dependably discriminate between nonuniform tooth thickness and variations of other tooth parameters.

GEAR LEAD MEASUREMENT

In helical gears, the lead is the axial advance of the helix for one complete turn of the gear. The lead angle is bounded by an imaginary tangent to the tooth surface, contacting it at the pitch circle, and by

the gear axis. The helix angle is the complement of the lead angle. The lead of a helical gear is constant, but the helix angle increases from the root to the tip of the tooth. These relationships are shown diagrammatically in Fig. 17-24.

The close matching of the leads of mating gears is important for assuring the uniform distribution of load across the width of the gear. Improper lead matching causes load concentrations at isolated areas of the teeth, with the probable consequences of noisy operation and gear damage. For that reason, tolerances are established for lead variations in gears pertaining to different quality numbers, and the measurement of the gear lead is an important process in the analytical inspection of gears.

For inspecting the lead of helical gears, the lead-measuring machine must impart to the sensing probe of the indicating or recording instrument an axial advance that is coordinated with the rotation of the work gear in a manner resulting in the theoretical lead of the gear being tested. Specially designed mechanisms are used for carrying out the advance of the slide, which holds the instrument probe, in a controlled relation to the rotating of the work gear. Commonly, that mechanism may consist of a sine bar arrangement or may have discs with variable pin radii, which can be set to correspond to the desired lead ranging from zero to infinity. The latter type of system is used in the lead-measuring machine shown in Fig. 17-25. In this case the radius setting of the pin is accomplished by means of end-measuring rods and large thimble micrometers. A master lead bar, with lead modification of known magnitude, can be supplied for checking the instrument's functions. The illustrated type of lead-measuring machine is built in two sizes, for work gears with maximum outside diameters of 14 inches and $25\frac{3}{4}$ inches. The pertinent maximum pitch diameters are 12 and 24 inches, respectively. The measuring head of these machines is designed to permit the checking

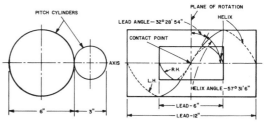

American Gear Manufacturers Association

Fig. 17-24. The interrelation of tooth inclination parameters designated by the terms lead, lead angle and helix angle.

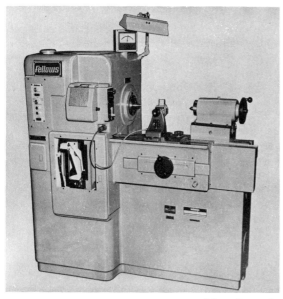

Fellows Corporation

Fig. 17-25. Gear tooth lead-measuring machine with built-in recorder instrument, adapted to test gears with 12-inch maximum pitch diameter; it has 120 degrees maximum rotation of the work spindle. Both right- and left-hand leads, as well as internal gears, can be inspected.

Fellows Corporation

Fig. 17-26. The arrangement of the sensing instrument and of its probe, used on the lead-measuring machine shown in Fig. 17-25, permits the checking of multiple workpieces on the same shaft.

Fellows Corporation

Fig. 17-27. Checking the lead of an internal gear on the lead-measuring machine shown in Fig. 17-25, using an optional pointer head of special design.

of the gear tooth lead of multiple workpieces on the same shaft, as shown in Fig. 17-26.

After having set the proper gear lead and having mounted the work gear on a shaft held between centers, the pointer of the measuring head is brought into contact with that gear tooth element along which the gear is checked. That is preferably coincident with the pitch circle, the location of which is ordinarily halfway between the tip and the root of the tooth. The movement of the pointer should be always in a direction normal to a plane tangent to the tooth at the point of measuring contact.

Most types of lead-measuring machines, including the model shown in Fig. 17-25, are adapted to check both external and internal gears, the latter with a special pointer mounted on the measuring head, in the manner shown in Fig. 17-27.

Both sides of the gear tooth can be inspected by changing the position and direction of deflection of the pointer, and both recordings can be made on the same chart paper, with the two tracings side-by-side, as shown in Fig. 17-28. The chart thus produced provides an easily appraisable record of the tooth helix. In addition to the lead, the crowning of the gear can also be measured, as shown in the lower portion of Fig. 17-28.

Another method of lead checking is by measuring the *base helix angle* of the gear. Several types of modern gear tooth profile-measuring machines, including those the linkage mechanisms of which are shown in Figs. 17-12 and 17-14, provide the means for carrying out and recording the results of such measurements. The base helix angle is, at the base

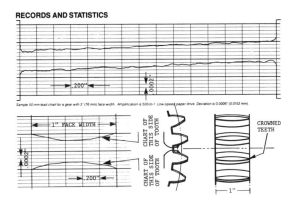

RECORDS AND STATISTICS

Sample 50 mm lead chart for a gear with 3" (76 mm) face width. Amplification is 500-to-1. Low-speed paper drive. Deviation is 0.0006" (0.0152 mm).

Fig. 17-28. Samples of gear lead tracings produced on rectilinear chart paper, displaying the lead deviations from the theoretical, as measured on both sides of the gear tooth. The lower chart illustrates how the crowning of the gear tooth can also be inspected, recording on the chart the gear tooth contour, which is shown in an exaggerated scale on the diagram on the right.

Fellows Corporation

cylinder of the involute gear, that angle which the tooth has in relation to the gear axis.

In addition to the specialized gear-measuring methods and machines shown thus far, coordinate measuring machines (see Chapter 12) can be used for gear inspection. When equipped with gear inspection software from Sheffield Measurement, the modern computer-controlled CMM is capable of performing the following checks: profile; lead; spacing and runout; and tooth topography. The results of these checks are then documented through both inspection data printouts and a series of graphic charts. When not being used for gear inspection, the CMM can be used for a wide range of other inspection tasks.

FUNCTIONAL GEAR CHECKING

The method determines the composite effect of gear errors, simulating the operational conditions by running two mating gears in mesh. In that test, also referred to as the *composite action method of gear inspection*, the gears are most commonly operated in tight mesh, using gear rolling instruments as equipment. The tight mesh engagement, which is used for most functional gear checking, produces contact on both flanks of the meshing gear teeth, and for that reason it is also termed double-flank or dual-flank testing. Another gear rolling test method, less commonly used, operated with fixed gear center distance, in a manner to limit the contact to a single flank on each of the meshing gear teeth; this method,

of relatively recent development, is termed single-flank gear testing, and will be described separately. The following discussions refer to the more commonly applied double-flank gear checking by tight mesh rolling.

The functional gear checking is carried out by rolling the gear being tested in tight mesh contact with a corresponding master gear of known accuracy, both gears being installed in a *gear rolling instrument*. The specimen gear, also termed the work gear, is mounted on a fixed arbor, while a parallel arbor for accepting the master gear is installed on a slide that is both adjustable for initial setup and has a limited free movement for accommodating center distance variations during the rolling of the meshing gears. Gear rolling instruments used for inspecting bevel, hypoid, or worm gears have mounting arbors at right angles to each other.

When particular parameters of the specimen gear vary from the design dimensions incorporated into the master gear, the rolling of the two gears in tight mesh causes variations in their center distance. Such potential variations from the designed dimensions of the gear being tested include the effects of incorrect profile, pitch, tooth thickness, runout and, in the case of helical gear, the effects of lead variations as well. The resulting changes in the center distance reflect the composite effect of these potential variations, a condition to which reference is made by the term composite action method of gear inspection quoted above.

In addition to the potential deviations of the listed parameters, other defective conditions can contribute to center distance variations, such as excessive surface roughness or waviness, nicks or similar surface blemishes.

The master gears to which the accuracy of the gear being inspected is compared must be of a quality class substantially better than that of the work gear. There are American Gear Manufacturers Association (AGMA) recommendations for the relationship between the quality numbers of the master gear and of the work gear, and also AGMA tolerance data for master gears of different quality numbers. These master gear tolerances specify the maximum allowable variations for total and tooth-to-tooth composite errors, as well as the tolerances of significant gear elements.

The center distance variations that occur during the rolling of the meshing gears can be observed.

1. By means of an indicator, which registers the movement of the slide holding the arbor of the master gear; or

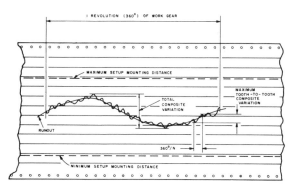

American Gear Manufacturers Association

Fig. 17-29. Chart produced by a recording type gear rolling instrument, displaying the gear tooth errors of a typical gear when run in conjunction with a corresponding master gear of known accuracy.

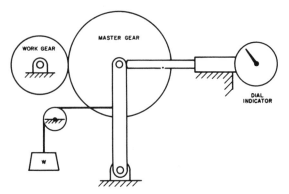

American Gear Manufacturers Association

Fig. 17-30. Schematic diagram of a typical indicator type gear rolling instrument. The weight acting on a string, which is guided by a pulley, is symbolic of an applied force commonly produced by spring action. (To simplify the diagram, the master gear is shown supported by a pivoting arm instead of the commonly used slide that assures a truly linear displacement of the master gear.)

2. By means of a graphical record produced in the form of a strip chart, which is adapted to detailed analysis and also provides a permanent record for future references.

The indicated or recorded center distance variations reflect the composite effect of both the tooth-to-tooth composite variations and the total composite variations, the latter also comprising the effect of runout. Figure 17-29 shows a typical strip chart record of a gear rolling test. In view of the diversity of defective conditions affecting the variations of center distance during a full rotation of the work gear, indicator readings are not adapted to analyze in detail the conditions that cause those variations. If the purpose of the testing, however, is only to determine whether the center distance variations are not exceeding an acceptable range, indicator-equipped gear rolling instruments, which are relatively inexpensive and permit quick testing, can prove adequate. In order to reduce the reliance on the skill and judgment of the inspector using an indicating gear rolling instrument, the indicator can have limit markers or may be complemented with light signals, announcing accept or reject conditions of the tested gear.

In the following, examples are presented of typical gear rolling instruments that are designed and widely used for gear inspection by the composite action method. The gear rolling instruments described here operate with gears in tight mesh (double-flank contact), and the effect of gear errors produces center variations.

Indicating Type Gear Rolling Instruments

Indicating gear rolling instruments operate by the principles shown in Fig. 17-30. One of the gears,

usually the work gear, is mounted on a fixed arbor, and the arbor for the other gear, the master, is installed on a slide permitting free linear movement in a direction radial with respect to the fixed arbor, toward which it is forced, usually by spring action.

The initial position of the movable arbor with respect to the fixed position can be adjusted corresponding to the designed assembly distance between the centers of the two gears to be rolled in tight mesh. Subsequently, when the gears are rolled, that center distance can vary, whereas the tight mesh condition is assured by a spring action. Variations in the position of the movable arbor are transmitted to an indicator displaying the center distance variations, which reflect the effect of several gear errors.

The indicator type gear rolling instruments are usually manually operated, built with simple designs permitting rugged construction, which is desired for production floor service where such inspection tools are widely used for on-the-spot quality control at the gear cutting machines. The indicators displaying the center distance variations commonly have 0.0005-inch graduations, and most types of these instruments are available either with stud arbors or in column types with adjustable center height for accommodating gears with integral spindles in various lengths. A typical model of such instruments has a center distance adjustment range of 2 to 9 inches, thus being adapted to a wide range of gear sizes. Indicator type manually operated gear rolling instruments of the described system are available in other size ranges as well and can be equipped with indicators having the desired degree of resolution.

Recording Type Gear Rolling Instruments

The most common gear errors that produce distinct center distance variations are gear runout and size variation. The tooth-to-tooth variations, however, may cause only a flickering of the indicator pointer, the magnitude of which is difficult to assess by visual observation. For that reason, when a more thorough analysis of the gear conditions is required, gear rolling instruments with mechanical or, more commonly, with electronic recorders are used. The chart tracings produced by such recorders also represent permanent records for future reference.

Figure 17-31 shows a widely used model of recording type gear rolling instruments, with that particular machine offering a center distance range of $3\frac{1}{8}$ to 15 inches, which can be further extended by using an optional center support. Work gears within the capacity range of the instrument can be mounted easily between centers and brought into alignment with the master gear by means of the elevating handwheel of the lower work spindle and a crank used for raising or lowering the upper work center. The cabinet containing the master gear spindle, its drive and the recording instrument can be moved by hand crank along a guideway to bring the master gear into mesh with the work gear. The force by which the master gear engages the work gear, termed gate pressure, is adjustable from 0 to 12 pounds. During the testing the master gear is driven by a variable-speed motor capable of rotating the master gear up to 10 rpm. The speed of the recorder chart drive motor is also steplessly variable, and the recording system provides six different magnifications, namely, 100×, 200×, 500×, 1000×, 2000×, and 5000×. The magnification normally used is 500 to 1, with each graduation line of the chart paper representing a 0.0002-inch departure from an ideal straight line that would result in the case of zero center variations.

The diagram in Fig. 17-29 illustrates the kind of information that can be obtained from chart tracings produced by recording type gear rolling instruments, including the model shown in Fig. 17-31.

While most types of spur gears either have bores and can be mounted on arbors or have integral shafts with centerholes, the mounting of other types of gears requires special fixturing. Figure 17-32 shows, as examples, the diagrams of such fixtures, which are available for the discussed model of gear rolling instrument.

Other makes of gear rolling testers are available in different sizes and center distance capacities. One versatile type has a center distance range of up to 23.6 inches for mounting work gears between centers and up to 31.5 inches for single-end mounting. That type of instrument, of modular design, can be operated with a wide variety of optional accessories, which are needed for testing internal gears, bevel gears, and even for the rolling test of worm gears against a master worm, with power drive for the worm. Expanded use of these instruments is assured by such accessories as a measuring head for single-error testing of concentricity, tooth spacing and tooth thickness, as well as attachments for checking and recording pitch errors on spur gears, helical gears, worm gears and bevel gears.

Gear Rolling Instruments for Fine-Pitch Gears

External and internal gears with fine pitch, typically in the range of 20 DP (diametral pitch) (about 1.25 module) and finer, are usually inspected on gear rolling machines specially designed for gears with such fine teeth, requiring light checking pressures only. A typical model of such instruments, designed for external spur and helical gears, accepts work gears up to 6 inches maximum pitch diameter, and offers a center distance range of $\frac{7}{8}$ to 4 inches. The

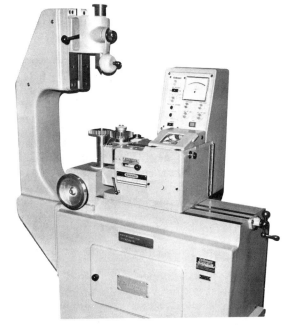

Fellows Corporation

Fig. 17-31. A widely used model of gear rolling machine for external and internal spur and helical gears, equipped with electronic indicating and recording instruments, for measuring composite errors resulting from variations in circular pitch, tooth profile and lead.

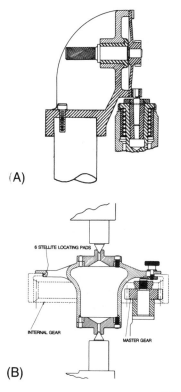

(A)

(B)

Fellows Corporation

Fig. 17-32. Diagrams of work gear mounting fixtures by which the gear rolling machine shown in Fig. 17-31 can be adapted for testing.
(A) Face gears that are mounted on a horizontal work spindle.
(B) Or, internal gears using a fixture with locating pads for mounting the work gear; this fixture is installed between the work centers of the basic machine.

trunnion-mounted master gear can be power-rotated at any of nine speeds for checking both spur and helical gears. The engagement pressure of the master gear is adjustable from 2 to 32 ounces. A sensitive dial indicator with 0.0001-inch graduations is provided for the visual inspection of the center variations of the gear being tested, and a recorder with four different magnifications produces charts for analysis and permanent record.

Universal Gear Testers

Instruments designated as universal gear testers are intended to carry out a wide variety of gear inspection operations, some of which are functional tests for determining composite action effects, while others are analytical measurements directed at specific gear dimensions. Universal gear testers are basically gear rolling machines with manual or motor drive and are equipped with recorders for charting

center distance variations in a manner easily associated with one complete revolution of the work gear. In addition, the basic instrument can be supplemented with a wide variety of accessory attachments for analytical measurements—checking tooth thickness, pitch, backlash, lead and involute profile. The analytical measurements are made with sensitive indicators using probes specifically adapted to the form and the location of the gear element to be measured.

While universal gear testers may substitute for several special-purpose instruments and thereby provide the means for all or most of the gear inspection operations of a particular manufacturing shop, the analytical measurements thus performed can be rather time-consuming, considering the need for installing and setting up different attachments.

SINGLE-FLANK GEAR ROLLING TEST METHOD AND EQUIPMENT

Composite action gear testing by the commonly used double-flank contact method, in which the work gear is rolled in tight mesh with the master gear, produces center distance variations by the composite action of various gear tooth errors. Although the double-flank method of gear rolling offers the advantage of simple instrumentation and operation, the information supplied does not represent conditions that result from a true simulation of the common meshing of gears in an assembly. In double-flank testing both flanks of the mating gears are in simultaneous contact along a portion of the tooth flank exceeding the area that is active in most types of actual gear operations.

Single-flank gear testing simulates the actual operating conditions of a gear pair running at a fixed—the designed—center distance, with the gear teeth contacting each other on one of their flanks only, thus allowing for backlash that is present in most gear couples. The single-flank gear tester measures the *transmission* error of a pair of gears, that is, the variation from a consistent angular velocity of the work gear that is driven by a master gear running at constant speed. The difference of the operating principles between the double-flank and the single-flank gear testing methods is shown in Fig. 17-33.

In single-flank contact testing the mating gears are installed at the designed center distance and, as indicated in diagram (B), circular optical gratings are mounted on both instrument shafts, which hold the work gear and the master gear. Each of the gratings gives a train of pulses whose frequency is the mea-

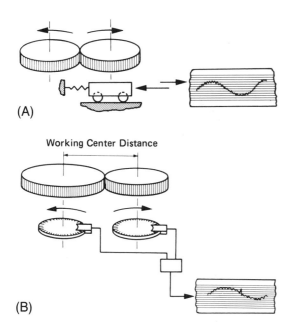

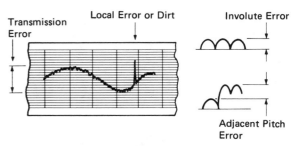

Goulder Mikron, Ltd.

Fig. 17-34. Typical chart tracing produced by a single-flank contact gear rolling tester, displaying total transmission variations, as well as errors of involute profile and adjacent pitch.

Goulder Mikron, Ltd.

Fig. 17-33. The operating principles of gear rolling test methods.
(A) Double-flank contact gear rolling with the gears in tight mesh. Variations of tooth profile and pitch produce center distance changes as composite effect.
(B) Single-flank testing with gears mounted at fixed center distance. Variations of gear tooth profile and pitch cause transmission errors, which are registered by an optical-electronic rotary resolver and recorded on a strip chart.

sure of the angular movements of the two shafts, and hence of the gear they carry. Because the numbers of teeth of the work gear and of the master are generally not equal, the trains of pulses are processed to produce a one-to-one ratio of the resulting signals. Thereby small differences between the modified reference frequency of the master and that of the work gear announce the single-flank, or transmission, error of the gear pair. The phase difference of the produced frequencies is measured electronically at a rate of 21,600 observations per revolution of the driven gear shaft and is recorded as a smooth wave form with high discrimination, which permits the detection of phase differences less than one second of arc.

Single-flank contact gear testing offers significant *advantages* in the following applications, which are listed as examples:

For indexing gears, in the case of which the direct measurement of transmission errors is much faster than the appraisal of findings derived from analytical measurements.

For power transmission gears, by detecting from the transmission error curve the sources of vibrations and noise.

For complex gear forms, such as bevel gears, worm gears, and offset conical gears, replacing the less-dependable inspection method based on contact patterns produced by coating the gears with a marking compound.

For determining the optimum center distance in the assembly of two product gears operated with critical transmission requirements.

The transmission error curve produced by single-flank testing, a typical example of which is shown in Fig. 17-34, resembles that resulting from gears rolled in tight mesh (double-flank testing), having tooth-to-tooth ripples superimposed on a waveform with a period of one revolution of the work gear. The transmission error curve produced by single-flank testing permits one to identify errors of profile and adjacent pitch of particular teeth, as well as the cumulative pitch error of the complete gear.

Examples of a few typical gear rolling instruments designed for single-flank and two-flank contact testing are shown in Figs. 17-35A and 17-35B. Figure 17-35B illustrates the work of a two-flank gear roll tester for the checking of rotational deviation, rotational tooth-to-tooth error, and torsional backlash on spur gears with straight and helical teeth, bevel and worm gears. Figure 17-35A shows the instrument setup with a measuring head equipped with two inductive and one coordinate measuring probes for use in single error tests (that is, radial runout, tooth thickness fluctuation, pitch and pitch-to-pitch errors).

The instrument shown in Fig. 17-36 is designed for worm and worm-wheel pairs with capacity for worm-wheels up to 30 inches maximum diameter, with a maximum center distance of 17 inches.

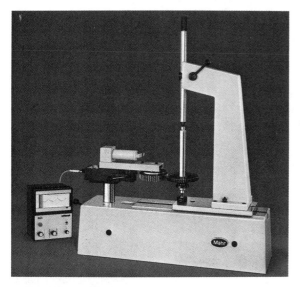

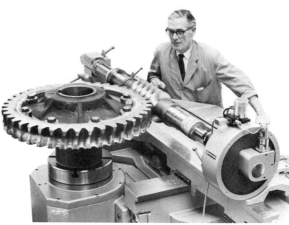

Goulder Mikron, Ltd.

Fig. 17-36. Single-flank contact gear rolling tester for heavy worm and worm-wheel pairs.

Mahr Gage Co.

Fig. 17-35A. Single-error testing probe type attachment for a two-flank roll gear tester.

Mahr Gage Co.

Fig. 17-35B. Workshop gear tester setup for two-flank roll testing of straight spur gears.

Two different arrangements of a special type of single-flank gear rolling tester are illustrated in Fig. 17-37. That particular machine is designed to permit changeover to two alternative arrangements: (A) for spur gears with center distances from 0 to 16 inches; (B) for bevel gears up to $16\frac{3}{4}$ inches in diameter. It is also possible to modify the positions on the machine table of the gear shaft bearings to increase the center distance in spur gear testing to $35\frac{1}{2}$ inches, thus accommodating spur gears to a maximum weight of 2200 pounds.

BACKLASH

Backlash is the clearance between mating tooth surfaces of a gear pair when assembled at the designed center distance. Backlash at the operating mounting distance is a desirable condition of most gear assemblies as a means of avoiding tooth jamming due to permissible errors in profile, pitch, concentricity, lead and so forth, and also due to thermal expansion of gears running at high speed. A certain amount of clearance between the mating tooth surfaces is also beneficial for the lubrication of the operating gears.

The required minimum backlash of operating gears can be provided by deliberately thinning the teeth during the manufacture of the gears, thus producing gears with teeth thinner at the reference circle than the spacing between the adjacent teeth. That condition of the gear can be inspected by measuring the tooth thickness directly, using one of the methods described earlier.

(A)

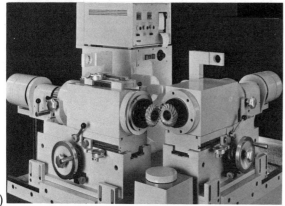

(B)

Klingelnberg Gear Technology, Inc.

Fig. 17-37. Single-flank contact gear rolling tester in flexible design, permitting its use in two alternative arrangements:
(A) For spur gears.
(B) For bevel gears.

Because of the permissible errors of gear parameters that affect the backlash of assembled gears, backlash variations are commonly present and may be a specified condition requiring inspection. The most commonly specified backlash tolerance refers to the minimum backlash, measured at the tightest point of mesh between mating gears. Generally backlash is measured by means of an indicator installed with its axis perpendicular to the tooth surface with which the indicator point is in contact. The mating gears are mounted at operating distance and one of the gears, the master pinion when checking a particular work gear, is held solidly and the gear being tested is turned in both directions. The range over which the work gear can be moved is the measure of backlash as the gear teeth in engagement. For determining the minimum backlash of the gear that test might have to be repeated over a complete or, in the extreme case, over several revolutions of the gear and pinion.

For checking the minimum backlash of large gear pairs that are too heavy to be rotated by hand feeler gages may be used; these measure directly the clearance between meshing teeth in various positions around the gear being inspected.

GEAR INSPECTION BY TOOTH CONTACT PATTERN EVALUATION

Tooth profile measurements by the methods already described are applicable to parallel-axis (spur and helical) gears, the teeth of which have equal profiles over the entire width of the gear. The measured profile permits the actual contact of the tooth with its mating element over the active area of the tooth surface to be evaluated.

The tooth design of bevel and displaced-axis (for example, hypoid) gears makes impractical the tooth contact evaluation by profile measurement because the shape of the bevel gear tooth varies along its entire width. Even if a series of individual profile measurements could be carried out on a particular tooth, those would still not reflect dependably the actual contact over the entire tooth surface.

Inspection by tooth contact pattern is a practical method, although it does not supply exact dimensional data. This method can be applied, however, to all types of gears, including spur, helical and, more importantly, bevel and hypoid gears.

Essentially, this method of inspection uses a technique that requires the application of a marking-compound coating over the entire surface of the gear teeth and then running the mating gears under light load for a few seconds. The method can be used for assembled gears, as well as for individual work gears mounted on a gear-testing machine.

The contact patterns developed on the gear tooth surface are observed visually, and permit the appraisal of the tooth profile and other tooth dimensions, as well as of the gear mounting conditions that affect the contact of the mating gear teeth. A uniform contact pattern, extending over the entire area of the active tooth surface, is the desirable condition, whereas other patterns are indicative of various tooth form or gear mounting errors.

Diagrams of typical tooth contact patterns, indicating different types of errors, have been published by the American Gear Manufacturers Association and serve as useful guides in the visual observation and evaluation of the gear tooth form and gear mounting conditions.

GEAR SOUND TESTING

Objectionable noise in gears results from either faulty design or manufacture. By detecting such noise before the gears are assembled, considerable time and expense can be saved. The sound testing of gears may also help to identify and avoid errors causing such undesirable operating conditions, considering that the combination of slight dimensional errors, even when each is within its tolerance limits, can result in objectionable gear noise, particularly in gears running at high speed.

The presence of specific types of gear noises is often indicative of particular errors that, when identified by sound testing, can be eliminated. Conditions affecting the intensity of various noise types are the speed, the degree of mesh (from tight to loose), and the amount of load; varying these conditions, or just establishing them at a level simulating that of actual service, are means of effective gear sound testing.

The machine shown in Fig. 17-38 is a very practical tool of gear testing for both the production shop and the gear laboratory; the machine is built to study the operation of gears at various loads and speeds. The machine not only screens out environmental sounds, but also amplifies gear noise by its specially designed acoustical sound box and horn. The test gears are mounted on the precision spindles of the machine; the center distance between these spindles is easily changed to accommodate gears of different sizes, and gage blocks may be used to assure the accurate setting of the center distance. A conveniently operated lever controls the loading and unloading mechanism. A brake is provided for producing the required gear load. A variable speed drive motor produces any of four forward and reverse speeds. Special accessories and machine models are available for testing internal gears or a set of three gears, the latter by means of a special third spindle and a supplementary brake.

Essentially the same machine model is available designed for semiautomatic sound testing, in which the sequential variations of load and direction of rotation are produced automatically, allowing the operator to concentrate on listening to the gear noise under varying conditions and to exchange the work gears when the machine stops automatically for work loading.

Fully automatic gear sound testers are used in high volume production where speed and economy of the testing process are important. The test gears are loaded and unloaded automatically, and a speeding cycle similar to that used in the semiautomatic

National Broach & Machine Co.

Fig. 17-38. Gear sound testing machine with sound box and horn, designed with adjustable center distance of the spindles and variable speed and gear load.
(Top) General view of the machine.
(Bottom) The cover opened, showing the arrangement of a pair of helical gears in mesh.

model is provided. In these machines, however, the human ear for appraising the gear sound level and human judgment for segregating rejects are replaced by an electronic sound discriminator, which handles those functions automatically. The discriminator causes a solenoid-operated diverter in the unloading chute to segregate the gears that are outside the sound tolerance to which the discriminator has been adjusted.

INSPECTION OF GEAR CUTTING HOBS

One of the most widely used and productive methods of cutting gear teeth is by hobbing, a continuous generating process capable of developing gear teeth with a high degree of dimensional accuracy. The utilization of these favorable process potentials is greatly dependent on the accuracy of the employed tools—the hobs—both in their original as-manufactured condition and during their entire service life, which comprises a series of resharpening operations. Incorrectly manufactured or resharpened hobs can adversely affect various important dimensions of the manufactured gears, such as the pressure angle and the profile of the teeth. Consequently, the thorough examination of the hobs in their manufacture and regular inspection in service, particularly following each tool sharpening, are important requirements for assuring the production of dependably accurate gears.

The inspection procedures for process control may include the following operations:

Inspecting the profile of the operating tooth face and of its position with respect to the hob axis.

Inspecting the flute lead.

Inspecting the spacing of the flutes.

Verifying the true running of the flanges and tooth tips.

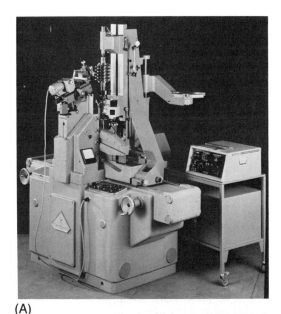

(A)

Klingelnberg Gear Technology, Inc.

Fig. 17-39. Bench type hob inspection instrument with adjustable work-mounting and indicator-holding elements, equipped with several indicators for inspecting on the shop floor the critical dimensions of hobs to a maximum of 12 inch diameter.

(B)

Klingelnberg Gear Technology, Inc.

Fig. 17-40. Hob measuring machine with optical setting and electronic recording device, for manual control or power drive by infinitely variable speed motor for continuous measurements. Adapted for inspecting with a high degree of accuracy all the important parameters of regular and special hobs.
(A) General view of the machine.
(B) Close-up of the operating area, showing the main slide, traversed manually or by power, over a distance that is displayed by an optical device (box in front, with lid open).

For the requirements of the shop floor these measurements can be carried out on bench type hob inspection instruments, such as the model shown in Fig. 17-39. That particular instrument uses gage blocks for accurate setting, and the various measurements are carried out by means of several indicators, with sensitivities corresponding to the typical tolerances of the hob dimensions for the inspection of which they are used. As examples: a high-precision sensitive indicator with 0.0001-inch resolution is used for checking the tooth face profile and flute lead, and two indicators with 0.0005-inch resolution serve for the inspection of the flute spacing.

For a more comprehensive measurement of hob parameters with a very high degree of accuracy, hob testing machines, such as the instrument shown in Fig. 17-40, are used. That machine permits the infinitely variable setting of all theoretical hob values, using a sensitive optical device for setting and measurement. For carrying out the measurements in a continuous process, the instrument is equipped with infinitely variable motor drive with jump feed control. An electronic recording unit produces charts with cumulative error diagrams of consecutive teeth; the thus-produced chart tracings clearly reveal the characteristics of the critical profile sections, without the interference of the more extended but not active intermediate sections, which are suppressed by the use of the jump feed control. The machine is also adapted to coordinate measurements of hobs and worms with special profiles, as well as to the pitch measurement of rack type gear shaping cutters.

AUTOMATIC GEAR INSPECTION MACHINES

Several different objectives may warrant the inspection of gears in an automatic process, such as:

1. Inspection during manufacturing, between two successive production steps, for avoiding the continued processing of defective workpieces.

2. Inspection of finished gears, which are produced in large volume and are intended for critical applications requiring the 100% checking of several specific gear dimensions.

3. Thorough analysis of a series of gear parameters measured with a high degree of accuracy, generally combined with permanent record production.

Many of the gear measuring instruments described in the preceding discussions have various automatic functions; two significant functions of those instruments, however, generally are not automated. They are:

1. The mounting and removal of the work gear being inspected;

2. The evaluation and decision-making based on the determined dimensional conditions.

Consequently, the operation of instruments that are not expressly designated as automatic gear measuring machines requires some degree of manual control, at least for handling the workpieces, including the starting of the measuring process and the assessment and decision-making by the inspector.

In the following, examples of various types of automatic gear inspection machines are described briefly, with emphasis on the purpose of their applications. In view of the diversity of the objectives, which such instruments are intended to meet, no categorization has been attempted, nor is any particular order being observed in the subsequent presentation of machines that are selected as characteristic representatives of different types of equipment.

Automatic Inspection of Gears Prior to Shaving or Roll Finishing

That inspection must assure that the semifinished gears fed into gear shavers or gear rolling machines have no malformed teeth, and that their tooth sizes, as well as outside diameters, are within tolerance limits. The purpose of that 100% inspection is to prevent damage to expensive shaving cutters and gear rolling dies.

The premachined gears are checked on the machine shown in Fig. 17-41 at the rate of up to 720 pieces per hour. The gears to be inspected are stored in a magazine and brought, one at a time, into mesh with a motor-driven rotating master gear and an OD contact roller. Electronic gaging units measure the movements of the master gear and of the OD contact roller, producing signals for operating trap doors in the exit chute. The semifinished gears are segregated, according to the measured stock allowance, for roll finishing or for rotary shaving.

Automatic Inspection Machine for Checking Two Different Internal Helical Gears

Figure 17-42 shows an automatic gear gaging machine that indiscriminately accepts and inspects internal helical automatic transmission gears in two different widths, checking seven variables at an hourly rate of 300 pieces. The gears have 57 teeth on a 3.2344-inch pitch diameter, and the helix angle of the teeth is $20\frac{1}{2}$ degrees. The gears are checked for size, eccentricity, tooth action, helix variation, nicks, and face runout. The inspected gears are fed out of

National Broach & Machine Co.

Fig. 17-42. Automatic inspection machine for checking seven variables of internal helical ring gears. The machine has several exit chutes for the good gears and for those with different categories of defects, segregating the salvageable gears.
(A) General view of the machine.
(B) Close-up of the operating area, displaying the gears being checked.

Automatic Gear Gaging Center for Checking Each End of Transmission Sun Gears

The automatic gaging machine shown in Fig. 17-43 checks size, eccentricity, nicks, helix angle, overall length and face flatness, and finally segregates rejects into six categories; all these operations are carried out at the rate of 600 gears per hour. The gear being checked is an automotive automatic transmission helical sun gear with a 2.130-inch diameter and a 2.393-inch length. The gear has 26 teeth and a 21 degree helix angle. Size is measured at the high point of eccentricity, independent of nicks and helix angle, while eccentricity is checked independent of size, nicks and helix angle. Other conditions and dimensions checked independently are nicks, helix angle on both ends and overall length over wobble high points of the end faces. The gear being inspected is held by a live expanding plug and cup arbor, locating the gear on the minor diameter of its internal spline teeth. Each work gear engages in tight mesh with two master gears on a gimbal-mounted inertia checking head and is power rotated through 360 degrees for the simultaneous checking of both ends. The exit chute for the inspected gears has trap doors for segregating the faulty parts into six specific

National Broach & Machine Co.

Fig. 17-41. Automatic inspection machine for semifinished helical gears from 1-inch to 3-inch outside diameter.
(A) General view of the machine.
(B) The operating area with a large master gear and a contact roller engaging the work gear, which is mounted on a pair of retractable arbors.

the machine in four chutes, used for (a) good gears, (b) reject gears, (c) gears with nicks, and (d) salvageable rejects. The different width gears are identified and segregated in each of the four chutes. In the checking position the gears, one at a time, are brought into mesh with a system of four master gears, two of which are fixed and locate the work gear, whereas the other two are spring loaded and connected with instruments for measuring the accuracy of the teeth. One of the fixed position master gears is driven, causing the work gear to make $1\frac{1}{2}$ revolutions for the inspection operation, which also includes checking of face runout.

National Broach & Machine Co.

Fig. 17-43. Automatic gaging center for checking both ends of transmission sun gears for six variables and segregating rejects. The machine is shown with covers removed to make visible the checking stations and the chutes for the segregated gears.

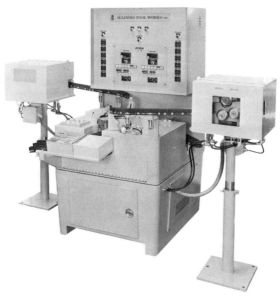

Illinois Tool Works, Inc.

Fig. 17-44. Automatic gear measuring machine with dual stations for the final inspection of transmission pinions. Before entering the inspection machine the pinions pass through a special unit, seen on both sides of the machine, for cleaning and an oversize check.

reject conditions. Individual lights on the control panel signal each gear characteristic that was found to be out of limits.

Dual Station Automatic Final Gear Inspection Machine

The automatic gear measuring machine shown in Fig. 17-44 is designed for the final inspection of automatic transmission pinions manufactured in high volume production. The machine, which has two checking stations, each with its own part feed channel, can inspect 1000 parts per hour, checking each pinion for pitch diameter size, then segregating the workpieces as under, good or over; runout and nicks are also checked. The inspected and segregated workpieces are discharged through the appropriate chutes in accordance with the determined condition. Before entering the feed channels of the machine, the pinions to be inspected pass through special units, mounted on adjustable columns and seen on both sides of the inspection machine. These special units serve both as burnishers, to clean the gears before inspection, and as a preliminary size check for keeping oversize gears, which could damage the measuring instruments, from entering the feed channels.

Such devices are also used on the production line, preceding machines using expensive tools, such as shaving cutters, which are sensitive to damage caused by gears with nicks or oversize pitch.

Functional Gear Element Analyzer

Functional gear checking on a gear rolling machine, by running the work gear in tight mesh with a corresponding master gear of known accuracy, reveals those composite errors of the work gear that cause center distance variations. By recording the variations of the center distance, the chart tracing representing one complete revolution of the work gear can be analyzed, discriminating center distance variations caused by particular types of gear errors, the extent of which can be evaluated individually. That analysis process for assessing the effect of various gear errors, however, is rather time-consuming, and is also dependent, to some extent, on the judgment of the inspector.

In gear rolling machines equipped with recorders, the speed at which the test gear can be run to produce dependable chart tracings is limited by the response speed of the recorder. Substantially, often 10 to 20 times faster response is possible when the chart recorder is replaced by velocity pickups in combination with electronic memory devices that can store the signals produced by center distance variations and then display the information and may even ac-

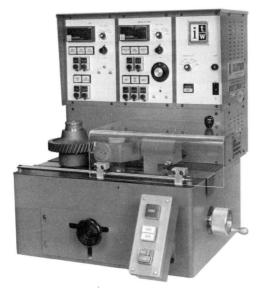

Illinois Tool Works, Inc.

Fig. 17-45. Functional gear element analyzer, which performs an automatically controlled gear rolling test and determines, at high speed, the extent of the thus-assessed variations of several significant gear characteristics, displaying that information in accordance with the tolerance setting of the individual parameters.

tuate gates that segregate the workpieces according to the determined conditions of error.

The machine shown in Fig. 17-45 was designed to accelerate the performance of the functional gear-measuring process, operating without dependence on personal judgment and providing prompt information as to which, if any, of several gear dimensions is outside the established tolerance limits. The operation of that gear element analyzer is very fast. Approximately 4 seconds after the beginning of the gear rolling cycle, a 1.3-second "readout" period begins; during that time, a final assessment of all the accumulated data is made and the results are displayed. That information remains visible until the next cycle is started; but after the elapse of the "readout" time, approximately 5.3 seconds following the beginning of the cycle, the drive clutch disengages and the gear rolling cycle ends.

All measurements are determined by sensing the center distance variations between the master gear and the work gear, and/or the wobble action of the head as generated by rotating the work gear mounted on the work arbor, while in tight mesh with the master gear. Errors in the following work gear dimensions are detected and evaluated in relation to the preset tolerance limits:

a. Pitch diameter size, measured by detecting the average center distance deviations from zero, representing the mean (specified) size of the gear.

b. Total pitch diameter runout is detected by measuring the maximum variation of center distance at the frequency per second equal to one revolution of the work gear. Filter networks prevent higher frequency signals, such as those caused by rolling tooth action and nicks, to interfere with this check.

c. Tooth-to-tooth rolling action is determined by measuring the average center distance variation at the frequency per second equal to the number of teeth multiplied by the revolutions per second of the work gear.

d. Nicks are detected by measuring the center distance variation at the frequency per second equal to the number of teeth multiplied by the revolutions per second of the work gear and will be determined as the intermittent "spikes" over and above the acceptable average tooth-to-tooth action.

e. Total composite error is the measured maximum variation of the center distance at the frequency per second equal to one full revolution of the piece part.

f. Lead: "lead average" and "lead variation" are sensed through the wobble action of the checking head. "Lead average" is detected by measuring the average lead deviation from zero representing a zero lead error or a lead variation (wobble) with peaks equally distant from zero. "Lead variation" is detected by measuring the peak-to-peak amplitude of wobble, independent of the amount of lead average.

g. Gimbal nicks are similar to the errors termed nicks, but are located near the tips of the gear teeth and are detected by measuring the intermittent "spikes" over and above the acceptable average tooth-to-tooth action.

Polar-Coordinate Measuring Center

The machine shown in Fig. 17-46 is a computer numerical control (CNC) gear measuring center designed for fully automatic checks of profile, helix angle, pitch and concentricity on external and internal spur gears with either straight or helical teeth. Automatic checks of positional and form errors can also be made on bevel gears, worm and worm gears, hobs and cams. For companies engaged in large-volume gear manufacturing or purchasing, a dedicated and

Mahr Gage Co.

Fig. 17-46. Polar coordinate gear measuring center with computer control, graphics screen and printer.

fully automatic gear measuring machine is advised over the more versatile, but less efficient, CMM.

Computerized Chart Analysis for Tooth Profile and Lead-Measuring Machines

The chart tracings produced by involute profile- and lead-measuring machines are well adapted to detailed analysis, by comparing the deviations of the actual tracing line to the theoretical path of the inspected gear tooth element and determining the value of the recorded variations by the scale of the chart.

That procedure, although accurate and dependable, can prove quite time-consuming when a large number of charts must be evaluated. Various types of profile- and lead-measuring machines, for example, the instrument whose mechanism is shown in Fig. 17-12, can be equipped with an additional device for computerized chart evaluation. The system operates by digitizing the analog values picked up by the measuring probe and feeding those values into the memory of the computer for calculation. The plotter produces a standard chart sheet with tracing lines corresponding to the scale of the chart, on which the determined values are also printed out in numerical values by the computer. Furthermore, it is possible to operate the computer with plotter as a terminal of a large computer and to use the determined values as data for statistical quality control.

The evaluation of the measured variations include the following characteristics:

1. For profile measurement—total profile error, profile angular error and profile form error;

2. For lead inspection—total lead error, lead angular error and lead form error.

18.
Process Control Gaging

Dimensional measurements provide the information needed to control all metalworking processes by which specific forms and sizes are imparted to the workpiece. The connection between dimensional measurement and the control of the metalworking process also usually involves evaluating the measurement results, decision-making regarding actions needed for correcting the detected imperfect condition, and performing the appropriate machine adjustments.

These functions, when carried out by the operator, must be aided by gages that readily supply work-size information with little or, preferably, no action on the part of the operator. In the high volume production of tightly toleranced parts on semiautomatic or fully automatic machine tools, the gaging must be complemented by the just enumerated supplementary functions, more specifically by

1. The evaluation of the momentary work size compared with the specified dimension and, optionally,

2. The operation of the appropriate machine controls for assuring that the workpiece size is in compliance with the pertinent specifications.

Gages that assist in assuring, or actually participate in controlling, the size of the processed workpiece are designated *process control gages*.

The need for machine adjustments to provide work-size control during processing depends on several circumstances, the specified dimensional tolerance of the part being an important factor. When producing parts with high accuracy requirements, even on dependably functioning machine tools, some degree of work-size control during machining might be needed, for any or several of the following reasons:

1. Tool wear, particularly in the case of tools that are subjected to rapid wear, such as grinding wheels;

2. Thermal deformation of machine tool members, such as originating from electrical equipment, hydraulic systems or mechanical sources due to friction;

3. Changes in the temperature of the coolant or of the environment;

4. Vibrations, elastic deformation and so forth.

The combined effect on the workpiece size of the types of variables listed above can be detected promptly by appropriate dimensional measuring instruments that are designed to supply information for initiating corrective action while the machining is still in progress. Because those listed above and similar causes usually affect the workpiece size only gradually, in a manner termed *drift*, detection and correction in the incipient state of such size changes generally avoid the production of out-of-tolerance parts.

In addition to preventing work-size deterioration during machining, either by guiding the operator or by initiating automatic machine adjustments, specific systems of process control gages also act directly on the machine controls that assure the size of the produced work. Measuring systems that control the size of each individual workpiece while it is being machined are known as *size control gages* or *machine control gages*.

Process control gages generally measure the workpiece during the actual machining process. There are conditions, however, that can prevent the direct measurement of the part during its processing, such as lack of access to the surface being machined or the short duration of a single part's processing. For process control purposes such workpieces are measured immediately after leaving the machine tool, applying a method that may be considered as indirect process control gaging. The commonly used terms for distinguishing between the direct and the indirect methods of gaging during machining are the *in-process gaging* and the *post-process gaging*.

Many machining processes that eject the finished parts in rapid sequence and that have drift characteristics that only affect the part size very slowly can assure dependable size control when the gaging is applied following the actual processing of the workpiece. Post-process gaging systems are of the indicating only or of the automatic work-sizing type. For added assurance against passing faulty parts, post-process gaging systems may be complemented with *sorting devices* for automatically discarding individual parts having sizes outside the tolerance limits.

To facilitate the survey of the various purposes served by process control gages, as well as of the different control operations performed by such gaging systems, the preceding aspects are recapitulated in Table 18-1.

GRINDING GAGES

Process control by gaging during machining first found extensive application in grinding, particularly in the precision grinding of cylindrical surfaces.

Several characteristics of precision grinding prompted the use of these gaging processes, such as the close tolerances commonly required for the finish ground part dimensions in combination with the relatively rapid wear of the grinding wheel, an inherent process property that is always present but not consistent in its effect on the size of the ground work surface. Even if the grinding wheel would have perfectly homogeneous structure, its breakdown, manifested in the wheel diameter, would progress at an increasing rate during the lifetime of that tool, owing to the gradual reduction of the wheel periphery (that affecting both the grain volume in momentary contact with the workpiece, as well as the effective cutting speed). In view of these conditions for all grinding processes, a preset in-feed depth, irrespective of whether it is manually or mechanically controlled, does not consistently produce uniform part size; that can only be assured by the direct gaging of the surface being ground.

The essential characteristic of a grinding gage is its capability of determining the size, most commonly the diameter, of the ground surface while the part is mounted on the grinding machine, that is, in a position where the grinding process can be either resumed promptly should the gage indicate the need for additional stock removal or performed continuously until the finish size is reached. In current grinding practice such gages are operated in continuous contact with the work, and for that purpose are equipped with indicators that permit the accommodation of workpieces even when they are grossly oversized (indicators with pretravel range). To permit the continuous gaging of the work being ground, the grinding gages are designed to operate without inhibiting the grinding process by avoiding interference with the grinding wheel while in contact with the work.

The engagement of the gage with the workpiece may be controlled (a) by manual placement; (b) by power, initiated by the operator; or (c) by fully automatic placement, controlled by other operating functions of the grinding machine. After the grinding operation has been terminated, the gage is retracted, again by manual or automatic control, to clear the work area of the grinding machine and to facilitate the unloading of the finished as well as the loading of the next workpiece.

The basic types of grinding gages, in general, those intended for use on cylindrical grinding machines, have common design characteristics, which can be distinguished on the instrument shown in Fig. 18-1, illustrating a typical manually operated grinding gage. By means of the mounting bracket the gage is installed on the wheelhead or, less frequently, on the tailstock of a cylindrical grinding machine.

The bracket holds the adjustable torque spring, which ensures the tracking contact of the gage with the engaged workpiece during the entire grinding process, and a hydraulic dashpot, which facilitates the swinging of the support arm, moved down for engagement and up for retraction. The actual gage frame is connected with the support arm by means of a ball bearing or roller bearing hub. The gage frame contains the caliper portion, the indicator instrument, and, for manually operated grinding gages, the lifting handle or knob. The upper contact point of the caliper head is attached to a plunger that acts on the indicator, thereby transmitting the size changes of the work surface during the grinding process.

Figure 18-2 shows a grinding gage installed on a cylindrical grinding machine (A) mounted on the tailstock and (B) attached to the headstock. After it has been brought into engagement with the cylindrical workpiece, the contacts of the caliper portion

TABLE 18-1. APPLICATIONS OF DIMENSIONAL MEASUREMENTS FOR CONTROLLING MACHINING PROCESSES

GAGING APPLIED	FUNCTIONS PROVIDED	APPLICATIONAL PURPOSE	CHARACTERISTIC INSTRUMENTS
In-Process — Gaging	Continuous indication of work size during processing	Gaging the workpiece during machining for continuously indicating the momentary size of the worked surface in relation to the attainable finish size to which the gage has been adjusted in the setup of the operation.	Grinding gages with the work-size sensing member acting directly on an indicator.
	Continuous work-size gaging combined with control functions for assuring correct product size	Monitoring the workpiece size during the process by appropriately designed contact members of a gage head, generating signals to be converted into dimensional values by a remotely located amplifier, which may also trigger switching functions for terminating the operation when finish size is reached.	Machine control gages with remote meters and work-size-related simple switching function.
	Work-size gaging directing the control of successive operational phases	The continuously monitored size changes of the processed workpiece, when reaching predetermined dimensional conditions with respect to the finish size, produce specific control functions, e.g., varying the feed rate in successive steps.	Machine control gaging systems with programmed control functions.
Post-Process — Gaging	Work-size indications, optionally with signals or even machine shut-off action	Workpieces machined in rapid sequence are gaged individually immediately following processing to provide continuous information on extent and trends of size variations which may call for machine adjustments. Gage indications may be supplemented with light or sound signals, optionally also with switches for stopping the process, thus allowing several machines to be attended by a single operator.	Post-process gages with meters and, optionally, with signaling and/or shut-off function.
	Work-size monitoring for corrective machine adjustments triggered by the gage	Signals generated by the post-process gage may also initiate machine adjustments for correcting automatically the developing work-size deviations in their incipient state, before actual rejects are produced. One typical application is through-feed centerless grinding with only gradual drift of the work size.	Post-process gages feeding work size data to the size-control device of the machine.
	Work gaging complemented with size-determined sorting of processed parts	As a further safeguard against delivering out-of-size parts intermingled with dimensionally correct parts, post-process gages are often complemented with sorting devices for removing the reject parts produced as strays or due to untimely machine adjustment. Discarded parts may be separated as over- and undersize, as determined by the gage.	Post-process gages complemented with part-sorting devices.

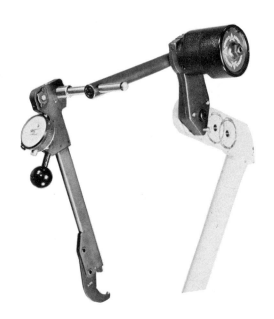

Arnold Gauge Co., Div. Lehr Precision Tools Inc.

Fig. 18-1. Manually operated grinding gage for outside diameters, with 0.0001-inch least graduation type mechanical indicator. The hydraulic mounting (at right) assures sensitive gage support during the grinding operation and then lifts the gage into a retracted position for workpiece exchange. The caliper portion can serve either a single diameter (as shown) or it can be adjustable.

of the gage, which have been adjusted to indicate zero in a position corresponding to the desired finish diameter of the work, stay in engagement during the entire grinding process.

The workpiece dimension controlled by grinding gages is most commonly, but not exclusively, the outer diameter of cylindrical parts. Grinding gages, composed of appropriate elements, may also be used for lateral dimensions, those measured in the axial direction of the workpiece in relation to a specific reference plane, or for both diametral and lateral dimensions, which can be gaged concurrently, with individual size indications. Examples of gaged dimensions are pointed out in Fig. 18-3.

Accordingly, grinding gages may be supplied in assemblies designed to measure the following:

1. Multidiameter workpieces using an instrument equipped with several gage units or with quick interchange pick-off hub assemblies for each diameter;

2. The distance between two opposite shoulders or shoulders facing in the same direction;

3. Tapers—by gaging the diameter at a specific distance from a radial reference plane; and

4. The diameters of noncontinuous cylindrical surfaces, namely, those interrupted by flats, keyways, or spline grooves, using gage calipers equipped with specialcontacts.

Grinding gages equipped with special narrow calipers are used for crankshaft regrinding, with the caliper portions of the gage fitting the limited clearance space between the check and the steady rest, which must be accommodated in these operations.

Figure 18-4 shows a grinding gage equipped with two calipers, each acting on its own indicator, for the concurrent gaging of the diameter and a lateral dimension of the workpiece.

The *capacity* of grinding gages, expressed as the diameter of the workpiece that can be inspected with the instrument, depends on the type of caliper used. The caliper parts of these gages are interchangeable and are commonly available for single diameters or in adjustable designs to cover work diameter ranges from $\frac{5}{16}$ to $1\frac{1}{2}$ inches to 8 to 12 inches, even up to 24 inches for special applications. Within the listed ranges, each of the calipers is adjustable to closely approximate the nominal diameter of the workpiece. The fine adjustment to the finish size of the gaged dimension is carried out at the indicator, using a master gage or workpiece, contacted by the caliper, as reference element.

The *indicating members* of grinding gages are commonly mechanical indicators with a sensitivity of 0.0001-inch or, for part dimensions with wider tolerances, 0.0005-inch per graduation. For applications that require higher indicator sensitivity, grinding gages may be equipped with the appropriate probes for air or electronic gaging, using remote-indicating instruments that are usually placed on a solid base mounted on the wheelhead slide of the grinder, in a position permitting easy observation by the operator. When an electronic gage is used as the sensing and indicating member, the equipment can be complemented with electrical switches acting on the machine controls, resulting in an automatic size control system. Grinding gages comprising electronic indicating and electrical switching elements are also usually equipped with hydraulic actuators that place the gage into the measuring position on the workpiece after it has been mounted on the grinder and the automatic operational sequence has been started by the operator. The operations of the gage and of the grinder proceed in an automatic sequence, beginning with the hydraulically actuated

(A)   (B)

Arnold Gauge Co., Div. Lehr Precision Tools Inc.

Fig. 18-2. Grinding gages with adjustable calipers installed on cylindrical grinding machine.
(A) Mounted on the tailstock in position behind the work.
(B) Attached to the headstock in forward position. Many other mounting arrangements are possible, including the most frequently used installation on the wheelhead.

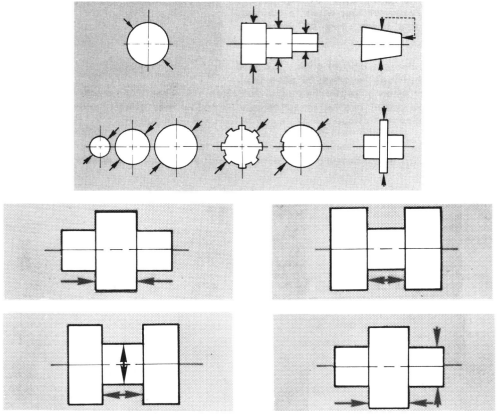

Arnold Gauge Co., Div. Lehr Precision Tools Inc.

Fig. 18-3. Diagrams of typical workpiece dimensions that can be measured during machining with appropriately equipped grinding gages. The outside diameters of cylindrical sections with plain or interrupted surfaces, as well as the precisely located diameters of tapered sections, are commonly gaged. The axial dimensions may be widths and lateral locations, gaged individually or in combination with a diameter.

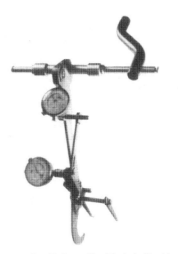

Arnold Gauge Co., Div. Lehr Precision Tools, Inc.

Fig. 18-4. Grinding gage equipped with two caliper members, each with its own indicator, for measuring concurrently the width of a flange and an adjacent outside diameter. Grinding gages can also be arranged for the measurement of diameters and lateral dimensions in many other positional combinations.

placement of the gage, followed by the control of the wheel feed, to reaching the finish size of the workpiece, when the wheel slide retracts and the gage returns into its initial position, permitting the unobstructed unloading of the finish ground part.

Grinding gages equipped with the complementary members just described represent the transition from gage-assisted but manually controlled grinding to that performed by automatic work sizing with machine control gaging systems, which are discussed next.

MACHINE CONTROL GAGING SYSTEMS

General Definition

The machine control gages operate by monitoring continuously the size of a specific part dimension during its machining and compare the momentary work size with a single or several successive nominal values; at the preset size the gaging system actuates switches for controlling particular functions of the machine tool.

The complete machine control gaging system comprises several distinct but interconnected members, namely, the following:

1. The *gaging unit*, often referred to as the gage head or the caliper;

2. The *amplifier* with the work-size display, which converts the signals originating from the gaging unit into analog or digital dimensional values, usually in relation to a preset size, which serves as the reference base;

3. The *control elements,* which direct the feed movements of the machine tool in correspondence with the momentary work size, until the preset finish size is reached. The control functions of the gaging system may comprise several steps, such as applying different feed rates for roughing and finishing.

The major members of machine control gaging systems are manufactured in a wide variety of designs, adapted to the characteristics of the workpiece, to the process, to the basic machine tool and to several other conditions. These are discussed in greater detail under separate headings for each of the major members listed.

Machine control gaging systems assure the automatic size holding of the machining process, and when applied on machine tools that are also equipped with mechanized work-handling devices for loading and unloading the workpieces, a completely automated processing is accomplished.

Advantages of Machine Control Gaging

Significant advantages are offered by machine control gaging systems over processes controlled manually or with rigid mechanical means, for example, cams. Examples of these advantages are listed below, noting, however, that the actual benefits of any or several of these potential advantages depend on the workpiece specifications and on pertinent process-related conditions.

Machine control gaging systems:

1. Assure size holding by the process to very tight tolerances, often with an accuracy exceeding that feasible by any other system of machine control;

2. Operate by continuously gaging the actual size of the workpiece, thus avoiding the deleterious effects of uncontrollable process variables;

3. Eliminate the dependence on the operator, thus providing both economic benefits and freedom from the consequences of human errors;

4. Permit optimum utilization of the production equipment by dispensing with interruptions for measurements and adjustments; may even participate in applying adaptive controls when using systems designed for such operation;

5. Control the actual size of each individual workpiece so that scrap is avoided and inspection of the finished parts is generally not required.

To derive significant benefits from these potential advantages, the application of machine control gaging must be warranted by tolerance specifications that, in the applicable processes, cannot be assured by purely mechanical means or, in the case of manual control, would require the frequent or continuous gaging of the workpiece in combination with appropriate machine adjustments. Consequently, the most rewarding field for the application of machine control gaging is the finish grinding of tightly toleranced parts that are produced in large volumes.

THE MAJOR MEMBERS OF MACHINE CONTROL GAGES

The *gaging unit* or *gage head* contains the sensing members, consisting of the probe fingers that act on the gaging cartridges or transducers. Most commonly, electronic gages are used, although in some cases air gages, which use air nozzles as the probe elements, are preferred. The sensing members are designed to continuously follow the size changes of the worked surface during the machining process, and must be adapted to the configurations of the workpiece and to the accessibility of the machined surface.

With regard to their size-measuring and referencing characteristics, three basic types of gaging contact systems are commonly available: (a) the single-contact system, which relies on a machine element to represent the fixed reference point from which the size of the work is measured; (b) the double-contact system using a single transducer, which senses the variations of the distance between the two contacts, both being in engagement with the actual work surface; consequently, its operation does not rely on an extraneous reference point; (c) the double-contact system using two transducers, one for each contact finger; the signals generated by these two transducers are then combined by the appropriate amplifier to produce a single value corresponding to the momentary diameter of the workpiece. The principles of design of these three contact systems are shown in Fig. 18-5, using as examples gage heads for external diameters. The same design principles can be applied to internal in-process gages.

The automatic operation of machine control gaging systems also requires that the sensing members

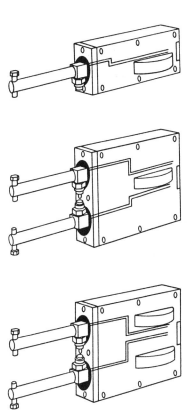

Marposs Gauges Corp.

Fig. 18-5. Gage heads of machine control gaging systems; principles of design of the generally used types. (Top to bottom) Single-contact, double-contact with common transducer, and double-contact with individual transducers.

retract at the end of the cutting part of each operation, thus avoiding interference with unloading and loading, and, subsequently, move back into the proper gaging position. Consequently, for the automatic operation of machines that produce work with size controlled by in-process gaging, the gage head and its probes must perform various movements synchronously with several other functions of the machine, such as wheel feed and work loading and unloading; these movements of the gage head comprise two separate, but coordinated sets of functions.

a. The approach and retraction movement of the gage head first advances the probe fingers into the area of (or actual) engagement with the work surface and then—at the end of the operation—clears that area to avoid interfering with the loading and unloading of the workpiece.

It is either a linear movement, usually in a radial direction to the workpiece, such as the jump-on gages used for external surfaces, or a swivel movement, such as that of the swing-in gage heads used for internal surfaces. A linear movement of the gage head in the axial direction is carried out by through-the-workpiece type internal gages.

b. The probe fingers of many types of gage heads have their own approach and retraction movements, coordinated with those of the gage head, which bring the probes into actual contact with the work surface, followed by the withdrawal of the fingers, which precedes the retraction of the gage had at the end of the operation. In certain types of machine control gages the movements of the probe fingers may comprise several steps, such as for bridging the shoulders of bores in recessed locations or a locking action for restricting the approach movement of the fingers when gaging interrupted work surfaces. References are made to several of these movements in the subsequent description of typical machine control gages.

Although the gage heads of machine control gages are manufactured in a wide variety of models for different applicational requirements, the most commonly used types of machine control gage heads are reviewed in Table 18-2. Subsequently, characteristic examples of various types of gage heads will be illustrated and discussed in greater detail. References are made in that table, by indicating the pertinent figure numbers, to the illustrations that accompany the descriptions of the selected gage head types.

The *amplifier* converts the signals generated by the gage head probes into meter readings, displaying the momentarily sensed work-size deviations from the preset reference dimension in standard units of measurement, that is, inches or millimeters, generally in very small increments of these basic values. For most general applications analog gages with meters are used, although in many cases digital display systems are preferred, particularly because of the wider range they provide, even when readouts in very fine increments are needed. The digital-display type amplifiers are manufactured with different degrees of resolution, up to those with least increments of 0.000010 inch. The limited range of the analog type meters can be compensated for by using two interconnected meters on the same amplifier, for example, one with a measuring range of 0.020 inch and the other, with an expanded scale, covering a range of 0.002 inch, both meters having the same reference dimension, which produces zero indication. A widely used model of an analog amplifier with two meters is shown in Fig. 18-6.

Marposs Gauges Corp.

Fig. 18-6. Analog amplifier with two interconnected meters having least graduations of 0.0005 and 0.00005 inch. The amplifier is installed on a cylindrical grinding machine and displays the momentary work size sensed by the double-contact gage head.

The meters guide the operator in manually controlled machining systems, while in the more commonly used automatic systems the size display has an informative role only, the actual control of the feed and other operational functions of the machine being actuated by the control elements of the system.

The *control elements* for those machine functions that must operate in accordance with the momentary size of the workpiece being processed are interconnected with the amplifier and are triggered by signals generated at the attainment of present work-size dimensions. A wide range of different control functions can be accomplished by the appropriate types of control elements, such as:

1. Starting the feed, preferably with a rapid approach movement to the position where the grinding wheel makes actual contact with the work surface;

2. Varying the feed rate, which may be high in the roughing phase and then lower in the finishing phase, and finally allowing a specific spark-out time in grinding, with the feed movement halted, or switching to a microfeed that can eliminate the need for spark-out;

3. Applying adaptive control, varying the feed to accomplish preset rates of stock removal in the subsequent phases of the process, the operation of that control system, however, remaining subordinated to the essential sizing function of the system;

4. Terminating the operation when the set finish size of the workpiece has been reached.

**TABLE 18-2.   COMMONLY USED TYPES OF GAGING UNITS FOR
THE IN-PROCESS CONTROL OF WORKPIECE DIAMETERS—1**

WORK AREA	SURFACE CONFIGU-RATION	GAGING UNIT TYPE		APPLICATIONAL CHARACTERISTICS	FIGURE REFER-ENCE
		HEADS	FINGERS		
External	Continuous	Single	One	A single finger, contacting the lowest point of the part diameter, does not have to be retracted for work loading.	18-9
		Single	Two	Most commonly used system, adapted for fingers or jaws in various designs, usually installed on a "jump-on" slide for approach and retraction.	18-10 18-11
		Double	One on each	Adapted for large work diameters, which exceed the capacity of a single head. Requires amplifier which combines signals originating from two independent transducers.	18-12
	Continuous or Interrupted	Single	One	Single finger gaging units, contacting the bottom of the workpiece, may be complemented with mechanisms preventing the fingers from "dropping" into the gaps on the work surface.	18-9
		Single	Two	Two different systems are used: (a) cam control of the fingers synchronized with the interruption on the work surface; (b) signal evaluating device which eliminates the effect on the gaged OD size caused by the interruptions of the envelope surface of the workpiece.	18-10 18-11
		Double	Two	Two individual heads mounted on a common bracket may have their signal-combining system complemented with a gap-bridging device for cancelling the signals caused by work surface interruptions.	18-12

Depending on the design system used by the manufacturers of machine control gages and on the complexity of the functions thus controlled, the control elements can be incorporated into the case of the amplifier or into a separate cabinet, which is often of modular design and interconnected with the gaging signals of the amplifier. The principles of interconnecting the above-mentioned members of a machine control gaging system installed on a cylindrical grinding machine are shown in Fig. 18-7. That diagram represents only one of various types of in-process gaging applications for performing the control of the machining process by monitoring the size of the workpiece during its machining.

The control elements of the gage system may act on the hydraulic feed movements of the machine tool or may be used to control an electronic feed system that has for actuating member a DC stepping motor through a reducer. An appropriate controller causes these motors to rotate by the number of steps corresponding to the amount of feed needed to reach a specific work size, as determined by the gaging

TABLE 18-2.   COMMONLY USED TYPES OF GAGING UNITS FOR THE IN-PROCESS CONTROL OF WORKPIECE DIAMETERS—2

WORK AREA	SURFACE CONFIGU-RATION	GAGING UNIT TYPE		APPLICATIONAL CHARACTERISTICS	FIGURE REFER-ENCE
		HEADS	FINGERS		
Internal	Continuous	Single	One	Single finger units, with swing approach and retraction, are used for internal surfaces when the space available alongside the operating grinding wheel is too small for two fingers and a dependable reference position of the workpiece is assured, such as in shoe-type grinding.	
		Single	Two	Commonly used type when the space in the bore while the grinding wheel is in engagement can accommodate two fingers contacting opposite points of the hole diameter.	18-13
	Continuous or Interrupted	Single	Two	For gaging bores with axial interruptions the preceding described type of two-finger contact gaging unit may be used when equipped with a synchronized locking device.	18-13
		Double	Two	Essentially similar to the two-head gages used for external diameters, but equipped with device for cancelling the effect of surface interruptions on the gaged hole diameter.	18-14
	Recessed	Double	Two	In addition to the withdrawing and repositioning of the fingers, usually by swing motion, these internal gages also have a radial movement, preceding the withdrawing or following the positioning of the fingers.	18-15
	Constricted	Single	Two	For very small diameter bores, which do not provide space for even a single finger alongside the operating wheel, gaging heads approaching the workpiece inside the workhead spindle of the grinder are used; these reciprocate synchronously with the wheel spindle and insert the contact fingers while the wheel clears the bore.	18-16

member of the system. The rate of the feed during the metal-cutting period of the process may be varied automatically in accordance with the actual dimensional conditions of the work that is being monitored by the machine control gaging system, applying in successive stages lower rates as the actual size of the worked surface approaches the specified final dimension.

MACHINE CONTROL GAGING UNITS FOR SIZING THE WORK DIAMETER

Machine control gaging is applied for the final operation in the manufacture of tightly toleranced workpieces that are predominantly, but not exclusively, of essentially cylindrical form; these are machined on their external or internal surfaces, which

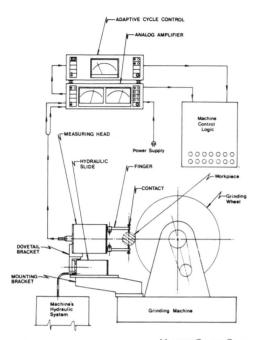

Marposs Gauges Corp.

Fig. 18-7. The principles of interconnection between the major members of a machine control gaging system installed on a cylindrical grinding machine. The machine's hydraulic system is used to approach and then to retract the measuring (gage) head; this system also comprises adaptive control regulating the feed to assure optimum stock removal rates, in accordance with the gaged size of the workpiece.

have varying degrees of accessibility. The following discussion first reviews the most common applications of machine control gaging, namely, those for the manufacture of generally round precision parts, which must have rigorously controlled outside or inside diameters.

For loading and unloading the workpiece in the machine tool, it is generally necessary that the gage head clears the work area by being retracted and, subsequently, moved back into the exact operational position where its sensing members contact the essential elements of the work surface. These retraction and approach movements can have mechanical, hydraulic or pneumatic actuation, and their direction is, most commonly, either along a straight line, radial with respect to the work (jump-on gage heads), used for external diameters, or along an arcuate path (swing-in gage heads), preferred for internal surfaces. The principles of these two basic

gage head movements are shown in Fig. 18-8, noting, however, that other types of approach/retraction movements are used in cases where neither of these basic motion directions is suitable.

The subsequent descriptions of typical gage heads generally follow the order selected for Table 18-2, complementing the concise information supplied in that general overview.

DIAMETER GAGING HEADS OF MACHINE CONTROL GAGING SYSTEMS

The gage heads described and illustrated in the following have been selected as representative models of several commonly used types of in-process gaging units. While the forms of execution of various makes of gage heads differ, any particular set of operational conditions requires essentially similar gage head characteristics. For that reason, concise indications of those design characteristics are used as subtitles in the subsequent discussion of typical gage heads serving the in-process control of external and internal diameters.

One-Finger Type Gaging Heads for Outside Work Diameters

The one-finger type of gaging head contacts the underside of the workpiece. It offers the advantage of dispensing with the retraction movement at the end of the operation; the unloading of the finished and the loading of a new part can be carried out while the gaging head remains in its operating position. The one-finger type gage heads can be mounted on a fixed bracket and usually have a heavy-duty guard to protect the finger from mechanical damage. Such gage heads are used for the high volume production of workpieces that are positively located in the machine tool, held either between centers or in a precise collet. A method of work holding must be used that dependably establishes the reference position of the work axis in relation to the machine tool, because the gaging head senses the radial distance of the contacted part surface from a plane associated with the workholding device of the machine tool. Consequently, any runout of the machine tool work spindle affects the gaged part dimension, a condition that must be considered when applying the one-finger type gaging head. This type of gaging head is manufactured for both continuous and interrupted work surfaces, such as the model shown in Fig. 18-9.

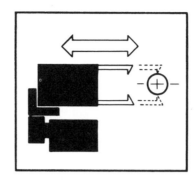

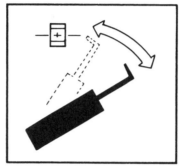

Fig. 18-8. Diagrams of the basic approach/retraction movements of gage heads used in machine control gaging system.
(Top) Jump-on gage head (used for external diameters).
(Bottom) Swing-in gage head (used for internal diameters).

Marposs Gauges Corp.

Marposs Gauges Corp.

Fig. 18-9. One-finger type gaging head for outside work diameters, installed on a fixed bracket for contacting the underside of a firmly mounted workpiece. The version of the gaging head shown is adapted for workpieces with interrupted surfaces by including a device that prevents the contact element from penetrating into the gaps of the work surface.

Two-Finger Type Gaging Heads for Outside Work Diameters

Figure 18-10 shows a typical form of execution of gaging heads with two fingers for measuring the size of workpiece outside diameters during machining. The head with two fingers or jaws is mounted on a slide that has hydraulic or pneumatic actuation, which retracts the gage head while the finished workpiece is unloaded and a new one is loaded, and then repositions the gage head by advancing it so that the tips installed on the fingers contact the workpiece in its diametral plane. The function of the approach/retraction movement is usually integrated into the general control system of an automatic machine tool, typically a cylindrical grinder, which advances the gage head after the rotation of the workpiece has started and withdraws it when the operation is finished.

The position of the fingers is adjusted to the size of the general workpiece diameter in the setup phase of the operation; that setup is complemented with a fine adjustment to assure that the meter of the amplifier indicates zero when the gaged diameter is equal to the design size of the work. Many models of two-finger type gaging heads have sufficiently large finger movement for contacting the workpiece in its unfinished state; in other models, however,

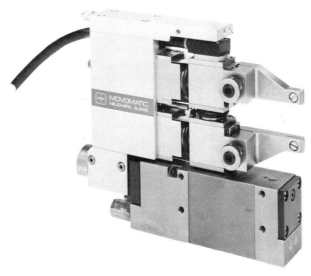

Meseltron/Movomatic

Fig. 18-10. The basic form of execution of a gaging head with two jaws, for external diameters, serving as the sensing unit of an automatic machine control gaging system. The gaging head is mounted on a retraction base and the jaws are designed to accept the appropriate type of contact tips, which are not shown.

means are provided to delay the contact with the work surface until its diameter has been brought close to the finished size. Other models of two-finger type gaging heads, such as the one shown in Fig. 18-10, are normally equipped with a jaw-retraction mechanism, commonly actuated hydraulically, that closes the fingers only after the gaging head has been advanced by its jump-on slide into the operating position. When the finished size of the part has been reached, the jaws open first, then the retraction slide withdraws the gaging head, clearing the workpiece area to permit unimpeded work loading.

The gaging head shown in Fig. 18-10 as a detached unit can be seen in Fig. 18-11 during its operation on a cylindrical grinder. A similar application of the two-finger type in-process gaging head can be seen in Fig. 18-6; that illustration shows a gaging head of different make, but with essentially identical operational characteristics.

The two-finger gaging heads represent the most commonly used sensing system for the in-process diameter gaging of cylindrical or, occasionally, tapered external work surfaces, because this type of gaging head with its two contact points provides assured mutual referencing in the boundary planes of the part's diameter, in distinction to one-finger gaging

heads, whose indications are based on an externally located reference plane.

Two-finger type gaging heads, similar in appearance to those serving continuous outside surfaces, are also manufactured for gaging the diameters of interrupted surfaces. Several design varieties are available, which differ with regard to the control of the finger movement and the evaluation of the produced signals. When gaging the diameter of rotating parts with interrupted surfaces, (a) the fingers must be prevented from dropping into the gaps of the work surface and from being damaged by the edges of the grooves that interrupt the continuity of the surface—this is usually accomplished by appropriate damping elements; and (b) the signals produced by the finger contacts must be processed in an amplifier to filter out the effect of the surface interruptions on the gaged outside diameter of the workpiece.

Most models of two-finger type gaging heads have a single transducer for sensing the variations in the distance of the opposite contact tips installed into the two fingers. In some other models each finger has its own transducer, and the signals originating from these two sources are combined by the appropriately designed amplifier to indicate, as a single value, the momentary separation of the contact tips, that distance representing the actual work diameter.

Meseltron/Movomatic

Fig. 18-11. The two-jaws type gaging head of an in-process gaging system installed—mounted on its pneumatically operated retraction unit—on a cylindrical grinding machine for monitoring the continuous diameter of a cylindrical workpiece.

Regarding the workpiece diameter capacity, the various current models of two-finger type gaging heads are designed to provide different adjustment ranges, which usually are interrelated with the available minimum center height of the grinder. A typical adjustment range of gaging heads used in applications where no particular space limitations interfere with the installation is of the order of ⅛ inch to 4 inches, commonly accomplished by means of interchangeable fingers.

Double-Head Type Gaging Units for Outside Diameters

Although the work diameter range for the monitoring of which the two-finger type gaging heads are adapted comprises a substantial portion of workpieces with dimensional tolerances warranting in-process gaging, for large work diameters, typically those exceeding about a 4-inch diameter, gaging units comprising two individual heads are needed. A characteristic representative of double-head type gaging units is shown in Fig. 18-12. The two heads, each with a single finger and interchangeable contact points, are mounted in adjustable positions on a common bracket, which is usually installed on a hydraulic slide for functioning as a jump-on gage.

The gage heads have individual transducers for originating signals, which then are combined in the amplifier, to produce indications corresponding to

the momentary size of the monitored work diameter. The adjustable distance between the two mechanically independent heads permits the use of this type of gaging unit for a substantially larger range of diameters than is possible with a single-head type device. The illustrated model has a maximum work diameter capacity of 7 inches, and in the version shown the gaging heads are adapted for both continuous and interrupted work surfaces.

One-Finger Type Gaging Heads for Internal Diameters

The use of one-finger type gaging heads for internal diameters generally is confined to operational conditions resulting from the grinding of small diameter bores, where the space cleared by the operating grinding wheel can only accommodate a single-contact finger. For internal grinding the one-finger type gaging head is mounted on a swinging, hydraulically actuated bracket, which operates synchronously with the grinding cycle.

As pointed out in conjunction with the one-finger type external gaging heads, a single-point contact of the work surface involves extraneous referencing for gaging the diameter. For that reason the reliability of diameter gaging by single-point contact is predicated on the running accuracy of the work spindle and, in the case of internal surfaces, also on the concentricity between the produced bore and the outside diameter of the part, which is generally the surface by which the workpiece is held in the machine tool.

For gaging bore diameters behind shoulders, such as in the case of ball bearing and roller bearing rings, single-finger gaging heads with pneumatically operated retraction movements are needed.

Two-Finger Type Gaging Heads for Internal Diameters

The two-finger type gaging head is the most widely used system for the in-process gaging of internal diameters because it is compact in design and, by providing its own referencing, assures the dependable monitoring of the developing bore diameter. Figure 18-13 shows a widely used representative of the two-finger type bore gaging heads in operation on an internal grinding machine. That particular model is applicable for a bore diameter range of 0.20 inch to 4.72 inches by means of interchangeable fingers, which are installed into the adjustable mounting guides of the head.

The head itself can be installed on the machine tool either by means of adjustable dovetails or in a swinging hydraulic support bracket, depending on

Marposs Gauges Corp.

Fig. 18-12. Double-head type gaging unit in use on a cylindrical grinding machine. Both heads are mounted on a common support bracket, which permits a wide range of adjustments of the distance between the contact points installed on the two individually supported fingers.

Marposs Gauges Corp.

Fig. 18-13. Two-finger type bore gaging head installed on an internal grinding machine for the in-process monitoring of the bore diameter, which, in this case, accommodates both contact fingers alongside the operating grinding wheel.

Marposs Gauges Corp.

Fig. 18-14. Double-head type gaging unit for the in-process gaging of a bore with diameter that exceeds the capacity of a single-head type sensing device. The two heads are installed on a common bracket, which permits one to adjust the distance between the heads over a wide range.

whether the contact fingers are to reciprocate in and out of the bore synchronously with the grinding wheel traverse or whether they are to remain in the bore during the grinding cycle. The illustrated type of internal gaging head is adaptable to both continuous or interrupted work surfaces and provides automatic finger retraction to assure the necessary clearance between the contact tips and the work surface during the positioning phase of the gaging.

Double-Head Type Gaging Units for Inside Diameters

For in-process gaging of workpieces with inside diameters exceeding the range accommodated by the single-head two-finger type gaging heads, gaging units comprising two heads, each with individual contact finger and transducer, are employed. A characteristic example of the double-head type internal gaging units is shown in Fig. 18-14. The two heads are mounted on a common bracket, designed to permit distance adjustment, whereby the illustrated model is adapted for bore diameters up to 7 inches.

Various types of mountings are available to permit the clearing of the work area for loading and unloading. The gaging head has automatic finger retraction, which operates during the positioning of the contact fingers inside the workpiece bore. The signals originating from the two separate transducers are combined in the amplifier into a single indication, which represents the deviation of the bore diameter from the design size to which the gaging unit has been adjusted in the setup phase of the operation.

Double-Head Type Gaging Units with Extra-Long Finger Retraction for Deep Groove Inside Diameters

Certain kinds of workpieces, the outer rings of ball and roller bearings with precision ground internal raceways being typical examples, require gaging heads with extra-long retraction movement of the fingers to permit them to clear the relatively high shoulders behind which the critical bore diameter must be gaged. A representative model of such gaging units, which is shown in Fig. 18-15, has two individual heads installed on a common swiveling bracket, which removes the contact fingers from the workpiece area during the loading phase of the operation. The retraction movement functions—withdrawing the fingers prior to their removal from inside the work bore by the swiveling bracket and releasing the contact fingers after their return into the gaging position inside the work bore—are coordinated with the swivel of the bracket. The travel of that retraction movement can be adjusted according to the height of the shoulder that must be cleared, up to a total of 0.400 inch.

In order to accommodate a wide variety of workpieces with deep internal grooves, this model of gaging unit can be used with two different brackets and accepts interchangeable fingers, producing a working capacity range of $\frac{7}{8}$-inch to 12-inch bore diameter.

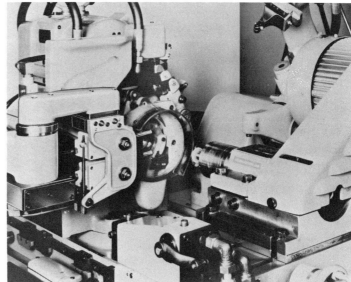

Fig. 18-15. Double-head type internal gaging unit with extended finger retraction movement, for gaging the diameter of internal surfaces behind shoulders, a design common for the raceways of ball bearing rings, such as that shown in the photograph and also indicated in the adjacent diagram.

Marposs Gauges Corp.

Two-Finger Type Gaging Heads Entering the Bore Through the Workhead Spindle

For the in-process gaging of bores of very small diameter in which the space between the operating grinding wheel and the bore wall is too narrow to admit even a single-contact finger or when a two-finger type gaging contact is needed for highly dependable sizing, a special type of internal gaging unit has been developed and is widely used for the described operational conditions.

This type of internal gaging unit, one well-known example of which is shown in Fig. 18-16, enters the bore of the workpiece through the workhead spindle of the grinding machine, namely, from the end of the workpiece bore that is opposite the side from which the grinding wheel is introduced. The illustration shows the gaging unit retracted to the setup position, while being adjusted to the design diameter of the work bore. The advancement and retraction of the gaging unit is coordinated with the operation cycle at the termination of which the retracted gaging unit will in no way interfere with the unloading and loading of the workpiece, procedures usually performed by an automatic work-handling system.

The through-the-spindle type in-process bore gaging units, which are preferably installed by the machine tool manufacturer when building the new equipment, can function either by staying in a fixed position inside the work bore during its grinding or by carrying out a reciprocating traverse synchronously with the similar movement of the grinding wheel. The latter mode of operation has the advantage of not requiring any clear space alongside the grinding wheel and also of scanning the work bore essentially over its entire length.

The illustrated model of through-the-spindle type internal gaging unit is adapted for both continuous

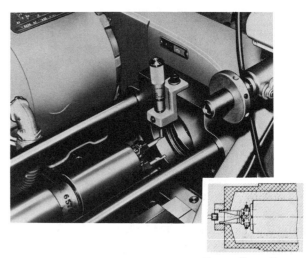

Marposs Gauges Corp.

Fig. 18-16. Through-spindle type in-process internal gaging unit shown in retracted position at the rear end of the work spindle of an internal grinding machine. In its retracted position, the gaging unit clears the workpiece entirely, as well as the area in front of it, thus avoiding interference with the operation of the automatic work-handling device. The adjacent diagram indicates the position of the contact fingers inside the work bore (not seen in the photograph), in relation to the operating grinding wheel.

and interrupted bore surfaces and can be used for internal diameters over a range of 0.315 inch to 2.283 inches.

MACHINE CONTROL GAGING SYSTEMS FOR DIVERSE WORK-SURFACE CONFIGURATIONS

In addition to the most common uses of in-process gaging, that is, the grinding of an essentially cylindrical surface that is accessible to the contact elements of the gaging unit during its machining, there are also various other important applications for machine control gaging. Each of these uses requires gaging units adapted to particular workpiece configurations and operational conditions. In the following, examples of those other applications of machine control gaging will be discussed, with brief descriptions of the pertinent gaging units.

Machine Control Gaging of Multiple Diameters

For controlling the sizes of multiple diameters on a single workpiece during the sequential grinding of these surfaces, by relying on the direct gaging of the diameter being ground, several alternative methods and equipment are available. The choice of these methods and instruments, which are pointed out in the following, depends, in addition to other considerations, on the characteristics of the available basic machine tool whose operation is to be controlled by the in-process gaging of the work surface being machined.

In Fig. 18-17 three diagrams illustrate the principles of multiple diameter cylindrical grinding operations, the sizing performance of which can be controlled by the different gaging systems reviewed here.

(A) Size control for the grinding of several different diameters in one setting. Two alternative meth-

ods, each requiring different instruments, can be applied: (a) Using a single gaging head, which originates signals for a measuring and control unit containing several independent potentiometers, each being set up to the appropriate zero position. These potentiometers are switched on sequentially, either by manual or automatic action. The adjustment capacity of the gaging head is a limiting factor with respect to the range of workpiece diameters for which this method is applicable. (b) Using for each diameter an individual gaging head with its own potentiometer, those being brought into action by a gaging head selector unit. Employing for each work diameter the appropriate model of gaging head, this method has practically no limitations regarding the range of diameters that must be gaged in a single setting of the workpiece.

(B) Size control for grinding several similar diameters on a workpiece with numerical input of the nominal sizes. A conventional multiphase machine control system, producing different feed rates for roughing and finishing, is complemented with a digital–analog converter. Programmed numerical inputs, corresponding to the nominal sizes of the sequentially ground diameters, produce the appropriate zero settings of the gaging and control system. A limiting factor is the adjustment range of the gaging head, typically about 0.079 inch with 40-microinch resolution.

(C) Size-control system for fully automatic, numerically controlled multidiameter grinding machines that raises the performance accuracy by taking over the control of the wheel in-feed during the finishing phase in the grinding of the individual work diameters. The servo-controlled gaging unit is programmed to be set to the nominal values of the sequentially ground different diameters and provides a setting range of about 7.9 inches. The sensing member of the system is a double-head type gaging unit, each head having its own electrohydraulic positioning device, servo-amplifier, and digital–analog converter for the signals originating from punched tapes or decade input switches, either one of these being used as a data carrier.

Gaging of Crankshaft Pins for Process Control

The grinding of crankshaft bearing surfaces with high-dimensional accuracy, by applying a machine control gaging system, requires gaging units of special design. For the main bearings of the crankshaft thin gaging heads are usually needed, because of the

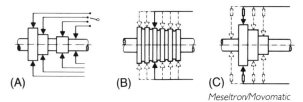

(A)   (B)   (C)

Meseltron/Movomatic

Fig. 18-17. Diagrams of different workpieces, each with several diameters, that are ground in succession to finished size by controlling the process with different types of gaging systems that automatically measure the actual size of the diameter being ground.

Marposs Gauges Corp.

Fig. 18-18. Gaging head for the in-process size control of crankshaft pin diameters. The head has interchangeable contact fingers of specially narrow and extended design for reaching work surfaces that only offer limited access.

narrow space resulting from the design of the workpiece and the application of steady rests. The in-process gaging of the pin diameters involves even further access difficulties, owing to the displaced axis of rotation of that surface with respect to the main axis of the crankshaft, a condition requiring gaging units with particularly long contact fingers. Figure 18-18 shows a gaging head of special design for crankshaft pin grinding applications. One of the two fingers is rigidly mounted and serves as the reference member for the function of the sensing finger. This type of gaging head is usually installed on a hydraulically actuated jump-on slide and has a pivoting mounting with spring load for assuring that the reference finger stays in continuous contact with the work surface during the actual grinding operation. The narrow, only 0.70-inch wide fingers of the gage head are both adjustable and interchangeable to accommodate an extended range of workpieces with different dimensions.

Laterally Contacting Gaging Heads

When finish ground on a cylindrical grinding machine preferably of the angular approach type, the faces and shoulders of workpieces have their locations determined by the axial position of the work with respect to the operating face of the grinding wheel. For controlling by means of direct gaging that

position of the workpiece, in-process gages of the laterally sensing type, which contact directly the face surface being ground, are frequently used. An example of the lateral gaging heads with a single-contact element for controlling the location of a face during its grinding is shown in Fig. 18-19. That gaging head is mounted on a swinging bracket, by means of which the head can be retracted or brought into contacting position during the successive phases of the grinding operation.

In addition to that application with a single sensing member, laterally contacting in-process gages are also used with two sensing members for controlling the lateral position of the surface being ground in relation to

a. A parallel surface of the workpiece, such as the distance of the face being ground from a similarly oriented face of an adjacent flange or the workpiece;

b. A surface in the opposite axial direction, such as the width of a flange while ground on one face in precisely controlled position with respect to the other face previously finished, of the same workpiece element.

Laterally contacting gaging heads are also used in conjunction with diameter gaging heads on plain or angular-approach cylindrical grinding machines, e.g., in operations that must produce a cylindrical section with accurate diameter and in a controlled

Marposs Gauges Corp.

Fig. 18-19. Lateral contact gaging head with a single sensing finger, for indicating the position of a face surface on the workpiece relative to a radial reference plane representing the operating face of the grinding wheel.

Meseltron/Movomatic

Fig. 18-20. Two-contact in-process gaging unit installed on an angular-approach type cylindrical grinding machine for the axial positioning by lateral contact of one side of the part (left) and concurrently gaging the diameter being ground (right). In the front are examples of typical workpieces finish ground with the aid of that gaging system.

axial position with respect to a finished shoulder of the groove of the workpiece. A two-contact gaging unit installed on an angular-approach type cylindrical grinding machine for laterally positioning one face of the part and for gaging the diameter being ground is shown in Fig. 18-20.

Controlling the Workpiece Thickness in Surface Grinding

The automatic grinding of flat surfaces on workpieces with close thickness tolerances requires the continuous monitoring of the critical work size, that being measured as the distance between the part's opposite side and that developed in the grinding process. In surface-grinding processes, that opposite or reference side of the workpiece is identical with the part side supported by the table or, more commonly, by the magnetic workholding plate of the reciprocating or rotary-table type surface-grinding machine.

Consequently, the gaging system must measure one dimension only and requires a single-contact gaging head. Exceptions to that rule are the double-faced disc grinders, on which both surfaces of a part are finished in a single pass; for such operations double-contact gaging heads are used.

Automatic surface-grinding operations are usually carried out on a batch of identical workpieces, mounted short distances apart along the rectangular table of the reciprocating type or around the circular table of the rotary type surface grinder. Consequently, the in-process gaging system must have the capability of measuring the size of a series of parts to be ground to the same thickness, carrying out the gaging as if a single continuous surface had to be measured. That requirement is met by gaging systems that bridge the interruptions of the signals caused by the gaps between the consecutively mounted parts or by the noncontinuous surface of individual workpieces.

Size-control gaging systems serving surface-grinding operations are equipped with gaging units incorporating automatic retraction of the sensing finger as soon as the set finish size of the work has been reached, and the system causes the termination of the grinding process. In a widely used model of size-control gages for surface grinders, that retraction movement lifts off the contact arm by $\frac{1}{8}$ inch from the finished workpiece, but for work with grinding allowances exceeding that clearance, an optional device is available for providing a lift-off distance of $1\frac{1}{8}$ inches.

Match Grinding by Size-Control Gaging

Component parts that must be assembled with tightly toleranced fits, particularly those requiring interference, are often manufactured efficiently by first finishing the bore of the female member of the assembly with an economically producible tolerance and then grinding the outside diameter of the mating part to a size corresponding to that specific bore diameter. Thereby, the inherently higher accuracy potentials of outside diameter grinding can be utilized to the benefit of critical fit conditions, avoiding the need for expensive and possibly wasteful segregation.

Size-control systems serving match grinding commonly consists of a stationary bench unit that accepts the female part of the assembly and a machine-mounted outside diameter gaging unit for the in-process monitoring of the male member's size while it is being finish ground. Both gaging units transmit their signals to a common amplifier-meter, which is designed to use the size signals originating from the bore gaging unit, modified by the set clearance or interference, to function as the reference (zero) value

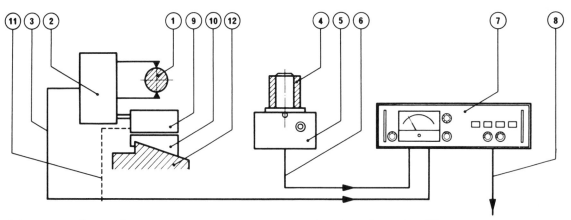

1 Component to be ground
2 External measuring head
3 Cable ext. meas. head – meas. and control amplifier
4 Component to be matched
5 Internal measuring head
6 Cable int. meas. head – meas. and control amplifier
7 Measuring and control amplifier
8 Main and control cable
9 Hydraulic retraction unit
10 Sole plate
11 Hydraulic pipes
12 Table of the machine

Meseltron/Movomatic

Fig. 18-21. Operational diagram of the members of a gaging system for match grinding, comprising a bore gaging unit, an in-process outside diameter gaging unit, and a common measuring and control amplifier.

for indicating the momentary size of the mating part's outside diameter.

Figure 18-21 shows the functional interrelation of the three principal members in an in-process gaging system for match grinding. That system comprises the stationary bore gaging unit (center) for determining the reference dimension of a particular mating part, the outside diameter of which is finish ground and its size is monitored by the machine-mounted in-process gaging unit (left). The signals originating from both gaging units are transmitted to a common gaging and control amplifier (right).

Figure 18-22 shows the stationary bore gaging unit while a work-locating arbor for the bore of a particular part is being installed. The internal gaging unit has interchangeable arbors, and the external gaging unit offers a wide adjustment range. Furthermore, both units are available in different models to accommodate workpieces in various size categories.

The repeat accuracy of this gaging system in maintaining the required fit conditions—clearance or interference—between the mating parts is stated to be on the order of 8 to 10 microinches.

SIZE CONTROL BY POST-PROCESS GAGING

Centerless through-feed grinding is a typical example of, and one of the most rewarding areas of ap-

Meseltron/Movomatic

Fig. 18-22. The stationary bore gaging unit of the match grinding size control system, shown in the setup stage, while a centering arbor, corresponding to the dimensions of the female part of the assembly, is being inserted. That arbor contains the mechanical contact point of the electronic bore gaging system.

plication for, process control gaging based on work-piece measurements immediately following the operation. That position of the work-size determination differs from the conventional use of size-control gaging, by which the size of the workpiece is monitored while the process, commonly finish grinding, is in progress, hence its post-process gaging designation.

Size control by post-process gaging may have to be applied because the workpiece, during its processing, is in a location that excludes access by the sensing contacts of the gaging system. The dependability of post-process gaging, however, is predicated on machining processes with inherently steady size-holding performance, that is, with variations in the size of the successively finished workpieces developing only very slowly, essentially as a gradual drift.

That latter requirement of steady size holding can be assured, in general, in finishing-type through-feed centerless grinding operations, a method selected to describe a characteristic size-control system operating by post-process gaging.

In other applications, where the limited accessibility of the workpiece during its processing necessitates the use of post-process gaging, although the steady-size-holding process is not assured in a dependable manner, post-process gaging stations may be complemented with segregating devices. The latter serve to separate those parts that were finished with out-of-tolerance size before the control signals issued by the post-process gage produced the necessary machine tool adjustments.

The widely used post-process gaging type, size-control system, whose operating members are pointed out in Fig. 18-23, comprises a linear actuation device of unique design, developed for assuring dependable size-control functions on centerless or double-disc grinding machines, which require compensating adjustments in two directions. On these categories of grinding machines in-process adjustments are needed, in addition to compensating for wheel wear, to eliminate the adverse effect on the workpiece size of thermal variations, the latter requiring machine adjustments occasionally in a direction opposite to that necessitated by wheel wear. That two-directional backlash-free machine slide adjustment, in very small increments, is accomplished by a special linear actuator using a magnetostrictive armature, with the property of magnetically induced length changes, operated in conjunction with two straddlingly arranged hydraulic clamping devices.

Such linear actuator devices permit the signal originating from the gaging system to be utilized ef-

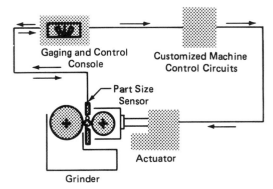

AirTronics, Div. of Am. Gage & Machine Co.

Fig. 18-23. Elements of a post-process gaging system with process control functions, for centerless grinding operations. (For reasons of clarity the workpiece and sensor are shown in a common plane with the wheels; actually, the sensing of the work size is effected as the parts emerge from engagement with the wheels, but still are riding on the work-rest blade.)

fectively, thereby maintaining a consistent work-size accuracy with a tolerance of 0.0001 inch. That tolerance, commonly plus or minus, may represent the total range of deviation in the centerless finish grinding of parts with uniform stock allowance. Figure 18-24 is a diagram of the system's gage indicator face with adjustable set points, representing different signal-originating positions, dependent on the measured work size.

For centerless grinding operations that do not require work-size holding on the order of 0.0001-inch tolerance, post-process gages can be used in a design permitting greater variations in the stock allowance, such as may be expected in drawn tubing. Figure 18-25 shows the diagram of the gaging unit serving the

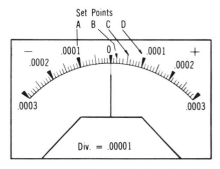

AirTronics, Div. of Am. Gage & Machine Co.

Fig. 18-24. Face of the gage indicator used in the process control system shown in Fig. 18-23. The four set points represent: (A and D) the undersize and oversize limits; (B and C) the corresponding fine compensation trigger points, in positions biased in the direction of "oversize," as a precautionary measure.

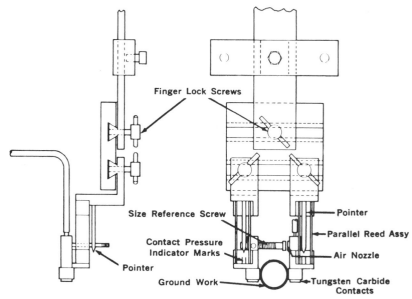

Fig. 18-25. Gaging unit of a post-process gaging system serving the centerless through-feed grinding of tubing in great length. The straddlingly arranged and flexibly mounted mechanical contacts transmit the momentary work-size conditions to an air gage with its nozzle directed against the head of an adjustable reference screw.

Finger Lock Screws

Size Reference Screw

Contact Pressure
Indicator Marks

Pointer

Pointer

Parallel Reed Assy

Air Nozzle

Ground Work

Tungsten Carbide
Contacts

AirTronics, Div. of Am. Gage & Machine Co.

centerless grinding of tubing in great lengths, operated in conjunction with a size-control system, which comprises a display console and a control panel with timers and relays, controlling the operation of a high-torque motor, which is coupled directly to the lead screw of the centerless grinder. The system is designed to control the diameter of the centerless through-feed ground work to 0.0005-inch tolerance, but under well-controlled operational conditions even higher work-size accuracy can be maintained.

The gaging unit of the system operates by air gaging the variations in the distance between two contacts straddling the rotating workpiece as it emerges from the area of the abrasive wheels. The tungsten carbide contacts are mounted on flexible reeds, permitting a floating position of the sensing members whose momentary separation, which represents the diameter of the contacted work, is monitored by the air gage. The nozzle of that air gage faces the head of a size reference screw, which can be adjusted quickly according to the desired work size.

Post-process gages of special design are also available for controlling the width of tightly toleranced parts, ground at high production rates on double-disc grinders. That method definitely excludes the contact of the workpiece during the actual stock removal, but because of the slow size variation of the consecutively ground parts, occurring as a gradual increase of width caused by grinding wheel wear, sensitive post-process gaging can detect that trend sufficiently in advance for initiating automatic machine adjustments, thus preventing the production of out-of-tolerance parts. Signals originating from the process control gaging system cause a stepping motor to move the wheel slide by the distance needed to compensate for the detected size variation when that is still well within the tolerance range specified for the width of the finished part.

Such post-process gages for double-disc grinders can be used for workpieces passing through and between the faces of the abrasive discs along a straight rail or for workpieces retained in rotary part carriers. The size control assured in automatic operation by such post-process gaging systems, mentioned as a typical part tolerance in the case of 0.010- to 0.015-inch stock removal, is +0.001 inch.

INDIRECT GAGING FOR PROCESS CONTROL

As pointed out earlier, through-feed centerless grinding, a highly productive and widely used method of finishing cylindrical workpieces with tightly toleranced diameters, represents a rewarding area of application for process control gaging. Rigorously specified work size is automatically obtained by the use of such a process control system, even though direct gaging of the workpiece during the actual machining is not practical because there is no access to the work in process by the gage sensor. Under such limitations, one way of accomplishing effective gaging for process control is by post-process gaging, using the gaging system described in the preceding section.

Another approach is by indirect gaging, monitoring with great sensitivity and rapid response the variations of the distance between the two wheels of the grinder, which determines the diameter of the centerless ground workpiece. The principles of operation of such indirectly gaging process control systems are displayed in Fig. 18-26 and described in the following.

A carbide pad rides on the surface of the operating grinding wheel. That pad is a part of a free-sliding assembly mounted on the grinding wheel housing, and the anvil attached to the pad represents the surface of the grinding wheel, transmitting any changes in the wheel diameter due to wear. Mounted on the regulating wheel housing is the air gage assembly with an extension arm at the tip of which is the air nozzle, facing the reference anvil. Any change

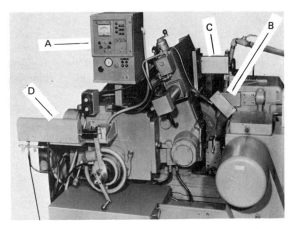

ITT Industrial & Automation Systems

Fig. 18-27. The indirectly sensing process control gage, whose operating principles are shown in Fig. 18-26, installed on a centerless grinding machine: *(A)* control panel; *(B)* gage reference anvil with its carbide pad; *(C)* gage extension arm with the air jet assembly; *(D)* in-feed actuator, which advances the position of the grinding wheel to compensate for wear detected by the air gaging system.

from wheel wear in the gap between the anvil and the air gage nozzle is sensed by a monitor/control system and results in signals causing an actuator to advance the wheel slide until the preset width of that gap is restored. The gage has a sensitivity threshold of 20 microinches and the in-feed actuation is effected in increments of 0.0001 inch. A small air-operated resonator is energized during the advance movement of the wheel slide to minimize the effect of friction-caused sticking (the adherence of mating way surfaces referred to, in shop language, as "sticktion"). Figure 18-27 shows the operating members of a centerless grinding machine with the elements of the indirectly gaging process control system installed.

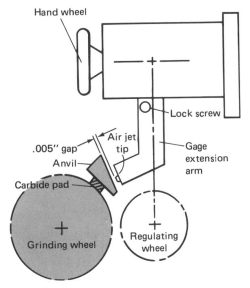

ITT Industrial & Automation Systems

Fig. 18-26. The principles of operation of an indirectly sensing process control gage developed for use on centerless grinding machines, where the position of the active wheel surface determines the diameters of the consecutively ground parts. Changes in the position of the grinding wheel periphery are transmitted to an anvil whose carbide pad is in uninterrupted contact with the wheel surface; that anvil faces the nozzle of an air gage that monitors the position of the wheel surface.

CONTINUOUS GAGING SYSTEM

In strip mills and wire mills producing rolled or drawn material at high traverse speed and in very great lengths, special gages are used to monitor the thickness and/or width of the manufactured material as it exits from the processing equipment. Such gages are usually equipped with mechanical contacts, typically floating arm mounted rolls, acting on fast response (linear variable differential transformer) (LVDT) transducers. The detected variations from the preset zero size are shown either on an analog meter or on a digital display. Gage setting accuracy is on the order of 0.0001 inch.

Advanced systems have push-button gage zeroing from the operator's station, which can be operated while the material rolling is in progress, and serves to eliminate temperature-related fluctuations. Generally oversize–undersize limit warning lights are also provided either on the meter panel or as a separate unit for complementing the basic gaging system. The light signal units of continuous gages frequently comprise time-delay relays, with timing range adjustable, for example, from 0.1 to 10 seconds. That capability's purpose is to prevent the control from responding to conditions such as dirt or welds, which could produce false oversize signals.

The more recently developed computerized continuous gaging systems also have interchangeable English unit/metric capability, an analog meter for continuous monitoring of deviation errors from the set nominal size, plus direct digital display of the absolute value of strip thickness; that latter display is operated on command.

DIGITAL READOUT SYSTEMS FOR CONTROLLING LINEAR DISTANCES

For process control purposes the size of a particular workpiece dimension is ordinarily gaged on the part itself, usually during machining. The gage indications, generally representing the momentary deviation from the required nominal size, serve to guide the manual or to control the automatic-feed movement of the involved machine tool member.

That procedure may be reversed for attaining comparable sizing results. Gaging the *actual* travel distance of that machine member whose position determines the size of the produced workpiece dimension can be used as the basis for controlling the process. Characteristic workpiece dimensions that are effectively produced by controlling the process based on indications of actual travel distances are the separation of hole centers or of particular work features, for example, shoulders or groove walls at a tightly toleranced distance from another element on the part's surface. These are listed as examples, but the sizes of many other types of work dimensions are also controlled with high accuracy by travel-distance measurement.

The gaging systems to be discussed in the following differ in a very important way from the commonly used travel-distance measuring devices of machine tool slides. These latter rely on the expected travel equivalent of an input action, typically the number of revolutions of a lead screw counted in full and fractional turns, the latter indicated by graduated dials. The accuracy of the movement thus produced depends on the precision of the involved mechanical machine tool elements, particularly of the lead screw and its nut. Thermal, frictional, elastic, and other conditions including play and backlash of mechanical couples as well as the stick-and-slip phenomenon often present in slide movements, also affect the exact correlation between input action and the resulting actual travel distance.

These limitations of the exact agreement between input action and the distance of the produced travel of machine tool members has long since been recognized and has led, in uses requiring very high distance accuracies, to the direct positional control of traveling machine tool members, using for that purpose gage blocks, dial indicators and so forth. These methods of direct positional measurement, however, are rather cumbersome to carry out, are prone to human errors and have many other shortcomings as well.

The recent significant advances in electronics, which have permitted the development and resulted in the wide adaptation of a new generation of direct travel-distance measuring systems, have contributed to the great benefits of digital readouts, which is also used as the general term of system designation. A brief review of that important category of process control gaging devices, the *digital readout systems* used on metalworking equipment, is, therefore, also included in this chapter.

It should be pointed out that digital displays are also used in certain types of equipment for indicating the linear equivalent of full and fractional turns of a lead screw, usually recorded by means of rotary encoders. Or, instead of counting the rotational increments of the lead screw, such increments, representing extremely small distances, may be dependably produced, for example, by stepping motors used in computer numerical controlled machine tools. Although such input-movement recording or producing systems may have the capabilities that are of comparable or even equivalent accuracy to those of the actual travel-distance measuring systems, from the viewpoint of these discussions they are not considered to represent instruments of dimensional measurement, and are therefore only mentioned without further discussion.

The first widely accepted applications of digital readout systems were for position indicating elements of measuring machines. These applications and several systems of digital readout devices serving that purpose are discussed in Chapter 12. The following discussions are about an area of application of much wider scope, namely, the use of digital readout systems for process control gaging as the measuring equipment of various types of machines tools. Some of these recently developed capability features of digital readout systems, such as those serving programmed operations, are now also used for measuring machines, and in that respect the following descriptions complement those in Chapter 12.

General Description and Function

The purpose of digital readout systems used for process control is to measure and to indicate in digital values the distance of displacement of the machine tool members whose position determines the size of specific workpiece dimensions that are developed in the machining process. To measure the actual distance of displacement the digital readout systems use pairs of sensing elements solidly mounted on each of the moving machine members, commonly slides, and their fixed position guideways. The relative displacement of the mating sensing elements is exactly identical with that of the associated machine members, whose distance of travel is thereby faithfully represented by that of the sensing elements.

The basic digital readout system comprises two principal members:

1. The *sensing element*, commonly designated as the *transducer*, which produces discrete pulses when the moving slide progresses over a specific number of distance increments; and

2. The *counter*, which counts and displays in digital values the number of distance increments over which the slide moved.

Before proceeding with the examination of these two principal members and their different types of execution, a few aspects of the application of digital readout systems for controlling machining processes are pointed out:

1. Frequently, more than a single slide of the machine tool must be moved over a specific distance to produce the desired workpiece dimension. Digital readout systems commonly are installed on all the slides whose positions participate in determining the critical workpiece dimension and, in general practice, the indications of these systems identified by the common designations of the axes in a system of rectangular coordinates: *A* for longitudinal (traverse), *B* for transverse, and *C* for vertical movement.

2. Most commonly, digital readout systems are used for measuring linear displacements, but systems are also available and are regularly used for measuring angular values in small fractional increments of rotary movements.

3. While in multidirectional applications each of the installed systems functions as an independent unit with its own transducer and counter, several concurrently used systems have their counters mounted into a common instrument housing, with their individual display screens contained in a single frame.

4. Digital readout systems offer many potential advantages over other types of displacement distance-measuring devices; several of these advantages are enumerated under the following title. In many cases these advantages definitely call for, or even require for technical reasons, the use of digital readout systems, while in other applications economic considerations—comparing the system's actual benefits over the conventional distance-measuring devices with the involved procurement cost—may have to direct the choice.

Properties of Digital Readout Systems

The following listing of system properties, which provide various operational advantages over the conventional indirectly measuring mechanical systems, is based on capabilities not necessarily available in all makes and models. That diversity of system design is warranted, considering that not all the enumerated properties can be utilized in every kind of application; therefore, it would be wasteful to incorporate into a device capabilities that would serve no useful purpose in actual operation.

Furthermore, the capabilities mentioned here may be available in instruments of various makes and models to different degrees, primarily with regard to the power of resolution (last increment of display) and the more sophisticated characteristics, such as programmed operation.

Finally, that listing, while extensive, is not entirely comprehensive; particular models, designed for special applications, offer capabilities beyond those listed subsequently as generally available properties.

Notwithstanding the preceding reservations, the following listing should prove useful in two respects: (a) in appraising the potential advantages, which can be derived from the use of digital readout systems, in general; and (b) assisting the future user in selecting a model that can provide the greatest benefits from a particular set of operational conditions and process requirements.

Here is the listing, in random order, of system properties that can offer significant advantages in many different applications:

a. *Direct distance measurement.* The distance measured is that of the actual translational movement of the machine tool member, in distinction to other systems that measure the input movement causing that travel.

b. *Freedom from wear.* The master scales (the transducer elements) of digital readout systems generally are noncontacted components; consequently, they are never exposed to wear. Wear does, however, affect systems comprising translation actuating members that concurrently serve as elements of the measuring device.

c. *Travel speed.* The permissible travel speed of machine members with translation distance measurement by digital readout systems is generally much faster than that of conventional systems, which have to avoid overstressing the actuating members in view of their role as elements of the measuring system.

d. *Cumulative errors avoided.* Cumulative errors, which can occur in conventional measuring systems, particularly over long travel distances and with several consecutive measurements, are practically nonexistent in most systems of distance measurement by digital readout.

3. *Digital display.* The essential characteristic of digital readout systems, that of displaying the traveled distance in digital values that are not subject to interpretation, assures error-free operation, greatly reducing potential uncertainties originating from the human element.

f. *Floating zero.* After a reference position from where the travel distanced is to be measured has been established, the counting device of the instrument can be set to zero by simply pressing a button. Subsequently, also, whenever a new reference position is chosen as the starting point of another distance to be measured, only the same simple procedure—pressing a button—is needed to reset the counter to zero. That capability of digital readout systems is termed "floating zero." The readout process can also be reversed, where more convenient, presetting the required distance on the counter, and then moving the slide until the zero display position has been reached.

g. *The resolution.* This is the finest increment of discrimination appearing as the value of the smallest fractional unit displayed by the digital readout system. It may not be equal to that of the most sensitive electronic gages used for comparative gaging, which display deviations from a present size on a meter; however, that indicating range is limited to 100 times, or much less, the least increment. Digital readout systems are available with resolutions satisfying most production needs, and that high degree of resolution is retained over the entire capacity range of the device, which may be 100,000 times, or often much more, the least increment of the measured displacement.

h. *Accuracy of repeat.* Because indications are based on incremental master scales, which are produced with an accuracy generally far exceeding the degree required for most industrial length measurements, and the system operates without making physical contact with those masters, the digital readout length-measuring systems are free of drift, periodic errors, and other potential sources of errors, including the effects of travel direction reversals, to which most other types of length measuring devices are exposed.

i. *Inch/metric conversion.* During the current period of gradual conversion from the inch to the metric system, means are often required for carrying out length measurements either in inches or in millimeters. The capability of incremental length-measuring systems with digital readout to convert inch increments into equivalent metric units, or vice versa, and to present a digital display of the measured distance in either of these two unit systems, as selected by the operator simply by pushing a button, can be of significant advantage in many applications.

j. *Interfacing capability.* Digital readout length-measuring systems, in their many different forms, can be interfaced with complementary devices, such as printers and computers, for uses including the production of printouts of the performed measurements, carrying out programmed inspection procedures, and many other functions serving advanced methods of inspection.

The Limitations of Digital Readout Systems

Notwithstanding their many advantages, digital readout systems are not the best or even applicable means for all conditions of length measurement. It may be well to point out several of that system's limitations for the purpose of assisting in a judicious choice of the method best suited for a particular length-measuring task.

a. Digital readout systems are not self-contained dimensional measuring instruments; they are complementary elements of machine tools and measuring instruments using slides with straight line movements along one or more axes. The readout system only registers and displays in numerical values the distances of these movements.

b. The controlled movements of the slide on which the workpiece is mounted do not establish the boundary points of distances between specific elements of the surface of the measured object; that must be determined by means of sensing devices operating by physical contact or optical targeting and mounted on a stationary member of the machine, or on a slide when that latter is being moved in relation to a stationary workpiece. When digital readout systems are used on machine tools, the sensing device is substituted by the cutting tool of known physical dimensions.

c. The accuracy of the readout system is only one factor, although an important one, of the aggregate accuracy of the length measurement. The precision of the slide movement, the controlled travel of which represents the distance being measured, is also a constituent factor, together with the sensitivity of the targeting device (for example, indicator) used for identifying or the precision of the tool used for developing the distance boundaries.

d. When used as a means of process control, installed on machine tools, the effective working accuracy of the slide positioning may be affected to an extent far exceeding that inherent in the digital readout systems, by such factors as

(1) Sag of the overhanging—not supported by guideways—portion of the machine slide, such as in the case of knee type milling machine tables;

(2) Error due to sag increases with the distance of the worked surface from the plane of the readout systems' master scale; and

(3) By thermal distortions in the slides and workpiece.

The above-mentioned limitations are related to the applicational conditions of digital readout systems. These conditions should be fully appraised and considered in order to derive the greatest benefits of the potential advantages available in properly applied digital readout systems.

The Transducers of Digital Readout Systems

The transducers are the means by which the travel distance of the moved machine member is registered in the form of pulses, each representing a specific distance increment. The transducers of digital readout systems consist of two members, one of which remains stationary, mounted adjacent to the guideways of the moving slide to which the mating transducer member is firmly attached. The relative displacement of these two complementary transducer members coincides with the measured travel distance of the moving machine element.

The functions of transducers serving digital readout systems can be accomplished by several different means, resulting in specific transducer types, each suitable for a particular, although not rigidly delineated, applicational condition. The commonly used transducer types and their characteristics are as follows.

1. *Ruled scales.* These are actually gratings with very closely spaced lines alternating with blank spaces and most commonly are made of chromium deposited on glass, a method that is adapted to very high accuracy of execution. The pulses produced by the mutual displacement of the mating scales are sensed optically, as moiré fringes, in a manner described in Chapter 12.

2. *Glass scales* for installation on machine tool slides are supplied in casings protecting the delicate grating surface from mechanical shocks and also sealing them against the penetration of coolant and scarf. A pair of glass scale sets, together with the pertinent counterdisplay, are shown in Fig. 18-28. The most frequently used resolutions for glass scales are 0.0005 and 0.0001 inch. Glass scales are manufactured and are usually readily available in many different sizes to suit the total lengths of the slides on which they are to be installed, but for practical reasons, the maximum length of glass scales is about 60 inches.

3. *Tape type transducers.* Tapes as transducer elements offer the advantages of extended travel length capacity and moderate cost. The practically unlimited travel distance is particularly advantageous for slide movements over 60 inches, where glass scales are only rarely used, and the moderate cost is beneficial when the somewhat lower general accuracy of the tape system is adequate.

Anilam Electronics Corp.

Fig. 18-28. The members of a digital readout system using glass scales, two of which, for longitudinal and crosswise slide movements corresponding to the *A* and *B* axes, are shown here, alongside the counter and digital display box. The box is installed remotely from the machine slide for convenient observations; the aluminum housings containing the well-protected glass scales and access by the pertinent reader heads are mounted on the machine slides and their guideways, respectively.

Two different systems of tape type transducers are available, the operating characteristics of which are indicated in the diagrams in Fig. 18-29.

(A) *The linear tape transducer*, which resembles a steel measuring tape without visible graduations, measures incrementally in a manner that produces pulses that are converted into digital values. This provides a much higher degree of resolution than the visually readable graduations. The unwinding or rewinding tape is in slippage-free contact with a plug-gage made with a bore for being pressed on the shaft of a rotary encoder, whose rotating grating disc produces the pulses corresponding in number to the travel length of the tape. The tape contacts, although over 270 degrees, only the gage plug, and it is re-

ceived by a separate constant-torque spring type wind-up; consequently, the diameter variations of the spool do not affect the accuracy of the measurement. A housing, containing the mechanical components and the rotary pulse generator as one unit, and the finger holding the free end of the tape as the mating unit are mounted on machine elements that move in relation to each other; that movement causes the tape to extend or to retract. The extended tape is shielded from dirt by an aluminum extrusion. A widely used make of linear tape type digital readout system is manufactured in two models; one allows a maximum travel of 120 inches and the other, with components in heavy-duty design, allows a maximum travel of 360 inches. Both types provide 0.001-inch and 0.0005-inch resolution. The maximum cumulative error of these systems is stated to be 0.0005 inch in 20 inches, a factor to be considered in determining the system's field of applications.

(B) *The endless tape transducer* is a product of more recent development and uses a precision stainless steel tape with ends connected to form a continuous circular tape. This system eliminates the use of a rewind mechanism and permits calibration prior to installation. The operating principles of that system can be seen in the diagram in Fig. 18-29B, and Fig. 18-30 shows several major members of an endless tape system, including the counter for two axes and the tape housing for one of the axes, together with the pertinent finger that must be attached to the moving machine slide. The illustrated make of end-

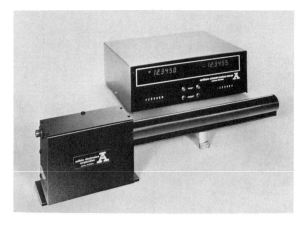

Anilam Electronics Corp.

Fig. 18-30. Endless tape system—several major members, including the counter box and one of the two tape housings with encased tape and the finger, which is to be attached to the moving machine tool member for its positive connection with the tape.

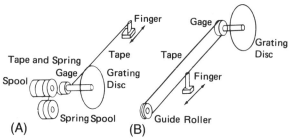

Anilam Electronics Corp.

Fig. 18-29. Operating characteristics of the two basic systems of tape type transducers.
(A) Linear tape transducer.
(B) Endless tape transducer.

less tape readout system is manufactured in two different models: one for long displacement distances up to a maximum travel of 160 inches, with 0.0005-inch highest resolution, and the other for higher-accuracy requirements, with 12-inch maximum travel and 0.000050-inch resolution.

The rack-and-pinion system is recommended particularly for applications with travel lengths exceeding the practical limits of glass scales. The transducer elements of that system comprise a rack installed on one member of the machine tool, and engaged by the pinions of the pulse-producing device mounted on the other machine tool member, the mutual movements of which are to be measured. Figure 18-31 shows the principal elements of an input system, with the racks mounted on a spar that is to be installed on a machine tool member, as well as the rotary encoder whose pinion engages the racks. Rack-and-pinion type transducer systems are effectively used on machine tools with very long slide movements, such as large lathes, jig borers, and horizontal boring mills, an example of which is shown in Fig. 18-32.

The racks are manufactured in 12- and 16-inch or longer sections and are installed in spars that may have lengths of 16 feet for a single unit; several of these units can be mounted abutted end to end to

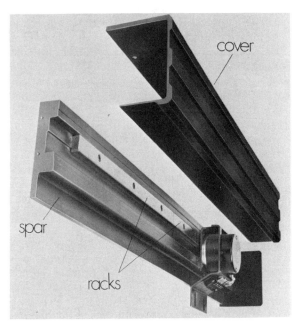

Elm Systems

Fig. 18-31. Principal elements of a rack-and-pinion type input system, shown with the protective cover removed. The rotary encoder (at right) carries a pinion in engagement with the rack, for registering by precise pulses the relative movements between the two machine tool members on which these mating components of the system are installed.

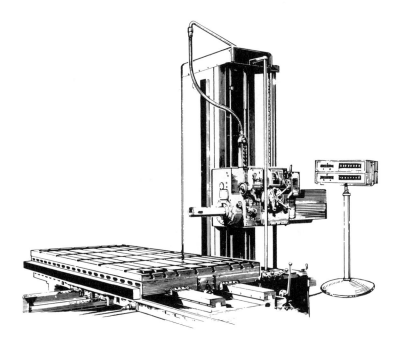

AirTronics, Div. of Am. Gage & Machine Co.

Fig. 18-25. Gaging unit of a post-process gaging system serving the centerless through-feed grinding of tubing in great length. The straddlingly arranged and flexibly mounted mechanical contacts transmit the momentary work-size conditions to an air gage with its nozzle directed against the head of an adjustable reference screw.

produce a total measuring length of 100 feet or more. The individual racks are usually precision ground to a high degree of accuracy, typically ±0.00015 inch, and the installation of the individual racks on the spars is carried out in a manner to compensate for even very small inaccuracies, thus holding the non-accumulating accuracy over the whole length of the system to typically ±0.0004 inch, at 68°F (20°C) envelope temperature. The calibration of a set of racks installed on a spar is preferably carried out by laser interferometers, which have an accuracy better than 0.5 part per million. That calibration procedure can also compensate for some types of errors of the machine on which the rack-and-pinion type transducer is used.

The optical encoders, on the spindles of which the pinions in engagement with the racks are mounted, usually have a very high response speed, permitting traverse rates on the order of 2400 inches per minute. Because the racks are made of steel they have practically the same coefficient of thermal expansion as the machine tools on which that system of travel-distance measurement is installed.

THE COUNTER DISPLAYS OF DIGITAL READOUT SYSTEMS

The counters of digital readout systems have the basic function of converting the discrete signals produced by the transducer into digital values that are immediately displayed in clearly visible numerals. In addition to that basic function, the counters are also designed to perform various processing functions, which most commonly include the following:

a. "Zero reset"—by which the reference point for a series of consecutive measurements can be established at any location of the monitored travel;

b. "Preset"—designating the ability to enter a desired location into the counter for future reference, specifically as the target position for the next movement;

c. "Inch/metric conversion"—permitting the numerical value of the monitored travel distance to be displayed in either inches or equivalent millimeters, a feature valuable particularly during the current, perhaps decades-long, period of conversion to metric values throughout the industry.

Examples of counter displays performing the listed functions are shown in Figs. 18-28 and 18-30.

For applications of digital readout systems where other functions of the counter can provide benefits

by reducing the positioning time and the incidence of operator's error, counter consoles complemented with microprocessors are available in different designs with capabilities satisfying a wide range of different process requirements. The microprocessors used in readout systems are essentially minicomputers with programs for performing the desired functions. The advantage of the microprocessor-equipped systems over the conventional hardwired logic systems is their ability to have their program changed in the field without the need for adding or exchanging a printed circuit board; a new program can be set simply by using the keyboard entry.

Here are a few examples of capabilities offered by a typical microprocessor digital readout system:

1. *Keyboard* for entry of dimensions retained by a programmable memory, which has the capacity of, regularly, 12 and, optionally, 99 different positions. The position entries are addressed to the axis selected by the operator and the entered positions can be recalled for visual check. The console can be supplemented with battery backup for protecting the program from loss in the case of power interruption.

2. The *absolute zero recall,* after it is once set at any point, permits one to carry out a number of incremental movements, zeroing the counter with each of those, and still display the total distance traveled simply by recalling the absolute zero.

Anilam Electronics Corp.

Fig. 18-33. Advanced type of microprocessor readout console for three axes with complementary recorder for off-line preparation of tape cassettes, each storing a particular positioning program.

3. *Add/subtract calculator* capacity is particularly useful for cutter offset compensation during programming from a blueprint. Automatic error compensation by entering a factor for known machine inaccuracies can be helpful by automatically correcting the consequences of such equipment deficiencies with respect to any particular axis.

The readout console shown in Fig. 18-33 offers even further capabilities, such as:

1. Tape cassette memory storage for permanently retaining a programmed memory that will have to be used again.

2. Off-line programming is also possible with a recorder added to the console, as shown in Fig. 18-33. Preferably, a separate console, independent of those installed on the machine tools, is used for such off-line tape cassette preparation, thus permitting an even better utilization of the machine tools, which can be operated without interruption for reprogramming when changing over to workpieces needing a different positioning program. This system has the capacity of 340 memory locations.

The preceding listing of a few selected capabilities of microprocessor-equipped digital readout systems is far from complete. Its purpose is only to indicate the great variety of capabilities that, as a result of recent advances in electronics, became available at reasonable cost for improving the efficiency of accurate distance setting on machine tools. Settings of both initial and final positions—which help to control travel distances with dependable accuracy—that are practically free from human error and that can be done with much less downtime than is usual when conventional movement controls are used, represent significant advances in the application of dimensional measurement in the production process.

19.
Automated Dimensional Measurements

INTRODUCTION

Before defining the concept of automated dimensional measurements it is well to review the major steps generally involved in the measurement of geometric dimensions. These steps are:

1. *Positioning* the object and instrument in relation to one another, also termed *staging* when the object is placed on the instrument. In most cases this step establishes the mutual reference position of the object and the instrument for the subsequent measurement of a single, multiple or series of dimensions.

2. *Travel*, or movement, over a controlled or predetermined distance. That travel may be over a fixed distance, corresponding to the nominal size of the inspected part dimension and relying on the instrument's indicator to signal the amount of eventual deviations, or to a point of coincidence where the significant surface element of the object coincides with the fixed target (for example, mechanical or optical) of the instrument. In the latter case the actual distance of travel represents the size of the measured dimension. Gages made or preset for specific dimensions incorporate the nominal distance, thus dispensing with travel as a step in the measuring process.

3. *Indications*, when measuring with indicating instruments, are produced by contact between the indicator's sensing member and the object after travel over the nominal distance has been carried out. The indications confirm the coincidence, or display the amount of deviation of the measured object dimension from its nominal value.

4. *Appraisal* of the measurement result with regard to the applicable size and tolerance specifica-

tions. In manually executed measurements this is usually a mental process, often complemented by manual recording.

5. *Decision-making and application*, such as accepting or rejecting the part or assigning it to a particular size category (segregation), executed by the appropriate disposal of the part.

In common measurement or gaging practice most of the steps listed are performed by the person carrying out that operation. In *automated dimensional measurements*, on the other hand, most or even all of these steps are performed by the instrument, without any human action.

Automated dimensional measurements, as a concept defined for the purpose of the subsequent discussions, do not have distinct boundaries. As an example, instrumentation serving process control gaging—the subject of the preceding chapter—comprises various types incorporating many automatic functions. In that case, however, the emphasis is on controlling the machining process of a particular workpiece, while the purpose of automatic measuring systems described in the following is primarily inspection for qualifying a product or for determining specific dimensional conditions.

Accordingly, this chapter will discuss the following:

1. *Automatic gaging and sorting*. These comprise processes applied in conjunction with the production of small- and medium-sized components, basically as a means of acceptance or rejection. The general system is frequently adapted to assigning the accepted parts to different size categories, such as would be needed for selective assembly with the corresponding categories of similarly inspected mating

parts, a dimensional sorting termed *segregation*. Instruments serving that system are designed to check either a single or a limited number of part dimensions in a characteristically fast sequence.

2. *Automatic multidimension gaging.* The inspection of dimensionally critical component elements, which are manufactured in high volumes for assembly into complex machinery products, such as internal combustion engines, is carried out at efficient rates and with dependable accuracy by special-purpose automatic gaging machines. Such component parts usually have several geometric parameters whose dimensional accuracy is critical for the proper operation of the assembled product and must be inspected, preferably in processes comprising the measurement of several individually toleranced and positionally interrelated dimensions; such processes are designated as multidimension gaging. Special gaging machines are used with fully automated operation, including, in addition to the gaging proper, a variety of auxiliary functions, such as material handling, object positioning, evaluation of the gaging results and the appropriate disposal of the inspected parts.

3. *Automatic gaging as a complementary operation.* The mechanized assembly of parts that require matching according to the sizes of their features, which must operate with closely toleranced fit conditions, must be preceded by the selection of the properly mating components. Often it is practical and economically advantageous to combine the complementary size matching and assembly operations in a single automatic gaging and assembly machine. Examples of automatic assembly machines complemented with automatic gaging and sorting functions are presented, with the emphasis of the discussion on the performance and equipment of the complementary operations.

4. *Automatic statistical analysis.* This constitutes a unique application of automation to gaging operations, in this case directed at the evaluation and recording phase of the process. While statistical analysis, being based on probability, is not equivalent in the reliability of its findings to 100% inspection by automatic means, it is still widely used in the manufacture of mass-produced parts when a very small ratio of undiscovered rejects has no critical consequences. The recording and evaluation phases of conventional gaging with statistical analysis are, however, time-consuming and prone to errors due to the human element. Recent advances in electronic data-processing equipment permit the convenient

combination of automatic recording and calculating functions with the gaging of each single piece, thereby producing statistical evaluation concurrently with the gaging of individual parts executed without human action, consequently, free from human errors.

5. *Computer-supported coordinate measurements.* The role of coordinate measuring machines throughout industry has increased considerably during the past decade, particularly as a corollary of the rapidly expanding use of computer numerical controlled (CNC) machine tools. For checking the correctness of the programs controlling the operations, as well as the actual performance of such automatic machine tools, the first piece produced in a production run must be inspected, a process for which coordinate measuring machines are particularly well suited. An essential requirement of inspection for that purpose is the rapid performance of the measuring process, preferably complemented with an inspection report on all the significant part dimensions. Coordinate measuring machines adapted to satisfy the indicated inspection requirement substantially reduce the downtime of expensive CNC machine tools, which would result from first-piece inspection performed by conventional measuring methods, including manually controlled and recorded coordinate measuring machine operations.

AUTOMATIC GAGING AND SORTING

General Process Characteristics

A large and continually growing number of assembly types produced by the engineering industry contain component parts whose functionally important dimensions must satisfy rigorous tolerance specifications. In many cases meeting that requirement is facilitated by the absolute range of acceptable sizes being significantly wider than the size variations allowed for parts contained in any particular assembly. These composite requirements of a wider, yet still exacting total size range and the much tighter tolerances for parts used in operating assemblies require individual gaging and sorting into size groups, in a process commonly termed *segregation*.

In addition to the functionally important part size, there are often other dimensional requirements as well, generally easier to control in the manufacturing process, but still of critical importance for the proper operation of the assembly; those will be referred to as supplemental dimensions. While most types of supplemental dimensions may be less prone to variations in the manufacturing process, they are often tightly toleranced for assembly or functional reasons; therefore, they also require 100% inspection for assuring the acceptability of the component part.

Considering that, in high-volume manufacturing, parts with critical primary and supplemental dimensions are produced in vary large quantities, the individual inspection and dimensional segregation of such component parts require equipment with commensurate production rates, occasionally as high as several hundred parts per minute inspected and sorted.

The inspection requirements outlined are satisfied by a particular piece of equipment, the automatic gaging and sorting machine, which performs in rapid sequence all essential phases of the inspection, namely, the following:

1. Material handling, including transporting of the individual parts to the gaging station, positioning or locating the parts for gaging, and subsequent transport of the inspected parts, in accordance with the instructions received from the gaging unit;

2. The actual gaging, including the checking of the primary and, when needed, supplemental dimensions;

3. Decision-making and issuance of instruction signals for the required material-handling functions.

Various aspects of equipment characteristics and capabilities are examined.

The gaging systems used by the manufacturers of automatic gaging and sorting machines are chosen to best satisfy the required degrees of discriminating sensitivity and operating speed, in compliance also with the space and access conditions of the workpiece, all these with regard to the initial and operating costs. The commonly used gaging systems and their operational characteristics are reviewed in the approximate order of their frequency of use.

Electronic gages offer the advantages of high response speed, which permits the performance of several hundred gagings per minute, and of very high sensitivity, which is needed for the specified accuracy level of the gaging. When called for by workpiece specifications, electronic gages used in automatic operations can have a discriminating capability of 20 microinches or, in extreme cases, even finer. An additional advantage of electronic gages is the availability of readily utilized signals for instructions to the material-handling system and, when needed, for recording purposes also.

Air-electronic gages are used when the configuration of the inspected part's feature interferes with access by the contact members—the probes—of the electronic gages, such as in the case of holes of very small diameter, and also when average size conditions have to be determined. The output of air gages used on automatic gaging machines is converted into electronic signals for directing the sorting operation of the material-handling system.

Electric and air-electric gages were used extensively in earlier automatic gaging and sorting machines; however, the principal justification for their choice—lower costs—has essentially been eliminated by advances in the design and manufacture of electronic instruments. With regard to gaging performance, electric gages have limited capabilities in several respects, such as their sensitivity (adapted only for tolerances not finer than about ± 0.00025 inch), lower operating speeds (limiting their use to not more than about 70 parts per minute), and restricted classifying capacity (for a maximum of about six size groups). Electric gages occasionally are combined with air gages for the sensing function, where the noncontact properties and small probe size of the air system are needed.

Mechanical gages, which do not emit signals to actuate sorting functions, are, in principle, not adapted for automatic gaging and sorting machines. An exception is the gaging and sorting of parts with plain geometric characteristics, such as balls and rollers, only requiring the gaging of a single dimension, typically but not exclusively the diameter. That part dimension can be associated conceptually with the width of a gap between supporting members and effecting the gaging by relying on gravity to pull the part through that gap whose width just corresponds to the part diameter. The succession of gap widths or, in practice, a gradually widening gap permits the part to drop down at different locations between the two supporting members, along which the parts travel.

Work-handling equipment of automatic gaging machines. The advanced automation of the gaging process requires the mechanized transport, positioning, and disposal of the parts being inspected. In addition to the economic and other benefits of automatic operations, inspection processes whose expected production rate exceeds the limits of consistent human performance, definitely require the application of mechanized work handling. The work-handling systems actually used may consist of purely mechanical devices or may operate in combination with the application of pneumatic or electromagnetic force.

The functions of work-handling mechanisms in automatic gaging and sorting machines generally comprise all the movement and holding phases of the part's progress through the machine, starting at the feeding from a hopper or other storage system, or, occasionally, directly from the production line, through conveyors to the discharge into the appropriate size group's container. Essential material-handling functions between these two boundary points of the travel through the gaging machine are the proper orientation of randomly positioned parts, transfer to the gaging station(s) timed to correspond to the movement of the gage probes, locating and retaining the part in the gaging station as well as actions termed staging and discussed subsequently.

The results of the gaging performed determine the instructions issued by the gaging unit to the material-handling system for the disposal of each individual part, depending on its gaged condition. Accordingly, each part is channeled into containers corresponding to its size group or reject condition.

The staging of the workpiece during gaging. The workpiece to be inspected can either be held for gaging in a stationary position or be gaged on-the-move. In addition to the characteristics of the gaging machine, particularly with regard to the response speed of its gaging and decision-making unit, and the speed with which the sorting instructions are carried out by the material-handling system, the configuration of the workpiece and of its inspected feature are also controlling factors. An obvious example is that it is not practical to gage on-the-move the diameter of a bore, which requires the penetration of a probe into a hole to be presented in a stationary position.

In other cases, when the part must be inspected in different places and/or orientations, the gaging station of the machine must provide the necessary movements of the part or gaging probe; these may be linear or rotary, which is an indexing or continuous rotation. Automatic gaging machines performing the gaging of workpieces in several positions usually are equipped with special instruments for evaluating the combined effect of signals produced by a succession of individual measurements or by continuous gaging.

The inspection of form or nongeometric conditions of the workpiece. Most commonly automatic gaging machines are designed and used for the inspection of linear dimensions, specifically diameter and length. Often, however, the gaging of other geometric parameters is also required and provided for by supplemental gaging stations, which assure additional inspection capabilities. Several types of geometric conditions are determined as automatically gaged, compared, and evaluated linear dimensions, such as for checking straightness, taper, or out-of-

roundness. Other additional inspection capabilities may be provided for workpiece conditions whose checking requires that the basic dimensional gaging capacities be supplemented by specific motions or force as, for example, the automatic inspection of the internal radial clearance in ball or roller bearings. In particular cases, automatic gaging machines may be equipped with special devices that, complementing the performance of dimensional gaging, inspect workpiece conditions not related to geometry, such as determining the weight of a part or checking with eddy currents for proper hardness or freedom from cracks. The results of these special measurements are also controlling factors for the further disposition—acceptance and sorting, or rejection—of the inspected parts.

In the following, examples of automatic gaging machines are presented, which are selected to illustrate the wide range of applications for that category of inspection equipment as well as the variety of capabilities that machines designed for particular gaging and sorting operations possess.

Sorting by Checking Mechanically a Single Part Dimension

Many types of workpieces with plain geometrical forms have only a single dimension that is critical for functional and assembly reasons. In manual inspection procedures, single diameters or lengths are often checked by using snap gages, which verify whether the part can pass through the gap between opposite gage jaws.

The principles of diameter checking with a snap gage can be applied to an automatically operating procedure adapted to explore the part's actual diameter by checking it, in succession, with gaps of different widths. Further requirements for accomplishing an automatic inspection process are:

a. Substituting gravitational pull for the manual testing of the parts with a snap gage;

b. Providing a series of different gage gap widths in sequence by passing the parts along a single gradually widening gap, which provides increasing gap widths at distinct sections along its length;

c. Assuring the proper orientation of the parts by imparting rotation, in opposite directions, to the work-supporting members of the instrument;

d. Moving mechanically the parts along the series of gaps provided by the inspection instrument.

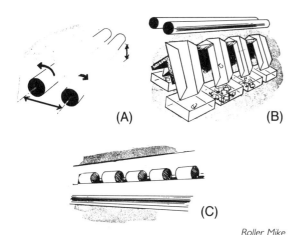

(A)   (B)

(C)

Roller Mike

Fig. 19-1. Operating principles of a purely mechanical size-checking and sorting instrument.
(A) Work-supporting rolls rotated in opposite directions, tilting upward to the feeding end and from there slightly diverging to produce a gap of increasing width.
(B) Closely adjacent chutes directed to alternate sides along and beneath the gap formed by the rolls, each chute leading into an individual container.
(C) The arrangement of the workpieces (for example, bearing rollers for diameter checking), which are supported by two nearly parallel rolls during their travel.

Roller Mike

Fig. 19-2. General view of a purely mechanical gaging and sorting instrument, named Roller Mike® by the manufacturer. Chutes, arranged alternately on both sides, receive the size-graded parts; oversized parts are discharged at the exit end of the instrument.

These requirements of an automatically operating single-dimension checking and grading instrument are accomplished by applying the design principles indicated in the diagrams of Fig. 19-1. To provide the gradually widening gap, two rolls, acting as straight edges, support the workpieces. These rolls are installed so as to diverge very slightly from parallel; thus the gap separating the work-support rolls is expanding by a small amount from one end, where the work is introduced, to the opposite end. The width of that gap at the entry end, as well as the amount of expansion along the entire length, are sensitively adjustable to correspond to the various grade sizes of the inspected part dimension.

The support rollers are gear driven, rotate in opposite directions, and to a sensitively adjustable degree tilt down toward the exit end. That rotational movement of the support rolls and their tilted position cause the workpieces introduced to proceed toward the exit end at a rate that can be controlled by the speed of the rolls (variable from 10 to 75 rpm) and the degree of tilt. Chutes, which lead into separate containers and are arranged under adjoining sections of the expanding gap between the two support rolls, receive the parts released by the rolls. That release occurs when the width of the gap, at a par-

ticular section of its length, just exceeds the gaged dimension, usually the diameter or length, of the inspected workpiece.

Figure 19-2 is the general view of a size-grading instrument operating by the gradually increasing gap between two work-supporting members. Regarding the capabilities and performance of the illustrated instrument, the following values are stated by the manufacturer:

1. Grading (sorting) accuracy complies with group incremental specifications of 10 millionths of an inch for high grade balls, 30 millionths for bearing needles, and 50 millionths for bearing rollers.

2. Hourly production rates vary from 10,000 for medium-sized bearing rollers to as much as 300,000 for small tungsten carbide balls used in pens.

Automatic Bore Gage with Interchangeable Air Plug

Automatic gaging of critical part dimensions between consecutive operations when combined with a sorting function avoid expending further work on reject parts and, by discovering dimensional discrepancies right at the source, also alert the operator to the need for machine adjustment.

The automatic gaging and sorting machine shown in Fig. 19-3 provides great applicational flexibility by permitting easy setting to specific workpiece di-

(A)

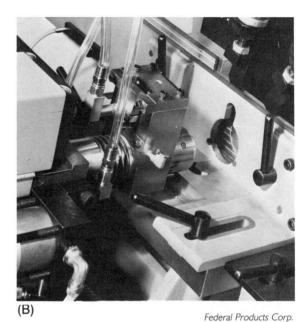

(B)

Federal Products Corp.

Fig. 19-3. Automatic bore gage with interchangeable air plug and adjustable material-handling elements, permitting cylindrical parts in a wide range of different sizes to be accommodated.
(A) General view of the gage.
(B) Close-up of the operating area.

Automatic Segregating Gage for Small Cylindrical and Tapered Parts

Selective assembly of products containing several component parts is generally done because in the manufacturing process critical workpiece dimensions often are easier to control when wider tolerances are permitted than those required within a particular assembly. Carrying out selective assembly is predicated on the availability of component parts segregated into specific classifications or size groups.

In large volume production such segregations by size of a great number of parts requires high speed gaging and sorting machines. Such machines combine the rapid and very accurate gaging of one or more dimensions of each individual part with co-ordinated material-handling functions for assigning to and actually transporting into individual classification containers each part in accordance with its measured size.

Typical workpieces that have very tightly toleranced dimensions, particularly diameters, and that are manufactured in substantial quantities are bearing rollers, both cylindrical and tapered; small shafts; and similar, essentially cylindrical parts. A

mensions within a very wide range. The machine is designed to inspect the bores of bearing rings or similar cylindrical parts up to 6-inch diameter and is generally used following the internal grinding operation. The gaging member is an interchangeable air plug used to inspect the workpiece bore by checking its diameter at two points as well as its length. The air gage indications produce signals that actuate the reject disposal gage for out-of-tolerance parts. Air cylinders are used for transporting and staging the parts; hydraulic action provides smooth, positive exploration of the bore by the air plug.

Various safety features are incorporated into this automatic gage, such as for diverting the parts from the gage should workpiece jam occur and for the case of electrical or pneumatic failure, which also actuates the automatic opening of the reject disposal gage.

(A)

(B)   *Federal Products Corp.*

Fig. 19-4. Automatic segregating gage equipped for measuring and classifying by diameter large-sized tapered bearing rollers at the rate of 1800 parts per hour.
(A) General view of the machine with automatically exchanged plastic tubes for collecting a specific number of parts assigned to a particular classification.
(B) Close-up view of the disposal mechanism with transport belt and gates controlled by the classifying unit at the gaging station.

characteristic example of a high speed gaging and segregating machine for such parts is shown in Fig. 19-4, that particular automatic gage being tooled for measuring the diameters and also for checking the taper angle of large tapered bearing rollers. The machine is gaging each individual roller, determining the correctness of the taper and the actual diameter (at a controlled axial level), the latter for assigning the part to one of eight size groups, each 0.0005 inch apart. Parts with diameters outside the range of the eight classifications or size groups, or with inadequate taper, are rejected. The gaging unit of the machine contains two opposed, matched electronic gage heads to measure the diameter and two matched heads for checking the taper.

The material-handling system of the machine positions each part individually into the gaging stations, holding it there in a stationary position for a very short time while the diameter and the taper are measured and that information is fed into a classifying unit. That unit then opens the appropriate disposal trap door along the conveyor belt, which moves the part following its gaging. The parts discharged through each of the trap doors are collected in tubes, which accept a specific number of rollers, a complement for a particular assembly, that being controlled by a preset counter. After the required number of parts has been collected by a tube, that tube is replaced by an empty one by an automatic indexing device.

Reject parts are also classified into groups of "Oversize Diameter," "Undersize Diameter," and "Reject Taper." The basic machine can be equipped for gaging and handling cylindrical parts up to 3-inch diameter, and its hourly production rate is up to 1800 parts.

Automatic Gaging Machine for Measuring Several Dimensions of Inspected Parts

In addition to plain linear dimensions, such as diameter and length, geometric characteristics, for example, cylindricity and roundness, are often important conditions for the proper functioning of mechanical parts. The manufacture of automobiles is one of the better known industries where mass-produced parts must satisfy exacting specifications.

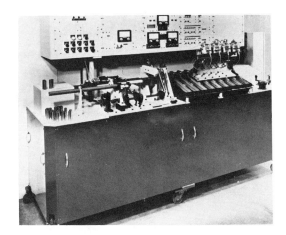

Federal Products Corp.

Fig. 19.5. Automatic gaging machine for the dimensional inspection and sorting of automotive wrist pins at speeds up to 4000 parts per hour. The machine determines the adequacy of several dimensional conditions, such as length, straightness and roundness, then measures the diameter for classifying the geometrically satisfactory parts into four good categories, as well as rejects for specific reasons.

While the geometric characteristics generally are defined as minimum requirements with specified values of permissible deviations, the functionally significant linear dimensions, particularly diameter, may be acceptable over a wider range, as long as the parts are segregated into classifications, each with specific, tightly toleranced diameters.

The automatic gaging machine shown in Fig. 19-5 serves for the dimensional inspection and sorting by size of automotive wrist pins and is equipped for measuring such geometrical characteristics as out-of-roundness, defined for this purpose as diameter variations, which must not exceed 0.0001 inch TIR (total indicator reading). To inspect that characteristic, the part must be rotated while being contacted by a sensitive indicator; this operation is performed at the "out-of-round" station of the machine, where the part rests in a Vee trough while being rotated by a belt.

The "good" parts, which satisfy the requirements of roundness, taper (not exceeding 0.0001 inch), length and diameter ranges, are classified into one of four groups, in increments of 0.0001 inch, and are discharged through individual channels, where the "good" parts are color marked to identify their classifications. The diameters are gaged at two places, 2.250 inches apart, and for the case of a diameter difference within the straightness limits, the larger diameter takes precedence.

Other channels serve to discharge the rejects, which are also classified, identifying the cause of rejection. All these measurements as well as the required material-handling and part-disposal functions are carried out at high speed, resulting in an hourly performance rate of up to 4000 parts.

Double-Track Automatic Measuring and Sorting Machine Complemented with Selective Weighing

The measuring machine shown in Fig. 19-6 is designed for the concurrent inspection of two different models of connecting rods, whose general outline can be seen in the inset. For performing these parallel inspection operations the machine has two work-handling tracks and separate gaging stations for the two workpiece sizes, as shown in the close-up view in Fig. 19-7.

Transporting the parts to consecutive inspection stations is done by cam-driven walking beams with individual carriers (pockets) for the two types of workpieces, which are automatically oriented and loaded at the entry end of the machine; sensors using infrared light beams verify the proper orientation of the loaded parts.

Conditions measured or checked at the consecutive gaging stations include the following:

1. Weights of pin end and crank end, and total weight;

2. Presence or absence (including blockage) of oil squirt holes, checked with an infrared sensor;

Federal Products Corp.

Fig. 19-6 Automatic measuring and sorting machine for the simultaneous inspection of two different sizes of connecting rods, with common general outline, as shown in the inset. The machine has a maximum throughput of 2400 parts per hour, carrying out a series of dimensional inspections and weighing, before accepting the parts.

Federal Products Corp.

Fig. 19-7. Close-up view of a pair of gaging stations to which two simultaneously inspected connecting rods of different models are transported by the walking-beam type handling system of the automatic measuring machine shown in Fig. 19-6.

Automatic Gaging Machines Complemented with Process Monitoring Function

Particular types of automatic gaging machines are designed to serve, in addition to dimensional inspection and sorting, as instruments assisting the size control of the process by signaling the occurrence of drift before the limit of the toleranced size has been reached. The gage signals alert the operator to the need for machine tool adjustment, or may even produce such adjustments automatically as feedback control, thereby eliminating or at least greatly reducing the incidence of reject parts. At the same time, should parts with defective size still be produced, such rejects are detected and disposed of by the basic gaging and sorting functions of the machine. Automatic gaging machines with such extended functions are often used with beneficial results at intermediate positions of a production process comprising several consecutive operations performed by different machine tools and for the final inspection following the last finishing operation on the part.

3. Bore diameter inspection at the pin and the crank ends with two gage plugs for each part, gaged at three consecutive levels;

4. Crank bore roundness measurement, combined with diameter gaging, by rotating the gage plug at one level and interpreting by data processing the individual indications of the four measuring contacts of the gage plug, which determine diameter, bend, twist, center distance, taper and roundness.

The gaging machine does not classify the "good" parts, that is, those that satisfy all the inspected conditions. Classification is applied, instead, to separate the repairable rejects, according to the reason for rejection.

The maximum throughput rate of the described automatic measuring machine is 2400 parts per hour. The meter panels of the machine offer digital readouts for the inspected parameters of both types of parts, providing values whose least significant digit is either 4 microinches, 0.1 gram or 1 degree. Below each readout is a two-digit thumbwheel selector, permitting the selection of whether the output on the digital readout shall display on-line data (the output of each measuring head), the preset tolerance or historical data (the output of each dimensional check for the previously completed gaging cycle).

The effectiveness of the indicated process monitoring functions provided by a separate automatic gaging machine is predicated upon carrying out the gaging immediately as the part leaves the production unit and in the sequence the parts are actually produced. One of the advantages of combining gaging and sorting with the emission of signals caused by defective parts is the avoidance of unnecessary or even harmful machine adjustments that might be prompted by a single out-of-size part, without that single piece being representative of the continuous size-holding performance of the production machine. Considering that defective parts are eliminated dependably by the automatic sorting functions of the gaging machine, out-of-size signaling or feedback action may be safely delayed from taking effect until a specific number of consecutively produced parts have the gaged defective size, thus representing a persistent condition or trend. Automatic gaging and sorting machines, which serve process monitoring as a supplemental function, must provide cycle rates at least equal to those of the pertinent production machines.

The diagram in Fig. 19-8 indicates in a simplified manner the operating principles of a typical process monitoring gaging machine, which combines the basic segregating function with the emission of signals for guidance in controlling the workpiece size. In this case, the machine issues warning signals to the operator when significant size drift persists, shutting

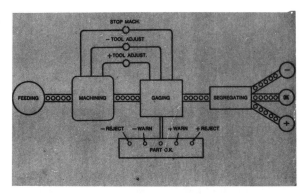

Giddings & Lewis Measurement Systems

Fig. 19-8. Diagrammatic listing of the major functions of an automatic gaging and sorting machine, which also provides process-monitoring signals. The size information obtained by gaging serves for both the classification of the parts to different good and reject categories and the signaling of size drift in consecutively gaged parts; or, when the tolerance limit is exceeded, the gage will shut off the production machine.

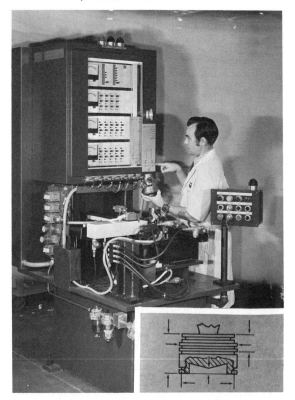

Giddings & Lewis Measurement Systems

Fig. 19-9. Automatic gaging machine for checking five dimensions of the engine piston shown in the inset. The machine is designed to be integrated into the production line, with interconnected part-handling systems, serving for both the inspection of the machined parts and the monitoring of the size-holding performance of the production unit.

off the production unit automatically when the tolerance limit is actually exceeded.

A characteristic example of an automatic gaging machine with process monitoring capacity is shown in Fig. 19-9. This machine receives machined parts through a conveyor line directly from the production unit at the rate of 300 parts per hour. The machine checks five dimensions, diameters and locations, as shown in the inset.

AUTOMATIC MEASURING MACHINES FOR INSPECTING MULTIPLE WORKPIECE DIMENSIONS

The automatic measuring machines described here are characterized by their capability of inspecting products with several critically dimensioned features and, in many cases, making qualifying decisions based on the results of those measurements. Although most of the gaging machines discussed in the preceding section are designed to measure and evaluate more than a single dimension, products with a multiplicity of features, each with rigorously toleranced dimensions are being measured with great accuracy by special inspection machines.

No strictly defined boundaries can be established between automatic gaging equipment assigned to one or the other of these sections. Nevertheless, the majority of the machines discussed serve purposes distinctly different with regard to their complexity and to the characteristics of the required equipment. Consequently, assigning the involved inspection processes and the pertinent gaging machines to separate categories should improve the clarity of the discussion.

Size Classification Preceded by Qualifying Inspection

Various types of component parts that are subsequently assembled by matching the sizes of particular features to the corresponding dimensions of the mating part (for example, the diameters of a bore with that of a cylindrical part) also must often satisfy rigorous specifications regarding several other dimensions and geometric conditions.

The measurements performed by automatic gaging and segregating machines serving the inspection of workpieces with a substantial number of tightly toleranced dimensions and form requirements, in combination with the segregation of the parts into several size classifications, are indicated in Fig. 19-10. That work flow and function diagram lists the essential inspection operations performed, in this

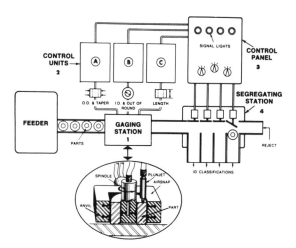

Giddings & Lewis Measurement Systems

Fig. 19-10. Work flow and function diagram of an automatic seg-regating gage, which also incorporates means for the concurrent inspection of several geometric conditions. The diagram indicates the operating principles only and does not represent the actual gaging machine, which provides several complementary functions and sorts the qualified parts into 10 size categories.

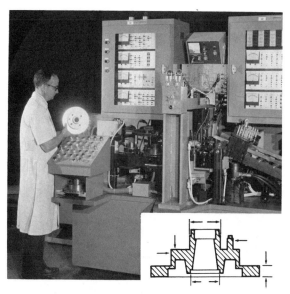

Giddings & Lewis Measurement Systems

Fig. 19-11. Automatic multiple-dimension gaging machine per-forming 20 dimensional measurements on disc-brake rotors, sev-eral of which are indicated by arrows on the part's cross-sectional diagram. The machine has a nominal cycle rate of 450 parts per hour, and produces signals that can be processed by an interfaced computer supplying printouts for various quality-control evalua-tions.

case at a single gaging station, although the mea-surements made include linear dimensions (inside and outside diameters and length), taper (registered as the difference between diameters at two levels), and roundness (that determined as diameter varia-tions around a bore measured by a rotating diameter gage).

The diagram is the simplified representation of the various material-handling, gaging and evaluating functions commonly incorporated into automatic measuring machines that serve the inspection of multiple dimensions of specific manufactured com-ponents. The equipment actually used must be de-signed in accordance with the general configuration, critical size and form characteristics, and other tol-eranced characteristics of the particular component parts.

Automatic Measuring Machine for the Inspection and Evaluation of Several Dimensional and Form Characteristics

The automatic measuring machine shown in Fig. 19-11, along with the diagrammatic outline of the workpiece to be inspected, is designed for the com-plete dimensional and geometric inspection of disc-brake rotors. Altogether 20 different dimensions are inspected, some of which are indicated in the part's diagram. The conditions measured include flatness, gaged as lateral runout of the inner and outer braking surfaces; thickness variations around the entire in-ner surface; radial squareness and parallelism; bolt circle runout; and bolt count. A special character-istic of the gaging machine is its adaptability for in-terfacing with a separate computer that periodically can produce printouts for the evaluation of such product quality aspects as dimensional trends and the incidence of rejects; this is in addition to mea-suring and qualifying the actual dimensions and geometric conditions of each inspected part.

Automatic Gaging and Classifying Machine for Automotive Crankshafts

The fully automatic multidimensional gaging and classifying machine, the gaging stations of which are shown in Fig. 19-12, serves the total dimensional in-spection of automotive V-8 crankshafts at the rate of 250 parts per hour. The crankshafts are delivered to the gaging machine by the in-plant transfer system, to be picked up by the gaging machine's own transfer mechanism, which carries the parts by indexing mo-tions to the successive gaging and marking stations of the machine. That transfer mechanism first lifts the crankshaft into a stationary gaging station and positions it there for measuring four pin bearing journal diameters in three separate places, as well as

Ex-Cello Corp., Micro-Precision Operations

Fig. 19-12. Workpiece area of an automatic gaging and classifying machine for automotive V-8 crankshafts, showing the locating and holding devices used in the two consecutive gaging stations into which the parts are transferred automatically and held during the gaging, in a stationary position at the first station and rotated at the second station.

the hub outside diameter, the pilot inside diameter and the seal outside diameter, the latter diameters in one place each, Subsequently, the transfer mechanism carries the crankshaft to and lifts it into the second gaging station, which rotates the part while gaging the five main bearing journal diameters in two places, the thrust bearing size, and runout, as well as the radius of the six crankshaft counterweights.

After the gaging has been completed the parts are carried to the paint marking station for color coding by spray jets to indicate acceptance or rejection. Furthermore, paint sprays, controlled by the gaging information retained by the machine, are applied to indicate the corresponding size classification (one out of three) of the five main bearing journals. Following that last inspection operation, the parts are off-loaded by the in-plant transfer system.

Altogether there are 57 gage heads in the two gaging stations. A valuable feature of this gaging machine is its automatic calibration system: After a minimum and maximum tolerance gage master has been installed in each gage station, all gage circuits are calibrated to their proper tolerance range by simply pushing the "Calibrate" button.

Automatic Gaging and Size Coding Machine for Engine Blocks

The automatic gage shown in Fig. 19-13 is designed to operate as part of an engine block transfer line. The gage measures the inside diameters of eight cylinder bores and ink stamps a size code adjacent to each bore. All the actions of the gage and its code stamping system are directed by a programmable logic controller (PLC).

The transfer of the engine blocks from the automatic production line and back onto it following the completion of the gaging and coding is performed by a hydraulically actuated walking beam, which also transports the parts through the five stations of the gaging machine and positions them for the gaging operation. Two of the machine stations are idle positions for receiving the blocks from the production line (entry) and holding the parts before moving them back to the line (exit). The actual gaging is carried out in two stations, one of each for four cylinders of the block and an additional station for code stamping.

Each bore is gaged by a measuring plug with two lever-mounted contacts; the levers bear on individual cartridge gage heads. The signals from each pair of gage heads are fed to a modular amplifier where they are combined to indicate the deviation of the actual bore size from a master setting that represents the nominal diameter.

When a block is positioned by the transfer mechanism in a gaging station, all four plugs at that station are hydraulically advanced into the bores to a predetermined depth. During that advance the contacts of the plugs are pneumatically retracted, to be released for engaging and measuring the bore as soon as the plugs have reached the measuring position. When the measurement is completed, the contacts are again retracted while the plug leaves the bore and the contacts enter their "home position." In that position the contacts are inside a minimum size master. Preceding the gaging of each consecutive part, the contacts are released for a few seconds to engage the master, thus producing size signals, which are registered by the PLC for automatic compensation when the signals deviate from zero. To facilitate the calibration of the gages for a specified tolerance range, a second master, that with maximum size, is mounted behind each minimum master on the measuring plug slide. By operating the machine in manual mode, and using appropriate push-buttons, the plug can be positioned in the maximum master for verifying the gage calibration.

At both measuring stations there are two measuring assemblies contained in the individual unit. The stamping station has one ink stamp coding unit on each side. The PLC directs the stamping of the proper code adjacent to each bore based on data accumulated from the eight gagings at the two mea-

Federal Products Corp.

Fig. 19-13. Automatic gaging machine for automotive engine blocks, measuring the diameters of the cylinder bores and using the accumulated data for directing the subsequent stamping of the corresponding diameter values adjacent to the individual bores. For each probe the machine has automatic zero setting and manually initiated tolerance range calibration.

suring stations in which the particular block has just been inspected.

From each measuring plug the signals are read out on analog meters with 0.0002-inch graduations, and two-digit readouts display the marking code for each bore measured.

AUTOMATIC GAGING MACHINE WITH PART-MATCHING FUNCTIONS

Frequently, the purpose of inspection by automatic gaging machines, in addition to determining the part's dimensional adequacy, is to segregate the generally acceptable parts into a specific number of size classifications that are used for selective assembly. Several of the automatic gaging machines described in the preceding two equipment categories actually perform part classifications, either by physical segregation or by applied marking for directing the subsequent selective assembly.

By combining the gaging and classification with successive operations in which the automatic machine performs the selection of mating parts with the matching size and, in some cases, actually assembles the components, significant advantages are provided. In addition to consolidating several complementary operations into a single unit, thereby realizing savings in operating cost and floor space, the segregated storage and transport and the marking of the parts are avoided, and the accompanying hazards of possible errors or mix-ups are also greatly reduced.

A few characteristic examples of automatic gaging and sorting machines that are complemented with equipment for selecting components of matching size and, in particular applications, for assembling them are described in the following.

Automatic Gaging and Matching Several Components of an Assembly

The fuel metering assembly, that product shown in outline in Fig. 19-14, consists of three components: the plunger, the barrel and the sleeve. Each of these components must satisfy very rigorous form specifications, particularly regarding taper and out-of-roundness, with a total tolerance of 0.000025 inch

(A)

Federal Products Corp.

Fig. 19-14. Operating elements of a multiple-parts gaging and size-matching machine for selecting members of a three-parts assembly that have very closely toleranced fits.
(A) Members of the assembly that have critical clearance specifications.
(B) Gaging station for shafts.
(C) Verifying station for upper collars, with classification racks in the background.

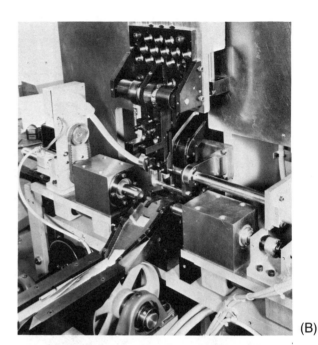

(B)

(C)

for the plunger and 0.00005 inch for the barrel and the sleeve bores. The total range of acceptable size variations of the diameters, outside for the plunger and inside for the other two components, is ±0.0003 inch from the nominal. Within any particular assembly, however, the diameters of the mating-part surfaces must not deviate by more than 30 microinches from their respective classification values, in order not to exceed the tolerance of the fit, specifying a nominal clearance of 50 microinches with a maximum tolerance of ±0.000030 inch.

The specifications of the total size range and the much more rigorous requirements for the fit conditions require first the classification of the generally acceptable parts into 20 size categories, 30 microinches wide each. Subsequently, component sets with matching sizes must be selected and separated for the consecutive assembly into a product with the specified fit conditions.

A special multiple-part gaging machine with matching and segregating functions, whose gaging stations in close-up views are shown in Fig. 19-14, serves to perform the described measurements and the selection of the matching component by the following operations:

a. Measures the inside diameters of the sleeves and collars, exploring each bore individually with the air plugs of air-electronic gages. The plugs, advancing along a spiral path, explore nearly the entire bore length for taper or out-of-roundness beyond tolerance limits, in addition to measuring the diameter. Parts that are geometrically correct, and within the total acceptable size range, are assigned to 1 of 20 good categories, each representing a 30-microinch wide band. The classified "good" parts, as well as those rejected for specific reasons, are collected in individual magazines.

b. Measures the shafts, each in two axial planes close to the ends, with the shaft rotated while being measured. The registered diameter values serve to discover excessive taper and out-of-roundness, in addition to the size of the geometrically acceptable parts. That portion of the system's operation is activated only after an adequate number of classified sleeves and collars has been accumulated in the storage magazines.

c. Based on the measured diameter of the acceptable shaft, the system selects a sleeve and a collar with matching size to be collected together with the just gaged shaft as elements of an assembly having the specified fit condition.

d. Each selected part of a particular set is subjected to a second verifying measurement before being released for subsequent assembly. That double checking is regarded as a necessary safeguard in view of the critical functional role of the required fit.

e. Temporarily discarded for a later reclassification are parts with either of the following size conditions:

(1) With diameters in one of the threshold bands (each 5 microinches wide) between the adjacent size classifications;

(2) Good shafts for which there are no matching size sleeves and collars available in the storage magazines, as determined by three consecutive attempts of the operating system.

f. The three matched components of an assembly are transferred to a nest on a conveyor that will transport the ready-to-assemble set into the assembly area.

Automatic Measuring Machines with Size-Matching and Assembly Functions

Assemblies whose rigorous fit specifications are met by selecting parts with matching sizes may be produced by consolidating into a single unit all the phases of the selective assembly process. The major functions performed are individual gaging of the component parts, segregation into different size categories, matching of the parts having mutually corresponding sizes, and actually assembling the selected components.

While the functionally significant sizes of the complementary components, as determined by gaging, represent the information directing the selection of the parts to be assembled, the execution of the fully automatic process also requires complex material-handling functions, controlled by logic utilizing the signals of electronic gages. When the parts are received directly from the production machines, the size classification must be preceded by dimensional inspection with regard to the geometric forms and the overall size range of the incoming parts. Those found to be not satisfactory in said respects must be diverted from the flow of components used for selective assembly.

As a typical example, equipment performing such combined gaging, size matching and assembly functions, which is used for selectively assembling automotive pistons, is now described.

The inspection and size classification of the pistons, followed by actual assembly with wrist pins, then their preparation for assembly into cylinder blocks, are carried out by two companion measuring and sorting machines, one of these also complemented with assembly functions.

The first of the two machines, shown in Fig. 19-15 together with the outline drawing of the workpiece, inspects the pin bore diameter of the incoming cylinders after they have been oriented by the built-in material-handling system for positioning the off-center pin bore in the proper direction. The pin bore

Federal Products Corp.

Fig. 19-15. Automatic measuring, classifying and assembling machine for automotive pistons, one of which is shown in the inset. Following measurements permitting qualifying decisions and classification by bore size, pistons with accepted geometry and bore diameters are assembled automatically with the matching wrist pins, which were gaged previously and are stored by size categories corresponding to the classified bore diameters.

diameters are measured in two planes at each end, using multisensor electronic probes for determining the following:

1. The out-of-roundness, toleranced at 0.0002 inch;

2. The taper, toleranced at 0.0002 inch; and

3. The diameter, which must be within a specific size range of 0.0004 inch, for being classified into one of four categories of 0.0001 inch each. Cylinders with out-of-size bores are rejected.

A close-up view of one of the pistons being measured at the gaging station is shown in Fig. 19-16. The indexing conveyor that brings the pistons to the gaging station transports them, after the gaging, to the stamping station, where the measured pin bore diameter is stamped on the "good" pistons.

The dimensionally satisfactory pistons are transported to one of four assembly stations in accordance with the size classification of their bores. At these stations previously measured and sorted wrist pins, corresponding in diameter to the pin bores of the assigned pistons, are stored. At the selected assembly stations a wrist pin of matching size is pushed, under controlled hydraulic pressure, into the pin bore of the piston.

Such an automatic selective assembly process—inserting matching size wrist pins into piston bores

Giddings & Lewis Measurement Systems

Fig. 19-17. Close-up view of the assembly area of an automatic measuring, size-matching and assembly machine, which measures the pin bores of pistons, retains that information and selects the matching diameter wrist pin out of five different sizes. In the upper corner of the left side are the magazines holding the pins; in the front is the conveyor transporting the assembled pistons with the inserted wrist pins.

that were measured at a preceding station of the machine—is shown in Fig. 19-17. This illustrates in close-up view the assembly area of an automatic measuring and assembly machine of a different manufacturer, operating similarly to the just-described equipment, but designed to select one out of five wrist pins, which are stored in individual magazines adjacent to the assembly station of the machine.

The companion gage to which the pistons assembled with wrist pins are transferred by conveyor measures the outside diameters of the parts, for inspecting the skirt characteristics of the pistons. These include the discovery of out-of-tolerance taper and ovality for rejection and the classification of the piston diameter into one of four categories, 0.0004 inch each, to mach-fit an equal number of cylinder bore size categories at the assembly of the engine blocks. The good pistons are stamped with an identifying letter, whereas rejects whose diameters are outside the acceptable size range or those with out-of-tolerance form conditions are diverted.

AUTOMATIC STATISTICAL ANALYSIS OF INDIVIDUAL LINEAR DIMENSIONS

The automation characteristics of the methods and equipment discussed in this section apply not to the measuring process proper, which is usually performed by manual work handling, but to the re-

Federal Products Corp.

Fig. 19-16. Close-up view of the gaging station in the measuring machine shown in Fig. 19-15. Here, two diamond-tipped multi-sensor electronic probes measure the bore diameter in two planes at both ends, supplying data for evaluating the form characteristics and determining the bore diameter.

cording and evaluation functions complementing the gaging of the parts.

Product size evaluation by statistical analysis can provide, in particular cases, an economical method of dimensional inspection. One-hundred-percent inspection may be substituted by measuring a specific sample lot and analyzing the individual size values by statistical methods; that procedure may be used when the presence of a very small number of out-of-tolerance parts and a degree of probability that is always less than absolute are acceptable conditions.

The reasons the inspection of sample lots may be substituted for the measurement of each individual piece are economic: savings in inspection costs and time. Although the actual measurement of all the manufactured parts requires greater effort and longer time than that of a limited number of parts in a sample lot, the statistical evaluation of the latter group by conventional methods, involving manual recording and significant calculation, may reduce substantially the benefits resulting from shortened measuring time. Furthermore, the manual recording dimensions are prone to human error even when skilled personnel are used; this circumstance is an additional factor affecting the economics of statistical quality control performed in the conventional manner.

Recent developments in computer technology and the availability of sophisticated microprocessors at moderate cost have opened up new applications for statistical quality control as a method of inspecting and qualifying the dimensional conditions of parts manufactured in substantial volumes. By using a regular electronic length-measuring gage that is interconnected with a special statistical analyzer, the inspector is entirely relieved of the responsibility of data collection and subsequent calculation procedures; all these functions are performed automatically and for all practical purposes simultaneously with the gaging of all individual parts within a particular sample group. The special analyzer instrument, which generally includes a printer, also supplies in the form of a comprehensive printout all the quality-control information needed for the evaluation of the lot.

Figure 19-18 illustrates a typical statistical analysis setup that includes a microcomputer, video monitor, alphanumeric keyboard, electronic gaging instrument and appropriate software. Scores of major gage manufacturing and computer software writing companies have developed sophisticated statistical analysis software packages that cost anywhere from $10.00 to tens of thousands of dollars. These soft-

Fred V. Fowler Co., Inc.

Fig. 19-18. Computer, monitor and keyboard with statistical analysis software. As an example, a bench stand with digital micrometer gage is connected to the computer.

ware packages, when used in conjunction with other necessary hardware, are capable of collecting, displaying and storing measurements for a variety of statistical analyses (that is, X-bar and R, X-bar and S, and attribute control charting). Data can be collected automatically or entered manually via the keyboard. Once the computerized statistical analysis is complete, a hard-copy printout is generated for review, as well as for a permanent record. Today, many manufacturers require that their suppliers show evidence of statistical process control for all parts produced as a term of the purchasing agreement.

COMPUTER-SUPPORTED COORDINATE MEASUREMENTS

The availability of compact computers with a wide variety of capabilities and at relatively low cost, resulting from spectacular scientific and technological advances, particularly during the past two decades, and that are still progressing, has permitted a substantial upgrading, even sophistication, of coordinate measurements.

Coordinate measuring machines, which were discussed in detail in Chapter 12, provide, even in their manually controlled operations, cost advantages and inspection capabilities greatly superior to the traditional layout type measurements on surface plates, which had been used for the inspection of coordinate dimensions. The capabilities of computers when assisting or controlling the inspection operations of coordinate measuring machines can provide great *advantages* in several respects, such as the following:

a. The speed of carrying out the inspection process. The short duration of even complex measuring operations involving a substantial number of dimensions can be an important source of direct savings in inspection costs. In many cases further benefits result from the speedy delivery of the inspection results by reducing substantially the downtime of expensive production equipment, particularly CNC machine tools requiring "first-piece" inspection prior to starting the regular production. Periodic inspection during the course of continuous production, when performed quickly, can provide early detection of faulty parts before the number of rejects accumulates.

b. Higher reliability of the measurements due to the substantial reduction or even elimination of human error.

c. The computer-supplied evaluation of the measurement results, such as the printed-out presentation of incorrect dimensions, possibly complemented with statistical analyses of the incidence and characteristic of defective parameters.

d. The calculating functions performed by various systems contribute to the further automation of the measuring process. Such functions are carried out by means of computer software—termed routines—which are usually available for many types of measurement-related calculations to simplify the setup of specimens and to develop significant dimensional information.

The actual availability and extent of benefits indicated as examples and in general terms are dependent on the category and type of computer equipment that complements the basic coordinate measuring machine. The selection of the computer system is usually guided by conditions related to several applicational aspects, such as the following:

1. The characteristics of the workpieces, the number of inspected dimensions, the degree of configurational complexity, the level of required accuracy and so forth.

2. Production process requirements, for example, with regard to allowable duration of the inspection operation and the variety of workpieces to be inspected consecutively or alternately.

3. Quality-control procedures for which information is needed: proving visual display only; simple or evaluated printouts; computed values, for example, to compensate for alignment errors of the specimen; or conveying the acquired measurement data to a central computer for further processing.

4. Measuring process requirements, such as automatically directing the travel movements of the coordinate measuring machine and the location of the data acquisition points by a program, either established individually or by a readily available routine.

Basic Categories of Dimensional Conditions Determined by Coordinate Measurements

The majority of dimensional conditions, for whose inspection coordinate measurements are advantageously applied, belong to or can be related to one of four subsequently listed categories. These are indicated by the diagrams in Fig. 19-19 that will help in visualizing the positional relationships among the significant points by which each particular dimension or geometric form is defined. To determine the exact location of these points in a three- (or occasionally four-) dimensional space, as well as the mutual relations of said locations, is the essence of the coordinate measuring process. The execution of that process, as well as the interpretation and evaluation of the measured values, are made much simpler, faster and more dependable when the coordinate measuring process is assisted by an appropriate computer.

Two Major Systems of Computerized Coordinate Measurements

To satisfy the equipment requirements resulting from the previously listed and similar applicational aspects, the computer-supported coordinate measuring machines are designed to operate by specific basic systems, and each machine type is usually available with various optional accessories. The two major systems of computerized coordinate measuring processes and equipment are the following.

1. Computer-aided measurements, in which the successive targetings, that is, contacting with the instrument-probe specific gaging points on the workpiece, are performed manually by the operator, either by direct action or by controlling the power movements of the machine. Information on the direction and distance of the actual travel between the successive target points is received by the computer

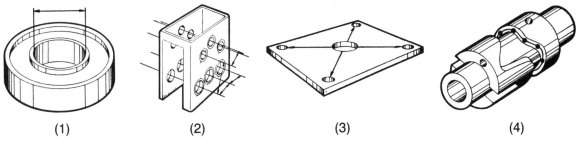

(1)   (2)   (3)   (4)

Brown & Sharpe Mfg. Co.

Fig. 19-19 Basic categories of dimensional conditions for the inspection of which computer-supported coordinate measurements can offer significant advantages over most alternate methods.
(1) *Size*, usually measured as a linear distance between specific boundary points.
(2) *Distance*, the separation of actual or virtual points (for example, the centers of holes) on the workpiece.
(3) *Location*, the position of a feature relative to reference elements on the part or to an external origin.
(4) *Shape*, is determined by the actual location of significant points on the work surface relative to their nominal positions in a multiple-axis coordinate system.

of the coordinate measuring machine for recording, storing and, frequently, evaluating based on a preestablished program.

2. Direct computer control relieves the operator from the manual targeting of the machine probe; a computer-directed servo-positioning system produces movements for establishing the mutual positions of the probe and the specimen exactly at the nominal locations of successive target points of the workpiece. The actual positions of those target points, as sensed by the machine probe, may or may not coincide with those defined in the program, thus producing gaging information that is retained by the computer. The retained gaging data are then processed and the inspection results—such as deviations from the nominal dimensions or forms, their characteristics and the relevant quality conditions—are evaluated. The acquired and processed inspection information appears as an alphanumeric display on a screen usually complemented by a printout.

Examples of coordinate measuring machines equipped to operate by these major systems of computer-supported measurements are presented later.

Data Acquisition and Interpretation Aided by Computer Processing

In dimensional inspection processes performed by coordinate measurements, the locations of specific points on the specimen surface are first determined by physically contacting the individual points with the probe of the coordinate measuring machine, an action that may be called target sensing. That action is followed by determining:

a. The coordinate location of the probe during the sensing contact; and

b. The indicated degree of coincidence between the probe position and the contacted point on the work surface, also taking into account the form and the physical dimensions of the probe.

That simplified definition of the process is predicated on several rigorously established conditions. These may be conceptual only, such as physically nonexistent feature elements, and hence must be reconstructed by adequate measuring procedures. Examples of basic or frequently required conditions for acquiring meaningful dimensional information by coordinate measurements are given in the following.

a. *Alignment of axes and planes.* The workpiece must be mounted on the worktable of the coordinate measuring machine in a manner that ensures the alignment of the specimen axes and planes with those of the probe (or table) movements of the machine.

b. *Origin establishment.* The workpiece must be in a position where the elements of its surface from which the part's dimensions are referenced will be in definitive relationship with the travel-distance indications of the measuring machine.

c. *Compensation for probe dimensions.* The actual sizes of the machine probes, unless all measurements are made by contacts with the same surface element of the probe, must be considered for the proper interpretation of the measured distances between the interrelated workpiece features.

d. *Deriving virtual points from physically present surfaces.* It is common practice in mechanical design to define dimensions by referencing from conceptual or virtual points, namely, those not physically present on the workpiece. A few examples are the axis of a solid cylinder, the center of a hole and the mutual boundary lines of intersecting planes that are not directly measurable because they are eliminated by edge rounding or covered by a fillet. A significant number of coordinate measurements complemented with calculations may be needed to substitute for the reference elements that cannot actually be contacted on the workpiece.

c. *Inspecting the form and size of basic geometric elements.* Mechanical parts other than those of very simple shape generally contain basic geometric elements. When inspected for form and size by conventional coordinate measurements, a substantial number of individual surface points must be contacted and their positions registered to produce meaningful measurement data, which will still require calculations for proper interpretation. Examples of such basic geometric elements are a straight line, a plane, a circle, a cylinder, a cone and a sphere.

f. *Frequently recurring geometric conditions.* Depending on the general configuration and design specifications of various mechanical parts with different degrees of complexity, particular geometric conditions require the position determination of a significant number of contact points on the part's surface, when inspected by conventional coordinate measurements. When such specific geometric conditions recur in the inspection practice of a plant, conventional coordinate measurements may prove to be less favorable economically than alternate methods of measurement, including the use of dedicated gages.

Randomly named examples of such recurring geometric conditions are

1. Radius of curvature of rounded surface sections;

2. Compound angles relative to a reference plane;

3. Roundness of external or internal features; and

4. True position location of various features, either regardless of size or at maximum material condition.

Computer assistance can significantly facilitate and accelerate the executions of the above-mentioned and many other types of coordinate measurement procedures, making that method the preferable one even for purely economic reasons. In addition, computer-assisted coordinate measurements provide many unique advantages owing to their high operating speed, great applicational flexibility and consistent accuracy, in addition to supplying displayed and printed measurement information, complemented with data developed for quality-controlled evaluations.

The role of computers or of special programmable calculators used for complementing coordinate measuring machines consists mainly in substituting appropriate types of programmed data-processing functions for many of the setup conditions commonly requiring time-consuming manual action, as well as for the tedious and lengthy calculations. Several categories of those functions, with particular regard to the purposes of their operation, are now described briefly.

Programmed Data-Processing Functions of Computer-Assisted Coordinate Measuring Machines

The inspection by coordinate measurements of mechanical parts with widely different configurations and complexity is greatly simplified when the actually measured values are analyzed and converted, by appropriate calculations, into the dimensional information sought. That conversion of the measured values, for interpretation in accordance with the required data, is performed by programmed calculators or desk-top computers to which the values of the individual measurements are automatically transmitted. Different calculations must be performed, depending on the dimensional information required, and those calculations are controlled by appropriate programs.

These programs are available as software in the form of magnetic discs (that is, $5\frac{1}{4}$ inch and $3\frac{1}{2}$ inch disks). Software capabilities are constantly being improved or upgraded to increase the efficiency of taking coordinate measurements. Several basic examples of the data supplied by different coordinate measurement analysis programs will now follow.

Automatic Axis Alignment

The axis of a workpiece mounted on the table of the coordinate measuring machine may be offset by tilt in one direction, producing a two-dimensional misalignment, or in two different directions, resulting in a three-dimensional misalignment. Such conditions of the workpiece position in relation to the reference axes of the coordinate measuring machine are shown in the diagrams in Fig. 19-20. The interference with the measurement results of these and other cases of workpiece misalignment can be elim-

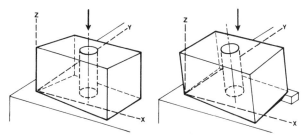

Giddings & Lewis Measurement Systems

Fig. 19-20. Examples of part misalignment in relation to the co-ordinate measuring machine axes *X-Y-Z*, causing offset angles. These conditions can be compensated by programmed calculations that convert the measurement data into the dimensions of the correctly aligned part.
(Left) Two-dimensional misalignment.
(Right) Three-dimensional misalignment.

inated by applying the calculations of an appropriate type of automatic-axis-alignment program.

When using such programs the setup and orientations of the part on the table of the coordinate measuring machine can be obviated, and only simple clamping is needed. There are several optional methods by which actual axis alignment of the specimen can be substituted by calculations. The optional methods, shown in the diagrams of Fig. 19-21, are the following:

(A) Two features are used in this program with the origin midway between the two points.

(B) The first two features determine the offset angle and the third feature becomes the origin, which, however, may be relocated to coincide with either of the first two features, or may be in any other location. This is the most frequently used alignment program.

(C) The offset angle is determined by the first two points, and a third feature is required for determining the origin, which is at the intersection of the *X* axis and a line perpendicular to it, while also passing through the third feature. The preferred method is to establish the origin at one corner of the workpiece. This method may be applied to two sets of measurements labeled 1–3 and 1′–3′.

(D) A method similar to (C), except it uses four features to calculate the origin. The origin is at the intersection of the *X* axis and a line through features 3 and 4.

(E) This method is used for referencing measurements to the center line of a bore. Both the origin and the alignment of the third axis are determined by diameter measurements at the top and the bottom.

Alignment of Planes

Basically, rectangular parts clamped on the worktable of a coordinate measuring machine in a position where the sides of the part are aligned with the machine's *X* and *Y* axes, respectively, usually have their top surface parallel with the machine ta-

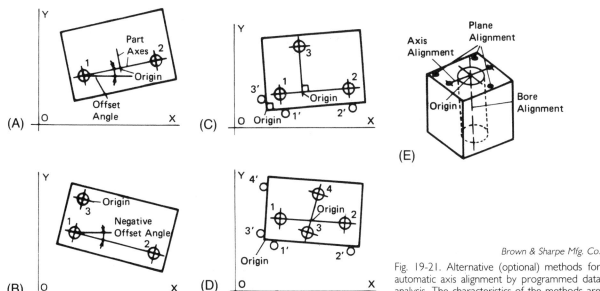

Brown & Sharpe Mfg. Co.

Fig. 19-21. Alternative (optional) methods for automatic axis alignment by programmed data analysis. The characteristics of the methods are described in the text.

ble. This parallelism is an important requirement when the top surface of the part serves as the measuring plane; consequently, that condition may have to be checked and corrected in the case of deficiency.

When the top surface, or any other surface needed for measuring planes, is not parallel with the table, and a correction of that condition by fixturing, shimming, or other mechanical procedures is to be avoided, the plane-alignment feature of automatic alignment may be used. It will provide compensation for this form of misalignment through a series of three measurements taken on the top surface of the part to produce a dependable X-Y plane; the measurements should be taken in a triangular pattern, as far apart as possible, in the manner indicated in Fig. 19-22A.

A ball probe is used for plane alignment, as well as all further measurements by which, in addition to determining the orientation of the X-Y plane, the procedure also establishes the third axis origin, that is, for axis Z in the X-Y plane. This third-axis origin lies in the center of the ball probe (see diagram) and the distance, equal to one half of the probe diameter above the contacted part surface, must be considered when taking measurements along the Z axis.

An alternate use for the plane-alignment routine is to compensate for X-Y misalignment when aligning a part in another plane, such as when the workpiece is clamped to the table for measurement on its front face in the Z-X plane. Before aligning the Z axis by using one of the other alignment programs, the plane-alignment routine is used on the front face of the part to compensate for any misalignment of the X axis in the X-Y plane, as indicated in Fig. 19-22B.

Deriving the Locations of Virtual Points from Related Actual Surfaces

In conventional coordinate measurements the location of a specific point on the part's surface is determined by a single contact of that point with the machine probe. Single-point contact may be used also to determine the location of a bore center (a point conceptually at equal distance from any point on the bore's periphery and in a plane at right angles to the axis of the bore), by using a tapered probe of adequate diameter (see the illustration on the left in Fig. 19-23).

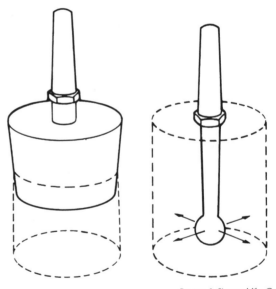

Brown & Sharpe Mfg. Co.

Fig. 19-23. Two techniques are used on coordinate measuring machines for determining the location of the center of a bore.
(Left) Single-point technique using a tapered probe. Applicable only to a limited range of bore diameters within the practical limits of the probe size.
(Right) Multipoint technique, requiring several measurements on the side surface of a feature for producing data from which the location of that feature's center can be determined by calculations generally performed by a programmed process. This method is applicable for large holes and also for outside surfaces (bosses).

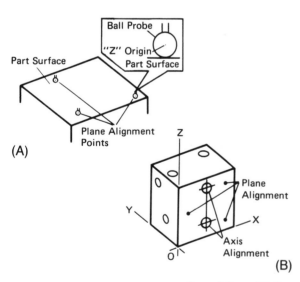

Brown & Sharpe Mfg. Co.

Fig. 19-22. Measurements needed to collect data for the application of plane-alignment programs.
(A) Alignment of the top surface of the part parallel with the table of the coordinate measuring machine. Inset shows the location of the Z-axis origin, relative to the contacted top surface of the part.
(B) Mode of compensation for X-Y when measuring on the front face of the part, in the Z-X plane.

That single-point technique is not feasible, however, for the case of a large diameter bore, or for determining the center of a solid cylindrical feature. A procedure known as the multipoint technique permits contacting in succession several physically present points of the pertinent part feature (see the illustration on the right in Fig. 19-23), the positions of these points being retained by the computer and used subsequently to determine, by programmed calculation, the derived position of the reference point sought. The multipoint technique can be applied to either internal or external surfaces of regular geometric shapes and may serve to determine, in the case of cylindrical forms, for example, either the co-ordinate position of the axis or the diameter of that feature.

Conversion of Cartesian into Polar Coordinates

To determine the position of a point on the workpiece by conventional coordinate measurements, the travel of the probe between the locations of the successive target points on the workpiece is carried out along the axes of a cartesian, that is, a rectangular system of coordinates, as indicated in the diagram on the left in Fig. 19-24. However, for the inspection of particular types of dimensional conditions, for example, the actual position of holes designed to be located with their centers on a circle and spaced at specific angular distances apart, measurements by polar coordinates offer the advantage of permitting direct correlation with the pertinent design specifications.

Programmed methods are available for converting the positions of the probe, which has been moved along the axes of a rectangular system of coordinates, into the corresponding polar coordinate values. As indicated in the diagram on the right in Fig. 19-24, the program determines the polar coordinate position of any individual point in a given plane by the length of the pertinent radius vector from a fixed origin and from the angle that the vector makes with a fixed line.

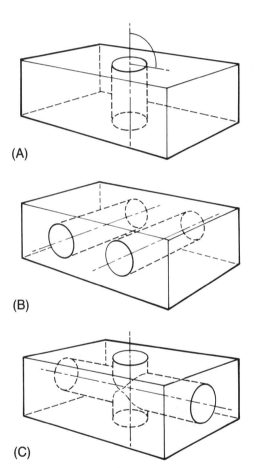

(A)

(B)

(C)

Giddings & Lewis Measurement Systems

Fig. 19-25. Examples of basic geometric factors in defined positions relative to the general reference elements of the main part. The actual location of the features and of their significant elements can be determined by measuring the coordinate positions of a few points on the interrelated surfaces and processing the obtained data by an appropriate analysis program.
(A) Perpendicularity of the bore axis to the plane of a part surface.
(B) Parallelism of bore axes.
(C) Location of the point of intersection of the axes of bores at right angles to each other.

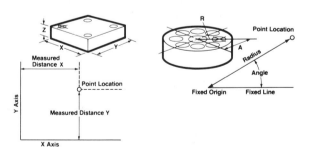

Brown & Sharpe Mfg. Co.

Fig. 19-24. (Left) Coordinate measuring machines determine the location of a point on the part in a rectangular (cartesian) system of coordinates.
(Right) That measured location of a point in a given plane can also be expressed in polar coordinates by means of calculations, generally by a programmed procedure, using as factors a fixed line and the radius of the point location from a fixed origin. The angle in degrees is also determined.

Dimensional Inspection of Interrelated Feature Elements

Figure 19-25 illustrates a few examples of basic geometric features whose actual dimensional conditions relative to reference elements of the workpiece are determined quickly and dependably when processed by computer specific measurement data. Obtaining meaningful dimensional information in that manner represents a great advance over manually controlled coordinate measurements in which a series of individual surface points must be measured subsequently correlating the obtained value by means of calculations.

The advantages of processing the individual measurement data by a programmed computer routine are particularly significant when a large number of geometric interrelations, either on a single part or in repetitive inspection operations, must be determined.

Examples of a few typical feature configurations common in manufactured parts are described as examples of inspection routines. These are performed using stored programs for automatically computing a small number of specific measurement data and thereby developing meaningful inspection information.

Linear Dimensions and Coordinate Locations

Diagrams of two application examples are shown in Fig. 19-26.

(A) Thickness of slots and walls. The same routine can be used for measuring the thickness of a flange, or the length or width of a slot. Each dimen-

sion can be determined from the measurement of three points, two on one side and one on the opposite side of the feature. The diameter of the probe must be entered as plus for the slot width (inside measurement) and as minus for the flange thickness (outside measurement).

(B) Coordinate location of a virtual point. A program used for determining the location of virtual points, defined as the intersections of extended lines or sides, but intersecting in areas that on the actual part are corners of irregular form or fillets on the inside or rounded off on the outside. Two measurements are made on each of the pertinent sides and the information supplied contains the included angle in degrees, in addition to the coordinate location of the point of intersection.

Coordinate Measurement of Angles

The principles of two application examples are shown in the diagrams in Fig. 19-27.

(A) Measurement of the angle of a feature in a two-dimensional plane, relative to the base axis of the selected working plane, for example, X for the X-Y plane, although any other plane, namely, Y-Z or Z-X, may also be used as the working plane. The procedure requires the locking of the third axis of the machine (axis Z for the X-Y plane) and measuring two points on the feature, as far apart as possible. Any probe form, with the exception of tapered probes may be used. The result is an angle ranging from −90 degrees to +90 degrees.

(B) Measurement of the included angle between two surfaces. The measurement can be performed in any position regardless of the working plane selected by using a ball probe to measure three points on each

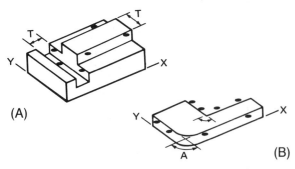

(A)

(B)

Brown & Sharpe Mfg. Co.

Fig. 19-26. Programmed measurement of linear dimensions and coordinate locations, by using the data of a limited number of actual point determinations.
(A) Thickness of slots and walls.
(B) Coordinate location of virtual point (defined, but not physically present on the part).

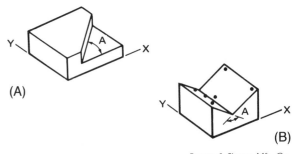

(A)

(B)

Brown & Sharpe Mfg. Co.

Fig. 19-27. Programmed coordinate measurement of angles.
(A) Angle measurement relative to a working plane.
(B) Angle between two surfaces, regardless of the selected working plane.

of the two surfaces, in a triangular pattern and as far apart as practical. The result is a measured angle ranging from 0 degrees to +180 degrees.

Programmed Coordinate Measurement of Geometric Forms

The true geometric form of particular part features often is functionally important and therefore is specified with close tolerances. The thorough inspection of form trueness, that is, the conformance of the actual feature shape to defined conditions, generally involves extensive measurements, such as by the tracing of characteristic form elements. In many cases, however, the determination of the form conditions by comparing the actual locations of specific points on the feature surface with the corresponding theoretical positions can provide sufficient proof of acceptable form.

That comparison of the actual point locations with the respective theoretical positions and determining any amount of deviation is greatly facilitated when processing the measured coordinated location data with the aid of appropriate programs, examples of which are indicated in the diagrams in Fig. 19-28 and described briefly in the following.

(A) Inspecting for flatness. Flatness may be defined as the condition of a surface, no point of which is outside a space bounded by two theoretically parallel planes set apart by the amount of the flatness

tolerance. Checking the compliance with that specification could require the positional determination of an undefined large number of points or surface elements containing those points. Such procedures, using tracing or optical techniques, are actually applied in the case of parts with very critical flatness requirements, for example, gage blocks. In general manufacturing practice, however, it may be sufficient to check the mutual locations of a limited number of surface points, as long as their combined positions are sufficiently representative of the flatness condition of the inspected surface.

The routine, whose operating principles are indicated in diagram (A), requires the measurement of a minimum of six points for calculating the plane that best describes all the measurement points. Subsequently, the routine calculates the distance between that plane and each measured point to determine the value of the deviation, which is the maximum distance on one side of that reference plane plus the maximum distance on its other side. Indicating the characteristics of deviation from flatness, the method may provide, by a simple procedure, useful information on the flatness condition of the inspected surface, as well as signal the need of more thorough testing when the computed deviation exceeds a specific limit.

(B) Calculated deviation from roundness. Roundness can only be determined with complete reliability by tracing the periphery of a nominally round feature with the stylus of a roundness tracing instrument. For general manufacturing purposes it may be sufficient to check the roundness condition of a feature by less than complete testing of its entire periphery. In such approximate roundness inspections a limited number of points, but not less than six, are measured on the periphery in a specific radial plane, that is, a plane at right angles to the axis. The measurement data produced are then entered into the computer for calculating the center and the radius of the feature. Subsequently, the programmed system calculates the radius from the center to each individually measured point to find the minimum and maximum values, whose difference is the measure of the approximate roundness deviation. That result is not equivalent to the departure from roundness determined by a stylus type roundness tracing instrument, but may be used to indicate the characteristics of roundness errors and also the need for further testing when the calculated deviation is large.

(C) Checking concentricity and determining the amount of eccentricity. To determine the amount of

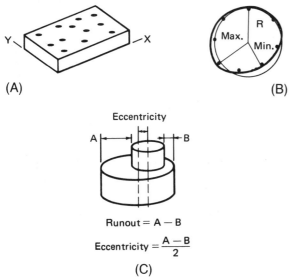

(A)

(B)

Eccentricity

Runout = A — B

$$Eccentricity = \frac{A - B}{2}$$

(C)

Brown & Sharpe Mfg. Co.

Fig. 19-28. Programmed coordinate measurement of geometric forms: (A) flatness; (B) roundness; (C) eccentricity.

eccentricity, that is, the mutual displacement of nominally coincident axes of two round features, the multipoint center and diameter method may serve as the starting step for producing the value of the diameter runout in terms of total indicator reading (TIR). The actual amount of eccentricity, that is, the displacement of the feature axes, is one half of the diameter runout, as indicated in diagram (C). The method may be used for outside diameters (bosses), as well as for inside diameters (holes), and the diameter of the second feature measured can be determined with or without maximum material condition (MMC).

The preceding examples are relatively simple programs and are described because they offer a wide range of applications. More complex programs are also available, such as for calculating the spatial positions or dimensions of various features, elements, and interrelated conditions, including the complete dimensional measurement of intricate shapes.

Design Characteristics of Automatic Coordinate Measuring Machines

The basic principles of dimensional inspection by CMM, as discussed in Chapter 12, remain the same, although the execution of complex and repetitive measurements becomes easier and faster through the supporting use of advanced electronics and automatic part loading. The availability of computers and dedicated microprocessors, which can be operated with a wide range of stored programs to perform specific measuring routines, greatly expanded the application potential and use of coordinate measuring machines.

The resulting wide variety of workpiece types for which computer-supported coordinate measurements became the preferred method of dimensional inspection has led to the expanding demand for that type of equipment. At the same time, such possible uses prompted significant developments in the design and manufacture of coordinate measuring machines. Also, the number of reputable manufacturers, domestic and foreign, specializing in the production of coordinate measuring machines grew, and the size ranges of their products increased substantially.

Development in design, in correspondence with the operational and applicational requirements, took place in several respects. These developmental directions have to differing extents been adapted and incorporated into various types of machines by the

several specialized manufacturers. Many of the new design features are, however, characteristic of a general trend, a few aspects of which are reviewed in the following.

Construction of the machines. The general characteristics are now those of a specific-purpose instrument and have become disassociated from machine tools, particularly jib borers, from which the earlier models of coordinate measuring machines have been developed. In coordinate measuring machines serving a specific purpose, the main spindle has no alternative role; it does not have to be rigidly supported and driven for metalcutting, as in jib borers; consequently, the probe spindle and its supporting members can be designed to carry out all the positioning movements along the three axes.

Range of sizes. This has been considerably expanded to satisfy the requirements of the widening field of products for which coordinate measuring machines are the preferred type of dimensional inspection equipment.

Giddings & Lewis Measurement Systems

Fig. 19-29. Five-axes automatic inspection center with pallet shuttle system, direct computer control, monitor and keyboard, printer and environmental enclosures for the CMM computer and printer.

Positioning movements. The distance of the probe travel along all three axes to specific positions is measured exclusively in increments producing electrical pulses that are displayed by digital readouts, a system that concurrently generates input signals for peripheral and supporting equipment.

Accuracy. A very high level of overall performance accuracy is obtainable, in the original delivery state and over extended use. This is less expensive to achieve, and, consequently, more generally producible in the case of equipment designed solely as measuring instruments that have stationary work supports and operate without the introduction of significant forces due to metalcutting or moving heavy machine members. Modern coordinate measuring machines have, in most cases, very rigid heavy granite blocks for work support, and the mov-

ing members of the probe assembly have frictionless guides, such as air bearings, which are practically free of wear and distortion. Modern pulse-producing scales have high resolution, such as 0.0001 inch, and retain their original accuracy indefinitely.

Equipment for data delivery and interpretation. Most automated coordinate measuring machines are equipped with data-processing control and part staging systems. These systems typically include a digital readout, computer, video display monitor, printer and pallet shuttle mechanism. Figure 19-29 shows a five-axes coordinate measuring machine with dedicated microprocessor, printer and basic machine. All are shown in environmental enclosures, and are complemented by a shuttle system that automatically loads palletized parts weighing up to 500 pounds.

Index